Mammals
OF THE
Neotropics

Mammals of the Neotropics, Volume 1, *The Northern Neotropics: Panama, Colombia, Venezuela, Guyana, Suriname, French Guiana*, by John F. Eisenberg, was published by the University of Chicago Press in 1989.

MAMMALS
OF THE
NEOTROPICS

The Southern Cone

VOLUME 2

Chile, Argentina, Uruguay, Paraguay

With plates in color and black and white by Fiona Reid

Kent H. Redford

and

John F. Eisenberg

The University of Chicago Press
Chicago and London

Kent H. Redford and **John F. Eisenberg** are both on
the faculty of the University of Florida: Mr. Redford is
assistant professor in the Center for Latin American
Studies and the Department of Wildlife and Range
Sciences and director of the Program for Studies in
Tropical Conservation. Mr. Eisenberg is the Katharine
Ordway Professor of Ecosystem Conservation at the
Florida Museum of Natural History. He is the author of
Mammals of the Neotropics, volume 1, and *The Mam-
malian Radiations*, both published by the University of
Chicago Press.

The University of Chicago Press, Chicago 60637
The University of Chicago Press, Ltd., London

© 1992 by The University of Chicago
Plates © 1992 by Fiona Reid
All rights reserved. Published 1992
Printed in the United States of America

01 00 99 98 97 96 95 94 93 92   54321

Library of Congress Cataloging-in-Publication Data
(Revised for vol. 2)

Eisenberg, John Frederick.
    Mammals of the Neotropics

    Vol. 2 by Kent H. Redford and John F. Eisenberg.
    Includes bibliographical references and indexes.
    Contents: v. 1. The northern Neotropics: Panama,
Colombia, Venezuela, Guyana, Suriname, French
Guiana—v. 2. The southern cone: Chile, Argentina,
Uruguay, Paraguay
    1. Mammals—Panama.   I. Redford, Kent Hubbard.
II. Title.
QL725.A1E38   1989   599.098   88-27479
ISBN 0-226-70681-8 (cloth : v. 2)
ISBN 0-226-70682-6 (paper : v. 2)

∞ The paper used in this publication meets
the minimum requirements of the American National
Standard for Information Sciences—Permanence of
Paper for Printed Library Materials, ANSI Z39.48-1984.

To *Pamela*, *Sofia*, and *Hugh*
and to *Jon David* and *Elisha*

Contents

Acknowledgments

Our work on the mammals of southern South America involved the cooperation of many colleagues. We offer our thanks to them all. In Santiago we were graciously hosted at the Museo Nacional by José Yáñez and Agustín Iriarte. We are also grateful to Angel Spotorno, Universidad de Chile, for permission to examine his collection. Throughout our time in Chile, Agustín Iriarte helped us in ways too numerous to mention. Our Chilean colleagues have been extremely supportive of our efforts in their backyard. The rodent and bat range maps were kindly reviewed by Agustín Iriarte, Luis Contreras, Angel Spotorno, José Yáñez, Fabian Jaksić, P. Marquet, Juan Carlos Torres, and Xavier Simonetti.

In Buenos Aires, Jorge Crespo hosted us at the Museo Nacional; his advice on the preliminary range maps for Argentina was invaluable. Our Argentine colleagues Rubén Bárquez, Ricardo Ojeda, Damián Rumiz, and Andrés Novaro reviewed the maps for several groups of Argentine mammals. In addition, Michael Mares has provided support and literature throughout the long gestation of this work.

In the United States, Bruce Patterson and Robert Timm of the Field Museum of Natural History, Chicago, generously allowed us to inspect the important collection of Chilean mammals. The mammals of Argentina were studied at the Museum of Vertebrate Zoology, Berkeley, where William Lidicker, Oliver Pearson, and James Patton were most helpful. Special thanks go to Oliver Pearson, who read the entire manuscript. Paraguayan mammals were studied at the University of Michigan, Museum of Zoology. We are indebted to Philip Myers for access to his collection and for reviewing portions of the text and the distribution maps. Jody Stallings and Andrew Taber both read portions of the Paraguayan mammal sections. Sydney Anderson kindly made available the collections of the American Museum of Natural History in New York. Access to the Carnegie Museum of Natural History, Pittsburgh, was given by Hugh Genoways and Duane Schlitter. The published works and unpublished dissertation of Jon C. Barlow on the mammals of Uruguay were extremely useful. Robert Brownell reviewed the cetacean and pinniped maps and species accounts.

The plates were prepared by Fiona Reid, and we are grateful for her unending patience and fine talent. The skulls were drawn by Chuck Gentry based largely on specimens at Harvard University's Museum of Comparative Zoology. Lynn Dorsey Rathbun and Sigrid James Bonner assisted with several of the figures, and Sigrid James Bonner also helped with many details on the range maps. Agustín Iriarte assisted us with plotting the locations on the maps using our museum data and the gazetteers assembled for the various countries. His knowledgeable eye has been very helpful. Many colleagues have aided us greatly by supplying literature references or reprints; to all of them many thanks.

Colleagues at the National Zoological Park, Washington, D.C., provided us with valuable data on life histories for several key species. Our thanks to Devra Kleiman, John Seidensticker, Steven Thompson, and Miles Roberts. Last, but by no means least, we owe an enormous debt of gratitude to Barbara Stanton, without whose great skill and endless patience this volume would never have been completed.

Introduction

A Historical Perspective

The early history of mammal studies in South America has been admirably documented by Hershkovitz (1987). Within the area covered by this volume the efforts of Don Felix de Azara loom large. Although not trained in biology, he described from the region of what is now Paraguay seventy-seven species of mammals, sixty-two of them unknown to science (Azara 1801). The early history of mammal studies in Chile is dominated by the efforts of Giovanni Molina (summarized by Osgood 1943). Claudio Gay (1847) prepared one of the best nineteenth-century descriptions of Chilean mammals. Argentina was investigated in detail by Charles d'Orbigny, together with the systematist Paul Gervais, and the mammal descriptions were published in volumes 2 and 4 of his serial publication *Voyage dans l'Amérique méridionale* (1835–47). It goes without saying that the observations and collections of Charles Darwin also contributed to our knowledge of mammalogy in southern South America, since he touched on the coasts of Uruguay, Argentina, and Chile (Darwin 1839). Together with the efforts of G. R. Waterhouse (1838–39) these collections were described and entered into the scientific literature. Hermann Burmeister described many new mammals from South America during the mid-nineteenth century in what now constitutes Bolivia and Paraguay. Oldfield Thomas was an active worker in mammalian systematics at the British Museum of Natural History from the late nineteenth century through the first quarter of the twentieth and received many collections from fieldworkers in South America. As a result, many type specimens of small marsupials and sigmodontine rodents from South America reside in London. This historical circumstance has made it difficult for recent workers to assign specimens to the correct taxa without making a "pilgrimage" there—not always, we may add, an onerous chore.

Research on Recent mammals in the twentieth century has been gaining momentum as more and more biologists have been recruited into the field. In the "southern cone" of South America, several names stand out in the years before 1950. These include Wilfred Osgood and Guillermo Mann for their efforts in Chile, and Angel Cabrera and Jorge Crespo for parallel endeavors in Argentina. Osgood (1943) produced the first detailed modern account for any mammalian fauna within the republics of South America. Mann (1978), building upon Osgood, produced a useful volume dealing with the natural history of the small mammals of Chile.

Angel Cabrera was a man of uncommon talents. A paleontologist as well as a neontologist, he was the first twentieth-century zoologist to attempt a systematic listing of the mammals of the South American continent (1957, 1960). Some of his students, suitably inspired, labored to fill the gaps in knowledge concerning natural history and ecology. Notable among this group is Jorge Crespo. Although we do not want to slight other workers in this brief introduction, let us merely say that a multitude of contributions have been forthcoming in the past thirty-five years. We hope most of them will be amply cited in the text.

The Fossil Record

Paleontology has had a long and distinguished history in the southern cone of South America. This statement is not meant to detract from the enormous contributions from Colombia, Venezuela, Bolivia, and Brazil. Rather, we wish to emphasize the historical perspective. Darwin (1839) called attention to the rich fossil deposits in Patagonia, but it was the Ameghino brothers of Argentina who began the most extensive efforts to recover their country's rich fossil deposits (Ameghino 1906). The history of paleontological research in South America has been reviewed by Simpson (1980).

Today one can scarcely appreciate the diversity and abundance of ungulates in South America before the completion of the Panamanian land bridge between North and South America during the Pliocene. Throughout the Tertiary the ungulate orders Notungulata and Meridungulata had evolved forms that resembled the elephants, rhinoceroses, horses, and hippopotamuses—so familiar to inhabitants of the more contiguous continental land masses. The extinction of this endemic fauna and its replacement from the north, only to have most of the northern immigrant megafauna collapse, constitutes an intriguing problem for biologists today (Stehli and Webb 1985).

Background for the Preparation of This Volume

In 1977 John Eisenberg was completing his book *The Mammalian Radiations* (Eisenberg 1981). As a footloose student with a year to spare before graduate school, Kent Redford began to help in compiling and analyzing the tables, graphs, and appendixes for the book. The generic names, the collecting localities, the habitats, and the names of previous investigators made the air heady with absorbing ideas, untested hypotheses, unknown animals, and unexplored places. Never before having considered South America as anything other than a source of photographs for Pan Am calendars, Redford was dazzled by thoughts of the Pantanal, the puna, and Patagonia.

Graduate school brought Redford a chance to go to Paraguay and spend three months in the Chaco working for Ralph Wetzel and Philip Myers. The purpose of the trip was to collect mammals and begin a research project on armadillos. It was a formative experience; the thorns, the rains, an abandoned truck, and the effects of hunting sent him across the border into Brazil. Even though he ended up working in Brazil, the lure of the southern fauna remained, and he welcomed the invitation from Eisenberg to team up and write volume 2 of this work, on the mammals of southern South America.

In 1978 Eisenberg traveled to Argentina with Richard W. Thorington, Jr., to join colleagues from CAPRIM in surveying the mammal faunas in the state of Chaco where it borders the Río Paraguay. Special attention was focused on *Alouatta caraya*, and a study was launched on the population biology of this species.

Although we have spent in the aggregate some six months in Argentina and Chile, neither of us has had extensive field experience in southern South America. However, owing to the widespread distri-

bution of many of the mammalian species of the region, we have had experience with them in other parts of South America or in the National Zoological Park, Washington, D.C. These taxa include the armadillos, the anteaters, many of the ungulates, several of the primates, and about 20% of the rodents, bats, and marsupials. Especially valuable in the zoological park was our access to a wide variety of hystricognath rodents, including some seventeen species from sixteen genera. Supplementing these experiences were visits to a number of museums in South America and the United States.

Organization of the Book

For the most part the text is organized taxonomically, with a chapter devoted to each order. This body of taxonomic chapters is wrapped in an introductory chapter and two concluding chapters. In chapter 1 we set the stage by orienting readers to the biogeographic history of South America and to contemporary habitats. The penultimate chapter deals with the community ecology of mammals in southern South America, and the final chapter discusses the effects humans have had on the mammalian fauna of the southern cone.

The taxonomic order of the volume follows Honacki, Kinman, and Koeppl (1982) for orders and families; within families the genera and species are arranged alphabetically. Within each chapter dealing with ordinal taxa the organization is similar. A diagnosis defines members of each order according to anatomical characteristics. The diagnoses are a guide to key anatomical features in the contemporary Neotropical context but are not exhaustive. The ordinal distribution is then presented and, if warranted, comments are made concerning taxonomy. A brief history of the evolution of the order is presented where appropriate.

Next each family within the order is considered in turn, repeating the diagnosis (field characters where possible), distribution, and remarks on natural history where warranted. Each genus is then taken up, with a more refined description, distribution for the genus, and information on natural history. Where there is only one species in a genus, the generic account is contained within the species account.

The species accounts are the most comprehensive of the taxonomic accounts. They include a set of external measurements, a physical description, geographical distribution, information on life history, notes on ecology and behavior, and usually a distribution map. At the end of each chapter we supply a list of references we used in writing the accounts. Citations of these works are mainly gathered at the end of each section. The hierarchical organization attempts to avoid repetition; if one turns to a species account and finds some information lacking, or.e should look under the appropriate subheading at the generic level, the family level, and finally the ordinal level.

How to Use This Book

Latin binomials are the only practical way to refer to species, since there is little agreement on common names. Even within a country there can be several regional names, and different countries rarely use the same common name. For many bats, rodents, and small marsupials there are no common names in any language. We have not attempted to coin new common names but have used those presented in Olrog and Lucero's (1981) treatment of Argentine mammals, supplementing when necessary from other sources. Common names and scientific names are listed in the indexes.

A bullet (•) before a species name indicates that the species was discussed in volume 1.

This book is based virtually completely on literature accounts and unpublished museum records. Neither of us is a taxonomist, and we have made no effort to revise taxonomic identifications. The taxonomy we use is based on Honacki, Kinman, and Koeppl (1982), with the following exceptions: Pearson (1984) for the fossorial sigmodontines; Eisenberg (1989) for *Puma*; Berta (1982) for *Cerdocyon*; Hershkovitz (1983) for *Aotus*; Reig (1987) for several annotations; and various references for *Ctenomys* as noted in the text.

We fully recognize that many of the taxa are unstable and may eventually be redescribed; some species will be subsumed and others will be split out. Where taxonomists have published their suspicions that such changes are needed, or where changes have been proposed but have not gained wide acceptance, we have included a note in the "comment" section of the species account. Some genera like *Ctenomys* and *Akodon* will probably be extensively revised, though most of the mammalian taxa in southern South America are likely to remain as described here.

For most species we include a distribution map with dots representing collection locations. These locations were obtained either from published literature, from museum records, or very rarely from personal communications from experts. When transferring the latitude and longitude to the base map we used a 1° × 1° grid, placing the dot in the center of the cell. Thus the locations are accurate only to within one degree. Not all recorded collecting locali-

ties were used; we showed all marginal locations and a broad sampling of interior localities. Unlike volume 1 (*The Northern Neotropics*), here no shading was used on the maps. This is because southern South America is much better collected than northern South America, and though there will certainly be exceptions to this generalization, if a species occurs in a broadly defined area it has likely been caught there.

Because this work is based on literature and museum records, there are undoubtedly some errors in distribution. We have tried to catch mistakes by having experts review the distribution maps. In some cases errors may be due to misidentification of the specimen by the original author, misidentification of the location by a later compiler, or an unrecorded (by us) specimen reidentification that has caused species to be combined. If authors have misidentified a species on which they report ecological or life history information that we used, then there will be errors in our accounts. We can provide a list of the authorities for each dot if readers have a particular interest in any of the maps.

A few keys are added to aid readers in identifying major taxa (families and genera). A complete key for the sigmodontine rodents is at present impossible, but excellent regional keys exist (see Pearson 1987 for small mammals of southwestern Argentina and Greenhall, Lord, and Massoia 1983 for Argentine bats).

We use abbreviations in many tables and also when referring to collections. The following refer to museums: USNMNH, United States Museum of Natural History; AMNH, American Museum of Natural History; CM, Carnegie Museum of Natural History; FM, Field Museum of Natural History; IEE-UACH, Instituto de Ecología y Evolución, Universidad Austral de Chile, Valdivia; MVZ, Museum of Vertebrate Zoology, Berkeley; UM, Museum of Zoology, University of Michigan, Ann Arbor; BA, Museo Argentino de Ciencias Naturales "Bernardino Rivadavia"; Santiago, Musco Nacional de Historia Natural; UConn, University of Connecticut, Storrs. "PCorps" refers to unpublished data from Peace Corps collections in Paraguay, made available courtesy of Jody Stallings and Mark Ludlow. Tracy Carter and Kim Hill kindly provided unpublished measurements.

The following abbreviations refer to linear measurements and weights: TL, total length; HB, head and body length; T, tail length; HF, hind foot length; E, ear length as measured from the notch to the tip; FA, forearm length (useful for bats); Wta, weight of the adult. Linear measurements are generally given in millimeters (mm), and weights are in grams (g) unless specified as kilograms (kg). Abbreviations in notations follow this pattern: TL 120–31 (= TL 120 to 131). When fewer than three individuals were measured, the measurements are included in the description. When three or more were measured, the measurements are displayed in tabular form with mean, standard deviation (S.D.) when given, minimum (Min.), maximum (Max.), sample size (*N*), location (Loc.) of the specimens measured, and source—the authority from which the measurements were taken or the museum from which we obtained specimen measurements. Locations are abbreviated as follows: A, Argentina; B, Bolivia; Br, Brazil; C, Chile; CR, Costa Rica; N, Nicaragua; P, Paraguay; U, Uruguay; V, Venezuela; Cap, in captivity.

Readers should be prepared for alternative presentations of measurements. The scheme is not uniform throughout the volume because of a variety of circumstances. First, the standard small-mammal measurements TL, HB, T, HF, and E are not useful across all orders. Among the Cetacea they often do not apply. Second, for some species only a limited number of specimens are available, and those collected before 1950 may not include weights or even ear measurements on the museum tags. A mixed presentation of data under species accounts should not be perturbing. Our knowledge of species varies.

Dental formulas correspond to a standard system; if a skull is viewed from one side, the upper and lower teeth are recorded in sequence from front to back: I, incisors; C, canines; P, premolars; M, molars. Thus the formula I 3/1, C 1/1, P 2/1, M 3/3 translates: viewing one side, there are three upper and one lower incisors, one upper and one lower canines, two upper and one lower premolars, and three upper and three lower molars.

In the descriptions of species we have referred to pelage colors using terms in common English usage. When working with a published description, we have used the author's terms, which often reflect the Ridgeway (1912) color terminology. We have tried to avoid this terminology in most accounts, since the color standards are not widely available and museum skins can change shade because of the preservatives employed in preparation and through exposure to light.

This book is not intended as a source of definitive species identifications. When in doubt the investigator must collect voucher specimens and have a taxonomist identify them. This is particularly important with smaller rodents, for which genetic, skull, and phallus characters are frequently required. Particular care should be taken with juvenile specimens, since they often do not precisely resemble the adult in either size or coloration.

Final Remarks

The mammalian fauna of Argentina and Chile is well known, thanks to the pioneering efforts of Mann, Osgood, Crespo, Oliver Pearson, and above all Cabrera. Recently Michael Mares, Ricardo Ojeda, Rubén Barquez, and several other outstanding Argentine mammalogists have expanded our understanding of mammal distributions in Argentina, and there have been parallel efforts in Chile by, among others, Luis Contreras, Angel Spotorno, and José Yáñez. The recent efforts of Jon C. Barlow in Uruguay, Michael Mares and colleagues in Argentina (see especially Mares, Ojeda, and Barquez 1989), and Philip Myers and the late Ralph Wetzel in Paraguay suggested to us that a synthesis for the southern portion of the continent might be possible. To this end we have produced a work, based largely on a compilation of the voluminous literature dealing with the mammalian fauna of Chile, Argentina, Uruguay, and Paraguay. We reserve Brazil, Bolivia, Peru, and Ecuador for a later collaborative effort.

We have tried to include most of the literature up to 1988. Our coverage of many of the Chilean and Argentine publications stopped in 1987, when the American Society of Mammalogy terminated its useful literature review. Since that time, we have included articles to which we could find references and reprints sent by colleagues.

We hope this book will prove useful, though we fully understand that the taxonomy for many groups is still in a state of confusion. Users of this book should realize that they have in their hands a progress report, a state-of-the-science work, that should be used as much to find out what is not known as what is known. We hope this volume will stimulate many of our colleagues, particularly younger ones, to fill in the gaps, correct the mistakes, and work toward a better understanding of the mammals of southern South America.

References

Ameghino, F. 1906. Les formations sédimentaires du Crétace supérieur et du Tertiare de Patagonie avec une parallèle entre leurs faunes mammalogiques et celles de l'ancien continent. *Anal. Mus. Nac. Buenos Aires* 14:1–568.

Azara, F. de. 1801. *Essais sur l'histoire naturelle des quadrupèdes de la provincia du Paraguay.* Translated from the original Spanish by M. L. E. Moreau-Saint-Mery. Vols. 1 and 2. Paris: Charles Pougens.

Berta, A. 1982. *Cerdocyon thous. Mammal. Species* 186:1–4.

Cabrera, A. L. 1958. Catálogo de los mamíferos de América del Sur. 1. Metatheria-Unguiculato-Carnivora. *Rev. Mus. Argent. Cienc. Nat. "Bernardino Rivadavia," Zool.* 4(1): 1–307.

———. 1960. Catálogo de los mamíferos de América del Sur. 2. Sirenia-Perissodactyla-Artiodactyla-Lagomorpha-Rodentia-Cetacea. *Rev. Mus. Argent. Cienc. Nat. "Bernardino Rivadavia," Zool.* 4(2): 308–732.

Darwin, C. R. 1839. *Journal of researches into the geology and natural history of the various countries visited by H.M.S. Beagle, under the command of Captain Fitzroy, R.N., from 1832 to 1836.* London: Henry Colburn.

Eisenberg, J. F. 1981. *The mammalian radiations: An analysis of trends in evolution, adaptation, and behavior.* Chicago: University of Chicago Press.

———. 1989. *Mammals of the Neotropics. Vol. 1. Mammals of the northern Neotropics: Panama, Colombia, Venezuela, Guyana, Suriname, French Guiana.* Chicago: University of Chicago Press.

Gay, C. 1847. *Historia física y política de Chile.* Vol. 1. *Zoología.* Museo de Historia Natural de Santiago [de Chile]. Paris: Malde y Renow.

Greenhall, A. M., R. D. Lord, and E. Massoia. 1983. *Key to the bats of Argentina.* Special Publication 5. Buenos Aires: Pan American Zoonoses Center.

Hershkovitz, P. 1983. Two new species of night monkeys, genus *Aotus* (Cebidae, Platyrrhini): A preliminary report on *Aotus'* taxonomy. *Amer. J. Primatol.* 4:209–43.

———. 1987. A history of the recent mammalogy of the Neotropical region from 1492 to 1850. *Fieldiana: Zool.,* n.s., 39:11–98.

Honacki, J. H., K. E. Kinman, and J. W. Koeppl, eds. 1982. *Mammal species of the world.* Lawrence, Kans.: Allen Press and Association of Systematics Collections.

Mann, G. 1978. *Los pequeños mamíferos de Chile.* Gayana: Zoología 40. Chile: Universidad de Concepción.

Mares, M. A., R. A. Ojeda, and R. M. Barquez. 1989. *Guide to the mammals of Salta province, Argentina.* Norman: University of Oklahoma Press.

Olrog, C. C., and M. M. Lucero. 1981. *Guía de los mamíferos argentinos.* Tucumán, Argentina: Ministerio de Cultura y Educación, Fundación Miguel Lillo.

Orbigny, A. D. d', and P. Gervais. 1847. Mamifères. In *Voyage dans l'Amérique méridionale (le Brésil, la république de Bolivia, la république du Pérou) exécuté pendant les années 1826, 1827, 1828, 1829, 1830, 1831, 1832, 1833,* ed. A. D. d'Orbigny, vol. 4, pt. 2. Paris: Pitois-Levrault.

Osgood, W. H. 1943. The mammals of Chile. *Field Mus. Nat. Hist., Zool. Ser.* 30:1–268.

Pearson, O. P. 1984. Taxonomy and natural history of some fossorial rodents of Patagonia, southern Argentina. *J. Zool.* (London) 202:225–37.

———. 1987. *Annotated keys for identifying small mammals living in or near Nahuel Huapi and Lanín National Park, southern Argentina.* Privately published.

Reig, O. A. 1987. An assessment of the systematics and evolution of the Akodontini, with description of new fossil species of *Akodon* (Cricetidae: Sigmodontinae). *Fieldiana: Zool.,* n.s., 39:347–99.

Ridgeway, R. 1912. *Color standards and color nomenclature.* Washington, D.C.: Privately published.

Simpson, G. G. 1980. *Splendid isolation.* New Haven: Yale University Press.

Stehli, F. G., and S. D. Webb, eds. 1985. *The great American biotic interchange.* New York: Plenum.

Waterhouse, G. R. 1838–39. Mammalia. In *The zoology of the voyage of H.M.S. Beagle, under the command of Captain Fitzroy, during the years 1832–1836: With notes by Charles Darwin.* Pt. 2. London: Smith, Elder.

Yáñez, J. L., J. C. Torres-Mura, J. R. Rau, and L. C. Contreras. 1987. New records and current status of *Euneomys* (Cricetidae) in southern South America. *Fieldiana: Zool.,* n.s., 39:283–88.

1 An Introduction to the Biogeography of Southern South America

Geography, Climate, and Vegetation

This volume details the distribution of mammals found in Chile, Argentina, Uruguay, and Paraguay (fig. 1.1). We selected these countries in part based on convenience, since their mammalian fauna is reasonably well known and documented. In addition, we can thus concentrate on the mammalian fauna of temperate South America and delimit the southern extent of the range of those mammals adapted to the tropics.

Compared with the Northern Hemisphere, the Southern Hemisphere has very little land; the southern half of the globe is dominated by oceans. More than a third of the land in the Southern Hemisphere is concentrated at the South Pole in the island continent of Antarctica. The effects of the extreme southerly position of southern South America and of the vast oceans that embrace it figure dramatically in the biology of its fauna. Readers should remember that Tierra del Fuego is over 20° farther south than the tip of Africa and 10° farther south than the tip of New Zealand. Another important factor to bear in mind is how rapidly the continent of South America attenuates at its southern end. In continental terms, the most southerly 15° constitutes a peninsula.

The dominant geological feature of this portion of South America is the Andes Mountains. This range is of recent origin, having begun to form in the Miocene, and active mountain building processes are still observable (fig. 1.2). The Andes and associated plateaus in southern South America run almost north-south and are set so far to the west on the continental landmass that the coastal lowlands on the Pacific side of the continent are very narrow. The mountain range, together with the variations in prevailing winds, vastly influences rainfall. For example, between 27° and 10° south latitude, the failure of westerly winds coupled with the cold Humboldt current produces an arid strip that in northern Chile is

termed the Atacama Desert. As one ascends the west face of the Andes, however, adiabatic cooling produces some precipitation. Thus steppes with low shrubs and grasses dominate at elevations above 1,000 m; and above 1,700 m, where plateaus occur between the various ranges of the Andes, the puna, a cool, high-elevation steppe, becomes the dominant biome. Farther to the south, on the west face of the Andes, where the westerlies again arrive laden with moisture from the Pacific, sclerophyll forest appears in a very narrow band between approximately 30° and 25° south latitude. This forest grades imperceptibly into temperate rain forest, which dominates the rest of the western face of the Andes to the south. On the eastern face, the altitudinal and latitudinal distribution of vegetation is similar. However, the extent of mesic vegetation is somewhat modified because of the rain shadow effect of the Andes themselves (see fig. 1.3). South of about 35° the high peaks stop, and from there to the end of the continent the Andes diminish in height and become less of a barrier between Argentina and Chile (Mann 1969).

Great rivers such as the Amazon and Orinoco, which profoundly influence mammal distributions in the north, do not exert the same influence in the south (see fig. 1.4). The major exception is the Río Paraguay, which demarcates two distinct zoogeographical regions in Paraguay (Myers 1982). This river joins with the Río Paraná to form the Río de la Plata, debouching at the great port of Buenos Aires. The Río Negro in the south of Argentina also appears to be a demarcation point for certain mammal distributions, but this zoogeographic limit may relate more to dramatic shifts in vegetation type and rainfall than to the river as any major barrier to dispersal.

It is possible to divide the region under consideration into discrete biomes as illustrated in figure 1.3. One can see that much of western Argentina is semi-

Figure 1.1. Map of southern South America showing the major cities and Argentine provinces.

arid, including the Monte Desert and the Patagonian steppes. To the east and northeast of these regions, slightly higher rainfall levels produce the scrub vegetation, so typical of the Chaco, that covers the entire portion of Paraguay west of the Río Paraguay. In more mesic regions around the Río de la Plata and over most of Uruguay occur the great grasslands of the Argentine and Uruguayan pampas. Multistratal tropical rain forest interdigitates with the drier regions along the courses of major rivers that provide relatively constant mesic conditions. It also occurs, as an extension of the Brazilian Atlantic forest, over Paraguay east of the Río Paraguay and into the Argentine province of Misiones. Another

tongue of tropical forest, an extension of the Yungas forest of Bolivia, stretches into the western Argentine provinces of Jujuy and Salta. Within these forests the ranges of many tropical mammals reach their most southerly extent. The Temperate Zone forms have suitable habitat at high elevations and at lower latitudes, but one should bear in mind the dominance of the Andes in stratifying vegetation forms by influencing the distribution of precipitation and providing a complicated mosaic of habitats suitable for both tropical and temperate species.

In his summary of floristic regions of the world, Takhtajan (1986) includes most of the southern cone of South America within the Holantarctic kingdom.

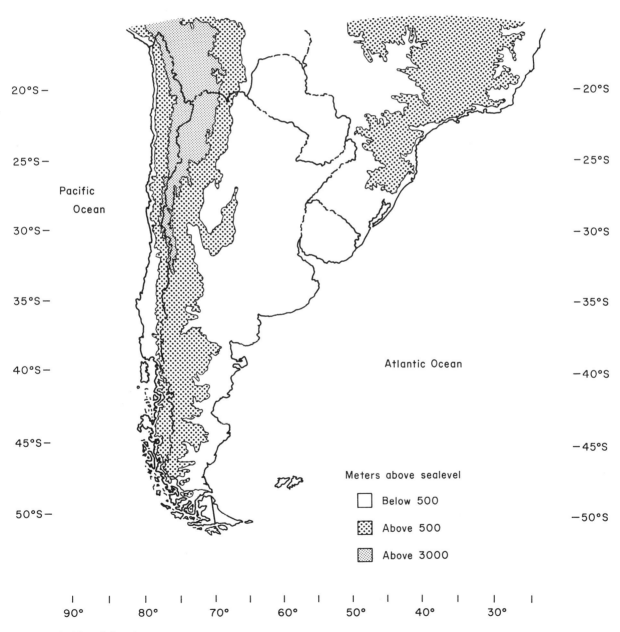

Figure 1.2. Map of elevations.

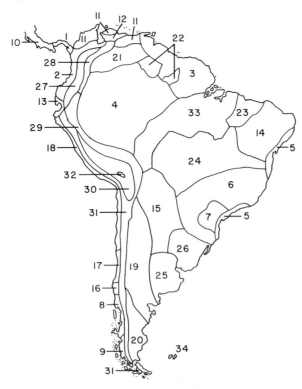

1 Panamanian
2 Colombian coastal
3 Guyanan
4 Amazonian
5 Serra do Mar
6 Brazilian rain forest
7 Brazilian planalto
8 Chilean *Nothofagus*
9 Valdivian forest
10 Central American
11 Venezuelan dry forest
12 Venezuelan deciduous
   forest
13 Ecuadorian dry forest
14 Caatinga
15 Gran Chaco
16 Chilean *Araucaria*
17 Chilean sclerophyll

18 Pacific desert
19 Monte
20 Patagonian
21 Llanos
22 Campos limpos—Amazo-
   nian savannas
23 Babaçu
24 Campos cerrados
25 Argentine pampas
26 Uruguayan pampas
27 Northern Andean
28 Colombian montane
29 Yungas
30 Puna
31 Southern Andean
32 Lake Titicaca
33 Madeiran
34 Insulantarctica

Figure 1.3. The major biomes of South America
(adapted from Udvardy 1978).

His scheme clearly unites the flora of southern South America with that of New Zealand and southeastern Australia. Within his classification, the flora of southern South America is termed the Chile Patagonian region. In the north of Argentina, elements of the Neotropical kingdom (Brazilian region) intersperse with the southern elements. Paraguay clearly lies within the Brazilian region of Neotropica. Numerous schemes of vegetation classification make subdivisions within the broad outline presented by Takhtajan; for example, Eiten (1974) lists fifteen vegetation forms for southern South America. Cabrera (1976) lists twelve phytogeographic provinces for Argen-

tina. The compromise approach to vegetation classification is portrayed in figure 1.3.

There are similarities in vegetation composition (in the systematic sense) and physiognomy when North and South America are compared. Glanz (1982) has published a useful essay comparing the mammalian fauna of "Mediterranean" Chile and California. Clearly there are similarities in vegetation between the two areas.

Broadly speaking, southern South America has tropical forms entering Paraguay, the extreme northwest and northeast of Argentina, and parts of Uruguay where the climate permits elements of the Brazilian floral region to dominate. Northern Chile is arid to Mediterranean in its climate, while central Argentina contains a spectrum from thorn forests to grasslands. At varying elevations southern Chile and a slender strip of southwestern Argentina contain the distinctive temperate, mesic-adapted flora termed "subantarctic" or "holantarctic."

The wet temperate flora of southern South America is distinctive, containing numerous endemic species and genera. For example, Takhtajan (1986) lists the following genera of trees as typical of, but not exclusive to, Chile and Argentina: *Araucaria*, *Podocarpus*, and *Nothofagus*. Some of these genera are shared with temperate South Africa and Australia. The biology of the southern beech forests (*Nothofagus*) of Argentina has been reviewed by Ward (1965). Some species are evergreen, others deciduous. All species seem to be abundant below 35° south latitude. It has been suggested that *Nothofagus* competes with the conifer *Araucaria*, but Veblen (1982) and Veblen et al. (1981) suggest that *Araucaria araucana* is an active colonizer of disturbance areas and that there is no general tendency for *Araucaria* to be displaced by species of *Nothofagus*. The small-mammal communities of these temperate forests have been analyzed by Pearson and Pearson (1982).

The wet grasslands of the pampas in Uruguay and Argentina have been the subject of numerous reviews (Soriano 1979), and many authors have noted the dominance of rodents in these habitats (Mann 1969; O'Connell 1982). These biotic regions have been vastly modified by humans through burning and through grazing livestock. The consequences of these induced changes will be explored in chapter 14.

Climatic Change and the Fauna of Southern South America

The mammalian fauna of southern South America was profoundly altered by the invasion of North American species at the completion of the Panamanian land bridge in the Pliocene (Stehli and Webb

Figure 1.4. Rivers of southern South America.

1985). We are also aware of the influences of more recent physical events on the biota of this vast southern continent. To cite a recently described phenomenon, consider the biological consequences of El Niño. Surface temperature changes in western Pacific waters can have profound effects on the primary productivity of the ocean off the coast of Peru. Furthermore, the rainfall patterns on continental South America can be severely altered (Barber and Chávez 1983). This event may be somewhat aperiodic, but given geological time considerations it has a short-term effect. On the other hand, the impact can be significant if rainfall levels are lowered in areas of marginal rainfall such as the Chaco of southern South America and the llanos of northern South America.

Longer-term events such as the Pleistocene glaciation cycles can have dramatic effects on mammal distributions. Recently Caviedes and Iriarte (1989) have proposed that the distribution of sigmodontine rodents in Chile has been influenced by glaciation in the southern Andes. Whereas the Andes served as a corridor for rodent dispersal from the north to the south, the Atacama Desert in northern Chile proved a formidable barrier to southern dispersal. The cold and arid climate of the Andes at 23° south latitude is also a barrier during nonpluvial periods (interglacial episodes). The depauperate nature of the Chilean rodent fauna derives in part from glacial episodes.

Eisenberg and Redford (1982) analyzed the mammalian fauna of southern South America and classified species according to their trophic strategy. The degree to which niches were saturated was tested by comparing the South American fauna with an appropriate temperate North American fauna. We concluded that the major feeding niches occupied by mammals in the North Temperate Zone are occupied by mammals of comparable size in temperate South America. But the number of species within each niche when the two areas are compared varies a great deal. In general there are fewer species for each major niche type in southern South America than in a comparable area of North America. We postulated that glaciation cycles in the Southern Hemisphere during the Pleistocene may have had a more profound impact in temperate South America because the land area of that continent decreases as one proceeds toward the Pole.

References

Barber, R. T., and F. P. Chávez. 1983. Biological consequences of El Niño. *Science* 222:1203–9.

Cabrera, A. L. 1976. Territorios fitogeográficos de la república Argentina. In *Enciclopedia Argentina Agricultura y Jardineria*. 2d ed., 1–85. Buenos Aires: Editorial Acme SACI.

Caviedes, C. N., and A. Iriarte W. 1989. Migration and distribution of rodents in central Chile since the Pleistocene. *J. Biogeogr.* 16:181–87.

Eisenberg, J. F., and K. H. Redford. 1982. Comparative niche structure and evolution of mammals of the Nearctic and southern South America. In *Mammalian biology in South America*, ed. M. A. Mares and H. H. Genoways, 77–84. Pymatuning Symposia in Ecology 6. Special Publication Series. Pittsburgh: Pymatuning Laboratory of Ecology, University of Pittsburgh.

Eiten, G. 1974. An outline of the vegetation of South America. In *Symposium of the Fifth Congress of the International Primatological Society*, 529–45.

Glanz, W. E. 1982. Adaptive zones of Neotropical mammals: A comparison of some temperate and tropical patterns. In *Mammalian biology in South America*, ed. M. A. Mares and H. H. Genoways, 95–110. Pymatuning Symposia in Ecology 6. Special Publication Series. Pittsburgh: Pymatuning Laboratory of Ecology, University of Pittsburgh.

Mann, G. 1969. Die Ökosysteme Sudamerikas. In *Biogeography and ecology in South America*, vol. 2, ed. E. J. Fitkau, J. Illies, H. Klinge, G. H. Schwabe, and H. Sioli, 171–229. The Hague: W. Junk.

Myers, P. 1982. Origins and affinities of the mammal fauna of Paraguay. In *Mammalian biology in South America*, ed. M. A. Mares and H. H. Genoways, 85–94. Pymatuning Symposia in Ecology 6. Special Publication Series. Pittsburgh: Pymatuning Laboratory of Ecology, University of Pittsburgh.

O'Connell, M. A. 1982. Population biology of North and South American grassland rodents: A comparative review. In *Mammalian biology in South America*, ed. M. A. Mares and H. H. Genoways, 167–86. Pymatuning Symposia in Ecology 6. Special Publication Series. Pittsburgh: Pymatuning Laboratory of Ecology, University of Pittsburgh.

Pearson, O. P., and A. K. Pearson. 1982. Ecology and biogeography of the southern rainforests of Argentina. In *Mammalian biology in South America*, ed. M. A. Mares and H. H. Genoways, 129–42. Pymatuning Symposia in Ecology 6. Special Publication Series. Pittsburgh: Pymatuning Laboratory of Ecology, University of Pittsburgh.

Soriano, A. 1979. Distribution of grasses and grasslands of South America. In *Ecology of grasslands and bamboolands in the world*, ed. M. Numara, 84–91. The Hague: W. Junk.

Stehli, F. G., and S. D. Webb, eds. 1985. *The great American biotic interchange*. New York: Plenum.

Takhtajan, A. 1986. *Floristic regions of the world.* Berkeley: University of California Press.

Urdvardy, M. D. F. 1978. *World biogeographical provinces.* Sausalito, Calif.: CoEvolution Quarterly. (Based on M. D. F. Urdvardy, International Union for the Conservation of Nature and National Resources Occasional Paper no. 18, 1975.)

Veblen, T. T. 1982. Regeneration patterns in *Araucaria araucana* forest in Chile. *J. Biogeogr.* 9: 11–28.

Veblen, T. T., C. Donoso Z., F. M. Schlegel, and B. Escobar R. 1981. Forest dynamics in south-central Chile. *J. Biogeogr.* 8:211–47.

Ward, R. T. 1965. Beech (*Nothofagus*) forests in the Andes of southwestern Argentina. *Amer. Midl. Nat.* 74:50–56.

Diagnosis and Comments on Reproduction

The dentition of marsupials is heterodont, with easily distinguished incisors, canines, premolars, and molars, and the number of teeth often exceeds the basic eutherian number of 44. The bony palate is fenestrated, a diagnostic character for most species when compared with the extant eutherians. The auditory bullae are formed principally by the alisphenoid. Epipubic bones are present in all extant species and are diagnostic for the order. The brain structure differs from that of eutherian mammals in that the corpus callosum is lacking.

Marsupial young have a brief intrauterine development and are born in an extremely undeveloped state. At birth the forelimbs are well developed, and the newborn crawls to the teat area, where it attaches to a nipple and remains some four to seven weeks. In many New World marsupials the teat attachment phase terminates when the female deposits the young in a nest, where she nurses them for another three to seven weeks until they are fully weaned. The teat area in many species is enclosed in a pouch. The biology of New World forms is reviewed in the volume edited by Hunsaker (1977a).

Distribution

At present, marsupials in the Eastern Hemisphere are confined to the Australian region, and they have undergone an extensive adaptive radiation in Australia and New Guinea. Some species have colonized the Celebes and the Moluccas. In the Western Hemisphere, the major adaptive radiation of marsupials occurred in South America. Most species are now found in South America, although many have extended their ranges into Central America. One species, *Didelphis virginiana*, has colonized temperate North America.

In the region dealt with in this volume, some tropical species are reaching the limits of their distribution, to be replaced in the south by more temperate-adapted forms. Genera typical of the tropical areas reach parts of northwestern Argentina in the provinces of Salta and Jujuy, eastern Argentina in the province of Misiones, and adjacent parts of eastern Paraguay. In the subtropics of Argentina, *Caluromys* and *Chironectes* are to be found, as well as some of the tropical species of *Marmosa*. Seven species of *Monodelphis* are recorded from southern South America, and one species, *M. dimidiata*, extends to the pampas region of Argentina. The truly temperate portions of Argentina and Chile are penetrated by *Didelphis albiventris* and the three genera endemic to the south—*Lestodelphys, Rhyncholestes,* and *Dromiciops.*

History and Classification

The oldest known marsupials are found in the mid-Cretaceous of North America. Cretaceous records from South America are late Cretaceous, but the radiation, as exemplified by the fossil record in South America, suggests that earlier finds of marsupials may await discovery (Marshall, De Muison, and Sige 1983). Throughout the Tertiary in South America marsupials radiated into insectivore, frugivore, and carnivore niches (see Marshall 1982a,b,c; Streilein 1982c for an extended account). Fossil marsupials first appear in Australia during the Oligocene, but clearly they must have arrived on the island continent before it was so distant from Antarctica. The recent discovery of fossil marsupials in the peninsula of Antarctica suggests that before the glaciation, when Antarctica, Australia, and South America were closer together, Antarctica may have served as the bridge for the earliest marsupials' transit from South America to Australia (Woodburne and Zinmeister 1982).

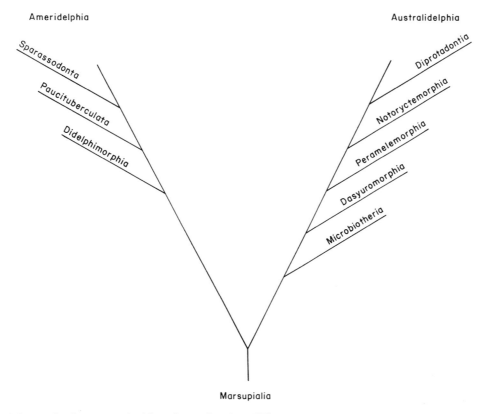

Figure 2.1. Cladogram for the Marsupialia (after Alpin and Archer 1987).

The order Marsupialia is classically divided into thirteen families, eight of which are extant. Ride (1964) proposed that the Marsupialia be divided into four orders. This view has been refined, and in this volume we follow the classification developed by Kirsch and Calaby (1977) and Alpin and Archer (1987). These authors have attempted to link an extant genus of South American marsupial (*Dromiciops australis*) with the major Australian radiation (see fig. 2.1).

FAMILY DIDELPHIDAE

Opossums, Comadrejas, Cuicas

Diagnosis

The dental formula tends to be conservative, with 50 teeth typical: I 5/4, C 1/1, P 3/3, M 4/4. The postorbital bar is not developed (fig. 2.2). The pouch may be either absent, as in the genera *Marmosa*, *Monodelphis*, and *Metachirus*, or well developed, as in *Didelphis*, *Philander*, *Chironectes*, and during lactation in *Caluromys*. There are five digits on each foot, and the hind foot has an opposable thumb. Members of this family have a long rostrum, large naked ears, and a tail that is often highly prehensile and usually almost hairless for at least the distal two-thirds. *Glironia* is exceptional in having its tail furred almost to the tip. This family exhibits a wide range of sizes. Some species may be as small as 80 mm in head and body length, while the larger species may reach 1,020 mm in total length. Pelage color is highly variable, ranging from gray to deep, rich brown. Some species exhibit a banding pattern of the dorsal pelage. Karotypic data have been summarized by Reig et al. (1977) and Seluja et al. (1984).

Comment

Certain characters of *Caluromys*, *Caluromysiops*, and *Glironia* lend credence to the tendency to place

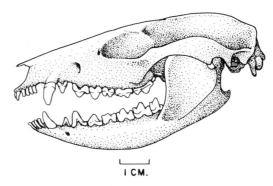

Figure 2.2. Skull of *Didelphis marsupialis*.

them in their own family or subfamily (Caluromyinae) (Creighton 1984, Reig, Kirsch, and Marshall 1987).

Distribution

Species of this family occur widely over the Neotropics and occupy almost every habitat type except extremely high elevations and the desert areas surrounding the Gulf of Venezuela, southwestern Peru, northwestern Chile, and the extreme temperate south.

Natural History

Most members of this family are nocturnal. All occupy omnivore, insectivore, frugivore/insectivore, or carnivore feeding niches. Many members are strongly arboreal and seldom come to the ground, especially those long-tailed forms that are specialized for multistratal tropical rain forests.

Marsupials are characterized by a unique mode of reproduction. All forms studied in the Western Hemisphere have only a yolk-sac placenta, and the shell membrane remains intact during the intrauterine growth phase. The young are born after a brief period of gestation, generally thirteen to fourteen days for the South American species. Upon passing out of the reproductive tract, the young climb unassisted to the nipple area, where each grasps a teat in its mouth and remains attached to it for a varying time during early growth and development. The mother transports the entire litter attached to her teats for approximately the first five to six weeks of their development. The young then detach and begin a nest phase where the mother does not continuously transport them but returns to the nest to nurse them. The teat area may be enclosed in a pouch, or no pouch may be present (e.g., *Marmosa*, *Metachirus*, and *Monodelphis*).

Eisenberg and Wilson (1981) measured the cranial capacities of several species of didelphid marsupials. When comparing their cranial capacities with those of eutherian mammals they found that several didelphid species (e.g., *Monodelphis* and *Didelphis*) had extremely small brains, but many others had brain sizes near the norm for small eutherian mammals. The wide variety of relative brain sizes within the family Didelphidae strongly suggests that we should reexamine generalizations concerning members' behavioral capacities based on *Didelphis*.

By comparing the relative proportions of limb segment length, trunk length, and tail length, we can classify didelphid marsupials into groups reflecting different locomotor adaptations. Terrestrial adaptations include a shortened tail (e.g., *Monodelphis*), and rapid terrestrial locomotion involves increasing the relative mass of the hind leg (e.g., *Metachirus*). Arboreal adaptation is reflected by a long tail with a proportionately high mass and relatively large, powerful forelimbs (Grand 1983).

McNab (1978) studied the energetics of Neotropical marsupials and concluded that, in contrast to most Australian species, their basal metabolic rates fall in line with rates measured for many insectivorous and frugivorous eutherian mammals. It is true that no didelphid marsupial has a high metabolic rate, but it is untrue to say that they have lower metabolic rates than comparable-size eutherians specialized for similar diets.

The community ecology of didelphids has been receiving more attention recently. Fleming (1972) studied three sympatric species in Panama. Charles-Dominique (1983) has completed an extensive comparative study of didelphid ecology in French Guiana. O'Connell (1979) investigated a didelphid community in northern Venezuela and has analyzed reproductive performance, habitat preference, and substrate utilization. Communication mechanisms were reviewed by Eisenberg and Golani (1977). The natural history of didelphid marsupials has been admirably reviewed in an article by Streilein (1982c) and in books by Hunsaker (1977a) and Collins (1973).

In the southern cone of South America the species of the genus *Monodelphis* appear to occupy the small terrestrial, insectivore, omnivore niche. Rodents of the genera *Oxymycterus*, *Geoxus*, and *Notiomys* are clearly providing some trophic competition with *Monodelphis*. One wonders, given the absence of the classical "Insectivora," to what extent rodents sustain competition with the "old endemic" insectivorous marsupials.

Genus *Caluromys* J. A. Allen, 1900
Woolly Opossum, Zorrito de Palo

Description

The dental formula of *Caluromys* is I 5/4, C 1/1, P 3/3, M 4/4. The first upper premolar is very small, situated directly behind the canine, and there is a distinct gap between the first premolar and the much larger second premolar. The fur of the body is very thick, hence the common name woolly opossum. The dorsal pelage varies from brown to gray brown, and a brown stripe runs from the muzzle to between the ears. The tail, furred for more than one-third of its length, is extremely long and fully prehensile. There are three recognized species (see plate 1).

Distribution

The genus is distributed from southern Veracruz in Mexico to Paraguay. The strong arboreal adaptations of this genus limit its distribution to areas of moist forest.

Life History and Ecology

The woolly opossums are distinctive when compared with other members of the family Didelphidae. Their litter size tends to be somewhat reduced, and they can live longer in captivity. *Caluromys* is typified by a comparatively large encephalization quotient (Eisenberg and Wilson 1981). A pattern of more extended maternal care is suggested by the studies of Charles-Dominique (1983) and Charles-Dominique et al. (1981). Trapping records and radiotelemetry indicate that these strongly arboreal animals come to the ground infrequently. They tend to be mixed feeders, eating the pulp of various fruits and supplementing this diet with nectar, invertebrates, and small vertebrates. Vocalizations of *Caluromys* are similar to those of other didelphids. A hiss is employed with an open-mouth threat during defensive behavior. Click sounds are made during male-female encounters and courtship and may grade into chirps (Eisenberg, Collins, and Wemmer 1975).

Caluromys derbianus and *Caluromys lanatus* appear not to overlap geographically, but *Caluromys philander* can co-occur with *C. lanatus*. The latter species is much larger than *C. philander*, and there is a strong suggestion that when congeners co-occur there is ecological segregation in feeding habits, indirectly indicated by the size differences (Eisenberg and Wilson 1981; Charles-Dominique et al. 1981).

• *Caluromys lanatus* (Illiger, 1815)[1]
Woolly Opossum, Cuica Lanosa

Measurements

	Mean	Min.	Max.	N	Loc.	Source[a]
TL	661.0	602	702	18	B, P, Peru	1
HB	273.3	201	319			
T	387.7	341	440			
HF	42.2	30	48	16		
E	34.9	30	40		P	
Wta	3,201.0					

[a](1) BA, FM, UM.

Description

The tail is long and heavily furred over 50% of its length, the fur gradually becoming thinner, with the distal 30% of the tail naked. All digits have well-developed pads and claws. This opossum has long, woolly fur that is light brown dorsally and darker, grading to orangish, on the shoulders, the front legs,

and the top of the head. The venter is lighter. The ears are uniformly darkish, and there are indistinct orangish eye-rings. The face is grayish white with a dark brown stripe originating at the nose and passing between the eyes to fade into orangish on top of the head (pers. obs.).

Chromosome number: $2n = 14$; FN $= 24$ (Reig et al. 1977).

Distribution

Caluromys lanatus is distributed from Colombia and Venezuela south to northeastern Argentina. In southern South America it is found only in eastern Paraguay and the Argentine province of Misiones

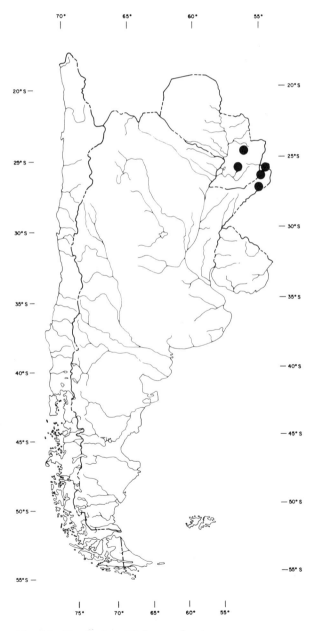

Map 2.1. Distribution of *Caluromys lanatus*.

---

1. Throughout this volume, a bullet before the species name indicates that the species was discussed in volume 1.

(Honacki, Kinman, and Koeppl 1982; Massoia and Foerster 1974; UM) (map 2.1).

Life History
In captivity estrus occurs at an average interval of twenty-eight days (Bucher and Fritz 1977).

Ecology
A specimen from Paraguay was obtained on a branch 10 m above the ground, and all indications are that this species is highly arboreal. In Venezuela all specimens taken were associated with multistratal evergreen tropical forests. This species is typical of rain forest and is at the extreme portion of its range in southern South America (Eisenberg 1989; Handley 1976; UM).

Genus *Chironectes* Illiger, 1811
• *Chironectes minimus* (Zimmerman, 1780)
Yapok, Water Opossum, Cuica de Agua

Measurements

	Mean	Min.	Max.	N	Loc.	Source[a]
TL	641.0	592	710	14	A, Br, P	1
HB	289.1	259	350			
T	351.9	310	386			
HF	64.5	60	69	10		
E	26.5	24	29			
Wta	386.7	550	650	3	P	

[a](1) Crespo 1974; Vieira 1949; BA, UM.

Description
The dental formula is I 5/4, C 1/1, P 3/3, M 4/4 (Marshall 1978c; Mondolfi and Medina Padilla 1957; Oliver 1976). The ears are moderately large, naked, and rounded. In addition to the usual distribution of facial hairs, the yapok has supernumerary facial bristles that are stout and long. The tail, which is longer than the head and body, is round and powerful, naked along most of its length, and coarsely scaled. Both sexes have a well-developed pouch. In males the scrotum is clearly visible because of its mustard-colored fur and is pulled into the pouch when the animal is in the water. The female is also able to seal her backward-opening pouch when in the water. The forefeet have expanded fingers with expanded fingertips, lack claws, and are used with great dexterity. The hind feet, webbed to the end of the toes, are used in swimming.

The peculiar color pattern and webbed feet make this, the only semiaquatic opossum, unmistakable (plate 1). It has short, dense, water-repellant fur, generally grayish white marbled with deep brown. The muzzle, the top of the head, and a band extending through the eye to below the ear are deep blackish brown. The dorsum is marbled gray or black, with rounded black areas coming together along the midline and extending from the top of the head to the base of the tail and expanding laterally into four broad transverse patches placed over the shoulders, center of the back, loins, and rump. The chin, chest, and belly are pure white, and the tail is black proximally and yellowish terminally.

Chromosome number: $2n = 22$ (Reig et al. 1977).

Distribution
The yapok is confined mostly to tropical and subtropical areas, from southernmost Mexico and Central America through Colombia, Venezuela, the Guianas, Ecuador, Peru, Paraguay, and along the eastern side of Brazil to northeastern Argentina. In southern South America it is found in Misiones, Ar-

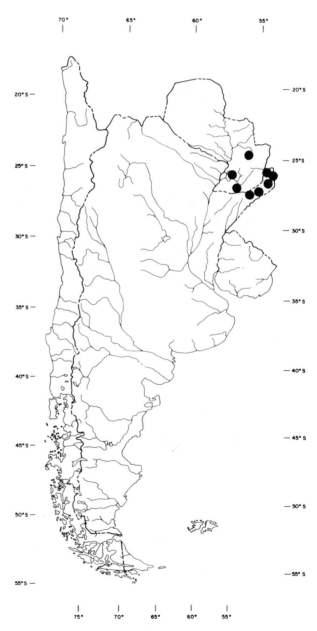

Map 2.2. Distribution of *Chironectes minimus*.

gentina, and in eastern Paraguay (Marshall 1978c; Massoia 1976; UM) (map 2.2).

Life History

Litter size in *Chironectes minimus* is one to five, with two to three most common. Animals from Venezuela have four or five nipples. The female keeps the young in her pouch when she swims. In captivity one female had her first estrous cycle at ten months (Marshall 1978c; Mondolfi and Medina Padilla 1957; Oliver 1976).

Ecology

The yapok is confined to areas of permanent water such as streams or rivers, usually within a forest. It is an excellent swimmer and diver, paddling with its hind feet and using its tail as a rudder. The den is usually a subterranean cavity, reached through a hole in the stream bank just above water level. Within this cavity it builds a nest; in captivity animals have been observed to transport nesting material with their tails.

Yapoks are carnivorous, eating small fish, crabs, crustaceans, insects, and frogs. Prey are captured with either the front feet or the mouth (Marshall 1978c; Mondolfi and Medina Padilla 1957; Oliver 1976; UM).

Genus *Didelphis* Linnaeus, 1758
Large American Opossum, Rabipelado, Comadreja

Description and Taxonomy

This is the large opossum commonly encountered over most of North and South America. The dental formula is I 5/4, C 1/1, P 2/3, M 4/4. Head and body length ranges from 325 to 500 mm and the tail from 255 to 535 mm. Animals may weigh up to 5.5 kg. The pouch is well developed in the female. The pelage consists of two hair types, a dense underfur and long guard hairs that are generally white tipped, giving a shaggy appearance. The almost naked tail is darkly pigmented from the base for approximately one-third of its length; the distal portion is generally white to pink. The forelimbs tend to be black, and the face is white to yellow. Depending on the species, the basic dorsal color may be gray with varying amounts of black; melanistic forms are known from several parts of its range (see plate 1).

There are three recognized species: *Didelphis albiventris* (= *azarae*), *D. marsupialis*, and *D. virginiana*. *D. virginiana* ranges from Massachusetts, in the United States, to Costa Rica (Gardner 1973). Its behavior has been studied in detail by McManus (1970). Life history strategies are reviewed by Sunquist and Eisenberg (1991).

Distribution

The genus is distributed from the northeastern United States to Patagonia in Argentina. The species *D. virginiana* has been introduced to the western United States, where it occurs from southern British Columbia to California.

• *Didelphis albiventris* Lund, 1840
White-eared Opossum, Comadreja Común, Overa

Measurements

	Mean	Min.	Max.	N	Loc.	Source[a]
TL	763.5	700	850	14	U	1
T	372.9	335	408			
HF	59.6	42	67			
E	54.1	49	61			
Wta	1,560.0	1,200	2,159			

[a](1) Barlow 1965.

Description

The white-eared opossum closely resembles its congener *D. marsupialis*, but adults are clearly separable by ear color. As in the black-eared opossum, there is considerable variation in body color within a population, with some individuals considerably darker than others. Another character that appears to separate the congeners is the length of the furred portion of the tail: *D. albiventris* has fur extending several centimeters up from the base, whereas the tail of *D. marsupialis* is mostly unfurred (see plate 1). Care must be taken in identifying juvenile *Didelphis*, because the white ear color of *D. albiventris* is much less pronounced in young animals (Crespo 1982; pers. obs.).

Chromosome number: $2n = 22$ (Reig et al. 1977).

Distribution

The white-eared opossum inhabits the subtropical and pampean zones from Mato Grosso state in Brazil south to central Argentina and the Andean zones from Venezuela to western Argentina. In southern South America *D. albiventris* is found throughout Paraguay and Uruguay, and in Argentina it occurs south to about Río Negro (about 40° S). In Jujuy it is found up to 2,600 m (Barlow 1965; Cerqueira 1984, 1985; Crespo 1982; Olrog 1979; Uconn (map 2.3).

Life History

In Uruguay nine females were found to have an average of 9.4 young, with a range of 4 to 12. In northeastern Argentina the mean number of young per litter was recorded as 7.1, with one to two litters produced between August and January. A study of opossum reproduction in northeastern Brazil concluded that mature males are always fertile and that female fertility is associated with the onset of rains.

This conclusion was also reached in a study of this species in Minas Gerais, Brazil, and may well apply to the Chaco as well (Barlow 1965; Cerqueira 1984; Crespo 1982; González 1973; Rigueira et al. 1987).

Ecology

Didelphis albiventris is a habitat generalist, apparently found everywhere except very high areas, very dry areas, and areas of dense woodland or forest. It is mainly terrestrial, though it can climb well. In Misiones province, Argentina, stomachs were found to contain large numbers of worms together with ants, small birds, eggshells, and vegetation. In northern Argentina it feeds on ripening grapes and fruit.

A study in Tucumán, Argentina, found an average home range of 5,700 m² for six animals (Barlow 1965; Cajal 1981; Crespo 1982; Fonseca, Redford, and Pereira 1982; Mares 1973; Mares, Ojeda, and Kosco 1981; UM).

Comment

Didelphis albiventris equals *D. azarae.*

• *Didelphis marsupialis* Linnaeus, 1758
Black-eared Opossum, Comadreja Grande

Measurements

	Mean	Min.	Max.	N	Loc.	Source[a]
TL	753.9	620	1,081	13	A, Br, P	1
HB	402.7	310	730			
T	351.2	310	390			
HF	54.6	45	60			
E	47.4	42	53			
Wta	1,012.5	725	1,400	4		

[a](1) Crespo 1974; Varejão and Valle 1982; UConn, UM.

Description

Didelphis marsupialis is similar in appearance to *D. albiventris*, but with darker dorsal pelage. The black ears separate it from *D. albiventris*. As in the white-eared opossum, there is considerable color variation within a population (Crespo 1974; pers. obs.).

Distribution

The species is found from Mexico south to northeastern Argentina. In southern South America it is found in eastern Paraguay and the Argentine province of Misiones (Crespo 1974; Cerqueira 1985; Honacki, Kinman, and Koeppl 1982; UM) (map 2.4).

Life History

Black-eared opossums are usually solitary, though two or more may be encountered together during the breeding season when males actively court females. The female builds a leaf nest in a tree cavity or burrow. Litter size varies with latitude, with the smallest litters near the equator.

In captivity, animals from southeastern Brazil had a mean litter size of 10.7 (*n* = 3), with a gestation period of fourteen to fifteen days. In the wild, animals from the same population produce two or three litters a year. Females from Rio de Janeiro state, Brazil, had an average of 7.2 young and had two litters a year, one in August and one in October (Eisenberg 1989; Motta, Carreira, and Franco 1983; Davis 1947).

Ecology

This species has been reasonably well studied in the northern portion of its range, but very little is known about its ecology in southern South America.

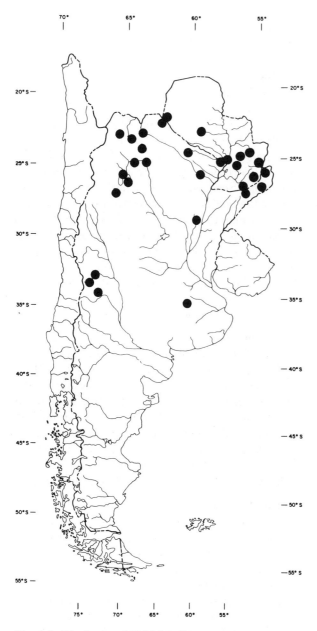

Map 2.3. Distribution of *Didelphis albiventris.*

Given adequate shelter and a sustained food supply, the home range of a lactating female may be rather stable, but the animals are opportunistic feeders and readily shift home ranges to adapt to fluctuating resources. In Venezuela, radio-tracking data indicate that extended home ranges are larger than estimates derived from trap, mark, and release studies. Mean home ranges in the llanos ranged from 123 ha for males and from 16 ha for females.

In a trapping study in the wet Atlantic forest of Rio de Janeiro state, Brazil, females were frequently recaptured in a small area, but males were very infrequently recaptured. In southeastern Brazil, *D. albiventris* can occur in microsympatry with *D. marsupialis* (C. M. C. Valle, pers. comm.), though in one study *D. marsupialis* was more typically found in moister habitats. *D. marsupialis* may be more sensitive to human disturbance than its congener (Cerqueira 1985; Davis 1945; Stallings 1988; Sunquist, Austad, and Sunquist 1987).

Genus *Lestodelphys* Tate, 1934
Lestodelphys halli (Thomas, 1921)
Patagonian Opossum, Comadrejita Patagónica

Measurements

	Mean	Min.	Max.	N	Loc.	Source[a]
TL	222.7	216.0	232.0	3	A	1
HB	134.7	132.0	139.0			
		132.0	144.0	—		2
T	88.0	81.0	99.0	3		1
		81.0	99.0	—		2
HF	16.5	15.7	17.7	3		1
		15.7	17.7	—		2
E	21.1	20.2	21.6	3		1
		18.0	22.0	—		2
Wta	76.0			1		3

[a](1) BA; (2) Marshall 1977; (3) MVZ.

Description

The dental formula is I 5/4, C 1/1, P 3/3, M 4/4, with a total of 50 teeth. The canines are exceptionally long. The ears are short, rounded, and flesh colored. The feet, markedly more robust than those of *Marmosa*, are equipped with claws that extend well beyond the terminal pads. The tail, much shorter than the head and body, is clearly used as a fat storage organ. This small opossum resembles *Marmosa elegans* in general dorsal color, being gray brown, almost pearl gray (plate 2). The fur is dense and very soft. The sides are paler, and the whitish to buff venter is sharply demarcated. The cheeks and a patch over the eyes are whitish, as are the hands and feet. The tail is dark grayish brown above and whitish below. In males the fur on the front of the throat is orange, and in females the fur around the nipples is orange (Marshall 1977; Crespo 1974; pers. obs.).

Distribution

The distribution of the Patagonian opossum is poorly known. It is found in the Argentine provinces of Río Negro, Santa Cruz, La Pampa, Mendoza, and Chubut. It occurs farther south than any other living marsupial (Marshall 1977; Crespo 1974; Thomas 1929; MVZ) (map 2.5).

Life History

Nothing is known of the biology of this rare animal. It has been reported to have nineteen mammae (Thomas 1929).

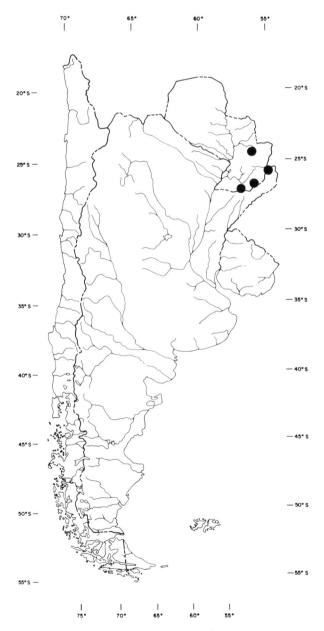

Map 2.4. Distribution of *Didelphis marsupialis*.

Ecology

The length of the canines and claws suggests a strongly carnivorous habit, and one specimen was trapped in a fox trap. Oliver Pearson (pers. comm.) reports that captive individuals killed live mice at lightning speed, eating everything—bones, teeth, and fur. A 70 g animal will eat an entire 35 g mouse in one night. The long, hard winters in the range of *L. halli* suggest that it may live like some of the north temperate weasels, foraging under the snow for small rodents and other animals, or may enter periods of torpor as suggested by data for *Marmosa elegans* in Chile (Marshall 1977; Thomas 1929).

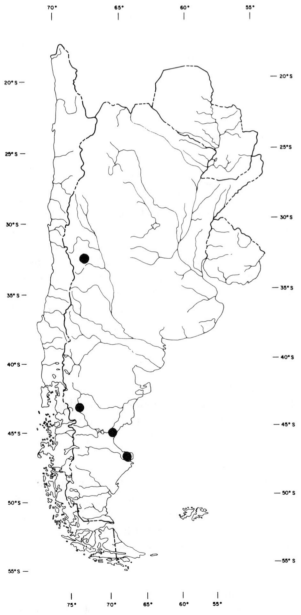

Map 2.5. Distribution of *Lestodelphys halli*.

Genus *Lutreolina* Thomas, 1910
• *Lutreolina crassicaudata* (Desmarest, 1804) Little Water Opossum, Thick-tailed Opossum, Comadreja Colorada

Measurements

	Mean	Min.	Max.	N	Loc.	Source[a]
TL	574.1	466	781	8	A, P, U	1
HB	289.4	197	378	17		
T	281.9	221	336	18		
HF	43.8	35	54			
E	26.3	22	38	16		
Wta	432.6	176	800	14		
	642.0	455	1,100	9	A	2

[a](1) Barlow 1965; Cajal 1981; Fornes and Massoia 1965; Mares, Ojeda, and Kosco 1981; CM, FM, UM; (2) Massoia 1973.

Description

The long weasel-like body and short, dense reddish or yellowish fur make this opossum distinctive. The tail is quite thick, naked at the tip and somewhat prehensile; it is furred for about 30% to 50% of its length. The ears are short and rounded and project only slightly above the fur. The limbs and feet are short and stout. There is disagreement about the extent of pouch development, but if it does occur, it is only slight. Some adults are only half the size of others. Males are often larger than females, suggesting a long period of continued growth in males.

There is considerable variation in coat color, but the upperparts are generally a rich soft yellow buff or dark brown, and the underparts are reddish ochraceous or pale to dark brown. Most of the tail is black, though the tip is often gray. The face is devoid of eye-rings or prominent markings, though the cheeks and chin may be lighter than the rest of the head (plate 1). It appears that coat color may vary with diet and climate (Barlow 1965; Lemke et al. 1982; Marshall 1978a; Ximénez 1967; UM).

Chromosome number: $2n = 22$ (Reig et al. 1977).

Distribution

Two apparently distinct populations of this species exist, one known only from a few specimens in Guyana and Venezuela, and the other occurring east of the Andes from Paraguay and southern Brazil south to central Argentina. It occurs in eastern Paraguay, in all of Uruguay, and in Argentina south to about Chubut province. In Jujuy province, Argentina, this opossum has been trapped up to 1,700 m elevation. The distribution of *L. crassicaudata* in Argentina may not currently connect across the center of the country (Barlow 1965; Marshall 1978a; Olrog 1976; Ximénez 1967) (map 2.6).

Life History
Very little information is available on this interesting opossum. One litter was recorded as seven and another as eleven (Lemke et al. 1982; Marshall 1978a; Ximénez 1967).

Ecology
This mesic-adapted animal is found along areas of permanent water. In Uruguay it is restricted to more extensive marshy and riparian habitats, and in Salto department it is most common along small, quiet brooks in dense forest. *L. crassicaudata* is somewhat weasel-like in shape and apparently in habits as well. It can swim and climbs well, and it is reported to be nocturnal, preying on small vertebrates, fishes, and insects. One stomach contained remnants of mollusk shells and sand, and this species has been caught in traps baited with mice. In captivity *Lutreolina* will eat insects and fruit and kills birds and mammals up to the size of *Microcavia*.

A trapping study found that two animals had an average home range of 800 m² (Barlow 1965; Cajal 1981; Marshall 1978a; Olrog 1976; Ximénez 1967).

Genus *Marmosa* Gray 1821
Mouse Opossum

Description
The dental formula is I 5/4, C 1/1, P 3/3, M 4/4. This genus is highly variable in size and includes some forty-seven species, nine of which occur in the area under consideration (R. Pine, in Collins 1973). Head and body length varies from 60 mm to over 200 mm, depending on the species, and the tail ranges from 100 to 281 mm and generally exceeds the head and body in length. The ear is usually slightly shorter than the hind foot. The tail is fully prehensile, and if its base is haired it is always for less than one-third the length of the tail. There are no white spots on the face; the eye has dark brown hairs around it so that from the front one sees a contrasting mask (plate 2). Dorsal pelage can vary from red brown to gray, and the venter is usually paler, varying from tan to cream.

Distribution
The genus is widely distributed from southern Veracruz, Mexico, to central Argentina. Different species often occupy discrete altitudinal zones, and some are adapted to the Andean foothills.

Natural History, Identification, and Systematics
The mouse opossums exhibit an interesting adaptive radiation, with many species apparently having narrow habitat requirements while others tolerate a wide range of habitat types. Arboreal ability varies from species to species; some are strongly arboreal and seldom come to the ground, while others show a more scansorial tendency (O'Connell 1979; Handley 1976). The female does not develop a pouch. The teats are arranged in varying symmetrical patterns in the posterior ventral area, and the number of nipples is variable both within and between species. For the genus, the nipple number ranges from nine to nineteen (Osgood 1921; Tate 1933); litter size is very high in some species. The young are born in an extremely undeveloped state after a thirteen- to fourteen-day gestation period and remain attached

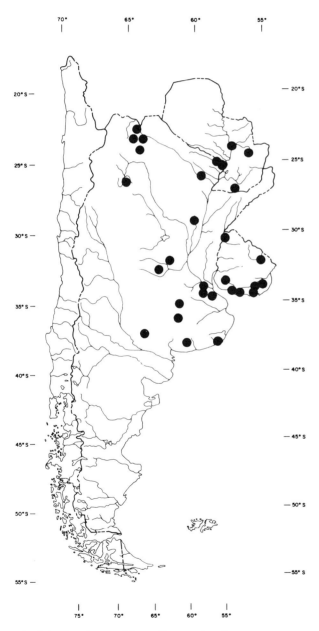

Map 2.6. Distribution of *Lutreolina crassicaudata*.

Table 2.1   Classification of *Marmosa* into Five Genera

Micoureus	*Marmosops* (continued)
M. cinereus	M. impavidus
M. constantiae	M. incanus
M. regina	M. invictus
M. alstoni	M. noctivagus
Marmosa	M. parvidens
M. andersoni	*Thylamys* (= *elegans* group)
M. canescens	T. elegans
M. lepida	T. macrura
M. mexicana	T. pallidior
M. murina	T. pusillus
M. robinsoni	T. velutinus
M. rubra	*Gracilinanus*
M. tyleriana	G. aceramarcae
M. xerophila	G. agilis
Marmosops (= *noctivaga* group)	G. dryas
M. cracens	G. emiliae
M. dorothea	G. marica
M. fuscatus	G. microtarsus
M. handleyi	

Source: Gardner and Creighton (1989).

to the teats for almost the first thirty days of life (Eisenberg and Maliniak 1967).

Creighton (1984) has proposed that *Marmosa* may be divided into at least four genera, whereas Reig, Kirsch, and Marshall (1987) would group the species into three genera. The recent classification by Gardner and Creighton 1989 is shown in table 2.1. Under the new scheme *Marmosa elegans* and allies would become *Thylamys* Gray 1943. Species of this genus have a woolly pelage and rounded premaxilla and store fat in their tails. Whereas most species of *Marmosa* have the teats arranged in a circular pattern, in members of the genus *Thylamys* the teats form the bilaterally symmetrical rows common in eutherians.

The genus *Micoures* Lesson, 1842, is distinctive in that most of the species are large, with head and body length exceeding 125 mm. The tail vastly exceeds the head and body length, averaging at least 1.3 times as long. The scales on the tail are rhomboid and coarse, with fourteen to sixteen rows per centimeter. The tail is never bicolored but is often whitish or mottled distally. The fur of the body extends at least 5 cm on the proximal portion of the tail. The tympanic bullae of the skull are small, and the postorbital processes are prominent.

The genus *Marmosops* Matschie, 1916, also has rhomboidal scales on the tail, but they are much finer, with twenty-two to twenty-eight per centimeter. The bullae are small, but the postorbital processes are lacking. The lower canines are premolariform, which is distinctive.

The genus *Marmosa* Gray, 1821, is subdivided

into the *Murina* group and the *Microtarsus* group; the latter could be elevated to the genus *Grymaeomys* Burmeister, 1854. In the *Microtarsus* group the tail scales are square and very fine grained, as many as forty per centimeter, and are annular rather than spiral. The tail tends to be weakly bicolor, and the bullae are large. The *Murina* group of the genus *Marmosa* has tail scales that are rhomboidal and coarse, at about fifteen to twenty per centimeter, and spiral as in *Micoures*. Although fur may extend on the proximal portion of the tail, it never exceeds 2.5 cm. The premaxillaries exhibit an acute outline. Postorbital processes are present and prominent.

The southern species of *Marmosa* often encounter extreme cold for brief periods. Some may be able to enter torpor for a few days. Fat storage on the body and in the tail may ameliorate nutritional stress during periods of reduced primary productivity. McNab (1978, 1982) has suggested that small marsupials may be near their energy limits when adapting to temperate climates. Their relatively long lactation and the obligate teat attachment phase of the young may place severe constraints on the reproductive female in successfully rearing a litter. It is no wonder that as marsupials reach temperate climates fat storage becomes a necessity.

Marmosa agilis
Marmosa Rojiza (Burmeister, 1854)

Measurements

	Mean	Min.	Max.	N	Loc.	Source[a]
TL	242.4	226	259	8	P	1
HB	100.9	91	109			
	75.4	70	84	13	A	2
T	141.5	128	162	8	P	1
	106.9	95	120	13	A	2
HF	17.4	16	19	8	P	1
	14.4	13	15	13	A	2
E	21.4	20	23	8	P	1
	15.9	15	18	13	A	2
Wta	27.0	23	34	8	P	1
	14.9	12	19	10	A	2

[a](1) UM; (2) Massoia and Fornes 1972.

Description

This medium-sized (note differences between samples) mouse opossum has a very long, bicolored tail equaling 140% of its head and body length. The ears are very large. The fur is close and smooth and uniform in length. The body color varies from dull, dusty brown to grayish brown, lighter on the sides and cream on the venter. The face is paler, with the cream of the belly extending well onto the cheeks, and adults have distinct black eye-rings (Massoia and Fornes 1972; Tate 1933; pers. obs.).

Distribution

Marmosa agilis is found in eastern Paraguay and several of the northeastern provinces of Argentina. It has not been recorded from Uruguay (Contreras 1982; Massoia 1970, 1980; Massoia and Fornes 1972; Tate 1933; UM) (map 2.7).

Life History

This mouse opossum is reported to have up to twelve young. Females lack a true pouch, and the teats remain hidden when the female is not lactating (Massoia and Fornes 1972; Nitikman and Mares 1987).

Ecology

Marmosa agilis is a characteristic inhabitant of the gallery forests of northeastern Argentina but has broad habitat tolerance. It has been caught under fallen trunks, in tree holes, and in moist woodland. It is reported to be an adept climber, and nests made of vegetation have been found 1.6 m off the ground. One such nest contained seven individuals. In eastern Paraguay it has usually been captured in vegetation, but sometimes has also been caught on the ground. In captivity these mouse opossums eat ground meat and drink a lot. They are frequently found in *Tyto* owl pellets.

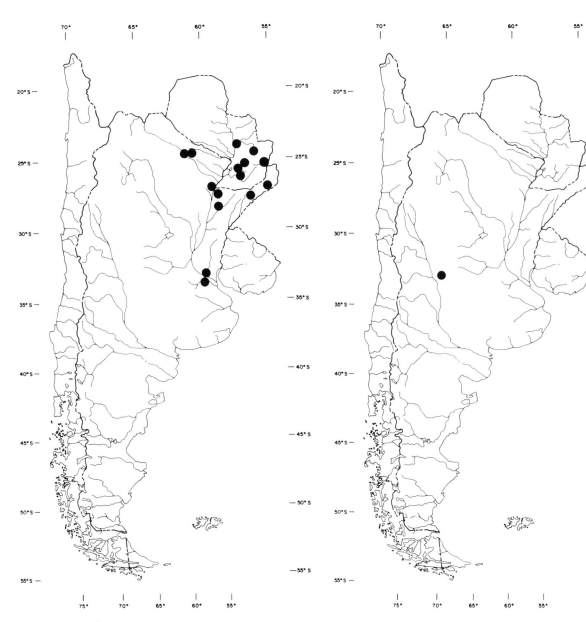

Map 2.7. Distribution of *Marmosa agilis*.

Map 2.8. Distribution of *Marmosa bruchi*.

Marmosa agilis is found throughout the Brazilian cerrado, usually associated with mesic areas such as gallery forests. In one gallery forest it was found at a biomass of 126 g/ha (Massoia and Fornes 1972; Nitikman and Mares 1987; UM).

Marmosa bruchi Thomas, 1921

Measurements

	Mean	Min.	Max.	N	Loc.	Source[a]
TL	170.7	159	185	7	A	1
HB	80.6	72	91			
T	90.1	84	91			
HF	12.4	11	13			
E	20.7	19	22			
Wta	14.9	11	19			

[a](1) UM.

Description

This medium to small *Marmosa* is predominantly gray dorsally, with a wash of warm brown. The brown increases laterally and is sharply demarcated from a white to cream belly. The face is paler, with narrow, dark eye-rings and white cheeks, and white hairs are also found at the base of the ears. The tail is faintly bicolored and serves as a fat storage organ (pers. obs.).

Distribution

In southern South America *M. bruchi* has been taken in the Argentine provinces of San Luis and La Pampa (MVZ, UM) (map 2.8).

Ecology

This species has been caught in traps on the ground as well as 1 m off the ground (UM).

Comment

G. K. Creighton (pers. comm.) considers *M. bruchi* a subspecies of *M. pallidor*, a species not listed in Honacki, Kinman, and Koeppl (1982). This group awaits formal revision.

• *Marmosa cinerea* (Temminck, 1824)
Marmosa Grande Gris

Measurements

	Mean	Min.	Max.	N	Loc.	Source[a]
TL	388.0	270.0	460.0	16	Br, A	1
HB	168.6	120.0	200.0			
T	219.4	150.0	260.0			
HF	25.1	22.5	29.5	8		
E	27.5	25.0	30.0	2		
Wta	109.9	56.0	194.0	36 m	Br	2
	99.1	53.0	230.0	28 f		

[a](1) Massoia 1972; Miranda-Ribeiro 1936; Vieira 1949; (2) Stallings 1988.

Description

This is the largest mouse opossum in Argentina and is readily identifiable by its size. In southeastern Brazil males are considerably larger than females. The ears are large to moderate in size and are gray; the tail is long and strikingly bicolored, with the distal portion cream or yellowish and the proximal portion black. Hair extends about 30 mm along the tail. The dorsum is some shade of gray, the sides are lighter, and the venter varies from buff to yellowish. The white of the venter extends to the entire chin and the lower cheeks. The eye is surrounded by a distinct dark eye-ring. *M. cinerea* is separable from its congener *M. agilis* by size and body color. The reddish dorsal coloration of live individuals fades soon after death (Husson 1978; Massoia 1972; Pine,

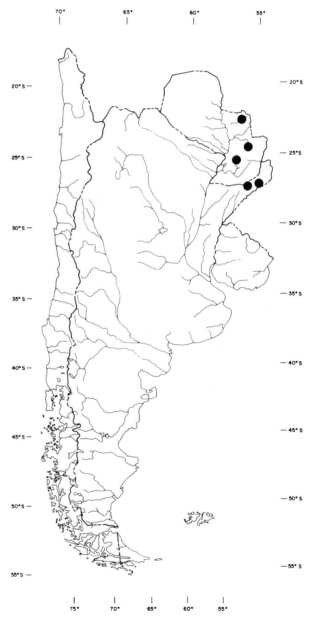

Map 2.9. Distribution of *Marmosa cinerea*.

Dalby, and Matson 1985; Stallings 1988; Tate 1933; pers. obs.).

Distribution

Marmosa cinerea has been trapped in northeastern Paraguay and in Misiones province, Argentina (Massoia 1972, 1980; UM) (map 2.9).

Life History

Females of this species do not have a pouch, and they have up to eleven teats. This mouse opossum has been studied in captivity, where it makes a nest with material it carries in its mouth. When detached from the teat the young utters a repetitive chirping cry, inducing the female to approach, grasp it with her forepaws, and push it under her venter, whereupon the young reattaches to the nipple.

In Venezuela breeding is tied to rainfall, with no reproduction during the winter dry season (Beach 1939; Eisenberg 1989; O'Connell 1979; Stallings 1988).

Ecology

Marmosa cinerea forages both arboreally and on the ground. In Venezuela it was trapped 47% of the time on the ground and 53% of the time in trees and bushes. In Minas Gerais, Brazil, it was found in brushy and forested habitats, though almost always off the ground. An examination of three stomachs showed only insects. It is nocturnal and constructs open, arboreal nests. From the limited data from southern South America, it appears that this species lives mostly in rich subtropical forest, though it has been captured arboreally in thorn forest in eastern Paraguay (Massoia 1972; Stallings 1988; UM).

Marmosa constantiae Thomas, 1904
Marmosa Grande Bayo

Measurements

	Mean	Min.	Max.	N	Loc.	Source[a]
TL	367.5	330	400	11	A, B, Br	1
HB	161.7	140	180			
T	205.7	190	220			
HF	23.5	21	27			
E	25.8	23	31			
Wta	90.0			1		

[a](1) Miranda-Ribeiro 1936; Olrog 1959; CM, FM, UM.

Description

The tail of this large *Marmosa* is about 125% the length of the head and body, proportionally slightly shorter than in *M. cinerea*. It is gray for most of its length, white on top for its final 20% and on the bottom for its final 40%. The dorsum is gray to gray brown; the venter is some shade of cream to yellow. There is frequently a thin line of bright tanish orange

separating the dorsal and ventral colors. The black eye-rings are well developed (Tate 1933; pers. obs.).

Distribution

This mouse opossum is known from Mato Grosso, Brazil, western Bolivia, and northwestern Argentina. In southern South America it has been recorded from Salta province, Argentina (Honacki, Kinman, and Koeppl 1982; Olrog 1959) (map 2.10).

Life History

Females are reported to have fifteen inguinal mammae (Tate 1933).

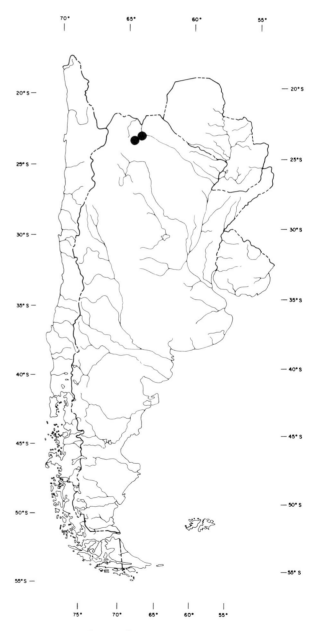

Map 2.10. Distribution of *Marmosa constantiae*.

Ecology

Marmosa constantiae is reported to be an arboreal inhabitant of moist forests (Olrog and Lucero 1981).

Marmosa elegans (Waterhouse, 1839)
Marmosa Elegante, Yaca

Measurements

	Mean	Min.	Max.	N	Loc.	Source[a]
TL	239.8	221.0	256.0	19	A	1
HB	114.1	106.0	121.0			
T	125.7	115.0	142.0			
HF	15.3	13.5	16.0			
E	22.5	21.0	24.6			
Wta	28.9	18.5	41.0	15		

[a](1) Cajal 1981; Mares, Ojeda, and Kosco 1981; Olrog 1959; CM.

Description

Marmosa elegans is a medium-sized mouse opossum whose tail, only slightly longer than the head and body, is finely haired throughout and frequently incrassated. The fur is very dense and soft, and the ears are large and naked. The color is generally grayish or light brownish, with lighter sides, a pure white, yellowish white, or gray white venter, and conspicuous black eye-rings that extend toward the nose. There is considerable variation in body color throughout the range. This is the only *Marmosa* found in Chile (Osgood 1943; pers. obs.).

Distribution

This wide-ranging mouse opossum is found in southern Peru, southern Bolivia, northwestern Argentina, and most of Chile, from sea level to at least 2,500 m. Except for one report eighty years ago, there are no verifiable specimens of *M. elegans* from south of the Bío-Bío River in Chile (Miller 1980; Tamayo and Frassinetti 1980) (map 2.11).

Life History

Females are reported to have nineteen nipples and up to fifteen young, though eight to twelve is a more common number. In Chile they reproduce from September to March, during which time two litters can be produced. There is little or no pouch development (Mann 1978; Schneider 1946; Thomas 1927).

Ecology

Marmosa elegans nests in various places—under rocks or roots, in trees and within cane patches, in rocky embankments, and in holes in the ground made by *Cavia*. This species stores fat, especially in the base of the tail, and animals hibernate during the winter.

The species has wide habitat tolerance: in Fray Jorge National Park in Chile it has been trapped in wet forest, brush, and riverine scrub. It is a typical element of the matorral scrub in Chile. In all habitats it seems to occur at fairly low densities. In Tucumán two animals were found to have an average home range of 289 m².

Marmosa elegans can be captured both arboreally and terrestrially. It is primarily an arthropod eater (arthropods and insect larvae made up 90% by volume of the diet in one study), though it also eats fruit, small vertebrates, and probably even carrion (Bruch 1917; Cajal 1981; Glanz 1977a,b; Mann 1978; Mares, Ojeda, and Kosco 1981; Meserve 1981a,b; Roig 1971; Schamberger and Fulk 1974; Schneider 1946; Simonetti, Yáñez, and Fuentes 1984; Thomas 1926).

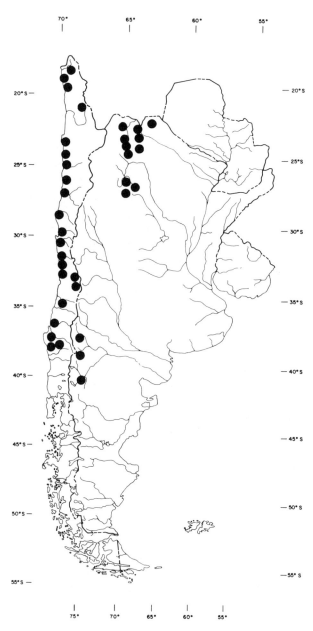

Map 2.11. Distribution of *Marmosa elegans*.

Marmosa formosa Shamel, 1930
Marmosa Enana

Measurements

	Mean	Min.	Max.	N	Loc.	Source[a]
TL	183.0	176	187	3	A	1
HB	90.0	83	94			
T	93.0	93	93			
HF	13.0	13	13			
E	20.7	20	21			

[a](1) Olrog 1959.

Description

This mouse opossum is *Monodelphis*-like in appearance, with short, stout feet, a short, tapering tail, and the dorsum a reddish brown. Ventrally it is buffy brown. The eye-rings are narrow and rather uniform, and the face is paler than the dorsum (Tate 1933; Olrog and Lucero 1981).

Distribution

This is another poorly known species of *Marmosa* and has been collected only from Formosa province in Argentina (Olrog 1959) (map 2.12).

Ecology

Marmosa formosa has been trapped in brush areas and moist forest (Olrog and Lucero 1981).

Comment

Marmosa formosa was formerly regarded as a subspecies of *M. velutina*.

Marmosa grisea (Desmarest, 1827)

Description

This small mouse opossum is known only from a few specimens. One specimen from UM measured TL 275; T 140; HF 17; E 25; Wta 54. One dried skin had a head and body length of 120 mm and a tail length of 155 mm. It has short, soft, velvety fur, mouse gray above, with a rufous tone on the shoulders and a pure white to creamy white venter. The eyes are clearly ringed with black, the ears are large and leafy, and the long tail is naked except for its basal centimeter. The tail is gray above for one-third to one-half its length, and the terminal half and underside are white (Goeldi 1894; Thomas 1894; pers. obs.).

Distribution

This mouse opossum is apparently confined to eastern Paraguay and southeastern Brazil. It has been trapped in eastern Paraguay and in the Serra dos Orgãos in Brazil (Goeldi 1894; Thomas 1894) (map 2.13).

Marmosa microtarsus (Wagner, 1842)

Measurements

	Mean	Min.	Max.	N	Loc.	Source[a]
TL	199.3	221	256	3	Br	1
HB	91.7	80	112			
T	107.7	95	130			
HF	15.0			1		
E	18.0			1		
Wta	31.0			1		2

[a](1) Miranda-Ribeiro 1936; Vieira 1949; (2) Stallings 1988.

Description

The tail is equal to 118% of the head and body length. The long pelage is usually rough or shaggy looking because of the presence of numerous guard hairs. This comparatively brightly colored *Marmosa*

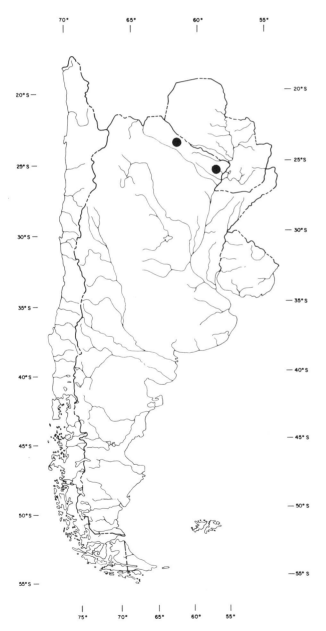

Map 2.12. Distribution of *Marmosa formosa*.

has a dorsum ranging from tawny to russet. The face is distinctly paler than the body, and the vibrissae are well developed and rather long. *M. microtarsus* is more vividly colored than *M. agilis* and has a rougher coat (Tate 1933).

Distribution

In southern South America *M. microtarsus* is known only from a few specimens from Rio de Janeiro and São Paulo states in Brazil and from Misiones province in Argentina (Davis 1947; Massoia 1980; Vieira 1949) (map 2.14).

Ecology

In moist Atlantic forest in eastern Brazil *M. microtarsus* was regularly caught on the ground as well

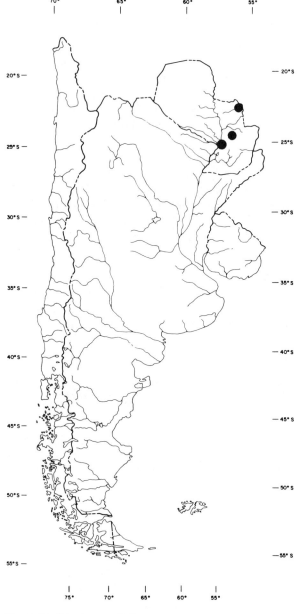

Map 2.13. Distribution of *Marmosa grisea.*

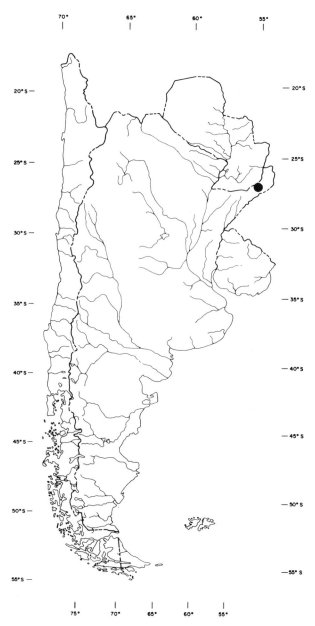

Map 2.14. Distribution of *Marmosa microtarsus.*

as in trees, in both virgin and second-growth forest, though its morphology suggests extensive arboreality. It will probably be found to be confined to areas of moderate to high rainfall (Davis 1947; Stallings 1988; Tate 1933).

Marmosa pusilla (Desmarest, 1804)
Marmosa Común

Measurements

	Mean	Min.	Max.	N	Loc.	Source[a]
TL	197.9	170.0	235.0	26	A, P	1
HB	94.3	75.0	120.0			
T	103.6	90.0	134.0			
HF	12.6	8.0	15.0			

E   21.7   18.0   24.1
Wta   18.3   12.1   29.5   12

[a](1) Daciuk 1974; Mares, Ojeda, and Kosco 1981; Olrog 1959;
BA, CM, UConn.

Description

The tail is strongly bicolored and is equal to about 110% of the head and body length. In some specimens the throat gland is well developed. The fur is thick and fine. Dorsally *M. pusilla* is brownish gray, and ventrally it is yellowish to white. The ventral color extends out onto the legs and cheeks, and the black eye-rings are poorly defined (Olrog and Lucero 1981; Tate 1933).

Distribution

This mouse opossum is widely distributed in Argentina, southwestern Bolivia, and Paraguay. In Paraguay it has been trapped in the eastern department of Amambay and throughout the Chaco, and in Argentina it has been recorded as far south as the province of Chubut. In Jujuy province it has been trapped up to 3,500 m (Daciuk 1974; Honacki, Kinman, and Koeppl 1982; Olrog 1959; Olrog and Lucero 1981; UConn, UM) (map 2.15).

Ecology

Within its range *M. pusilla* is frequently found in very dry areas. In Salta province, Argentina, it inhabits much drier areas than its congener *M. elegans* and has been trapped on rocky xeric hillsides, in dry thorn scrub, and along watercourses in dense vegetation. In the Paraguayan Chaco this mouse opossum was frequently trapped in and near thorn forest. One burrow inhabited by this species was under a cactus at 1,950 m, and in Salta *M. pusilla* was observed active even when there was snow on the ground (Lucero 1983; Mares 1973; Mares, Ojeda, and Kosco 1981; Olrog 1979; UConn).

Genus *Metachirus* Burmeister, 1854
• *Metachirus nudicaudatus* (E. Geoffroy, 1803)
Brown Four-eyed Opossum, Cuica Común

Measurements

	Mean	Min.	Max.	N	Loc.	Source[a]
TL	570.0	469	638	11	A, Br, P	1
HB	261.7	210	309			
T	308.4	242	360			
HF	44.1	39	52	8		
E	29.6	22	37	10		
Wta	350.0			1		
	281.5	102	480	35	Br	2
	235.9	91	345	51		

[a](1) Crespo 1950; Miranda-Ribeiro 1936; BA, UM; (2) Stallings 1988.

Description

This medium-sized opossum is often mistaken for *Philander*, the gray four-eyed opossum. However,

the difference in body color makes *Metachirus* unmistakable (plate 1). The tail is considerably longer than the head and body (118%) and has fur extending only 5–25 mm over its base, much less than in *Philander*. The scaly part of the tail is black and white but not as sharply demarcated as in *Philander*. There is no pouch. The dorsum is grayish brown, and the head has a striking color pattern: a dark band extends from the tip of the snout through the eyes and along the base of the ears, reaching a point midway between the ears and on some individuals stretching several centimeters past them. The dark color forms eye-rings. The ears are large, rounded, and unlike those of *Philander*, entirely dark. Also unlike that of *Philander*, the entire venter is white

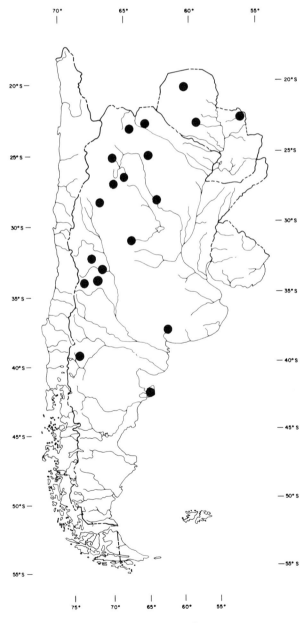

Map 2.15. Distribution of *Marmosa pusilla*.

or cream (Husson 1978; pers. obs.).

Chromosome number: $2n = 14$ (Reig et al. 1977).

Distribution

This widespread species is found from Nicaragua to Paraguay and northeastern Argentina. In southern South America *M. nudicaudatus* has been recorded from eastern Paraguay and Misiones province, Argentina (Crespo 1950; Honacki, Kinman, and Koeppl 1982; Massoia 1980; UM) (map 2.16).

Life History

Litter size ranges from one to nine, with a mean of five. The average teat number for a female is nine (Osgood 1921).

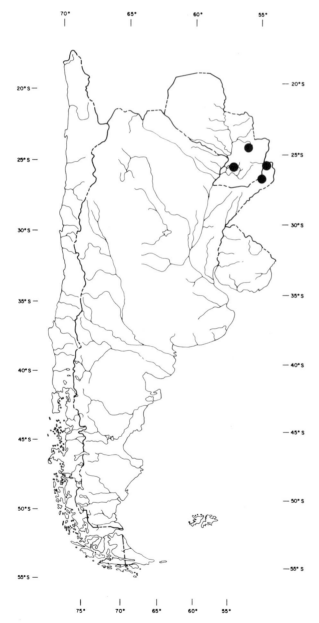

Map 2.16. Distribution of *Metachirus nudicaudatus*.

Ecology

Metachirus nudicaudatus is omnivorous, including fruits, small vertebrates, and invertebrates. In Minas Gerais state, Brazil, this species was caught in all forested habitats and found to be strongly terrestrial (Osgood 1921; Stallings 1988).

Genus *Monodelphis* Burnett, 1830
Short-tailed Opossum

Description

The dental formula is I 5/4, C 1/1, P 3/3, M 4/4. Head and body length ranges from 110 to 140 mm, and tail length from 45 to 65 mm. The tail is approximately half the head and body length, and this characteristic immediately distinguishes the short-tailed opossums from any other opossums within their range (fig. 2.3). Color varies widely depending on the species, from gray to chestnut brown. Some species have black dorsal stripes (plate 2).

Distribution

The genus *Monodelphis* is widespread from southwestern Panama to Argentina.

Life History and Ecology

Some species of this genus may be both nocturnal and diurnal. Species of the genus *Monodelphis* are predominantly terrestrial and crepuscular in their activity patterns. The litter size may be high in some species, and Pine, Dalby, and Matson (1985) suggest that *Monodelphis dimidiata* may exhibit a pattern of reproduction similar to that of the Australian *Antechinus stuartii*. In this form the males die shortly after a sharp seasonal reproductive period, and the adult females, after rearing their litters, do not survive to reproduce in the following year. The data of Pine, Dalby, and Matson are provocative, and appropriate field and laboratory studies could shed some light on this proposed reproductive pattern.

Monodelphis americana (Muller, 1776)
Three-striped Short-tailed Opossum,
Colicorto Estriado

Measurements

	Mean	Min.	Max.	N	Loc.	Source[a]
TL	153	142	159	3	Br	1
HB	107	100	111			
T	46	42	48			

[a](1) Vieira 1949.

Description

This distinctive short-tailed opossum has three black dorsal stripes, the middle one starting at the nose and extending to the base of the short tail, with the flanking ones shorter. The venter is heavily washed with orange (pers. obs.).

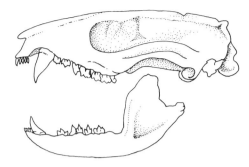

Figure 2.3. Skull of *Monodelphis* sp. (redrawn from Mares, Ojeda, and Barquez 1989) (typical skull CB = 25 mm).

Distribution

Monodelphis americana is distributed from Brazil to northern Argentina. In southern South America it has been recorded only from Misiones province, Argentina (Honacki, Kinman, and Koeppl 1982; Massoia 1980) (map 2.17).

Ecology

In the Atlantic forest of Brazil this species was found to make nests in the forks of trees or in bushes. In the Brazilian cerrado it was found to be terrestrial and active during the day (Davis 1947; Nitikman and Mares 1987).

Monodelphis dimidiata (Wagner, 1847)
Eastern Short-tailed Opossum, Colicorto Pampeano

Measurements

	Mean	Min.	Max.	N	Loc.	Source[a]
TL	164.3	114.0	231	23	Br, P	1
HB	99.6	55.0	151	16		
T	55.8	37.0	80			
HF	14.8	11.0	27	17		
E	10.9	7.1	14			
Wta	51.5	40.0	84	7		

[a](1) Barlow 1965; Fornes and Massoia 1965; Massoia and Fornes 1967; Miranda-Ribeiro 1936; Reig 1964; BA.

Description

This short-tailed opossum resembles the least weasel (*Mustela rixosa*) in general body form. The ears are very short. The tail, thick at the base and sparsely haired, is dark gray above and yellowish below. In animals from Uruguay the short, dense fur is ash colored from the center of the forehead back over the entire dorsum. Animals from Argentina have an olive brown dorsum, rufous sides, and a bright orange tan venter. On the sides of the head, flanks, and legs the fur is yellowish orange, paling on the venter (Barlow 1965; Miranda-Ribeiro 1936; pers. obs.).

Distribution

Monodelphis dimidiata is found in central Brazil, through Uruguay to the pampean region of Argen-

tina and as far west as Salta province, Argentina (Honacki, Kinman, and Koeppl 1982; Ojeda and Mares 1989) (map 2.18).

Life History

In Buenos Aires province, Argentina, litters were found in December–January, with a maximum of sixteen young. Litter sizes ranging from eight to fourteen have been reported. Neonates weighed between 0.08 and 0.11 g ($n = 3$). Breeding is done only by young of the previous year. There is great sexual dimorphism, and it has been suggested that this species may be semelparous like some of the Australian species of *Antechinus* (Pine, Dalby, and Matson 1985).

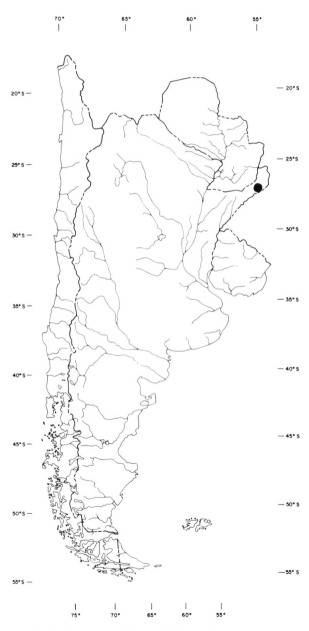

Map 2.17. Distribution of *Monodelphis americana*.

Ecology

Shrewlike in behavior, this short-tailed opossum appears to be confined to areas of grassland and marshes; it has also been caught in cultivated and uncultivated fields and along stream banks. It has been called both nocturnal and diurnal, perhaps varying its temporal pattern with season. *M. dimidiata* builds nests of vegetation and is reported to feed on invertebrates, small vertebrates, and plant material. It occurs at low densities; one study determined a density of less than two individuals per hectare (Barlow 1965; Fornes and Massoia 1965; Massoia and Fornes 1967; Pine, Dalby, and Matson 1985; Talice, Lafitte de Mosera, and Machado 1960; Villafañe et al. 1973).

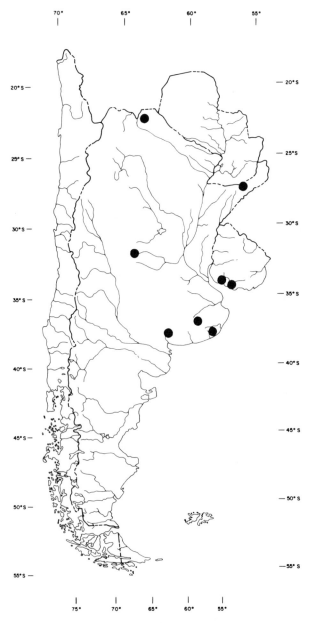

Map 2.18. Distribution of *Monodelphis dimidiata*.

Comment

Monodelphis fosteri was based on a juvenile specimen of *M. dimidiata* (Reig 1964).

Monodelphis domestica (Wagner, 1842)
Gray Short-tailed Opossum, Colicorto Gris

Measurements

	Mean	Min.	Max.	N	Loc.	Source[a]
TL	212.3	178	270	18	Br, P	1
HB	143.2	123	179			
T	69.1	46	91			
HF	17.7	14	22			
E	19.8	14	25	17		
Wta	71.4	58	95	11		

[a](1) Miranda-Ribeiro 1936; UConn, UM.

Description

This short-tailed opossum is a more or less uniform grayish brown dorsally with a whitish venter washed with orange. Some individuals have olive speckling on the dorsum and are lighter on the sides (Fadem et al. 1982; Olrog and Lucero 1981; pers. obs.).

Distribution

Monodelphis domestica is distributed in eastern and central Brazil, Bolivia, and Paraguay. It has been trapped in eastern Paraguay and throughout the Paraguayan Chaco (Honacki, Kinman, and Koeppl 1982; Myers and Wetzel 1979) (map 2.19).

Life History

Animals of this species obtained from the caatinga of Brazil were used to start a captive breeding colony. Extensive work on colony animals has made *M. domestica* one of the best studied of the South American small opossums. In captivity this species breeds throughout the year, and some females produce four litters a year. The female builds a compact, complicated nest, carrying nesting material with her tail.

Gestation lasts fourteen or fifteen days; young are born at about 0.10 g; litter size is three to fourteen, with an average of seven; and the estrous cycle is twenty-eight days. Young are attached to the nipple for about two weeks and then enter a nest phase. The female does not have a pouch but will transport young on her back. Young eat solid food at four to five weeks, can be separated from the female at seven weeks, and can reproduce at fifteen months. In captivity males often weigh considerably more than females.

In the Brazilian caatinga *M. domestica* reproduces through much of the year; it has at least two litters a year and maybe up to six. The shortest time between pregnancies is seven to eight weeks, and lactation lasts six to eight weeks. Age of first reproduction is five to seven months, and litters range from 6 to 11 with an average of 8.4. This reproductive pattern is probably the same in the Chaco. This species does not exhibit a near-semelparous mode of reproduction (Fadem et al. 1982; Streilein 1982a,b; Unger 1982).

Ecology

Primarily found in xeric situations, this species has been trapped in grassy areas, brush piles, and among jumbled rocks in a dry riverbed. It is also tolerant of man-made clearings. It is an accomplished predator, feeding primarily on invertebrates (Myers and Wetzel 1979; Streilein 1982a).

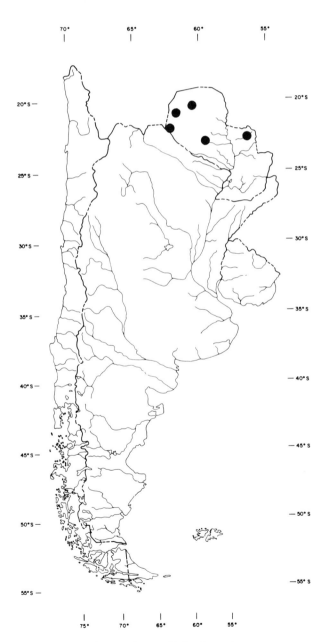

Map 2.19. Distribution of *Monodelphis domestica*.

Monodelphis henseli (Thomas, 1888)
Hensel's Short-tailed Opossum, Colicorto Rojizo

Measurements

	Mean	Min.	Max.	N	Loc.	Source[a]
TL	144.8	119.0	168.0	4	Br	1
HB	93.8	78.0	108.0			
T	51.0	36.0	62.0			
HF	15.2	15.0	15.5	3		
E	8.8	5.3	11.0			

[a](1) Miranda-Ribeiro 1936; FM.

Description

Monodelphis henseli is a small *Marmosa* with small ears. It is dark olivaceous brown to grizzled dark gray dorsally, with the venter, shoulders, and cheeks washed with rufous. The tail is like the dorsum above and rusty below (Miranda-Ribeiro 1936; Thomas 1888; pers. obs.).

Distribution

This species is distributed in southern Brazil, southeastern Paraguay, and northeastern Argentina (Honacki, Kinman, and Koeppl 1982; FM) (map 2.20).

Life History

Monodelphis henseli is recorded as having the impressive number of twenty-five nipples (Thomas 1888).

Comment

This species could be a junior synonym of *M. sorex* (Honacki, Kinman, and Koeppl 1982).

Monodelphis scalops (Thomas, 1888)
Colicorto de Cabeza Roja

Measurements

	Mean	Min.	Max.	N	Loc.	Source[a]
TL	208.0	206	210	2	Br	1
HB	145.5	145	146			
T	62.5	60	65			
HF	20.5	20	21			
E	16.0	16	16			
Wta	741.0					

[a](1) FM.

Description

This good-sized *Monodelphis* has distinctive markings. The head is a bright rufous extending to well behind the ears; from the shoulders to the rump the color is an agouti gray grading into dark rufous brown on the rump and tail base. The venter is a lighter version of the middorsal agouti gray flecked with yellow. There is a pronounced rufous color on the front legs and the chin as well as on the head (pers. obs.).

Distribution

This species is found in eastern Brazil, although recently it has been reported from Argentina (Honacki, Kinman, and Koeppl 1982; Olrog and Lucero 1981) (map 2.21).

Monodelphis sorex (Hensel, 1872)

Measurements

	Mean	Min.	Max.	N	Loc.	Source[a]
TL	114.7	112	118	3	Br	1
HB	70.7	67	73			
T	44.0	41	46			
HF	121.0					
E	5.2					

[a](1) Miranda-Ribeiro 1936.

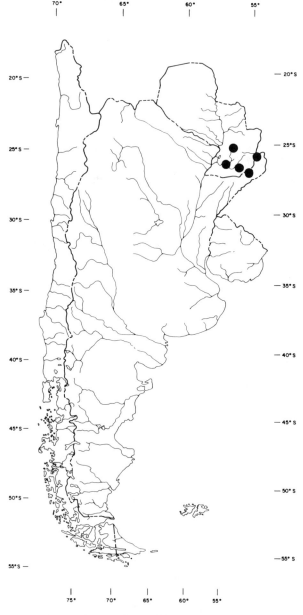

Map 2.20. Distribution of *Monodelphis henseli*.

Description

This species has a very dark brown dorsum, very lightly washed with rufus. The venter is a lighter shade, washed with yellow. On the venter the hair is sparse (pers. obs.).

Distribution

Although the species is thought to be confined to southeastern Brazil, specimens have been recorded from Misiones province, Argentina (Honacki, Kinman, and Koeppl 1982; FM).

Monodelphis touan (Shaw, 1800)

Description

One specimen from the Carnegie Museum measured TL 230; T 76; HF 19; E 17. The dorsum is grayish, with rosy to orange sides and an olive tan to cinnamon venter (Olrog and Lucero 1981; pers. obs.).

Distribution

This poorly known form is apparently found from Suriname south to Paraguay and northern Argentina. In Argentina it is reported to occur in Misiones province (Honacki, Kinman, and Koeppl 1982; Olrog and Lucero 1981) (map 2.22).

Comment

Monodelphis touan may be a junior synonym of *M. brevicaudata* (Honacki, Kinman, and Koeppl 1982) and is so treated in Eisenberg (1989).

Genus *Philander* Tiedemann, 1808
Gray Four-eyed Opossum, Chucha Mantequera, Zorro de Cuatrojos
• *Philander opossum* (Linnaeus, 1758)
Gray Four-eyed Opossum, Guaiki

Measurements

	Mean	Min.	Max.	N	Loc.	Source[a]
TL	558.6	437.0	620.0	12	A, Br, P	1
	525.6	500.0	546.0	5	Peru	2
HB	265.6	202.0	302.0	12	A, Br, P	1
T	293.0	235.0	320.0	12	A, Br, P	1
	277.8	253.0	299.0	5	Peru	2
HF	42.0	39.0	46.0	5	Peru	2
E	35.2	34.0	36.0	5	Peru	2
Wta	444.4	256.3	674.5	13	N	3

[a](1) Crespo 1950; Vieira 1949; BA, PCorps; (2) Gardner and Patton 1972; (3) Phillips and Knox Jones 1969.

Description

The most striking characters of this opossum are a sharply defined white spot above each eye (as in *Metachirus*) and a bicolored tail with the dark proximal part usually sharply separated from the much shorter white distal part (plate 1). The fur is short, soft, and woolly. The dorsum is dark gray grizzled with white; usually the central part is darker than the sides. The color is more blackish and less brownish than in *Metachirus*. Another pale spot is present in front at the base of the ears. The cheeks are whitish, and the large, rounded ears are whitish with black rims. The venter is cream colored, and the tail is furred for about 15% to 25% of its length (Husson 1978; pers. obs.).

Distribution

Philander opossum is distributed from southern Mexico south to northern Argentina, in Ecuador, eastern Peru, Bolivia, and Brazil. In southern South America it has been collected in the Chaco region and in Formosa and Misiones provinces, Argentina

(Crespo 1974; Honacki, Kinman, and Koeppl 1982; Massoia 1970; PCorps) (map 2.23).

Life History

In Nicaragua the average litter size of twenty-one females was 6, with a range of 3 to 7; in Suriname the number of pouch young averaged 2.8 (range 1–5; $n = 6$ females); in southeastern Brazil the average litter size of seven females was 4.5; and in northeastern Argentina litter size ranged from 4 to 6. In both Rio de Janeiro state, Brazil, and Misiones province, Argentina, breeding occurs from August to February.

Compared with *Caluromys*, *Philander* has a higher reproductive rate. Sexual maturity is attained

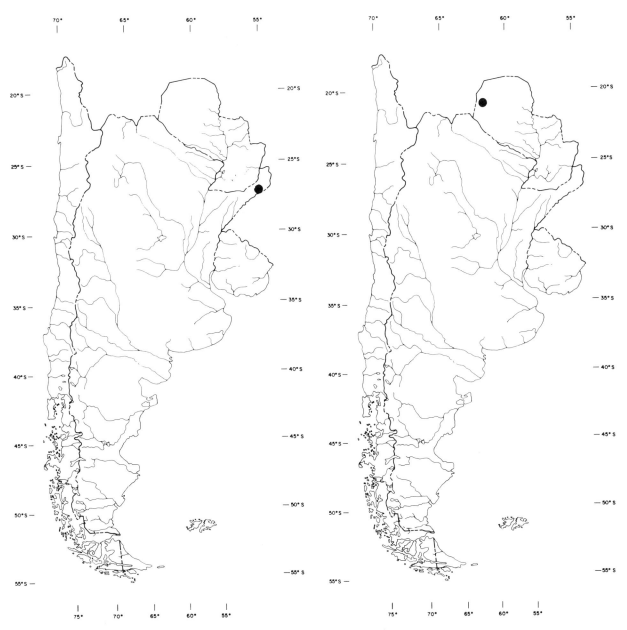

Map 2.21. Distribution of *Monodelphis scalops*.

Map 2.22. Distribution of *Monodelphis touan*.

earlier, often at less than seven months, and the rearing phase of a *Philander* female is somewhat shortened. The phase of teat attachment by the young is as short as sixty days, and the nest phase lasts eight to fifteen days before dispersal (Charles-Dominique 1983; Crespo 1982; Davis 1947; Husson 1978; Phillips and Knox Jones 1969).

Ecology

Philander uses clicks, chirps, and hisses in communication, comparable to the sounds noted for *Caluromys* and *Marmosa*. *Philander* has a relatively large brain, approximating the condition of *Caluromys*. In French Guiana *Philander* has a home-range

pattern characteristic of didelphids, with broad overlap among the ranges of neighboring adults. There is no clear-cut defense of a territory; home-range stability is a function of the availability of adequate resources. Apart from mating, contact among adult animals is minimal.

In southeastern Brazil *P. opossum* was found most commonly in moist areas but wanders through nearly all vegetation types. In French Guiana the stomach contents of four individuals contained 85% animal matter and 15% fruit and seeds (Charles-Dominique 1983; Davis 1947).

Comment

Philander has sometimes been referred to *Metachirops* (see Honacki, Kinman, and Koeppl 1982).

FAMILY MICROBIOTHERIIDAE

Diagnosis

This family contains a single living species, *Dromiciops australis*. The dental formula is I 5/4, C 1/1, P 3/3, M 4/4. The premaxillary bone is elongated, and the auditory bullae are very large; this latter character alone serves to distinguish this family from all other extant New World marsupials (fig. 2.4). The female has a well-developed pouch containing four teats.

Chromosome number: $2n = 14$.

Distribution

The extant species is found in south-central Chile and adjacent Argentina. See species account.

Genus *Dromiciops* Thomas, 1894
Dromiciops australis (Philippi, 1894)
Monito del Monte

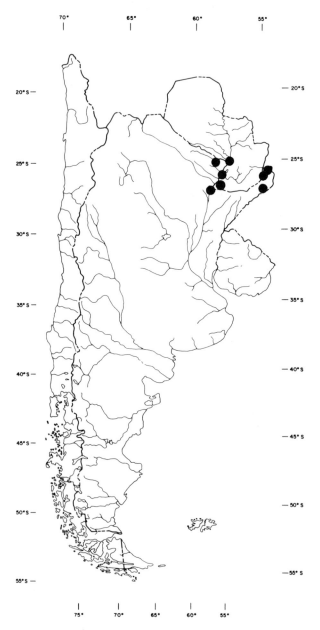

Map 2.23. Distribution of *Philander opossum*.

Measurements

	Mean	Min.	Max.	N	Loc.	Source[a]
TL	215.0	182.0	246.0	17	C	1
HB	107.1	86.0	122.0			
	107.9	83.0	113.0	17–20		2
T	107.9	93.0	132.0	17		1
	99.7	90.0	115.0	17–20		2
HF	17.0	16.0	19.0	17		1
	19.3	16.0	20.0	17–20		
E	16.4	12.0	19.0	17		1
	18.7	17.0	20.0	17–20		2
Wta	22.3	16.7	31.4	17–20		
	28.6	16.0	42.0	11		1

[a](1) Gollan 1946; Pine, Miller, and Schamberger 1979; FM, IEE-UACH; (2) Greer 1966.

Description

This small opossum has a distinct color pattern and is distinguishable from the sympatric *Marmosa* by its much smaller, sparsely haired ears, from the

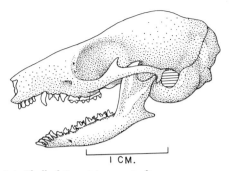

Figure 2.4. Skull of *Dromiciops australis*.

sympatric *Rhyncholestes* by its longer tail, and from both by its mottling. The pelage is dense and long. The tail, thickened at the base and used for fat storage, is thickly haired except for a narrow naked area on the underside of the tip and is moderately prehensile. The female has a small but distinct pouch.

The general dorsal color is fawn gray, considerably darker than the sides, which have large whitish patches, grading subtly with the dorsal color. These light patches are joined along the length of the body but are most pronounced before and behind the shoulders and over the hips (plate 2). The venter is light gray to yellowish white. The face is pale gray with distinct black eye-rings, and the neck is rufous brown (Mann 1978; Marshall 1978b; Osgood 1943; pers. obs.).

Chromosome number: The chromosome number is $2n = 13$ for males and $2n = 14$ for females. Males exhibit a sex-chromosome mosaicism (Gallardo and Patterson 1987) (fig. 2.4).

Distribution

Dromiciops australis is confined to southern South America, being found only from Concepción south to Chiloé Island and east to slightly beyond the Argentine border near Lago Nahuel Huapi (Marshall 1978b) (map 2.24).

Life History

Sexual maturity is reached in the second year of life. There are four phases to raising young: the pouch phase; the nest phase; riding on the mother's back; and traveling at night with littermates. The litter size is undoubtedly constrained by the number of nipples (four) (Mann 1958).

Ecology

Dromiciops australis appears to favor cool, moist forests, above 350 m, particularly with thickets of *Chusquea* bamboo. They are also found in *Araucaria* and *Nothofagus* forests. They appear to be rare wherever they occur, reaching densities of only 0.5 per hectare in Patagonian forests. They are excellent climbers and build nests both on the ground and at

about 1 m in bamboo thickets. This nest is spherical, about 20 cm in diameter, with a 3–4 cm opening and often is made of *Chusquea* leaves.

It appears that *Dromiciops* enters hibernation or periods of torpor in winter because of low temperatures and lack of food. Trapped animals enter torpor, perhaps in response to low temperatures within the trap. This species is omnivorous, with a marked preference for invertebrates, and is nocturnal. In one study thirty-eight *Dromiciops* stomachs contained 58.6% by volume mature arthropods. Collectively, invertebrates accounted for 71.8%. Annelids and invertebrates other than arthropods were not eaten. Some seeds and vegetation were present

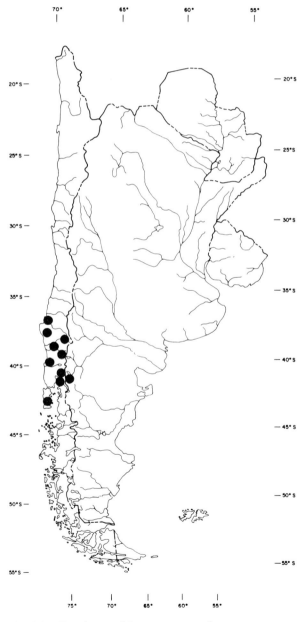

Map 2.24. Distribution of *Dromiciops australis*.

(Greer 1966; Mann 1958, 1978; Marshall 1978b; Meserve, Lang, and Patterson 1988; Pearson 1983; Pearson and Pearson 1982).

FAMILY CAENOLESTIDAE
Rat Opossums

Diagnosis
This family contains three genera and seven species. The dental formula is I 4/3, C 1/1, P 3/3, M 4/4. Members are easily distinguished from members of the family Didelphidae in that the first lower incisors are enlarged and projected forward. The remaining incisors are very reduced and show a single cusp (fig. 2.5). Rat opossums are ratlike in appearance, with head and body length ranging from 90 to 135 mm and the tail from 65 to 135 mm. The rostrum is long and narrow, and the eyes are small. The tail is covered with short hairs.

Distribution
The species of this family are confined to the high montane paramos of the Andes Mountains or the moist, temperate lowlands of Chile and Argentina. They range from western Colombia to southern Chile.

The seven recognized species within the family are found in the montane areas of Colombia, Venezuela, Ecuador to central Peru and Chile. They are confined to alpine meadows in the Andes and the lowland, temperate forests of Chile.

Natural History
Little material is available concerning the behavior of these animals in the field. The family has an ancient history, being recognizable in the Eocene of South America. It underwent an adaptive radiation in the Oligocene with numerous species that subsequently became extinct (Marshall 1982a,b,c). The extant forms are relicts in the high montane meadows of the Andes and the temperate lowlands of

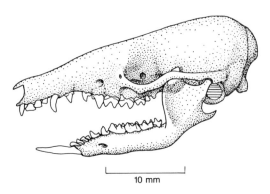

Figure 2.5. Skull of *Caenolestes* sp.

Chile (Osgood 1924). In these habitats they use runways, where they feed upon small invertebrates.

Genus *Rhyncholestes* Osgood, 1924
Rhyncholestes raphanurus Osgood, 1924
Chilean Caenolestid, Chilean Shrew Opossum

Measurements

	Mean	Min.	Max.	N	Loc.	Source[a]
TL	187.3	174.0	215	18	C	1
HB	108.5	97.0	128			
T	78.8	65.0	88			
HF	21.5	19.5	24			
E	11.9	10.0	13	17		
Wta	25.9	20.5	32	12		

[a](1) Osgood 1943; Pine, Miller, and Schamberger 1979; FM, IEE-UACH, Santiago.

Description
This small marsupial is unmistakable because of its soricine form and uniform dark color; both the dorsum and venter are dark brown or gray (plate 2). The tail, shorter than the head and body, is a uniform color, is sparsely haired, and serves as a fat-storage organ. The ears are short and lightly haired. There is a distinctive loose, fleshy lateral flap of skin on both sides of both the upper and lower lips. The female has no pouch (Osgood 1924, 1943; Pine, Miller, and Schamberger 1979; pers. obs.).
Chromosome number: $2n = 14$ (Gallardo and Patterson 1987).

Distribution
This caenolestid is found only in south-central Chile, Chiloé Island, and the adjacent mainland to 1,000 m. It may be found to exist in neighboring Argentina and perhaps more widely in Chile (Honacki, Kinman, and Koeppl 1982; Osgood 1943; Patterson and Gallardo 1987; Tamayo and Frassinetti 1980) (map 2.25).

Life History
There are seven teats and no trace of a pouch. In Osorno province reproductively active females were captured only in summer. Males are thought to be reproductively active throughout the year (Meserve et al. 1982; Patterson and Gallardo 1987).

Ecology
Rhyncholestes inhabits dense, moist forests with a dense understory. In one study, eleven individuals were caught near burrow entrances at the bases of trees or under fallen logs in dense forest, and all but one of the captures were at night. Though the species was previously thought to be extremely rare, this study concluded that it may in fact be highly localized in distribution. Like both *Dromiciops* and *Marmosa elegans*, *Rhyncholestes* appears to store fat

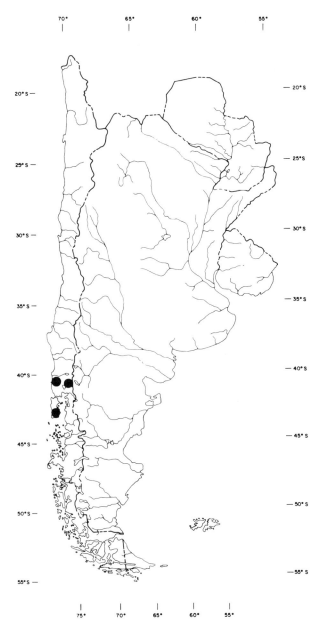

Map 2.25. Distribution of *Rhyncholestes raphanurus.*

in the tail during the winter. Although primarily insectivorous, this species also eats fungi, seeds, and earthworms. In one study, thirty-one *Rhyncholestes* stomachs contained invertebrates (54.7% by volume, including 7.9% annelids) and vegetation and fungi (39.8%) (Gallardo 1978; Kelt and Martínez 1989; Meserve et al. 1982; Meserve, Lang, and Patterson 1988; Patterson and Gallardo 1987; Pine, Miller, and Schamberger 1979).

References

Alpin, K. P., and M. Archer. 1987. Recent advances in marsupial systematics with a new syncretic classification. In *Possums and opossums: Studies in evolution*, ed. M. Archer, xv–lxxii. Chipping Norton, N.S.W., Australia: Surrey Beatty.

Barlow, J. C. 1965. Land mammals from Uruguay: Ecology and zoogeography. Ph.D. diss., University of Kansas.

Beach, F. A. 1939. Maternal behavior of the pouchless marsupial *Marmosa cinerea. J. Mammal.* 20:315–22.

Bruch, C. 1917. La comadrejita *Marmosa elegans. Rev. Jardin Zool. Buenos Aires* 13(51, 52): 208–12.

Bucher, J. E., and H. I. Fritz. 1977. Behavior and maintenance of the wooly opossum (*Caluromys*) in captivity. *Lab. Anim. Sci.* 27(6): 1007–12.

Cajal, J. L. 1981. Estudios preliminares sobre el área de acción en marsupiales (Mammalia-Marsupialia). *Physis*, sec. C, 40(98): 27–37.

Cerqueira, R. 1984. Reproduction de *Didelphis albiventris* dans le nord-est du Brésil (Polyprotodontia, Didelphidae). *Mammalia* 48(1): 95–104.

———. 1985. The distribution of *Didelphis* in South America (Polyprotodontia, Didelphidae). *J. Biogeogr.* 12:135–45.

Charles-Dominique, P. 1983. Ecology and social adaptations in didelphid marsupials: Comparison with eutherians of similar ecology. In *Advances in the study of mammalian behavior*, ed. J. F. Eisenberg and D. G. Kleiman, 395–422. Special Publication 7. Shippensburg, Pa.: American Society of Mammalogists.

Charles-Dominique, P., M. Atramentowicz, M. Charles-Dominique, H. Gérard, A. Hladik, C. M. Hladik, and M. F. Prévost. 1981. Les mammifères frugivores arboricoles nocturnes d'une forêt guyanaise: Interrelations plantes-animaux. *Rev. Ecol. (Terre et Vie)* 35:342–435.

Collins, L. R. 1973. *Monotremes and marsupials: A reference for zoological institutions.* Washington, D.C.: Smithsonian Institution Press.

Contreras, J. R. 1982. Mamíferos de Corrientes. 1. Nota preliminar sobre la distribución de algunas especies. *Hist. Nat.* 2(10): 71–72.

Creighton, G. K. 1984. Systematic studies on opossums (Didelphidae) and rodents (Cricetidae). Ph.D. diss., University of Michigan, Ann Arbor.

Crespo, J. A. 1950. Nota sobre mamíferos de Misiones nuevos para Argentina. *Comun. Inst. Nac. Invest. Cienc. Nat., Mus. Argent. Cienc. Nat. "Bernardino Rivadavia," Zool.* 1(14): 1–14.

———. 1974. Comentarios sobre nuevas localidades para mamíferos de Argentina y de Bolivia. *Rev. Mus. Argent. Cienc. Nat. "Bernardino Rivadavia," Zool.* 11(1): 1–31.

———. 1982. Ecología de la comunidad de mamíferos del Parque Nacional Iguazú, Misiones. *Rev. Mus. Argent. Cienc. Nat. "Bernardino Rivadavia," Ecol.* 3(2): 45–162.

Daciuk, J. 1974. Notas faunísticas y bioecológicas de Península Valdés y Patagonia. 12. Mamíferos colectados y observados en la Península Valdés y zona litoral de los Golfos San José y Nuevo (provincia de Chubut, república Argentina). *Physis,* sec. C, 33(86): 23–39.

Davis, D. E. 1945. The annual cycle of plants, mosquitoes, birds and mammals in two Brazilian forests. *Ecol. Monogr.* 15:244–95.

———. 1947. Notes on the life histories of some Brazilian mammals. *Bol. Mus. Nac., Zool.* 76:1–8.

Eisenberg, J. F. 1989. *Mammals of the Neotropics.* Vol. 1. *Mammals of the northern Neotropics: Panama, Colombia, Venezuela, Guyana, Suriname, French Guiana.* Chicago: University of Chicago Press.

Eisenberg, J. F., L. R. Collins, and C. Wemmer. 1975. Communication in the Tasmanian devil (*Sarcophilus harrisii*) and a survey of auditory communication in the Marsupialia. *Z. Tierpsychol.* 37:379–99.

Eisenberg, J. F., and I. Golani. 1977. Communication in Metatheria. In *How animals communicate,* ed. T. Sebeok, 575–99. Bloomington: Indiana University Press.

Eisenberg, J. F., and E. Maliniak. 1967. Breeding the murine opossum *Marmosa* in captivity. *Int. Zoo Yearb.* 7:78–79.

Eisenberg, J. F., and D. E. Wilson. 1981. Relative brain size and demographic strategies in didelphid marsupials. *Amer. Nat.* 118:1–15.

Fadem, B. H., G. L. Trupin, E. Maliniak, J. L. VandeBerg, and V. Hayssen. 1982. Care and breeding of the gray, short-tailed opossum (*Monodelphis domestica*). *Lab. Anim. Sci.* 32(4): 405–9.

Fleming, T. H. 1972. Aspects of the population dynamics of three species of opossums in the Panama Canal Zone. *J. Mammal.* 53:619–23.

Fonseca, G. A. B., K. H. Redford, and L. A. Pereira. 1982. Notes on *Didelphis albiventris* (Lund, 1841) of central Brazil. *Cienc. Cult.* 34(10): 1359–62.

Fornes, A., and E. Massoia. 1965. Micromamíferos (Marsupialia y Rodentia) recolectados en la localidad Bonaerense de Miramar. *Physis* 25(69): 99–108.

Gallardo, M. H. 1978. Hallazgo de *Rhyncholestes raphanurus* (Marsupialia, Caenolestidae) en el sur de Chile. *Arch. Biol. Med. Exp.* 11(4): 181.

Gallardo, M. H., and B. D. Patterson. 1987. An additional fourteen-chromosome karyotype and sex-chromosome mosaicism in South American marsupials. *Fieldiana, Zool.,* n.s., 39:111–15.

Gardner, A. L. 1973. *The systematics of the genus Didelphis (Marsupialia: Didelphidae) in North and Middle America.* Special Publications of the Museum 4. Lubbock: Texas Tech University Press.

Gardner, A. L., and G. K. Creighton. 1989. A new generic name for Tate's (1933) *Microtarsus* group of South American mouse opossums (Marsupialia: Didelphidae). *Proc. Biol. Soc. Wash.* 102(1): 3–7.

Gardner, A. L., and J. L. Patton. 1972. New species of *Philander* (Marsupialia: Didelphidae) and *Mimon* (Chiroptera: Phyllostomidae) from Peru. *Louisiana State Univ. Occas. Papers Mus. Zool.* 43:1–12.

Glanz, W. E. 1977a. Comparative ecology of small mammal communities in California and Chile. Ph.D. diss., University of California, Berkeley.

———. 1977b. Small mammals. In *Chile-California Mediterranean scrub atlas: A comparative analysis,* ed. N. J. W. Thrower and D. E. Bradbury, 232–37. Stroudsburg, Pa.: Dowden, Hutchinson and Ross.

Goeldi, E. A. 1894. Critical gleanings on the didelphidae of the Serra dos Orgãos, Brazil. *Proc. Zool. Soc. London* 1894:457–67.

Gollan, J. S. 1946. La comadrejita enana *Dromiciops australis australis* (F. Philippi). *Holmbergia* 4(9): 191–95.

González, J. C. 1973. Observaciones sobre algunos mamíferos de Bopicuá (Dpto. de Río Negro, Uruguay). *Comun. Mus. Mun. Hist. Nat. Río Negro, Uruguay* 1(1): 1–14.

Grand, T. I. 1983. Body weight: Its relationship to tissue composition, segmental distribution of mass, and motor function. 3. The Didelphidae of French Guyana. *Austral. J. Zool.* 31:299–312.

Greer, J. K. 1966. Mammals of Malleco province, Chile. *Publ. Mus., Michigan State Univ. (Biol. Ser.)* 3(2): 49–152.

Handley, C. O., Jr. 1976. Mammals of the Smithsonian Venezuelan project. *Brigham Young Univ. Sci. Bull., Biol. Ser.* 20(5): 1–90.

Honacki, J. H., K. E. Kinman, and J. W. Koeppl, eds. 1982. *Mammal species of the world.* Lawrence, Kans.: Allen Press and Association of Systematics Collections.

Hunsaker, D., ed. 1977a. *The biology of marsupials.* New York: Academic Press.

———. 1977b. The ecology of New World marsupials. In *The biology of marsupials,* ed. D. Hunsaker, 95–156. New York: Academic Press.

Husson, A. M. 1978. *The mammals of Suriname.* Leiden: E. J. Brill.

Kelt, D. A., and D. R. Martínez. 1989. Notes on distribution and ecology of two marsupials endemic to the Valdivian forests of southern South America. *J. Mammal.* 70:220–24.

Kirsch, J. A. W., and J. H. Calaby. 1977. The species of living marsupials: An annotated list. In *The biology of the marsupials,* ed. B. Stonehouse and D. Gilmore, 9–26. London: Macmillan.

Lemke, T. O., A. Cadena, R. H. Pine, and J. Hernandez-Camacho. 1982. Notes on opossums, bats, and rodents new to the fauna of Colombia. *Mammalia* 46:225–34.

Lucero, M. M. 1983. Lista y distribución de aves y mamíferos de la provincia de Tucumán. Ministerio de Cultura y Educación, Fundación Miguel Lillo, *Miscelánea* 75:5–53.

McManus, J. J. 1970. The behavior of captive opossum *Didelphis marsupialis virginiana. Amer. Midl. Nat.* 84:144–69.

McNab, B. K. 1978. The comparative energetics of Neotropical marsupials. *J. Comp. Physiol.* 125:115–28.

———. 1982. The physiological ecology of South American mammals. In *Mammalian biology in South America,* ed. M. A. Mares and H. H. Genoways, 187–208. Pymatuning Symposia in Ecology 6. Special Publication Series. Pittsburgh: Pymatuning Laboratory of Ecology, University of Pittsburgh.

Mann, G. 1958. Reproducción de *Dromiciops australis. Invest. Zool. Chil.* 4:209–13.

———. 1978. *Los pequeños mamíferos de Chile.* Gayana: Zoología 40. Santiago: Universidad de Concepción.

Mares, M. A. 1973. Climates, mammalian communities and desert rodent adaptations: An investigation into evolutionary convergence. Ph.D. diss., University of Texas at Austin.

Mares, M. A., R. A. Ojeda, and R. M. Barquez. 1989. *Guide to the mammals of Salta province, Argentina.* Norman: University of Oklahoma Press.

Mares, M. A., R. A. Ojeda, and M. P. Kosco. 1981. Observations on the distribution and ecology of the mammals of Salta province, Argentina. *Ann. Carnegie Mus.* 50(6): 151–206.

Marshall, L. G. 1977. *Lestodelphys halli. Mammal. Species* 81:1–3.

———. 1978a. *Lutreolina crassicaudata. Mammal. Species* 91:1–4.

———. 1978b. *Dromiciops australis. Mammal. Species* 99:1–5.

———. 1978c. *Chironectes minimus. Mammal. Species* 109:1–6.

———. 1982a. Calibration of the age of mammals in South America. *Geobios,* memo. spec., 6:427–37.

———. 1982b. Evolution of South American Marsupialia. In *Mammalian biology in South America,* ed. M. A. Mares and H. H. Genoways, 251–72. Pymatuning Symposia in Ecology 6. Special Publication Series. Pittsburgh: Pymatuning Laboratory of Ecology, University of Pittsburgh.

———. 1982c. Systematics of the South American marsupial family Microbiotheriidae. *Fieldiana: Geol.,* n.s., 10:1–75.

Marshall, L. G., C. DeMuizon, and B. Sige. 1983. Late Cretaceous mammals (Marsupialia) from Bolivia. *Geobios* 16:739–45.

Massoia, E. 1970. Contribución al conocimiento de los mamíferos de Formosa con noticias de los que habitan zonas viñaleras. *IDIA* 276:55–63.

———. 1972. La presencia de *Marmosa cinerea paraguayana* en la república Argentina, provincia de Misiones (Mammalia-Marsupialia-Didelphidae). *Rev. Invest. Agropecuarias,* INTA (Buenos Aires), ser. 1, *Biol. Prod. Anim.* 9(2): 63–70.

———. 1973. Observaciones sobre el género *Lutreolina* en la república Argentina (Mammalia-Marsupialia-Didelphidae). *Rev. Invest. Agropecuarias,* INTA (Buenos Aires), ser. 1, *Biol. Prod. Anim.* 10(1): 13–20.

———. 1976. *Fauna de agua dulce de la república Argentina.* Buenos Aires: Fundación para la Educación, la Ciencia y la Cultura.

———. 1980. Mammalia de Argentina. 1. Los mamíferos silvestres de la provincia de Misiones. *Iguazú* 1(1): 15–43.

Massoia, E., and J. Foerster. 1974. Un mamífero nuevo para la república Argentina: *Caluromys lanatus lanatus* (Illiger), (Mammalia-Marsupialia-Didelphidae). *IDIA* 311–14:5–7.

Massoia, E., and A. Fornes. 1967. El estado sistemático distribución geográfica y datos etoecológicos de algunos mamíferos neotropicales (Marsupialia y Rodentia) con la descripción de *Cabreramys,* género nuevo (Cricetidae). *Acta Zool. Lilloana* 23:407–30.

———. 1972. Presencia y rasgos etoecológicos *Marmosa agilis chacoensis* Tate en las provincias de Buenos Aires, entre Ríos y Misiones (Mammalia-Marsupialia-Didelphidae). *Rev. Invest. Agropecuarias,* INTA (Buenos Aires), ser. 1, *Biol. Prod. Anim.* 9(2): 71–81.

Meserve, P. L. 1981a. Resource partitioning in a Chilean semi-arid small mammal community. *J. Anim. Ecol.* 40:747–57.

———. 1981b. Trophic relationships among small mammals in a Chilean semiarid thorn scrub community. *J. Mammal.* 62:304–14.

Meserve, P. L., B. K. Lang, and B. D. Patterson. 1988. Trophic relationships of small mammals in a Chilean temperate rainforest. *J. Mammal.* 69:721–30.

Meserve, P. L., R. Murua, O. Loppetegui N., and J. R. Rau. 1982. Observations on the small mammal fauna of a primary temperate rain forest in southern Chile. *J. Mammal.* 63:315–17.

Miller, S. D. 1980. Human influences on the distribution and abundance of wild Chilean mammals: Prehistoric–present. Ph.D. diss., University of Washington, Seattle.

Miranda-Ribeiro, A. 1936. Didelphia ou mammalia-ovovivipara. *Rev. Mus. Paulista, São Paulo* 20: 245–427.

Mondolfi, E., and G. Medina Padilla. 1957. Contribución al conocimiento del "Perrito de Agua" (*Chironectes minimus* Zimmerman). *Mem. Soc. Cient. Nat. La Salle* 17:140–55.

Motta, M. de F. D., J. C. de A. Carreira, and A. M. R. Franco. 1983. A note on reproduction of *Didelphis marsupialis* in captivity. *Mem. Inst. Oswaldo Cruz* 78(4): 507–9.

Myers, P., and R. M. Wetzel. 1979. New records of mammals from Paraguay. *J. Mammal.* 60:638–41.

Nitikman, L. Z., and M. A. Mares. 1987. Ecology of small mammals in a gallery forest of central Brazil. *Ann. Carnegie Mus.* 56:75–95.

O'Connell, M. A. 1979. Ecology of didelphid marsupials from northern Venezuela. In *Vertebrate ecology in the northern Neotropics*, ed. J. F. Eisenberg, 73–87. Washington, D.C.: Smithsonian Institution Press.

Ojeda, R. A., and M. A. Mares. 1989. *A biogeographic analysis of the mammals of Salta province, Argentina: Patterns of species assemblage in the Neotropics.* Special Publications of the Museum 27. Lubbock: Texas Tech University Press.

Oliver, W. L. R. 1976. The management of yapoks (*Chironectes minimus*) at Jersey Zoo, with observations on their behaviour. In *Jersey Wildlife Preservation Trust, Thirteenth Annual Report,* pp. 32–36. Jersey: Trinity.

Olrog, C. C. 1959. Notas mastozoológicas. 2. Sobre la colección del Instituto Miguel Lillo. *Acta Zool. Lilloana* 17:403–19.

———. 1976. Sobre mamíferos del noroeste argentino. *Acta Zool. Lilloana* 32:5–12.

———. 1979. Los mamíferos de la selva húmeda, Cerro Calilegua, Jujuy. *Acta Zool. Lilloana* 33: 9–14.

Olrog, C. C., and M. M. Lucero. 1981. *Guía de los mamíferos argentinos.* Tucumán, Argentina: Ministerio de Cultura y Educación, Fundación Miguel Lillo.

Osgood, W. H. 1921. A monographic study of the American marsupial *Caenolestes. Field Mus. Nat. Hist., Zool. Ser.* 14:1–162.

———. 1924. Review of living caenolestids with description of a new genus from Chile. *Field Mus. Nat. Hist., Zool. Ser.* 14(2): 165–73.

———. 1943. The mammals of Chile. *Field Mus. Nat. Hist., Zool. Ser.* 30:1–268.

Patterson, B. D., and M. H. Gallardo. 1987. *Rhyncholestes raphanurus. Mammal. Species* 286: 1–5.

Pearson, O. P. 1983. Characteristics of a mammalian fauna from forests in Patagonia, southern Argentina. *J. Mammal.* 64:476–92.

Pearson, O. P., and A. K. Pearson. 1982. Ecology and biogeography of the southern rainforests of Argentina. In *Mammalian biology in South America*, ed. M. A. Mares and H. H. Genoways, 129–42. Pymatuning Symposia in Ecology 6. Special Publication Series. Pittsburgh: Pymatuning Laboratory of Ecology, University of Pittsburgh.

Phillips, C. J., and J. Knox Jones, Jr. 1969. Notes on reproduction and development in the four-eyed opossum, *Philander opossum*, in Nicaragua. *J. Mammal.* 50:345–49.

Pine, R. H., P. L. Dalby, and J. O. Matson. 1985. Ecology, postnatal development, morphometrics, and taxonomic status of the short-tailed opossum, *Monodelphis dimidiata*, an apparently semelparous annual marsupial. *Ann. Carnegie Mus.* 54(6): 195–231.

Pine, R. H., S. D. Miller, and M. L. Schamberger. 1979. Contributions to the mammalogy of Chile. *Mammalia* 43:339–76.

Reig, O. A. 1964. Roedores y marsupiales del partido de General Pueyrredón y regiones adyacentes (provincia de Buenos Aires, Argentina). *Pub. Mus. Mun. Cienc. Nat. Mar del Plata* 1(6): 203–24.

Reig, O. A., A. L. Gardner, N. O. Bianchi, and J. L. Patton. 1977. The chromosomes of the Didelphidae (Marsupialia) and their evolutionary significance. *Biol. J. Linnean Soc.* 9:191–216.

Reig, O. A., J. A. W. Kirsch, and L. G. Marshall. 1987. Systematic relationships of the living and Neocenozoic American "opossum-like" marsupials. In *Possums and opossums: Studies in evolution*, ed. M. Archer, 1–89. Chipping Norton, N.S.W., Australia: Surrey Beatty.

Ride, W. D. L. 1964. A review of Australian fossil marsupials. *J. Proc. Roy. Soc. West Austral.* 47:97–131.

Rigueira, S. E., C. M. de Carvalho Valle, J. B. M. Varejão, P. V. de Albuquerque, and J. C. Nogueira. 1987. Algumas observações sobre o ciclo reprodutivo anual de fêmeas do gambá *Didelphis albiventris* (Lund, 1841) (Marsupialia, Didelphidae) em populações naturais no Estado de Minas Gerais, Brasil. *Rev. Brasil. Zool.*, *São Paulo* 4(2): 129–37.

Roig, V. G. 1971. La presencia de estados de hibernación en *Marmosa elegans* (Marsupialia-Didelphidae). *Acta Zool. Lilloana* 28:5–12.

Schamberger, M., and G. Fulk. 1974. Mamíferos del Parque Nacional Fray Jorge. *Idesia* (Chile) 3:167–79.

Schneider, C. O. 1946. Catálogo de los mamíferos de la provincia de Concepción. *Bol. Soc. Biol. Concepción* 21:67–83.

Seluja, G. A., M. V. DiTomaso, M. Brun-Zorilla, and H. Cordoso. 1984. Low karyotypic variation in two didelphids. *J. Mammal.* 65:702–7.

Simonetti, J. A., J. L. Yañez, and E. R. Fuentes. 1984. Efficiency of rodent scavengers in central Chile. *Mammalia* 48:608–9.

Stallings, J. R. 1988. Small mammal communities in an eastern Brazilian park. Ph.D. diss., University of Florida, Gainesville.

Streilein, K. E. 1982a. The ecology of small mammals in the semiarid Brazilian caatinga. 1. Climate and faunal composition. *Ann. Carnegie Mus.* 51(5): 79–107.

———. 1982b. The ecology of small mammals in the semiarid Brazilian caatinga. 3. Reproductive biology and population ecology. *Ann. Carnegie Mus.* 51(13): 251–69.

———. 1982c. Behavior, ecology, and distribution of the South American marsupials. In *Mammalian biology in South America*, ed. M. A. Mares and H. H. Genoways, 231–50. Pymatuning Symposia in Ecology 6. Special Publications Series. Pittsburgh: Pymatuning Laboratory of Ecology, University of Pittsburgh.

Sunquist, M. E., S. N. Austad, and F. Sunquist. 1987. Movement patterns and home range in the common opossum (*Didelphis marsupialis*). *J. Mammal.* 68:173–76.

Sunquist, M. E., and J. F. Eisenberg. 1991. Reproductive strategies of female *Didelphis*. In *Current mammalogy*, vol. 3, ed. H. H. Genoways. New York: Plenum. In press.

Tálice, R. V., S. Lafitte de Mosera, and T. Machado.

1960. Observaciones sobre *Monodelphis dimidiata*. *Congreso Sudamericano Zool. Actas y Trabjos* (La Plata 12–24 October 1960), 4:149–56.

Tamayo, M., and D. Frassinetti. 1980. Catálogo de los mamíferos fósiles y vivientes de Chile. *Mus. Nac. Hist. Nat.*, *Chile* 37:323–99.

Tate, G. H. H. 1933. A systematic revision of the marsupial genus *Marmosa*. *Bull. Amer. Mus. Nat. Hist.* 66:1–250.

Thomas, O. 1888. Diagnoses of four new species of *Didelphis*. *Ann. Mag. Nat. Hist.*, ser. 6, 1: 158–59.

———. 1894. On *Micoureus griseus* Desm., with the description of a new genus and species of Didelphidae. *Ann. Mag. Nat. Hist.*, ser. 6, 14:184–88.

———. 1926. Two new mammals from north Argentina. *Ann. Mag. Nat. Hist.*, ser. 9, 17(99): 311–13.

———. 1927. On a further collection of mammals made by Sr. E. Budin in Neuquén, Patagonia. *Ann. Mag. Nat. Hist.*, ser. 9, 19(114): 650–58.

———. 1929. The mammals of Señor Budin's Patagonian expedition, 1927–1928. *Ann. Mag. Nat. Hist.* 4(10): 35–45.

Unger, K. L. 1982. Nest building behavior of the Brazilian bare-tailed opossum *Monodelphis domestica*. *J. Mammal.* 63:150–62.

Varejão, J. B. M., and C. M. C. Valle. 1982. Contribuções ao estudo da distribução geográfica das espécies do gênero *Didelphis* (Mammalia, Marsupialia) no Estado de Minas Gerais, Brasil. *Lundiana* 2:5–55.

Vieira, C. 1949. Xenartros e marsupiais do Estado de São Paulo. *Arq. Zool.*, *São Paulo* 7:325–62.

Villafañe, G. de, F. O. Kravetz, M. J. Piantanida, and J. A. Crespo. 1973. Dominancia, densidad e invasión en una comunidad de roedores de la localidad de Pergamino (provincia de Buenos Aires). *Physis*, sec. C, 32(84): 47–59.

Woodburne, M. O., and W. J. Zinmeister. 1982. Fossil land mammals from Antarctica. *Science* 218:284–86.

Ximénez, A. 1967. Contribución al conocimiento de *Lutreolina crassicaudata* (Desmarest, 1804) y sus formas geográficas (Mammalia-Didelphidae). *Comun. Zool. Mus. Hist. Nat. Montevideo* 9(112): 1–7.

Diagnosis

The name Xenarthra is used here because we concur with Emery (1970) that the Palaenodonta are not closely related or ancestral to the Xenarthra. In spite of the older ordinal name Edentata, not all members lack teeth; however, the tooth number is often re- duced and tooth structure is simplified. Among the existing families the incisors are absent as well as the canines, but the two-toed sloth has a presumptive premolar that is caniniform. The premolars and mo- lars, when present, do not have enamel. The tym- panic bone is ring shaped in many forms. All forms

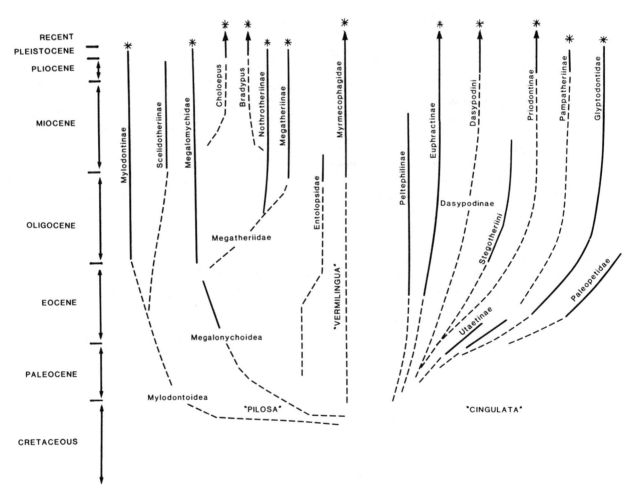

Figure 3.1. Phylogenetic tree of the Xenarthra (modified from Patterson and Pascual 1972). In their scheme the Chlamyphorini are subsumed in the Euphractinae and the Tolypeutini within the Priodontinae. Asterisk means survived to the Pleistocene.

have trunks reinforced with broadened ribs and by accessory zygapophyses on the thoracic and lumbar vertebrae, hence the ordinal name Xenarthra (narrow joints). Many of the living forms show great specialization for vastly different ways of life, making a simple diagnosis for the order difficult.

Distribution

The extant edentates are entirely confined to the New World. One species, the nine-banded armadillo (*Dasypus novemcinctus*) has extended its range in Recent times to the southeastern United States and the southern portions of the Great Plains. Other southern species, if they range across the Panamanian land bridge, generally do not extend farther north than southern Mexico. To the south three species of armadillos range to Patagonia, and *Zaedyus pichiy* ranges to the Strait of Magellan.

History and Classification

Edentates appear to have originated in South America and to have undergone their adaptive radiation there (Reig 1981). Although the suborder Palaenodonta is first noted in the Paleocene of North America, this is now believed to be either an early offshoot from South American stock or an entirely separate taxon. The more recent edentates first appear in the Paleocene of South America. These specimens are believed to belong to the armadillo family, the Dasypodidae. Six families of the Xenarthra are now extinct, including the giant ground sloths of the families Megalonychidae and Megatheriidae (Patterson and Pascual 1972) (see fig. 3.1). The glyptodonts (Glyptodontidae) were somewhat similar to armadillos in having bony scutes forming a turtlelike carapace, but they were an herbivorous, independent offshoot that first appeared in the Eocene and persisted to the Pleistocene (Gillette and Ray 1981). The glyptodonts, as well as some of the giant ground sloths, crossed from South America to North America in the late Miocene or early Pliocene at the completion of the Panamanian land bridge. They subsequently became extinct in North America at the close of the Pleistocene (Webb and Marshall 1982).

The biology of the Xenarthra is admirably reviewed in the volume edited by Montgomery (1985a). Captive maintenance has been summarized by Merrett (1983). The extant xenarthrans are included in four families, the Myrmecophagidae, the Bradypodidae, the Choloepidae (Megalonychidae), and the Dasypodidae.

Comment

The genera *Bradypus* and *Choloepus* were formerly subsumed under Bradypodidae, but see Honacki, Kinman, and Koeppl (1982) and Webb (1985).

FAMILY MYRMECOPHAGIDAE

Anteaters

Diagnosis

Teeth are totally lacking, and the rostrum of the skull is extremely elongated (fig. 3.2). The long tongue, extended for feeding on ants and termites, is coated with a copious sticky saliva produced from greatly enlarged submaxillary glands. The forefeet have an enlarged third digit with a strong claw, and the other digits are reduced or absent. The hind feet have four or five digits. The body is covered with hair; the tail is long and also haired to varying degrees.

Distribution

Species of anteaters occur from southern Mexico to northern Argentina. Ranges for the species are reviewed by Wetzel (1985a).

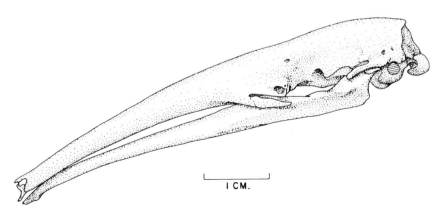

I CM.

Figure 3.2. Skull of *Myrmecophaga tridactyla*.

Natural History

Members of this family are highly adapted for feeding on ants and termites. The strong claws are used to open the nests, and the long, extensible tongue aids in feeding. Females produce only a single young at a time, and parental care is highly developed. The young is frequently carried on the mother's back when she changes resting areas or shortly before weaning, when it accompanies her to feeding sites. Relative to their body size anteaters have a low metabolic rate, apparently an adaptation for feeding on ants and termites, a ubiquitous resource of low nutritional quality (McNab 1982; Redford 1987a,b).

Genus *Myrmecophaga* Linnaeus, 1758
•*Myrmecophaga tridactyla* Linnaeus, 1758
Giant Anteater, Oso Hormiguero, Yuru Mi

Measurements

	Mean	Min.	Max.	N	Loc.	Source[a]
HB	1,265.5	1,100.0	2,000.0	16	Whole range	1
T	734.0	600.0	900.0			
HF	165.0	150.0	180.0	13		
E	46.7	35.0	50.0	9		
Wta	32.9 kg	22.0 kg	39.0 kg	5		
	32.1 kg	26.4 kg	36.4 kg	12 m	Br	
	29.2 kg	25.5 kg	31.8 kg	12 f		

[a](1) Redford, n.d.

Description

The giant anteater is one of the most distinctive mammals in South America, with its large size, long tail with long, coarse hair, and greatly elongated snout. The ears and eyes are very small. The front claws, particularly the third, are greatly enlarged and carried curled back as the animal walks on its "wrists" (fig. 3.3). The dorsum and tail are dark brown or black; the forelegs are mostly white, with black bands at the wrists and above the claws; and thin white bands pass from just below the ears back and up to well above the shoulders, descending to where the forelegs meet the body and enclosing a broad band of black. The tail, which can be nearly as long as the body, is uniformly brown with very coarse hair.

Distribution

Myrmecophaga is found from Belize and Guatemala south through the Paraguayan Chaco to the northernmost Argentine provinces of Misiones, Formosa, Salta, Jujuy, and probably Chaco and Santiago del Estero. This species is probably now extinct in Uruguay (Wetzel 1982a) (map 3.1.).

Life History

A single young weighing 1.1 to 1.6 kg is born after a gestation of 183–90 days. Its eyes open after about six days, and the lactation period is six to eight weeks or longer. The young is carried on its mother's back for about six to nine months, though it may be left in a "nest" while the female feeds. In captivity the age of first reproduction varies from 2.5 to 4.0 years; animals breed throughout the year, and the interbirth interval can be as low as nine months. In northeastern Argentina this species is reported to reproduce between September and March, though in central Brazil young are seen throughout the year. Giant anteaters can live at least sixteen years in captivity (Bickel, Murdock, and Smith 1976; Byrne 1962; Crespo 1982; Hardin 1976; Merrett 1983; Redford, n.d.; Shaw, Machado-Neto, and Carter 1987; Smielowski, Stanislawski, and Taworksi 1981).

Ecology

Giant anteaters are found in a large variety of habitats from tropical forest to the xeric Chaco. They appear to be most abundant in open vegetation formations with an abundance of ants and termites and can be active throughout the day and night depending on temperature and rainfall. In southeastern Brazil *Myrmecophaga* occurred at a density of be-

Figure 3.3. *Myrmecophaga tridactyla.*

tween 1.3 and 2.0 per square kilometer. In south-central Brazil female giant anteaters have average home ranges of 3.7 km² (*n* = 4), while male home ranges averaged 2.7 km² (*n* = 4)—not a statistically significant difference. In the Venezuelan llanos home ranges were much larger, with one report of a 25 km² home range.

Myrmecophaga appears to have very poor eyesight and fairly poor hearing, relying on smell to locate prey. Ants and termites compose most of its diet. The distribution of the social insect fauna available to a giant anteater appears to dictate to what extent it feeds on ants versus termites. It finds insects either by rooting with its nose or by digging into nests or mounds with its large, powerful front claws. The giant anteater is a very selective feeder, visiting many termite mounds or ant nests in the course of feeding. Cessation of feeding seems to be determined by the time it takes for the soldier caste of ants or termites to arrive at the breach and defend the nest, generally by noxious chemical secretions or by biting.

Giant anteaters are solitary and, except during the breeding season, seem to ignore one another. Adults probably have no serious predators; the major sources of mortality are humans and fire. The long, coarse hair is highly flammable, and anteaters are frequently found burned to death after a severe grass or forest fire (Eisenberg 1989; Montgomery 1985b; Montgomery and Lubin 1977; Redford 1985a, 1987b; Schmid 1938; Shaw, Machado-Neto, and Carter 1987; Shaw, Carter, and Machado-Neto 1985).

Genus *Tamandua* Gray, 1825
• *Tamandua tetradactyla* (Linnaeus, 1758)
Southern Tamandua, Oso Melero, Caguare

Measurements

	Mean	Min.	Max.	N	Loc.	Source[a]
TL	1,002.4	905.0	1,047.0	8	P	1
HB	590.3	522.0	635.0			
	640.5	594.0	692.0	6	Br	2
T	412.1	370.0	436.0	8	P	1
	458.5	425.0	498.0	6	Br	2
HF	88.3	57.0	105.0	8	P	1
	94.5	80.0	105.0	6	Br	2
E	46.5	40.0	50.0	8	P	1
	46.6	41.0	51.0	6	Br	2
Wta	5.12 kg	3.8 kg	8.5 kg	5	P	1
	6.20 kg	4.9 kg	7.0 kg	5	Br	2

[a](1) UConn, UM; (2) Redford, n.d.

Description

The tamandua is easily separated from the giant anteater by its smaller size, its coloration, and the shape of its tail. Most tamanduas in southern South America are golden brown with a black vest covering the dorsum and venter, crossing the shoulders in a black band, but on some individuals the vest may be greatly reduced or even absent. The tail is prehensile and only sparsely haired, the snout is considerably less elongated than that of the giant anteater, and the ears are proportionally longer. The claws on the front feet are enlarged but are not as long proportionally as those of *Myrmecophaga* (see plate 3) (Redford, n.d.; Wetzel 1975, 1985a).

Distribution

Tamandua tetradactyla is found from Venezuela south through Paraguay to northern Uruguay and the northern Argentine provinces of Santa Fe, Chaco, Salta, and Jujuy (Wetzel 1975, 1982a, 1985b) (map 3.2).

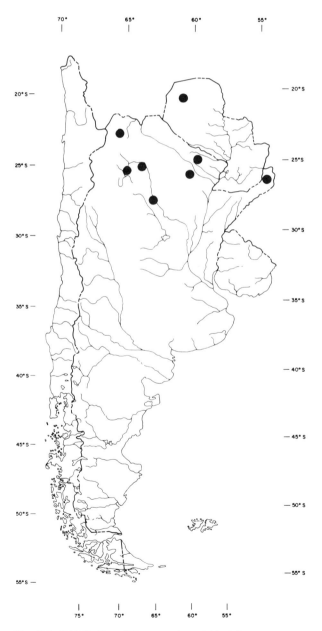

Map 3.1. Distribution of *Myrmecophaga tridactyla*.

Life History

The southern tamandua produces a single young after a gestation period of about 160 days. The young is carried on the mother's back or left in a nest. When it is old enough, it accompanies its mother as she forages (Crespo 1982; Meritt 1975; Montgomery and Lubin 1977).

Ecology

Tamandua tetradactyla is found in a wide variety of habitats, including tropical forest, dry scrub forest, and open grassland. With its prehensile tail it is an excellent climber, though it is often found well away from trees. The animal shelters in hollows in trees as well as abandoned holes in the ground such as those made by armadillos.

This medium-sized anteater feeds on both ants and termites, depending on the available prey. There is extensive individual variation in prey choice. The young, feeding with its mother, apparently learns her food preference, which may account for observed individual preferences as adults. There is also intrapopulation variation in activity cycles; some individuals are diurnal and others are nocturnal.

Tamanduas have been reported as prey for ocelots and jaguars, and the young are probably vulnerable to other felines and foxes. They are killed by humans for their meat as well as for their tough hide (Mares, Ojeda, and Kosco 1981; Montgomery 1985b; Montgomery and Lubin 1977; Redford 1983).

FAMILY BRADYPODIDAE
Three-toed Sloths

Diagnosis

The exact identification of the tooth types remains in some doubt; thus the dental formula is usually listed as 5/4–5. The teeth are cylindrical and grow throughout life (fig. 3.4). Head and body length ranges from 400 to 800 mm and weight from 2.25 to 5.50 kg; the stout tail is approximately 68 mm long. The neck contains eight or nine cervical vertebrae. The forelimbs are slightly longer than the hind limbs, and each forefoot has three long claws. The hair is long and rather stiff; individual hairs may support algae and thereby have a blue-green color. In addition to algal symbionts, moths have been recorded in the fur of members of this family (Waage and Best 1985).

Distribution

Three-toed sloths extend from eastern Honduras across South America to northern Argentina.

Natural History

Three-toed sloths are active both diurnally and nocturnally. They are highly specialized as selective browsers on the leaves of trees; the stomach is compartmentalized, and fermentative reduction of leaf structure occurs both in the stomach and in the in-

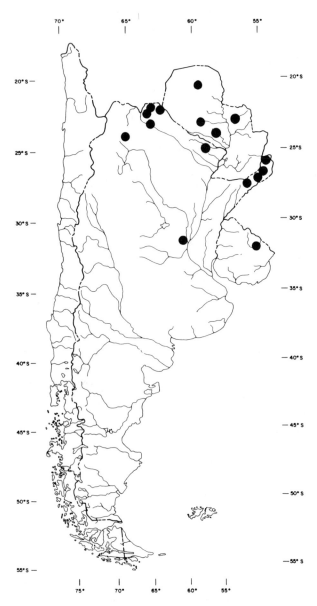

Map 3.2. Distribution of *Tamandua tetradactyla*.

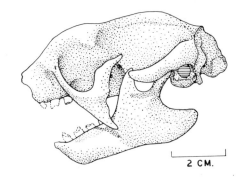

2 CM.

Figure 3.4. Skull of *Bradypus* sp.

testines. Three-toed sloths are characterized by a low metabolic rate and a low core body temperature, apparently an adaptation for feeding on leaves that are abundant but have a low nutrient content. They locomote slowly while suspended beneath a tree branch by all four limbs. A single young is produced after a gestation period of approximately six months (McNab 1978).

Genus *Bradypus* Schinz, 1825
• *Bradypus variegatus* Schinz, 1825
Brown-throated Three-toed Sloth, Perezoso Bayo

Measurements

	Mean	Min.	Max.	N	Loc.	Source[a]
HB	521.6	413.0	700.0	100	Whole	
T	57.7	38.0	90.0	101	range	1
HF	121.6	90.0	180.0	?		
E	13.3	8.0	22.0	41		
Wta	4.34 kg	2.25 kg	5.5 kg	25		

[a](1) Wetzel 1985a.

Description
This species is unmistakable (plate 3). The pelage on the shoulders, neck, throat, and sides of the face is brown. There is considerable variation in the pattern of lighter spots or splotches on the back. The throat and sides of the face are brown, and there is a prominent dark brown forehead and suborbital stripe outlining a paler area of the face. Males have a mid-dorsal yellow to orange speculum (Wetzel 1985a; Wetzel and Kock 1973).

Distribution
Bradypus variegatus is distributed from Honduras south to northernmost Argentina. It is known from the Argentine province of Misiones; two individuals in the Buenos Aires zoo were reported to be from Jujuy province, and one specimen in the Field Museum was collected in Jujuy. The southern boundary of this species' range is poorly known, though it has been reported in the Argentine provinces of Formosa, Chaco, and Misiones (Massoia 1980; Onelli 1913; Wetzel and Avila-Pires 1980; Wetzel 1982a, 1985a; FM) (map 3.3).

Life History
A single young is born after a gestation period of between 120 and 180 days and is nutritionally dependent on the female for about four weeks. It remains with its mother for about six months, after which it occupies a portion of its natal home range (Montgomery and Sunquist 1978).

Ecology
The sloth is strictly arboreal, regularly descending to the ground only to defecate. Because they are largely folivorous, sloths are found only in areas where most trees keep their leaves, though they are

reported to be able to survive subfreezing temperatures for a short period.

In some areas they can be important components of the mammalian biomass: on Barro Colorado Island (Panama) they made up an estimated 70%. In Panama animals were found to occupy home ranges averaging 1.6 ha. They are active both day and night, are strictly folivorous, and favor some species of trees over others, with this preference varying between individuals. Young sloths show the same preference for food and tree species as their mothers, learned as they are carried around on their mothers' backs. One of the criteria used in choosing trees is a crown that receives abundant sun: sloths behaviorally thermoregulate while resting by moving in and out of the sun (Luederwaldt 1918; Montgomery

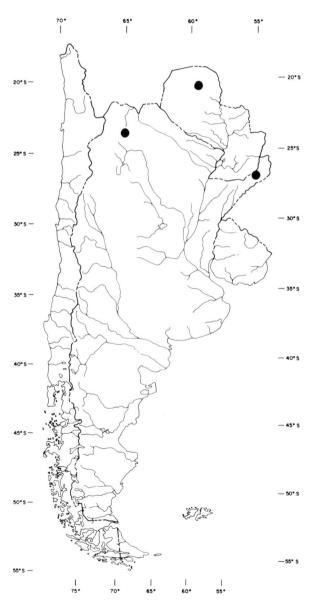

Map 3.3. Distribution of *Bradypus variegatus*.

and Sunquist 1975, 1978; Sunquist and Montgomery 1973).

Comment
Bradypus variegatus includes *B. infuscatus* and *B. boliviensis* (Honacki, Kinman, and Koeppl 1982).

FAMILY DASYPODIDAE
Armadillos

Diagnosis
The dental formula is highly variable and will be given under the generic accounts (see also Wetzel 1985b). The rostrum of the skull is moderately long (see figs. 3.5 and 3.6). Members of this family are characterized by numerous bony dermal scutes in regular arrangements, forming movable bands in the midsection and on the tail as well as immovable shields on the forequarters and hindquarters. The bony scutes are covered by horny epidermal scales. Sparse hairs appear between the bands and on the underside of the animal, which is not armored (see plate 3).

Distribution
The armadillos are distributed from Oklahoma in the United States, south to the Strait of Magellan in Chile and were recently introduced into Florida. The armadillos originated in South America and radiated extensively. The rich fossil record shows that they entered North and Central America during the late Miocene and Pliocene as the Panamanian land bridge neared completion (Webb and Marshall 1982; Webb 1985).

Natural History
The dermal scutes apparently aid in antipredator defense, perhaps most importantly by allowing animals to escape by running through dense, thorny vegetation. Armadillos' dietary habits vary greatly depending on the species. Some, such as *Cabassous* and *Priodontes* feed almost exclusively on ants and termites; others (*Euphractus*) eat fruits, invertebrates, and small vertebrates. All have extremely

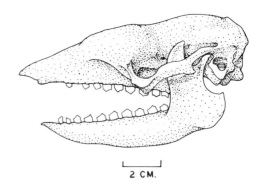

Figure 3.6. Skull of *Euphractus sexcinctus*.

well developed claws and are capable burrowers. Members of the genus *Dasypus* frequently forage on the surface, whereas others, such as *Cabassous*, are adapted for a fossorial existence.

The antipredator behavior patterns of armadillos have been related to their morphology. *Dasypus* can run fast and generally flees to a burrow or into dense brush if disturbed. *Euphractus* has a similar antipredator strategy but will sometimes bite. *Priodontes* and *Cabassous*, with their stout claws and low carapaces, will attempt to flatten their bodies and dig rapidly if attacked. *Tolypeutes* can and does roll into a ball when attacked. The hard carapace will deter most mammalian predators, but this passive defense renders members of the genus extremely vulnerable to human predation (Krieg 1929; Kühlhorn 1936).

Those species adapted for a fossorial life and specialized for feeding on ants and termites (e.g., *Priodontes* and *Cabassous*) have very low metabolic rates. The distribution of most armadillos is strongly governed by the mean low temperature of an area, since armadillos have a high thermal conductance and readily lose body heat at low temperatures. Though burrowing protects an individual from the extremes of ambient temperatures, only four species of armadillos have significantly occupied the warmer portions of the temperate zone. Further information in this widely varying group will be deferred for the species accounts (McNab 1980, 1982; Redford 1985b).

Genus *Cabassous* McMurtrie, 1831
Naked-tailed Armadillo, Cabasú

Description
The four species within this genus differ little in external morphology except for size (Wetzel 1980). The number of teeth is highly variable, ranging between 7/8 and 10/9. There are no teeth in the premaxillary bone. Head and body length ranges from 300 to 490 mm and the tail from 90 to 200 mm. The

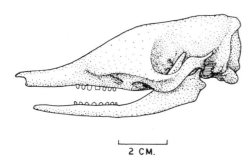

Figure 3.5. Skull of *Dasypus novemcinctus*.

snout is very short and broad, the ears are moderately large and funnel shaped, and the eyes are extremely small. The forefeet have five claws, the middle one extremely large and sickle shaped. The dorsal plates are arranged in transverse rows for the entire length of the body. The slender tail is distinctive, with either no armor or small, widely spaced thin plates; hence the common name naked-tailed armadillo. The tail alone serves to distinguish this genus from all other armadillos (see plate 3).

Distribution

The genus *Cabassous* is distributed from southern Guatamala south to northern Argentina. It prefers rather moist habitats with well-drained, loose soil (Wetzel 1980).

Life History

The species of *Cabassous* are specialized for feeding on ants and termites. They are strongly fossorial and may forage underground. The female gives birth to a single young.

Cabassous chacoensis Wetzel, 1980
Chacoan Naked-tailed Armadillo, Tatú-ai Menore

Description

Cabassous chacoensis is the smallest of the species of *Cabassous* and can be distinguished by its ears, which are strikingly smaller than in any other species in this genus. Measurements of two specimens were HB 300–306; T 90–96; HF 61; E 14–15. The ears are unique in having a fleshy expansion of their anterior margins (Wetzel 1980).

Map 3.4. Distribution of *Cabassous chacoensis*.

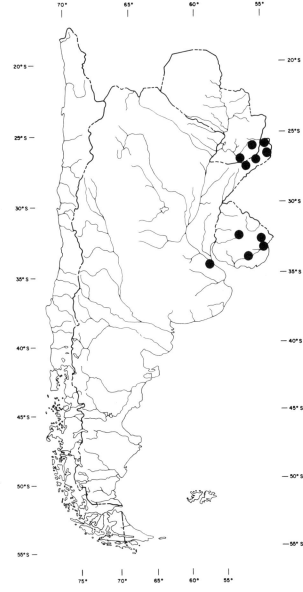

Map 3.5. Distribution of *Cabassous tatouay*.

Distribution

This species is found in the Gran Chaco of north-western Argentina, in western Paraguay, in south-eastern Bolivia, and possibly into adjacent areas of Brazil (Wetzel 1980) (map 3.4).

Ecology

This species appears to be confined to the xeric Chaco of Argentina and Paraguay and adjoining countries. Nothing is known of its natural history. When handled, males emit a loud grunt, but females remain silent (Meritt 1985; Wetzel 1982b).

Comment

This species has been mistaken for *C. unicinctus* (*loricatus*) (Olrog and Lucero 1981).

Cabassous tatouay (Desmarest, 1804)
Greater Naked-tailed Armadillo

Measurements

	Mean	Min.	Max.	N	Loc.	Source[a]
HB	457.8	410.0	490.0	5	Br	1
T	179.0	150.0	200.0			
HF	82.2	80.0	86.0			
E	41.7	40.0	44.0			
Wta	5.35 kg	3.4 kg	6.4 kg	3		

[a](1) Carter, unpubl.; Redford, n.d.

Description

Cabassous tatouay is much larger than any of the other species of *Cabassous* in southern South America: it is twice as heavy as the next largest species, *C. unicinctus*. It is further separable from *C. unicinctus* by the size of its ear: *C. tatouay* has a much larger, funnel-shaped ear that extends well above the top of the head (Redford, n.d.; Wetzel 1980).

Distribution

This species is found from Brazil south through Uruguay, in southeastern Paraguay, and in Argentina in Misiones province and perhaps as far south as Entre Ríos province (Ximénez and Achaval 1966; Wetzel 1982a) (map 3.5).

Ecology

Cabassous tatouay appears to be a species of open areas of vegetation, but as with all species of *Cabassous*, its highly fossorial habits make information on this species scarce. In southeastern Brazil, *C. tatouay* dug single-entrance burrows only into active termite mounds and was never found to reuse a burrow (Carter and Encarnação 1983).

Genus *Chaetophractus* Fitzinger, 1871
Hairy Armadillo, Peludo

Description

These armadillos are of intermediate size, ranging from 200 to 400 mm in head and body length. The head has a shield of dermal ossicles that almost covers the nose. The dorsum is made distinctive by a linear series of about eighteen bands of dermal scutes, the middle seven or eight bands being flexible. The venter is haired (see plate 3). Sometimes confused with *Euphractus*, members of the genus *Chaetophractus* have longer ears (see species accounts).

Distribution

The current distribution of the genus includes western Bolivia to Paraguay and thence to central Argentina. Outlying populations have been recorded from southern Peru and northern Chile.

Chaetophractus nationi (Thomas, 1894)
Andean Hairy Armadillo, Quirquincho Andino

Description

Chaetophractus nationi is intermediate in size between *C. villosus* and *C. vellerosus*. One specimen from Bolivia measured HB 268; HF 52; E 30. The hair on the shell varies from tan to buffy white, is up to 72 mm long, and can be sparse or thick (Wetzel 1985b).

Distribution

The distribution of this species is poorly known because it has frequently been confused with *C. vellerosus*. It is found in the puna of Bolivia and extends into the altiplano of Chile (Cabrera 1958; Mann 1978; Tamayo and Frassinetti 1980; Wetzel 1985b) (map 3.6).

Life History

One or two young are born in the summer (Mann 1978).

Ecology

This armadillo inhabits high-altitude areas with brushy vegetation. It deposits fat seasonally and hibernates during the winter. It is omnivorous, preferring insects and other invertebrates (Mann 1978).

Comment

More specimens are needed to either support the validity of this species or establish it as a high-altitude subspecies of *C. vellerosus* (Wetzel 1985b).

Chaetophractus vellerosus (Gray, 1865)
Small Hairy Armadillo, Small Screaming Armadillo, Quirquincho Chico, Piche Llorón

Measurements

	Mean	Min.	Max.	N	Loc.	Source[a]
TL	376	328	400	76 m	A	1
	368	265	419	71 f		
T	114	84	131	76 m		
	112	77	138	71 f		
HF	49	44	53	76 m		
	48	31	56	71 f		

E	28	22	31	76 m
	27	22	31	71 f
Wta	860	543	1,329	76 m
	814	257	1,126	71 f

[a](1) Greegor 1974.

Description

This species is the smallest and slenderest of the hairy armadillos and differs from other *Chaetophractus* and from *Zaedyus pichiy* in having much longer ears. The hair on the dorsum is usually tan (pers. obs.; Wetzel 1985b).

Distribution

Chaetophractus vellerosus is found in the Gran Chaco of Bolivia, western Paraguay, and Argentina, extending south to Mendoza province and the middle latitude of Buenos Aires province. In Buenos Aires province its distribution is disjunct (Carlini and Vizcaíno 1987; Wetzel 1982a) (map 3.7).

Ecology

This armadillo is found primarily in xeric areas from low to high altitudes and is not found in rocky soils, where burrow construction is impossible. It is usually found in areas with yearly rainfall between 200 and 600 mm, but a population exists in eastern Buenos Aires province, Argentina, where the annual rainfall is 1,000 mm.

It is nocturnal in the summer and diurnal in the winter and can go for long periods without drinking

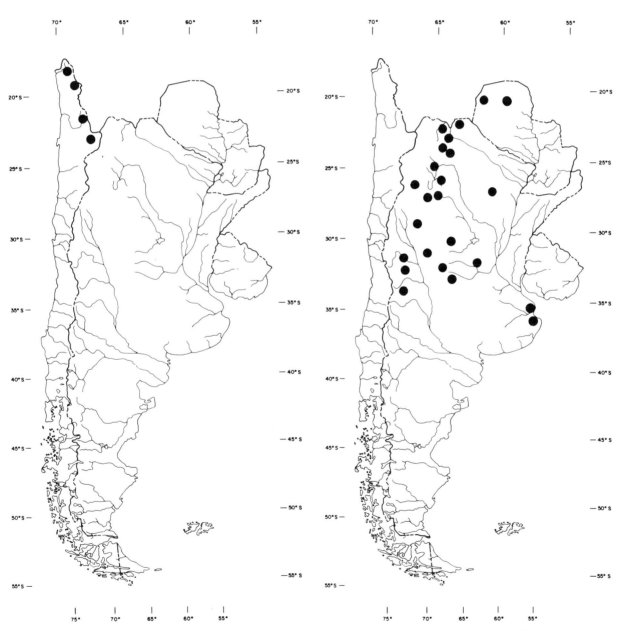

Map 3.6. Distribution of *Chaetophractus nationi*.

Map 3.7. Distribution of *Chaetophractus vellerosus*.

water. Burrows have diameters ranging from 8 to 15 cm, can be several meters long, may have multiple entrances, and are often at the base of shrubs. One animal uses several burrows throughout its home range. Apparently no nest is built in the burrow. The entrance is usually sealed when an animal is inside. The minimum home-range area for one individual was 3.4 ha.

When not in the burrow, *C. vellerosus* spends most of its time foraging. The diet of this species varies seasonally. During the summer the major food item was insects (46% by volume; $n = 48$ stomachs), whereas during the winter it was plant material (50.7% by volume; $n = 36$ stomachs), especially pods from the *Prosopis* tree. A significant percentage of the diet is vertebrates (27.7% by volume in summer; 13.9% by volume in winter), including anurans, lizards, birds, and the mice *Eligmodontia typus* and *Phyllotis griseoflavus*. It ingests a great deal of sand while foraging, and sand can compose 50% of the volume of a single stomach. In the winter males and females are up to 10% heavier than during the summer because they have a 1–2 cm layer of subcutaneous fat. When handled, this species frequently emits loud cries of protest (Carlini and Vizcaíno 1987; Crespo 1944; Greegor 1974, 1975, 1980a,b, 1985; Mares, Ojeda, and Kosco 1981; Myers and Wetzel 1979).

Chaetophractus villosus (Desmarest, 1804)
Larger Hairy Armadillo, Quirquincho Grande

Measurements

	Mean	Min.	Max.	N	Loc.	Source[a]
TL	436.7	386.0	486.0	10	P	1
HB	291.1	261.0	344.0			
T	145.6					
HF	61.7	57.0	66.0	9		
E	24.0	22.0	31.0	10		
Wta	1.32 kg	1.0 kg	1.40 kg	4		
	3.42 kg	3.2 kg	3.65 kg	2	A	2

[a](1) Wetzel 1985b; CM, PCorps, UConn; (2) Atalah 1975.

Description

Chaetophractus villosus is the largest *Chaetophractus*. It has a larger head shield than *C. vellerosus* and shorter ears. Some individuals have three to four holes in the pelvic shield that open to shallow glandular pits. It is a dark-colored armadillo with long black hair rather than the tan hair of *C. vellerosus* or the white of *C. nationi* (Pocock 1913; Wetzel 1985b).

Chromosome number: $2n = 60$; FN $= 90$ (Benirschke, Low, and Ferm 1969).

Distribution

This species is found in the Chaco of Paraguay and Argentina (in Argentina south to at least Santa Cruz province) and in Chile along the eastern edge from the Province of Bío-bío south to Aisén province (Wetzel 1982a; see Atalah 1975) (map 3.8).

Life History

In captivity mounting by the male was observed in every month of the year, and the birth season appears to be from February to December. The female builds a nest by pulling leafy material under her body and then kicking the pile backward into the nest mound. The gestation period is between sixty and seventy-five days, and lactation lasts about fifty-five days. If the nest is disturbed, the female will emerge and growl. Sexual maturity is reached about nine mouths (Roberts, Newman, and Peterson 1982).

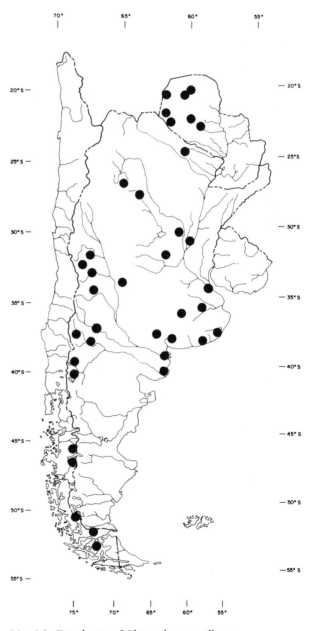

Map 3.8. Distribution of *Chaetophractus villosus*.

Ecology

In captivity, *C. villosus*, like the other euphrac-
tine armadillos, deposits large quantities of fat (Roberts, Newman, and Peterson 1982).

Genus *Chlamyphorus* Harlan, 1825

Description and Taxonomy

The genus *Burmeisteria* is included in this account. There are two extant species. These extremely small armadillos have a head and body
length usually less than 150 mm. The dorsal plates
are almost free from the body, being attached at the
head and loosely along the spine and at the pelvis.
These animals are strongly fossorial and appear to
occupy an almost molelike niche. The tail is short,
and the rump is abruptly squared off, so it looks like
a plug when the animal is burrowing (see plate 3).
The small size, large claws on the forefeet, reduced
ear, and diminutive eye render these armadillos so
distinctive that they cannot readily be confused with
any other species.

Distribution

The genus is at present confined to the more xeric portions of Bolivia, Paraguay, and northern
Argentina.

Chlamyphorus (Burmeisteria) retusus (Burmeister, 1863)
Chacoan Fairy Armadillo, Greater Fairy Armadillo,
Pichiciego Grande

Description

Chlamyphorus retusus is distinguishable from *C.
truncatus* by its larger size. One specimen from Paraguay measured TL 158.5; HB 116; T 42.5; HF 39; E
4.5 (PCorps). In addition, the species has the carapace completely attached to the skin of the back and
head, not underlain by fur. The head shield is wider,
extending laterally and ventrally to the level of the
eye, and the tail is not spatulate at the tip (see plate
3). The dorsum is golden yellowish, and there is
short gray white hair on the belly (Myers and Wetzel
1979; Wetzel 1985b; pers. obs.).
Chromosome number: $2n = 64$ (Benirschke, Low,
and Ferm 1969).

Distribution

Chlamyphorus retusus is found in the Gran Chaco
of southeastern Bolivia, western Paraguay, and
northwestern Argentina (Wetzel 1982a) (map 3.9).

Chlamyphorus truncatus Harlan, 1825
Lesser Pink Fairy Armadillo, Pink Fairy Armadillo,
Pichiciego Menor

Measurements

	Mean	Min.	Max.	N	Loc.	Source[a]
TL	129.6	111	148	10	A	1
HB	98.4	84	117			
T	31.2	27	35			

[a](1) Minoprio 1945.

Description

The shell is very flexible and attached to the body
only along the spine and on the butt plate. The shell
is underlain by long, soft fur. The tail is spatulate at
the end. Females have two nipples (Miniprio 1945,
1951).

Distribution

Chlamyphorus truncatus is found in the Argentine provinces of Mendoza, San Luis, La Pampa,

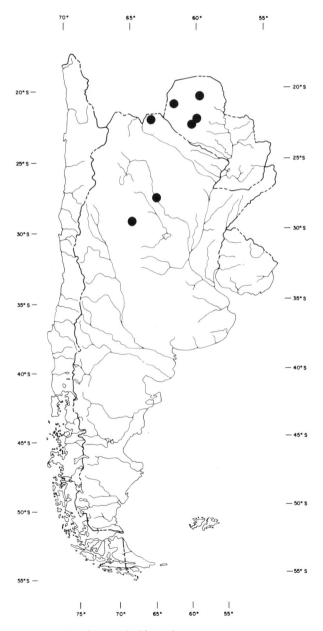

Map 3.9. Distribution of *Chlamyphorus retusus*.

eastern La Rioja, southern Catamarca, Córdoba, and western Buenos Aires (Wetzel 1982a) (map 3.10).

Ecology

This almost completely fossorial armadillo is confined to sandy soils in areas of little rainfall. It digs with its enlarged claws, kicks the soil back, and packs the soil with the butt plate. Burrows are never left open. In captivity *C. truncatus* is active underground day and night and typically comes aboveground for only a few minutes at a time to feed. It sleeps underground, in a curled position. Its spatulate tail supports the animal as it grooms and digs and leaves a telltale track when the animal ventures

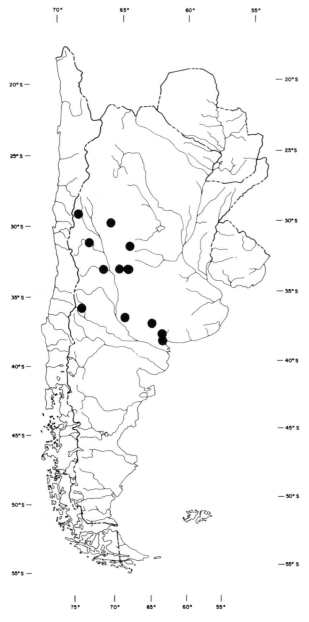

Map 3.10. Distribution of *Chlamyphorus truncatus*.

aboveground. Food is found by smell, apparently often aboveground, and five stomachs were found to contain mostly adult beetles and larvae (Minoprio 1945; Rood 1970; White 1880).

Comment

The neuroanatomy was described by Jakob (1943). He speculated that the pink fairy armadillo believes in a divine being and believes there can be no concept of God without the existence of a devil.

Genus *Dasypus* Linnaeus, 1758
Long-nosed Armadillo, Cachicame, Mulita, Tatú

Description

The rostrum is long (see fig. 3.5), and the dental formula is 7–9/7–9. The dark brown carapace comprises scapular and pelvic shields, with six to eleven movable bands separating the two shields. The ears are long and have no scales or scutes. The long tail, generally exceeding 55% of the head and body length, tapers to a slender tip. The proximal two-thirds of the tail is covered with rings, each formed by two or more rows of scales and scutes. The forefoot bears four long claws, the longest two on the second and third digits. The hind foot bears five claws, with the longest on the third digit (plate 3).

Distribution

The genus *Dasypus* is distributed from the south-central United States to the Río Negro in Argentina.

Life History

Species of *Dasypus* are insectivorous but will opportunistically eat small vertebrates and fruit. They construct burrows and transport nesting material such as grass and leaves by raking it under the body, pressing it against the abdomen with the forefeet, and hopping on the hind legs to the burrow.

Females are monovular but produce litters of two to eight. The young are genetically identical, since the zygote cleaves to produce several blastocysts before implantation (see *D. novemcinctus*) (Eisenberg 1961; Redford 1987a.)

Dasypus hybridus (Desmarest, 1804)
Southern Lesser Long-nosed Armadillo, Mulita Orejuda

Measurements

	Mean	Min.	Max.	N	Loc.	Source[a]
TL	459.5	397.0	498.0	15	U	1
T	168.4	132.0	191.0			
HF	67.3	55.0	75.0			
E	25.1	23.0	28.0			
Wta	1.5 kg	1.09 kg	2.04 kg	14		

[a](1) Barlow 1965.

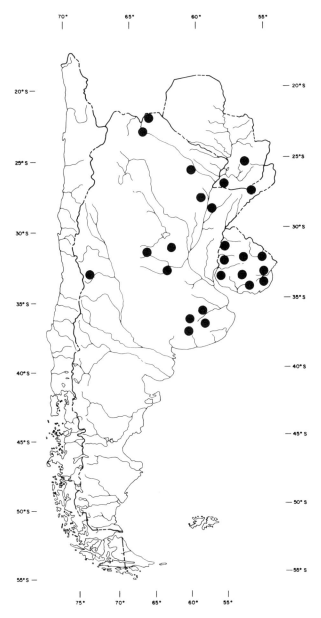

Map 3.11. Distribution of *Dasypus hybridus*.

Description

This small *Dasypus* usually has seven movable bands. It has smaller ears but a larger body than *D. septemcinctus*. The very lightly haired carapace and tail are dull brownish gray, and the belly is grayish to pink (Barlow 1965; Wetzel and Mondolfi 1979). Chromosome number: $2n = 64$ (Benirschke, Low, and Ferm 1969).

Distribution

Dasypus hybridus is found from eastern Paraguay, eastern Argentina, southern Brazil, and Uruguay west through northern Argentina to Jujuy province and south to Mendoza and Río Negro provinces (Wetzel 1982a) (map 3.11).

Life History

Embryo counts have ranged from seven to twelve, with eight the most common number of viable embryos (Crespo 1982; Galbreath 1985).

Ecology

Dasypus hybridus is an animal of the grasslands, found up to 2,300 m in Jujuy province, Argentina. It apparently digs burrows only in grassland or other very open vegetation, unlike *D. novemcinctus*, whose burrows are frequently found in the forest. Burrows are about 2 m long, with a single entrance less than 25 cm in diameter. Burrows are usually in sandy soils on flat or gently sloping ground, in banks, under rocks, or among the roots of trees.

This armadillo feeds much like *D. novemcinctus*, moving rapidly across the ground, sniffing constantly and digging shallow foraging holes. In Uruguay it often digs into ant and termite nests, and one stomach contained mostly ants and termites as well as Orthoptera, Lepidoptera, other invertebrates, and the remains of a small rodent (Barlow 1965; Olrog 1979).

Comment

Dasypus hybridus includes *D. mazzai* (Honacki, Kinman, and Koeppl 1982).

• *Dasypus novemcinctus* Linnaeus, 1758
Common Long-nosed Armadillo, Nine-banded Armadillo, Mulita Grande, Cachicamo, Tatú-hu

Measurements

	Mean	Min.	Max.	N	Loc.	Source[a]
TL	645.7	615.0	671.0	9	P	1
HB	354.6	324.0	381.0			
	465.9	431.0	508.0	15	Br	2
T	291.1	254.0	332.0	9	P	1
	312.5	211.0	366.0	13	Br	2
HF	86.3	61.0	114.0	9	P	1
	88.7	82.0	98.0	15	Br	2
E	40.8	37.0	51.0	9	P	1
	45.5	33.0	54.0	15	Br	2
Wta	2.54 kg	2.0 kg	3.0 kg	5	P	1
	4.10 kg	3.6 kg	4.7 kg	15	Br	2
	4.51 kg	2.2 kg	6.5 kg	51 m	P	3
	4.50 kg	2.0 kg	5.7 kg	51 f		

[a](1) PCorps, UConn; (2) Redford, n.d.; (3) Hill, unpubl. (only animals over 2 kg included).

Description

Dasypus novemcinctus is the largest of the *Dasypus* species occurring in southern South America. It also has more movable bands than either of the other species (mean, 8.3; range, 8–9). Unlike *D. hybridus* and *D. septemcinctus*, this species has extensive yellow triangular scales on the sides of the movable bands, a proportionally smaller ear, and a longer rostrum (Redford, n.d.; Wetzel 1985b). Chromosome number: $2n = 64$ (Benirschke, Low, and Ferm 1969).

Distribution

This species ranges from the southern United States south to Uruguay, eastern and western Paraguay, and Argentina south to the provinces of Santiago del Estero, Santa Fe, and Entre Ríos (Wetzel 1985b) (map 3.12).

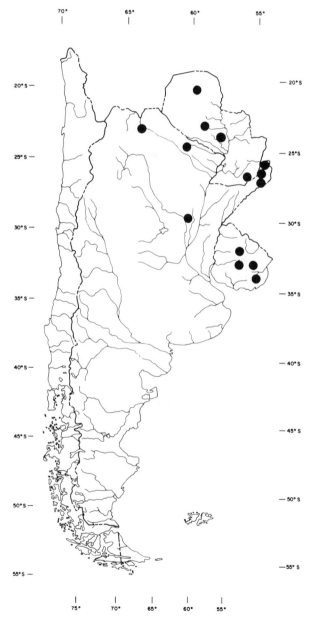

Map 3.12. Distribution of *Dasypus novemcinctus*.

Life History

Like other members of the genus, *D. novemcinctus* produces multiple young from a single fertilized egg. For this species the number of young is almost always four; the female has four nipples. The gestation period is eight to nine months, which includes a three to four month embryonic diapause. Four same-sex young are born, weighing about 85 g each. In the wild, weaning takes place after about three months. Littermates often remain together up to nine months. In the southern United States both sexes reach adult size at between three and four years, and females usually ovulate for the first time at two years. Females probably ovulate only once a year, but males produce sperm year round. In captivity females can bear young thirteen to twenty-four months after capture and subsequent isolation from males (Galbreath 1982, 1985; Storrs, Burchfield, and Rees 1988).

Ecology

Dasypus novemcinctus has the most extensive range of any of the edentates. It is found from the dry regions of the southern United States through the moist tropical forests of Amazonia to the grasslands of central South America. Though found in this great range of habitats, it seems most common in areas with reasonable rainfall and warm temperatures. The northern limits of its range are determined by cold, drought, and lack of insects, factors that probably control the southerly limits as well.

Even though it is found in grasslands, this species seems to prefer forested areas, where it usually digs its burrows, often near streams. These burrows may have more than one entrance and may contain a nest made of vegetation. Frequently a burrow containing an animal will be plugged with vegetation. In some areas, periodic flooding forces this species to construct nests aboveground.

Adults may be active day or night depending on temperature, rainfall, and degree of hunting. Although armadillos tend to forage alone, there is considerable home-range overlap. Though several individuals may use the same burrow, they may be members of a family group. Home-range size in *D. novemcinctus* varies considerably depending on the carrying capacity of the habitat and may be as small as 3.4 ha. At lower carrying capacities it can exceed 15 ha. In the southern United States *D. novemcinctus* is largely insectivorous, though it eats fruit, carrion, and small vertebrates. It will eat virtually any food item it can find and ingest. In South America this species feeds to a greater extent on ants and termites. These armadillos have very poor eyesight and locate food by smell. Most food items are probably taken from the ground surface or just beneath it. The strong claws are used to dig short, triangular feeding holes.

Dasypus novemcinctus is heavily hunted throughout its range for its delicate white meat. Because of its relatively large litter size and tolerance of humans, it can probably coexist with humans in rural areas (Barlow 1965; Eisenberg 1989; Galbreath 1982; González and Ríos 1980; Humphrey 1974; Redford 1987a,b; Wetzel and Mondolfi 1979).

Dasypus septemcinctus Linnaeus, 1758
Brazilian Lesser Long-nosed Armadillo,
Mulita Común

Measurements

	Mean	Min.	Max.	N	Loc.	Source[a]
HB	260.5	240.0	305.0	8	Br	1
T	147.5	125.0	170.0	7		
HF	60.0	45.0	72.0	2		
E	30.9	30.0	38.0	7		
Wta	1.63 kg	1.45 kg	1.8 kg	2		

[a](1) Hamlet 1939; Wetzel and Mondolfi 1979.

Description
Dasypus septemcinctus is the smallest of the species of *Dasypus* in southern South America. It does, however, have longer ears than *D. hybridus*. It has six or seven movable bands. Its carapace is dark, with little of the yellow found on *D. novemcinctus* (Redford, n.d.; Wetzel 1985b).

Distribution
This species is found from Amazonian Brazil south to the northern Argentine provinces of Salta, Formosa, and Chaco (Wetzel 1982a) (map 3.13).

Ecology
Dasypus septemcinctus is apparently a grassland species, although in southeastern Brazil it is reported to prefer gallery forests. It frequently expands burrows dug by other species of armadillos. In captivity young animals build nests at low temperatures (Block 1974; T. S. Carter, unpubl. data; Wetzel 1985b).

Genus *Euphractus* Wagler, 1830
• *Euphractus sexcinctus* (Linnaeus, 1758)
Yellow Armadillo, Six-banded Armadillo, Gualacate

Measurements

	Mean	Min.	Max.	N	Loc.	Source[a]
TL	616.4	556.0	700.0	13	P	1
HB	395.7	341.0	445.0	12		
T	220.2	200.0	255.0			
HF	83.5	80.0	89.0	13		
E	35.2	24.0	41.09	12		
Wta	3.95 kg	3.0 kg	5.9 kg	6		
	4.68 kg	3.2 kg	6.5 kg	14	Br	2

[a](1) PCorps, UConn; (2) Redford, n.d.

Description
This species is the largest of the "hairy" armadillos. At the anterior margin of the scapular shield there is no movable band. There are two to four holes in the pelvic shield. The carapace is characteristically yellow or tan and sparsely covered with pale hair (Redford and Wetzel 1985).

Chromosome number: $2n = 58$; FN = 102 (Be-

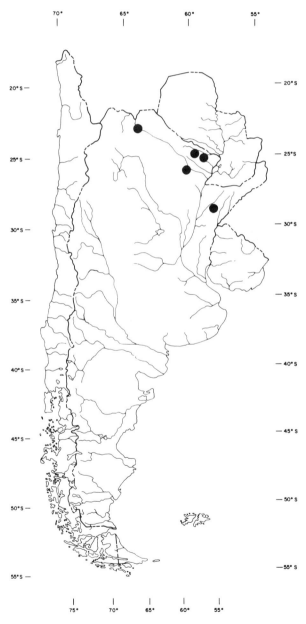

Map 3.13. Distribution of *Dasypus septemcinctus*.

nirschke, Low, and Ferm 1969; Jorge, Meritt, and Benirschke 1977).

Distribution

Euphractus sexcinctus is found in the savannas of Suriname and adjacent Brazil and south to Uruguay, eastern and western Paraguay, and Argentina south to Buenos Aires province (Redford and Wetzel 1985; Wetzel 1982a) (map 3.14).

Life History

Litter size ranges from one to three, and litters include both sexes. In captivity the female builds a nest before giving birth, and young weighing 95–

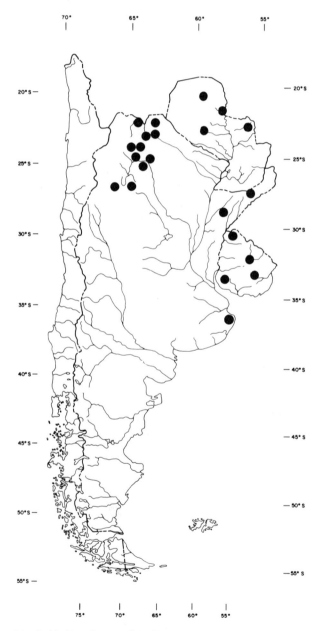

Map 3.14. Distribution of *Euphractus sexcinctus*.

115 g are born after a gestation of sixty to sixty-four days. The eyes open after twenty-two to twenty-five days, and the young reach sexual maturity at nine months. Two pregnant females were found in September and October in central Brazil and another in January in Uruguay (Barlow 1965; Gucwinska 1971; Kühlhorn 1954; Redford and Wetzel 1985; Sanborn 1930).

Ecology

The yellow armadillo is most commonly found in savannas, other open vegetation formations, and forest edges. It appears to use higher, drier habitats, though in Uruguay it is reported as most common in ecotonal areas, especially near streams. In the Brazilian Pantanal, in all vegetation types the biomass of *Euphractus* was estimated at 18.8 kg/km².

This species is largely diurnal, though occasionally it is active at night. It is a good digger and builds burrows with a single inverted U-shaped entrance. Unlike many other species of armadillos, it frequently reuses burrows. Captive animals are reported to mark the corners of their cages with the secretions from the pelvic shield scent gland, so this gland is probably used to mark burrows.

Like other hairy armadillos, *Euphractus* stores fat, and animals in captivity can weigh up to 11 kg. It is omnivorous and eats a broad range of animal and plant foods including carrion, small vertebrates, insects, bromeliad fruits, tubers, and palm nuts. Plant material can compose a significant proportion of the diet. In captivity the yellow armadillo will kill and eat large rats; however, they are very inefficient predators, unable to effect a killing bite. Members of *Euphractus* are active, alert animals with poor eyesight, and they use smell to locate food. Unlike many armadillos, they run to escape and bite when handled (Barlow 1965; Carter and Encarnação 1983; Redford 1985b; Redford and Wetzel 1985; Roig 1969; Schaller 1983).

Genus *Priodontes* F. Cuvier, 1825
• *Priodontes maximus* Kerr, 1792
Giant Armadillo, Tatú Carrera, Tatú-guaca

Measurements

	Mean	Min.	Max.	N	Loc.	Source[a]
HB	895.5	832.0	960.0	4	Whole	1
T	528.2	510.0	550.0		range	
HF	190.7	185.0	200.0	3		
E	53.8	45.0	60.0	5		
Wta	26.8 kg	18.7 kg	32.3 kg			

[a](1) Redford, n.d.

Description

The giant armadillo is unmistakable because of its size. The only armadillos it can be confused with are those of the genus *Cabassous*. However, *Priodontes* is much larger, with a darker shell, sharply marked laterally by a buffy border, and a well armored tail. Like *Cabassous* it has a rounded, blunt muzzle, a carapace with many narrow bands, and large scimitar-shaped foreclaws, the third of which is greatly enlarged (Redford, n.d.).

Chromosome number: $2n = 50$; FN $= 76$ (Benirschke, Low, and Ferm 1969).

Distribution

Priodontes is found from Colombia and Venezuela south to Paraguay and northern Argentina (Wetzel 1982a) (map 3.15).

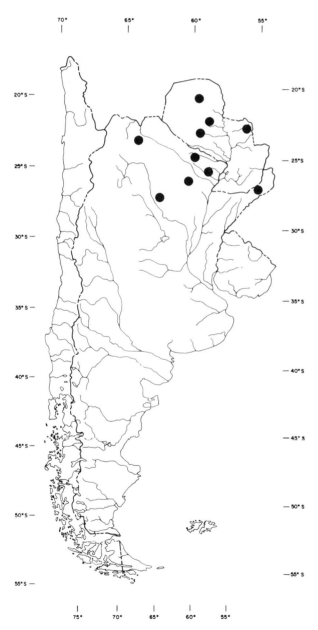

Map 3.15. Distribution of *Priodontes maximus*.

Life History

The litter size is one or two (Krieg 1929).

Ecology

Giant armadillos range over much of South America and are found in tropical forest and open savanna. They are largely nocturnal and highly fossorial and therefore are seldom seen. They are extremely powerful diggers, and their very large foraging and sleeping holes are unmistakable. Burrows tend to be clumped and are usually found in active or dead termite mounds. Individuals usually remain in their burrows for at least twenty-four hours and may stay there several days.

Priodontes is probably the most myrmecophagous of the armadillos and has been recorded as eating virtually nothing but ants and termites. Unlike other armadillos, *Priodontes* often destroys a termite mound while feeding (Barreto, Barreto, and D'Alessandro 1985; Carter 1983; Carter and Encarnação 1983; Redford 1985b).

Genus *Tolypeutes* Illiger, 1811
Tolypeutes matacus (Desmarest, 1804)
Southern Three-banded Armadillo,
Quirquincho Bola

Measurements

	Mean	Min.	Max.	N	Loc.	Source[a]
HB	250.7	218	273.0	17	A, Br	1
T	63.7	60	80.0	15		
HF	42.3	38	47.0	15		
E	22.8	21	32.0			
Wta	1.1 kg	1 kg	1.15 kg	10		

[a](1) Redford, n.d.

Description

Tolypeutes is probably the most distinctive of the armadillos and is unmistakable because of its small size, its hard, inflexible carapace, and its habit of curling up. Members of *Tolypeutes* are the only armadillos that can roll into a ball (plate 3) (pers. obs.).

Distribution

This species is found from southeastern Bolivia south through the Paraguayan Chaco to the Argentine province of Santa Cruz (Wetzel 1982a, as amended by O. Pearson, pers. comm.) (map 3.16).

Life History

Tolypeutes matacus gives birth to a single young after a gestation period of 120 days. The young opens its eyes after about twenty-two days and suckles for approximately ten weeks. The birth season is apparently between October and January (Krieg 1929; Meritt 1971; Sanborn 1930).

Ecology

Tolypeutes matacus prefers dry vegetation formations and is abundant in the most xeric parts of the Paraguayan Chaco. It apparently does not dig its own burrows. In fact, unlike most other armadillos except *Euphractus*, it will often run when chased rather than dig. *T. matacus* can be active throughout the day and night, though its major activity peaks are probably dictated by temperature and rainfall. It seems to feed primarily from the ground surface, occasionally digging shallow foraging holes. It is largely myrmecophagous but will take other soft-bodied invertebrates (Redford 1985b; Sanborn 1930; Schaller 1983; pers. obs.).

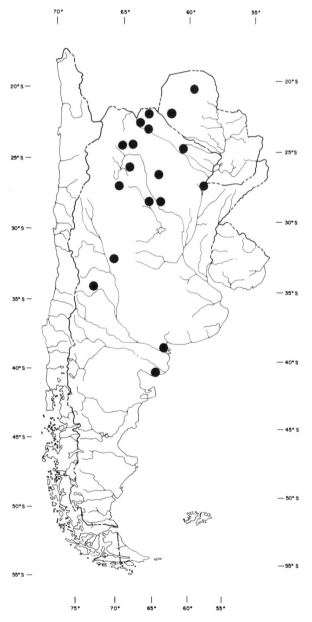

Map 3.16. Distribution of *Tolypeutes matacus.*

Genus *Zaedyus* Ameghino, 1889
Zaedyus pichiy (Desmarest, 1804)
Pichi

Measurements

	Mean	Min.	Max.	N	Loc.	Source[a]
TL	401.8	390.0	416.0	4	A, C	1
HB	277.0	250.0	300.0			
T	124.8	116.0	140.0			
HF	56.0	44.0	63.0			
E	14.7	14.0	16.0	3		
Wta	1.02 kg	1.25 kg	2.35 kg	6	A	2

[a](1) Allen 1905; Texera 1973; (2) Roig 1971.

Description

Zaedyus is a small hairy armadillo with sharply pointed marginal scutes. It differs from *C. vellerosus*, the smallest hairy armadillo, in this characteristic as well as in having a proportionally more slender head shield and much shorter ears (plate 3) (Wetzel 1985b).

Distribution

Zaedyus pichiy is found in Argentina from the provinces of Mendoza, San Luis, and Buenos Aires south to Río Santa Cruz and west to the puna of the Andes, and in Chile south from the province of Aconcagua to the Strait of Magellan (Wetzel 1982a) (map 3.17).

Life History

The gestation period is approximately sixty days, and the litter size is one to three. The young weigh 95–115 g. They are fully weaned by six weeks and reach sexual maturity at between nine and twelve months. In the wild the birth season is January and February (Meritt 1972; Merrett 1983).

Ecology

The pichi occurs farther south than any of the other armadillos, mostly in open vegetation formations. In the laboratory *Z. pichiy* shows a variable body temperature and a propensity to enter torpor. In the wild it is reported to hibernate.

Like other hairy armadillos, the pichi is omnivorous, eating carrion, insects, especially ants, invertebrates, and plant material, especially pods of the *Prosopis* tree (McNab 1980; Merrett 1983; Redford 1985b; Roig 1971; Wetzel 1982a).

References

Allen, J. A. 1905. Mammalia of southern Patagonia. In *Reports of the Princeton University expeditions to Patagonia, 1896–1899,* vol. 3, *Zoology.* Stuttgart: Schweizerbart'sche.

Atalah G., A. 1975. Presencia de *Chaetophractus villosus* (Edentata, Dasypodidae) nueva especie para la región de Magallanes, Chile. *Anal. Inst. Patagonia, Punta Arenas* (Chile) 6(1–2): 169–71.

Barlow, J. C. 1965. Land mammals from Uruguay: Ecology and zoogeography. Ph.D. diss., University of Kansas.

Barreto, M., P. Barreto, and A. D'Alessandro. 1985. Colombian armadillos: Stomach contents and infection with *Trypanosoma cruzi*. *J. Mammal.* 66: 188–93.

Benirschke, K., R. J. Low, and V. H. Ferm. 1969. Cytogenetic studies of some armadillos. In *Comparative mammalian cytogenetics*, ed. K. Benirschke, 330–45. New York: Springer-Verlag.

Bickel, C. L., G. K. Murdock, and M. L. Smith.

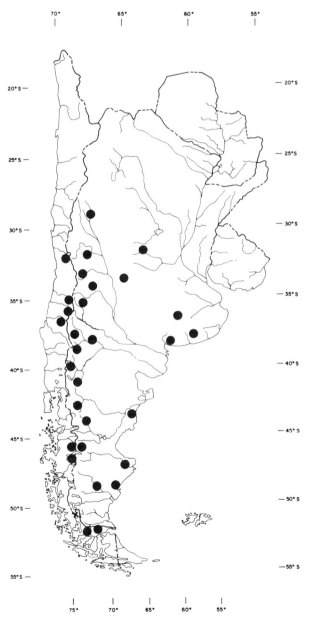

Map 3.17. Distribution of *Zaedyus pichiy*.

1976. Hand-rearing a giant anteater *Myrmecophaga tridactyla* at Denver Zoo. *Int. Zoo Yearb.* 16: 195–98.

Block, J. A. 1974. Hand-rearing seven-banded armadillos *Dasypus septemcinctus* at the National Zoological Park, Washington. *Int. Zoo Yearb.* 14: 210–14.

Byrne, P. S. 1962. Giant ant-eaters born in zoo. *J. British Guiana Mus. Zoo Roy. Agric. Commerc. Soc.* 36: 28–29.

Cabrera, A. L. 1958. Catálogo de los mamíferos de América del Sur. 1. Metatheria-Unguiculato-Carnivora. *Rev. Mus. Argent. Cienc. Nat. "Bernardino Rivadavia," Zool.* 4(1): 1–307.

Carlini, A. A., and S. F. Vizcaíno. 1987. A new record of the armadillo *Chaetophractus vellerosus* (Gray, 1865) (Mammalia, Dasypodidae) in the Buenos Aires province of Argentina: Possible causes for the disjunct distribution. *Stud. Neotrop. Fauna Environ.* 22(1): 53–56.

Carter, T. S. 1983. The burrows of the giant armadillos, *Priodontes maximus* (Edentata: Dasypodidae). *Säugetierk. Mitt.* 31: 47–53.

Carter, T. S., and C. D. Encarnação. 1983. Characteristics and use of burrows by four species of armadillos in Brazil. *J. Mammal.* 64: 103–8.

Crespo, J. A. 1944. Contribucion al conocimiento de la ecología de algunos dasipódidos (Edentata) argentinos. *Rev. Argent. Zoogeogr.* 4(1–2): 7–16.

———. 1982. Ecología de la comunidad de mamíferos del Parque Nacional Iguazú, Misiones. *Rev. Mus. Argent. Cienc. Nat. "Bernardino Rivadavia," Ecol.* 3(2): 45–162.

Eisenberg, J. F 1961. The nest-building behavior of armadillos. *Proc. Zool. Soc. London* 137: 322–24.

———. 1989. *Mammals of the Neotropics. Vol. 1. Mammals of the northern Neotropics: Panama, Colombia, Venezuela, Guyana, Suriname, French Guiana.* Chicago: University of Chicago Press.

Emery, R. J. 1970. A North American Oligocene pangolin and other additions to the Pholidota. *Bull. Amer. Mus. Nat. Hist.* 142(6): 457–510.

Galbreath, G. J. 1982. Armadillo, *Dasypus novemcinctus*. In *Wild mammals of North America: Biology, management and economics*, ed. J. A. Chapman and G. A. Feldhamer, 71–79. Baltimore: Johns Hopkins University Press.

———. 1985. The evolution of monozygotic polyembryony in *Dasypus*. In *The evolution and ecology of armadillos, sloths, and vermilinguas*, ed. G. G. Montgomery, 243–46. Washington, D.C.: Smithsonian Institution Press.

Gillette, D. D., and C. E. Ray. 1981. *Glyptodonts of North America*. Smithsonian Contributions to

Paleobiology 40. Washington, D.C.: Smithsonian Institution Press.

González, J. C., and C. Ríos. 1980. Refugios epideos del "tatu" *Dasypus n. novemcinctus* Linne (Mammalia: Dasypodidae). *Res. J. C. Nat. Montevideo* 1:129–30.

Greegor, D. H., Jr. 1974. Comparative ecology and distribution of two species of armadillos, *Chaetophractus vellerosus* and *Dasypus novemcinctus*. Ph.D. diss., University of Arizona.

———. 1975. Renal capabilities of an Argentine desert armadillo. *J. Mammal.* 56:626–32.

———. 1980a. Preliminary study of movements and home range of the armadillo, *Chaetophractus vellerosus. J. Mammal.* 61:334–35.

———. 1980b. Diet of the little hairy armadillo, *Chaetophractus vellerosus*, of northwestern Argentina. *J. Mammal.* 61:331–34.

———. 1985. Ecology of the little hairy armadillo *Chaetophractus vellerosus*. In *The evolution and ecology of armadillos, sloths, and vermilinguas*, ed. G. G. Montgomery, 397–405. Washington, D.C.: Smithsonian Institution Press.

Gucwinska, H. 1971. Development of six-banded armadillos *Euphractus sexcinctus* at Wroclaw Zoo. *Int. Zoo Yearb.* 11:88–89.

Hamlet, G. W. D. 1939. Identity of *Dasypus septemcinctus* Linnaeus with notes on some related species. *J. Mammal.* 20:328–36.

Hardin, C. J. 1976. Hand-rearing a giant anteater *Myrmecophaga tridactyla* at Toledo Zoo. *Int. Zoo Yearb.* 15:199–200.

Honacki, J. H., K. E. Kinman, and J. W. Koeppl, eds. 1982. *Mammal species of the world*. Lawrence, Kans.: Allen Press and Association of Systematics Collections.

Humphrey, S. R. 1974. Zoogeography of the nine-banded armadillo (*Dasypus novemcinctus*) in the United States. *BioScience* 24(8): 457–62.

Jakob, C. 1943. El pichiciego (*Chlamydophorus truncatus*) estudios neurobiológicos de un mamífero misterioso de la Argentina. 1. Parte. *Folia Neurobiol. Argent.* 2:1–106.

Jorge, W., D. A. Meritt, Jr., and K. Benirschke. 1977. Chromosome studies in Edentata. *Cytobiosis* 18:157–72.

Krieg, H. 1929. Biologische Reisestudien in Südamerika. 9. Gürteltiere. *Z. Morph. Ökol. Tiere* 14(1): 166–90.

Kühlhorn, F. 1936. Die anpassungstypen der Gürteltieren. *Z. Säugetierk.* 12:245–303.

———. 1954. Säugetierkundliche Studien aus Süd-Mattogrosso. 2. Edentata, Rodentia. *Säugetierk. Mitt.* 2:66–72.

Luederwaldt, H. 1918. Observações sobre a preguica (*Bradypus tridactylus* L.) em liberdade eno captivero. *Rev. Mus. Paulista* 10:795–812.

McNab, B. K. 1978. The comparative energetics of Neotropical marsupials. *J. Comp. Physiol.* 125: 115–28.

———. 1980. Energetics and the limits to a temperate distribution in armadillos. *J. Mammal.* 61: 606–27.

———. 1982. The physiological ecology of South American mammals. In *Mammalian biology in South America*, ed. M. A. Mares and H. H. Genoways, 187–208. Pymatuning Symposia in Ecology 6. Special Publication Series. Pittsburgh: Pymatuning Laboratory of Ecology, University of Pittsburgh.

Mann, G. 1978. *Los pequeños mamíferos de Chile*. Guyana: Zoología 40. Chile: Universidad de Concepción.

Mares, M. A., R. A. Ojeda, and M. P. Kosco. 1981. Observations on the distribution and ecology of the mammals of Salta province, Argentina. *Ann. Carnegie Mus.* 50(6): 151–206.

Massoia, E. 1980. Mammalia de Argentina. 1. Los mamíferos silvestres de la provincia de Misiones. *Iguazú* 1(1): 15–43.

Meritt, D. A., Jr. 1971. The development of the La Plata three-banded armadillo *Tolypeutes matacus* at Lincoln Park Zoo, Chicago. *Int. Zoo Yearb.* 11:195–96.

———. 1972. The behavior of armadillos in nature. *Yearbook of the American Philosophical Society, Report for 1971–1972*. Philadelphia: American Philosophical Society.

———. 1975. The lesser anteater *Tamandua tetradactyla* in captivity. *Int. Zoo Yearb.* 15:41–45.

———. 1985. Naked-tailed armadillos *Cabassous* sp. In *The evolution and ecology of armadillos, sloths, and vermilinguas*, ed. G. G. Montgomery, 389–91. Washington, D.C.: Smithsonian Institution Press.

Merrett, P. K. 1983. *Edentates*. La Villiaze, Saint Andrew's, Guernsey: Zoological Trust of Guernsey.

Minoprio, J. L. 1945. Sobre el *Chlamyphorus truncatus* Harlan. *Acta Zool. Lilloana* 3:5–58, v–xxii.

———. 1951. Fenómenos de similitud existentes entre *Tolypeutes matacus* Desm. y *Chlamyphorus truncatus* Harlan. *Anal. Soc. Cient. Argent.* 151:43–48.

Montgomery, G. G., ed. 1985a. *The evolution and ecology of armadillos, sloths, and vermilinguas*. Washington, D.C.: Smithsonian Institution Press.

———. 1985b. Movements, foraging and food habits of the four extant species of Neotropical vermilinguas (Mammalia: Myrmecophagidae). In *The evolution and ecology of armadillos, sloths, and vermilinguas*, ed. G. G. Montgomery, 365–377. Washington, D.C.: Smithsonian Institution Press.

Montgomery, G. G., and Y. D. Lubin. 1977. Prey influences on movements of Neotropical anteaters. In *Proceedings of the 1975 Predator Symposium*, ed. R. L. Philips and C. Jonkel, 103–31. Missoula: Montana Forest and Conservation Experiment Station, University of Montana.

Montgomery, G. G., and M. E. Sunquist. 1975. Impact of sloths on Neotropical forest energy flow and nutrient cycling. In *Tropical ecological systems*, ed. F. B. Golley and E. Medina, 69–98. New York: Springer-Verlag.

———. 1978. Habitat selection and use by two-toed and three-toed sloths. In *The ecology of arboreal folivores*, ed. G. G. Montgomery, 329–59. Washington, D.C.: Smithsonian Institution Press.

Myers, P., and R. M. Wetzel. 1979. New records of mammals from Paraguay. *J. Mammal.* 60:638–41.

Olrog, C. C. 1979. Los mamíferos de la selva húmeda, Cerro Calilegua, Jujuy. *Acta Zool. Lilloana* 33:9–14.

Olrog, C. C., and M. M. Lucero. 1981. *Guía de los mamíferos argentinos*. Tucumán, Argentina: Ministerio de Cultura y Educación, Fundación Miguel Lillo.

Onelli, C. 1913. Biología de algunos mamíferos argentinos. *Rev. Jardín Zool.* 9:77–142.

Patterson, B., and R. R. Pascual. 1972. The fossil mammal fauna of South America. In *Evolution, mammals, and southern continents*, ed. A. Keast, F. C. Erk, and B. Glass, 247–309. Albany: State University of New York Press.

Pocock, R. I. 1913. Dorsal glands in armadillos. *Proc. Zool. Soc. London* 72:1099–1103.

Redford, K. H. 1983. Lista preliminar de mamíferos do Parque Nacional das Emas. *Brasil Florestal* 55:29–32.

———. 1985a. Feeding and food preferences in captive and wild giant anteaters (*Myrmecophaga tridactyla*). *J. Zool.* (London) 205:559–72.

———. 1985b. Food habits of armadillos (Xenarthra: Dasypodidae). In *The evolution and ecology of armadillos, sloths, and vermilinguas*, ed. G. G. Montgomery, 429–37. Washington, D.C.: Smithsonian Institution Press.

———. 1987a. Dietary specialization and variation in two mammalian insectivores. *Rev. Chil. Hist. Nat.* 59(2):201:8.

———. 1987b. Patterns of ant and termite eating in mammals. *Current Mammal.* 1:349–400.

———. n.d. Sinopse dos edentados (tamanduás, preguiças etatús) do Brasil central. Unpublished manuscript.

Redford, K. H., and R. M. Wetzel. 1985. *Euphractus sexcinctus*. *Mammal. Species* 252:1–4.

Reig, O. 1981. *Teoría del origen y desarrollo de la fauna de mamíferos de América del Sur*. Monografie Naturae. Mar del Plata, Argentina: Museo Municipal de Ciencias Naturales Lorenzo Seaglia.

Roberts, M., L. Newman, and G. Peterson. 1982. The management and reproduction of the large hairy armadillo, *Chaetophractus villosus*, at the National Zoological Park. *Int. Zoo Yearb.* 22:185–94.

Roig, V. G. 1969. Termorregulación en *Euphractus sexcinctus* (Mammalia, Dasypodidae). *Physis* 29(78):27–32.

———. 1971. Observaciones sobre la termoregulación en *Zaedyus pichiy*. *Acta Zool. Lilloana* 28:13–18.

Rood, J. P. 1970. Notes on the behavior of the pygmy armadillo. *J. Mammal.* 51:179.

Sanborn, C. C. 1930. Distribution and habits of the three-banded armadillo (*Tolypeutes*). *J. Mammal.* 11:61–68.

Schaller, G. B. 1983. Mammals and their biomass on a Brazilian ranch. *Arq. Zool. São Paulo* 31(1):1–36.

Schmid, B. 1938. Psychologische Beobachtungen und Versuche an einem jungen männlichen Ameisenbären (*Myrmecophaga tridactyla* L.), Z. *Tierpsychol.* 2:117–26.

Shaw, J. H., T. S. Carter, and J. C. Machado-Neto. 1985. Ecology of the giant anteater *Myrmecophaga tridactyla* in Serra da Canastra, Minas Gerais, Brazil: A pilot study. In *The evolution and ecology of armadillos, sloths, and vermilinguas*, ed. G. G. Montgomery, 379–84. Washington, D.C.: Smithsonian Institution Press.

Shaw, J. H., J. Machado-Neto, and T. S. Carter. 1987. Behavior of free-living anteaters (*Myrmecophaga tridactyla*). *Biotropica* 19:255–59.

Smielowski, J., P. Stanislawski, and T. Taworski. 1981. Breeding the giant anteater. *Int. Zoo News* 28/5(174):1–6.

Storrs, E. E., H. P. Burchfield, and R. J. W. Rees. 1988. Superdelayed parturition in armadillos: A new mammalian survival strategy. *Lepr. Rev.* 59:11–15.

Sunquist, M. E., and G. G. Montgomery. 1973. Activity patterns and rates of movement of two-toed and three-toed sloths (*Choloepus hoffmanni* and *Bradypus infuscatus*). *J. Mammal.* 54:946–54.

Tamayo, M., and D. Frassinetti. 1980. Catálogo de

los mamíferos fósiles y vivientes de Chile. *Mus. Nac. Hist. Nat., Chile* 37:323–99.

Texera, W. A. 1973. *Zaedyus pichiy* (Edentata, Dasypodidae) nueva especie en la provincia de Magallanes, Chile. *Anal. Inst. Patagonia, Punta Arenas* (Chile) 4(1–3): 335–37.

Waage, J. K., and R. C. Best. 1985. Arthropod associates of sloths. In *The evolution and ecology of armadillos, sloths, and vermilinguas*, ed. G. G. Montgomery, 297–311. Washington, D.C.: Smithsonian Institution Press.

Webb, S. D. 1985. The interrelationships of tree sloths and ground sloths. In *The evolution and ecology of armadillos, sloths, and vermilinguas*, ed. G. G. Montgomery, 105–12. Washington, D.C.: Smithsonian Institution Press.

Webb. S. D., and L. G. Marshall. 1982. Historical biogeography of Recent South American land mammals. In *Mammalian biology in South America*, ed. M. A. Mares and H. H. Genoways, 39–52. Pymatuning Symposia in Ecology 6. Special Publication Series. Pittsburgh: Pymatuning Laboratory of Ecology, University of Pittsburgh.

Wetzel, R. M. 1975. The species of *Tamandua* Gray (Edentata, Myrmecophagidae). *Proc. Biol. Soc. Washington* 88:95–112.

———. 1980. Revision of the naked-tailed armadillos, genus *Cabassous* McMurtrie. *Ann. Carnegie Mus.* 49:323–57.

———. 1982a. Systematics, distribution, ecology, and conservation of South American edentates. In *Mammalian biology in South America*, ed. M. A. Mares and H. H. Genoways, 345–76. Pymatuning Symposia in Ecology 6. Special Pub-

lication Series. Pittsburgh: Pymatuning Laboratory of Ecology, University of Pittsburgh.

———. 1982b. The mammals of the Chaco of Paraguay. *Nat. Geog. Soc. Res. Repts.* 14:679–84.

———. 1985a. The identification and distribution of Recent Xenarthra (= Edentata). In *The evolution and ecology of armadillos, sloths, and vermilinguas*, ed. G. G. Montgomery, 5–21. Washington, D.C.: Smithsonian Institution Press.

———. 1985b. Taxonomy and distribution of armadillos, Dasypodidae. In *The evolution and ecology of armadillos, sloths, and vermilinguas*, ed. G. G. Montgomery, 23–46. Washington, D.C.: Smithsonian Institution Press.

Wetzel, R. M., and F. D. de Avila-Pires. 1980. Identification and distribution of the Recent sloths of Brazil (Edentata). *Rev. Brasil. Biol.* 40(4): 831–36.

Wetzel, R. M., and D. Kock. 1973. The identity of *Bradypus variegatus* Schinz (Mammalia, Edentata). *Proc. Biol. Soc. Washington* 86:25–34.

Wetzel, R. M., and E. Mondolfi. 1979. The subgenera and species of long-nosed armadillos, genus *Dasypus* L. In *Vertebrate ecology in the northern Neotropics*, ed. J. F. Eisenberg, 43–64. Washington, D.C.: Smithsonian Institution Press.

White, E. W. 1880. Notes on *Chlamydophorus truncatus*. *Proc. Zool. Soc. London* 8:8–11.

Ximénez, A., and F. Achaval. 1966. Sobre la presencia en el Uruguay del tatú de Rabo Molle, *Cabassous tatouay* (Desmarest) (Edentata-Dasypodidae). *Comun. Zool. Mus. Hist. Nat. Montevideo* 9(109): 1–5.

4 Order Chiroptera (Bats, Murciélagos)

Diagnosis

This order includes the only true flying mammals. Some other mammals glide, but the bats really fly. A wing membrane extends from each side of the body and hind leg to the forearm, where it is supported by the fingers, which are elongated to create a large surface area. There may be an additional membrane between the hind legs, sometimes enclosing the tail. The clavicle is well developed. The hind leg has become rotated to support the wing membrane, so that the knee is directed laterally and backward. The sternum is usually keeled for attachment of the massive pectoral muscles used in the power stroke during flying (see fig. 4.1).

In his classification of bats Miller (1907) used the morphology of the humerus as an indicator of specialization for flight. Family diagnoses often refer to the relative development of the greater and lesser tuberosities (trochiter and trochin) on the proximal end of the humerus.

Distribution

At present bats are found on all continent except Antarctica. The families Pteropodidae (= Pteropidae), Rhinopomatidae, Nycteridae, Megadermatidae, Rhinolophidae, Hipposideridae, Craesonycteridae, Mystacinidae, and Myzopodidae are confined to the Old World (Honacki, Kinman, and Koeppl 1982 include the Hipposideridae within the Rhinolophidae), while the Thyropteridae, Natalidae, Furipteridae, Mormoopidae, Phyllostomidae (= Phyllostomatidae), and Noctilionidae are confined to the New World. The families Emballonuridae, Vespertilionidae, and Molossidae are worldwide in their distribution. Bats show their greatest species richness and numerical abundance in the subtropical and tropical regions of the world (see fig. 4.2).

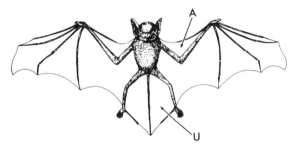

Figure 4.1. External anatomy of a bat. A = antebrachial membrane (propatagium); U = uropatagium (interfemoral membrane) (redrawn from Husson 1978).

History and Classification

Bats are poorly represented in the early fossil record. Their small size and delicate bones apparently reduce the probability of preservation. The earliest bat is known from the Eocene of North America (Jepsen 1970). This fossil clearly indicates that the bat was completely volant; thus the early ancestors of bats showing preflight adaptations are as yet unknown or unrecognized in the fossil record. The earliest records of the phyllostomid bats are from the Miocene of Colombia.

The bats are typically divided into two suborders, the Megachiroptera and the Microchiroptera. The megachiropterans are entirely Old World and are represented by a single family, the Pteropodidae. These are the Old World fruit bats, specialized for feeding on fruits, pollen, and nectar. Their ecological equivalent in the New World tropics is the family Phyllostomidae. The latter family, together with sixteen other families, is included in the Microchiroptera (see fig. 4.2.).

The megachiropterans are chiefly distinguished from the microchiropterans by the following characters: the second finger of a megachiropteran is capable of some independent movement and generally has a claw; and the postorbital process of the skull is well developed, helping to support the very large eyes that are typical of megachiropterans.

In the microchiropterans the second finger lacks a claw and is not capable of independent movement. Usually a median fold of skin termed the tragus projects into the pinna, or external ear. All microchiropterans show specializations of the auditory nerve and larynx that correlate with their highly evolved ability to echolocate, which serves them in orientation and prey capture. Only one genus of the Megachiroptera, *Rousettus*, employs echolocation for orientation. Bats of this genus produce the echolocating pulse by tongue clicks, not from the larynx as do the Microchiroptera. The standard reference works for the Chiroptera are Wimsatt (1970a,b, 1977) and Kunz (1982). A key to the South American families is given in table 4.1.

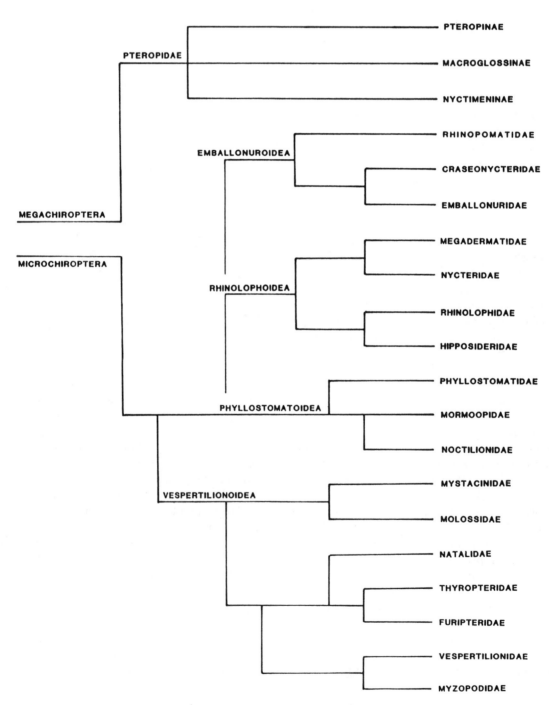

Figure 4.2. Cladogram indicating the relationships of the families of the Chiroptera (from Eisenberg 1981).

Table 4.1   Key to the Families of New World Bats

1 Nose leaf or conspicuous folds or plates of skin on chin present 2
1′ Nose leaf and folds or plates of skin absent 3
2 Folds of skin on chin; no nose leaf present .... Mormoopidae
2′ Nose leaf present; incisors 2/2, 2/1, or 2/0 .... Phyllostomidae[a]
3 Tail perforating uropatagium and enclosed in sheath for half its length; upper lip not noticeably cleft; claws of
   hind feet not noticeably long; most species have glandular sacs in the wing or tail (see text) .... Emballonuridae
3′ Tail and other characters not as above 4
4 Conspicuous suckers on forefeet and hind feet .... Thyropteridae
4′ Sucker disks absent 5
5 Tail usually longer than head and body but completely enclosed in interfemoral membrane; ears funnel
   shaped; tragus triangular .... Natalidae
5′ Tail and other characters not as above 6
6 Thumb (first finger) very reduced and enclosed in membrane or absent .... Furipteridae
6′ First finger present 7
7 Upper lip divided; tail shorter than interfemoral membrane but perforating uropatagium; claws of hind feet
   very long .... Noctilionidae
7′ Not as above 8
8 Tail extending beyond interfemoral membrane; incisors 1/2–3; tooth number 28–32 .... Molossidae
8′ Not as above; incisors 1–2/2–3, upper incisors widely spaced at base; tooth number 28–38 .... Vespertilionidae

[a]The nose leaf is vastly modified in *Desmodus*, *Centurio*, and *Sphaeronycteris* (see text). The Desmodontinae have a postcanine dental formula less than
4/4, and the upper incisors are bladelike (see text).

Natural History of Bats

Bat Form and Function

Comparing bat morphologies can often give us clues to habits and adaptation (Findley and Wilson 1982), since wing form is correlated with differences in flight patterns. The long, flat, narrow wings of the Molossidae correlate with rapid, long-range flight. *Myotis* has short, broad wings that reflect its brief periods of slow, maneuverable flight (Vaughan 1959). Jaw morphology reflects dietary adaptations: nectar and pollen feeders have a relatively long rostrum, and bats that habitually feed on large fruits usually have a very short rostrum and wide gape. Freeman (1979) suggested that the tooth and jaw structure of molossid bats reflect specializations for either "hard" prey such as beetles (e.g., *Molossus*) or soft prey such as moths (e.g., *Nyctinomops*). In the latter group the jaws are relatively thin and the teeth are smaller.

Many species of bats exhibit size dimorphism; frequently the female is larger than the male. Ralls (1976) advanced the idea that the larger size of females in most bat species could reflect selection through competition among females for scarce resources, but selection probably favors larger females because they can produce larger young that presumably have a higher probability of surviving. Myers (1978) concluded that large size in females may relate to problems of flight during pregnancy, since larger size and a longer forearm permit a larger wing surface to provide extra aerodynamic lift.

Bat Physiology

The basal metabolic rates of all mammals tend to decrease with increasing body size. Nevertheless, there are two broad subdivisions of bat species when basal metabolic rate is regressed against body weight. In one group, which includes the Vespertilionidae and the Molossidae, basal metabolism tends to be somewhat low. The second grouping of bats, with higher basal metabolic rates, includes most of the phyllostomids and pteropodids. In spite of their lower resting metabolic rate, the vespertilionids actively colonize and occupy habitats with lower mean annual temperatures than the phyllostomids require. The northern species of vespertilionids also hibernate during winter. The lower metabolic rate of the small vespertilionids may be viewed as an energy-conserving mechanism and an overall adjustment on the part of most to lower their body temperatures when roosting during the day. The larger phyllostomid bats tend to maintain a more constant body temperature over a range of environmental temperatures (McNab 1969). Small phyllostomids are unable to maintain a constant body temperature in extreme cold (McNab 1982).

Tropical environments do exhibit seasonality in abundance of both fruit and insects. Thus frugivorous and insectivorous bats will tend to adjust their reproductive patterns to times of the year when their food supply is abundant (Fleming, Hooper, and Wilson 1972; Wilson 1979; Bradbury and Vehrencamp 1977b). Bats adapted to the Temperate Zone hibernate during months of low food supply, an option not taken by the tropical bats.

Activity Patterns and Nocturnal Movements

Although bats are commonly known to be active at night, nonchiropterologists seldom realize that the time of active flight may vary significantly between species (Brown 1968). *Carollia perspicillata* shows an abrupt onset of activity in the early evening, which declines to a low level about midnight, and it has a second burst of activity just before dawn. *Sturnira lilium* has a low activity level early in the evening and a higher level later on, before dawn. A similar pattern is shown by *Artibeus lituratus* (Erkert 1982). Different activity rhythms among bats in a community may in fact have resulted from an effort to reduce interspecific scramble competition for food (e.g., *Arbiteus jamaicensis* and *A. lituratus*; Bonaccorso 1979). Or differences in activity rhythms might relate to predator avoidance. In small insectivorous species, however, the bimodal activity rhythm suggests two bouts of feeding, with the first meal digested during a resting interval and the second meal digested during the day. This implies an attempt to maintain an energy balance throughout a twenty-four-hour cycle.

During their nocturnal forays microchiropterans employ echolocation. The insectivorous and carnivorous bats have the most highly refined echolocation system both in call form and in the neurological adaptations needed to perceive the echoes from their intended prey. The literature on echolocation in bats is vast (see Novick and Dale 1971; Gould 1977; Griffin 1958; Griffin, Webster, and Michael 1960; Fenton 1982). Foraging in frugivorous species seems to place less stringent requirements on the precision of the echolocation system.

Frugivorous bats typically expend energy in searching for appropriate fruit trees, but once they find them they establish a night roost some distance from the feeding tree, which they exploit intermittently throughout the night. The night roost is at a distance from a fruiting tree to reduce competitive interactions with other bats and probably also to lessen exposure to predators that may otherwise be attracted to the tree. The demonstration by Morrison (1978) that *Artibeus jamaicensis* is less active under a full moon strongly suggests that predation is an ever present risk during bats' foraging journeys. Fleming (1982) reviews the literature on foraging strategies of frugivorous bats.

Roosting Sites

Bats rest in both diurnal and nocturnal roosts, opportunistically using caves, cracks, and tree cavities. More than one species may use the same cavity. Bats that roost in foliage or on branches or tree trunks frequently are solitary, but those that roost in dense foliage sometimes are found in larger groups, especially if they roost high in the canopy. The habit of roosting on smooth surfaces like the insides of large

leaves has often enhanced selection for foot disks such as those of *Thyroptera discifera*.

Many phyllostomids are "tent-making" bats. By biting leaves in particular patterns, they cause the leaf to fold over and thereby provide a shelter. Tent-making bats include *Ectophylla alba*, *Artibeus cinereus*, and *Uroderma bilobatum*.

Night roosts are used between foraging bouts and are usually some distance from the feeding site. Such night roosts are carefully chosen to minimize predation risk, so, thick-crowned trees are often preferred.

Bat Diets

Bats have specialized for different foraging strategies and different prey. Broadly speaking, they can be classified according to the scheme developed by Wilson (1973b). Foliage gleaners prey on insects that are feeding, resting, or moving on vegetation. Aerial insectivores catch flying insects. Frugivores feed extensively on fruit but also take insects at certain seasons (Ayala and D'Alessandro 1973). Nectarivorous bats have long tongues and usually long rostra. Although they feed on pollen and nectar, some of these species are important plant pollinators, and a co-evolved system of bats and plants has been discerned (Howell 1974; Sazima 1976). Pollinating bats also feed on fruits and insects. Carnivorous bats eat vertebrates such as frogs, lizards, rodents, birds, and other bats. Sanguivorous bats feed on blood and are found only in the Neotropics (see Desmodontinae below). Data concerning the feeding habits of Phyllostomidae are reviewed by Gardner (1977a).

Reproduction and Life Span

Mating systems are highly variable within the Chiroptera. Species such as *Saccopteryx bilineata*, *Artibeus jamaicensis*, and *Phyllostomus hastatus* tend to form harems, with males actively defending temporary groups of females against incursions by other males. Although vespertilionid bats living in temperate and subtropical latitudes mate in autumn and delay fertilization, this is not true of vespertilionids in the tropics. On the other hand, delayed embryonic development does occur in some species, and the delay length is under the control of the female. Fleming (1971) showed that in *Artibeus jamaicensis* mating occurs at the end of the dry season in Panama and births occur four months later, coinciding with a peak in the abundance of fruits. The female has a postpartum estrus, but the blastocysts implant and then enter a diapause for three months before developing normally for four months, so that the young are born at the end of the dry season just before a peak in the availability of large fruits. The delay in development allows the birth to occur when mother and young have optimum conditions for foraging.

In seasonally arid habitats the timing of reproduction is influenced by rainfall. With the onset of rains both insects and fruits become abundant. In less seasonal habitats bats may reproduce in any month, but subtle environmental factors may favor reproduction during a particular time, thus causing a marked peak. Other effects can also influence the timing of reproduction (Arata and Vaughan 1970; Wilson 1973b, 1979).

Insectivorous bats such as vespertilionids and molossids usually have a briefer gestation and a shorter lactation period than the phyllostomid frugivores (Kleiman and Davis 1979). Most species of bats produce a single young, but there are exceptions: *Eptesicus fuscus* and *Lasiurus cinereus* commonly produce twins, and *Lasiurus borealis* may have triplets. Most vespertilionids attain sexual maturity in less than a year; some, such as *Eptesicus fuscus*, mature as early as four months. The larger phyllostomids probably do not reach sexual maturity until they are nearly a year old. Vespertilionid bats, in spite of their small size, can attain a considerable age. *Eptesicus fuscus* has lived at least nineteen years, the vampire bat eighteen years, and *Arbiteus jamaicensis* seven years (Tuttle and Stevenson 1982).

Relative Brain Size

Bat brains vary considerably in comparative size. The relatively large brains of the New World phyllostomid bats are convergent with the patterns shown by the Pteropodidae. Foraging strategies that involve finding rich food resources isolated in small pockets are correlated with a large brain weight relative to body mass. The highly specialized aerial insectivores have the lowest brain/body weight ratios (Eisenberg and Wilson 1978).

Concluding Remarks

The natural history of New World bats has been described in numerous publications. Useful summaries are contained in Goodwin and Greenhall (1961), Husson (1962, 1978), and Baker, Knox Jones, and Carter (1976, 1977, 1979). Kunz (1982) contains much useful information on the ecology of the Chiroptera. Koopman (1982) has summarized distribution patterns for the South American species, and Baker et al. (1982) have summarized the data on karyotypes. A key to the bats of Argentina is offered in the publication by Greenhall et al. (1983).

Comment

Argentine bat distributions were modified in partial accordance with Bárquez (1987) and Bárquez and Lougheed (1990).

FAMILY EMBALLONURIDAE
Sac-winged Bats, Sheath-tailed Bats

Diagnosis

A distinguishing but not unique feature of this family of bats is that the tail is enclosed in a sheath of the interfemoral membrane, and the tip perforates the upper surface of the interfemoral membrane to lie free upon; hence the name sheath-tailed bats. The Noctilionidae also exhibit this feature. The dental formula is highly variable in the family, from I 2/3, C 1/1, P 2/2, M 3/3 in *Emballonura* to I 1/2, C 1/1, P 2/2, M 3/3 in *Taphozous*. The premaxillary bones do not meet anteriorly. Many species have an opening on the antebrachial membrane over a gland field, and scent is produced during some of their wing-waving displays. Other species (the diclidurines) have a gland field surrounding the tail in the uropatagium. During resting the third finger exhibits a reflexed proximal phalanx (see fig. 4.3). Although the trochiter is well developed, it is smaller than the trocin and does not articulate with the scapula (Sanborn 1937).

Distribution

This family is widely distributed in both the Old World and New World tropics and includes some twelve genera and fifty species. In the Neotropics it extends from Sonora, Mexico, south through the Isthmus to southern Brazil and Paraguay.

Natural History

These bats are entirely nocturnal, and most are specialized for feeding on insects. Strategies of insect capture are variable, but many species seem to be adapted for aerial pursuit (see Bradbury and Emmons 1974; Bradbury and Vehrencamp 1976a,b, 1977a,b). The usual number of young is one.

Genus *Peropteryx* Peters, 1867
• *Peropteryx macrotis* (Wagner, 1843)
Lesser Sac-winged Bat

Measurements

	Mean	S.D.	N	Loc.	Source[a]
TL	61.3	2.0	15 m	Br	1
	64.1	3.8	7 f		
T	14.2	1.6	15 m		
	14.1	2.1	7 f		
HF	6.6	0.5	15 m		
	6.6	0.8	7 f		
E	14.2	0.6	5 m		
	14.4	1.0	7 f		
FA	42.0	0.9	15 m		
	43.6	1.1	7 f		
Wta	4.2	0.6	15 m		
	4.6	0.9	7 f		

[a](1) Willig 1983.

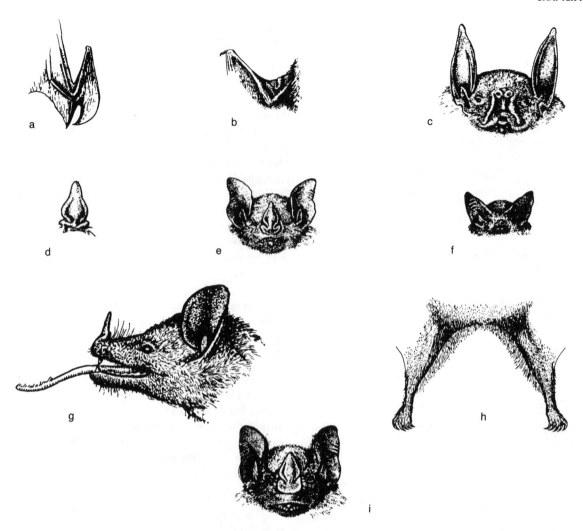

Figure 4.3. Some external features of the Chiroptera (part 1): (*a*) proximal flexure of emballonurid phalanges; (*b*) scent glands of *Peropteryx;* (*c*) face of *Noctilio;* (*d*) nose leaf; (*e*) face of *Carollia;* (*f*) face of *Peropteryx macrotis;* (*g*) head of *Glossophaga soricina;* (*h*) tail membrane of *Sturnira;* (*i*) face of *Artibeus jamaicensis.*

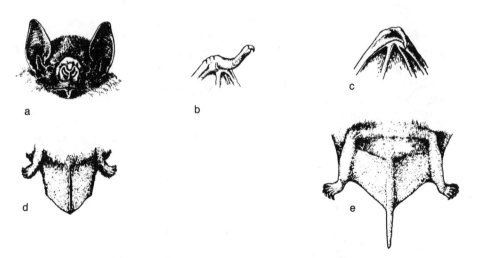

Figure 4.4. Some external features of the Chiroptera (part 2): (*a*) *Desmodus rotundus* face; (*b*) thumb of *Desmodus;* (*c*) absence of thumb in Furipteridae; (*d*) furred interfemoral membrane of *Lasiurus;* (*e*) free-tailed bat, *Molossus*—relation of tail to uropatagium.

Description

The dental formula is I 1/3, C 1/1, P 2/2, M 3/3. Females are larger than males. This small, dark bat has long, thick dark brown dorsal fur, with the venter the same color. Some individuals are washed with ash gray. The wings and ears are dark brown. The underside of tail is lightly furred in parallel lines (Willig 1983) (see fig. 4.3).

Chromosome number: $2n = 26$; FN $= 48$ (Baker et al. 1982).

Distribution

Peropteryx macrotis is widely distributed from southern Mexico to Peru, Paraguay, and southern Brazil. In southern South America it has been collected only from Paraguay (Bárquez 1983a; Honacki, Kinman, and Koeppl 1982; Myers and Wetzel 1983) (map 4.1).

Life History

In the caatinga of Brazil, this bat has been found roosting in colonies of up to ten individuals; there is always only one male, probably reflecting the existence of small harems (Willig 1983).

Ecology

In the Brazilian caatinga this species roosts in rock piles and culverts, and in Paraguay it uses limestone caves and crevices (Myers and Wetzel 1983; Willig 1983).

FAMILY NOCTILIONIDAE

Bulldog Bats, Murciélagos Pescadores

Diagnosis

This family includes only one genus and two species. The dental formula is I 2/1, C 1/1, P 1/2, M 3/3 (see fig. 4.5). There is no prominent nose leaf, and the upper lip is medially divided (see fig. 4.3). The tail is considerably shorter than the interfemoral membrane, and its tip protrudes through the membrane on the dorsal side. There are three phalanges on the third finger, and the claws of the feet are greatly enlarged. This suite of characters serves to distinguish the family from other Neotropical bats.

Distribution

This family is distributed from Sinaloa, Mexico, to northern Argentina.

Natural History

The species *Noctilio leporinus* is known as the fishing bat because it flies near the surface of water and seizes minnows in its claws. The other species, *N. albiventris*, is not as specialized for fishing but feeds primarily on aquatic insects.

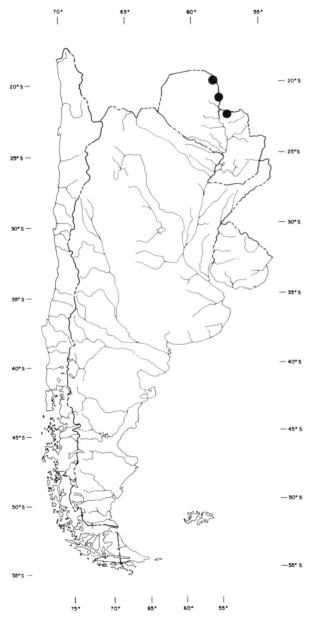

Map 4.1. Distribution of *Peropteryx macrotis*.

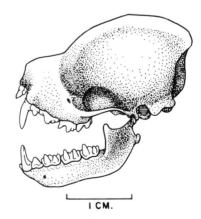

Figure 4.5. Skull of *Noctilio leporinus*.

Genus *Noctilio* Linnaeus, 1766
• *Noctilio albiventris* Desmarest, 1818
Lesser Bulldog Bat, Southern Bulldog Bat,
Pescador Menor

Measurements

	Mean	Min.	Max.	N	Loc.	Source[a]
TL	90.1	80.0	100.0	22	P	1
	98.6	93.0	104.0	10 m		2
	93.1	84.0	97.0	17 f		2
HB	70.7	64.0	77.0	22		1
T	19.4	16.0	29.0	22		1
	19.4	17.0	23.0	10 m		2
	18.6	16.0	20.0	17 f		2
HF	16.5	15.0	18.0	22		1
	19.2	18.0	20.0	10 m		2
	17.6	16.0	19.0	17 f		2
E	21.9	12.0	24.0	22		1
	26.6	25.0	28.0	10 m		2
	24.6	22.0	27.0	17 f		2
FA	60.7	57.0	64.0	22		1
	62.8	59.6	65.5	10 m		2
	61.2	57.7	63.4	17 f		2
Wta	27.7	21.0	36.0	22		1
	49.7	45.0	55.0	9	A	3

[a](1) PCorps; (2) Myers and Wetzel 1983; (3) Crespo 1974.

Description

The feet are large and robust, but not as much
as in *N. leporinus*. The large ears are narrow and
pointed, and the chin has raised cutaneous ridges.
Both dorsal color and ventral color in the lesser bulldog bat vary greatly, both geographically and individually. The dorsum ranges from grayish brown to
reddish brown, and the venter varies from whitish
to dark orange. Often, but not always, there is a
whitish median dorsal streak. The membranes are
dark gray to black (Hood and Pitocchelli 1983; Husson 1978).
Chromosome number: $2n = 34$; $FN = 58–62$.

Distribution

Noctilio albiventris ranges from Honduras to the
Guianas, eastern Brazil, northern Argentina, and
Peru. In southern South America it has been col-
lected in Paraguay and northern Argentina (Honacki,
Kinman, and Koeppl 1982; Myers and Wetzel 1983;
Olrog and Lucero 1981) (map 4.2).

Life History

The litter size is one, and there is probably only
one litter a year. Young do not fly until thirty-five to
forty-four days of age, and they have a long period of
maternal attachment before weaning, which occurs
at seventy-five to ninety days (mean = 80.5). The
young eats solid food for the first time at about forty-
five days and is fed masticated food from the mother's cheek pouches. When juveniles first leave the
roost, they fly with their mothers and continue to
nurse. The females nurse only their own young,

which they recognize by vocal signals (Brown,
Brown, and Grinnell 1983; Hood and Pitocchelli
1983).

Ecology

Noctilio albiventris roosts in hollow trunks or
branches and in buildings, often near water. The
bats are captured over lakes or large, slow streams,
where they hunt insects both aerially and on the surface of the water. They capture prey with the feet as
well as in the interfemoral membrane. An early peak
in activity about dusk is followed by a second peak
after midnight. This species primarily inhabits mesic
forests with surface water, and its distribution can be

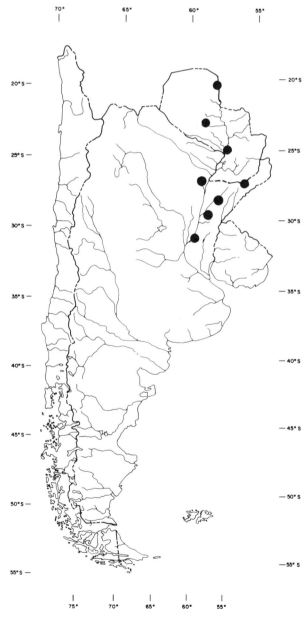

Map 4.2. Distribution of *Noctilio albiventris*.

highly localized (Brown, Brown, and Grinnell, 1983; Crespo 1974; Hood and Pitocchelli 1983; Hooper and Brown 1968; Myers and Wetzel 1983).

• *Noctilio leporinus* (Linnaeus, 1758)
Mexican Bulldog Bat, Pescador Grande

Measurements

	Mean	S.D.	Min.	Max.	N	Loc.	Source[a]
TL	119.8		114.0	129.0	4 m	P	1
	116.5		111.0	120.0	11 f		
T	28.0		25.0	30.0	4 m		
	26.0		21.0	29.0	11 f		
HF	32.3		32.0	33.0	4 m		
	29.3		27.0	31.0	11 f		
E	33.5		31.0	36.0	4 m		
	31.1		29.0	33.0	11 f		
FA	88.8		87.2	90.3	4 m		
	85.9		82.4	88.5	11 f		
Wta	69.2	11.5			20 m	Br	2
	61.3	4.1			20 f		

[a](1) Myers and Wetzel 1983; (2) Willig 1983.

Description

The pelage of *N. leporinus* is extremely short, and the chin has well-developed cross ridges, imparting the "bulldog" appearance. The muzzle and nose have excrescences. There is considerable geographic variation in size, with the largest individuals found in the north and south. The large ears are narrow and pointed. The dorsum varies from pale orange to dark orange and even grayish brown, and a distinct dorsal stripe, whitish to orangish depending on dorsal color, extends from between the ears to the rump. The venter ranges from whitish to bright orange (see plate 4). The wing and tail membranes are brownish and nearly naked (Hood and Knox Jones 1984; Husson 1978).
Chromosome number: $2n = 34$; FN $= 58–62$.

Distribution

The Mexican bulldog bat, or greater bulldog bat, occurs from western and eastern Mexico to northern Argentina. Discontinuously distributed within this large area, it is restricted mostly to nonarid lowland and coastal areas and to major river basins. In southern South America *N. leporinus* is found in Paraguay and across the northern and north-central portions of Argentina (Bárquez 1987; Hood and Knox Jones 1984; Myers and Wetzel 1983; Olrog and Lucero 1981) (map 4.3).

Life History

Females are monovular, and the litter size is one. There is one reproductive peak, with breeding in autumn and winter. Females apparently form nursery colonies, and breeding may be highly synchronous in strongly seasonal habitats (Goodwin and Greenhall 1961; Hood and Knox Jones 1984).

Ecology

Bats of this species are most commonly found in tropical lowland habitats; they are observed and collected most frequently over ponds and quiet streams, but they also frequent bays, lagoons, and estuaries of major rivers. In Paraguay they are common over slow streams and scattered temporary ponds of the lower Chaco.

This is one of the few bats to evolve fish-eating habits, but fish are not its only prey; it takes insects as well, and there is one record of its eating a frog. In the Brazilian caatinga groups of up to thirty were found roosting in large hollow trees (Hood and Knox Jones 1984; Hooper and Brown 1968; Myers and Wetzel 1983; Willig 1983).

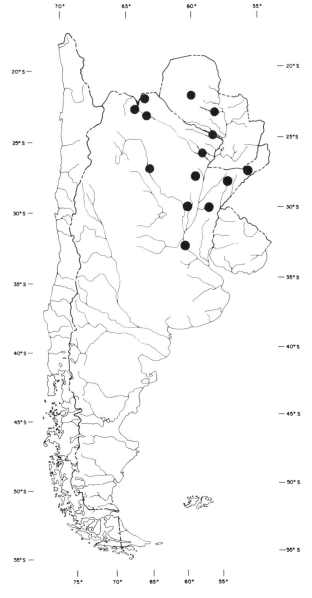

Map 4.3. Distribution of *Noctilio leporinus*.

FAMILY PHYLLOSTOMIDAE
(= PHYLLOSTOMATIDAE)
American Leaf-nosed Bats

Diagnosis

For this description the subfamily Desmodontinae (the true vampire bats) and some stenodermines are aberrant, since the nose leaf is very reduced, but their morphology is so distinctive that they cannot be mistaken for other species of bats (see Desmodontinae below). The New World leaf-nosed bats are characterized by a skull without a postorbital process. The third finger has three complete bony phalanges. The most distinctive feature is the nose leaf that uniquely defines almost all members of this family (fig. 4.3). The tooth number is highly variable from twenty-six to thirty-four, reflecting in part the diverse feeding specializations exhibited by the family as a whole.

Distribution

This family is confined to the Western Hemisphere and is distributed from southern California and Arizona, as well as the Gulf coastal plain of Texas, south through the Isthmus to northern Argentina. It includes some 51 genera and over 140 species that are for the most part adapted to lower elevations and the tropical and subtropical areas of the New World.

Natural History

These bats show a variety of feeding specializations, though many feed on fruit (Carvalho 1961). According to Wilson (1973b) most genera, because they include a great deal of fruit in their diet, may be considered frugivores. Some are mixed feeders, however, eating fruit as well as insects, and their search for insects seems to involve close inspection of leaves; thus they are termed foliage gleaners. Genera with this feeding habit include *Micronycteris*, *Tonatia*, *Mimon*, and *Phylloderma*. Some species are specialized for feeding on nectar and pollen: most notable are *Glossophaga*, *Anoura*, and *Lionycteris*. Two genera belonging to the subfamily Desmodontinae are specialized for feeding on blood, including *Diphylla* and *Desmodus*. Some bats are active carnivores, preying on lizards, birds, and small mammals; these include *Phyllostomus*, *Trachops*, *Chrotopterus* and *Vampyrum*. *Artibeus*, *Carollia*, and *Sturnira* are specialized for frugivory, as are many stenodermine species (Gardner 1977a).

Because many phyllostomid bats are foliage gleaners, nectarivores, or frugivores, their food does not require precision localization by means of echolocation, in contrast to that of the aerial insectivores. The echolocating pulse of phyllostomids generally is of low amplitude, rather brief, and highly frequency modulated. Because of these acoustical characters, these bats are sometimes referred to as "whispering bats."

Most species of phyllostomids produce a single young. Some species are highly seasonal in the timing of births, especially in areas subject to seasonal aridity. Birth often occurs at a time of maximum fruit or insect abundance (Wilson 1979).

An analysis of bat communities in Panama by Bonaccorso (1979) and Humphrey and Bonaccorso (1979) indicates that abundance varies considerably depending on the habitat. Some species are specialists in second-growth habitats, others are specialized for feeding along streams or creeks, and still others are specialists in mature forests. Regardless of the habitat preference, only three or four species make up the most abundant forms. Over 70% of all bats netted in a mature forest in Panama may belong to just four species: *Artibeus jamaicensis*, *A. lituratus*, *Glossophaga soricina*, and *Carollia perspicillata*.

Within any habitat type, the species of bats belonging to the same trophic category seem to avoid direct competition by specializing for different vertical strata or different foods. It appears from the work of Gardner (1977a) and Bonaccorso (1979) that feeding categories are not as easily demarcated as previously thought. Wilson's trophic categorization into frugivores, foliage gleaners, and so on, has been refined in Bonaccorso's publication and further refined in the article by Humphrey, Bonaccorso, and Zinn (1983). Many of the so-called frugivores are highly opportunistic feeders, eating insects at one time of year and fruit at another. Opportunism seems to be more characteristic of most medium-sized phyllostomids (Humphrey, Bonaccorso, and Zinn 1983).

The biology of phyllostomid bats has been elegantly summarized in three volumes edited by Baker, Knox Jones, and Carter (1976, 1977, 1979). The monograph on *Carollia perspicillata* by Fleming (1988) is considered a classic. A key to the subfamilies of the Phyllostomidae is included in table 4.2. Chromosomal data included in the species accounts are from Baker et al. (1982).

SUBFAMILY PHYLLOSTOMINAE

Diagnosis

Bats of the subfamily Phyllostominae have a long rostrum and long ears. Most species retain the tail and a well-developed uropatagium, but others show tail reduction and varying degrees of reduction of the uropatagium.

Table 4.2   Field Key to the Subfamilies of the Phyllostomidae

1   Nose leaf rudimentary, face not wrinkled; calcar absent; wing attached near knee joint; upper incisors broader
    than canines . . . . Desmodontinae
1' Nose leaf prominent or face exhibiting prominent wrinkles; upper incisors smaller than canines 2
2   Muzzle long and narrow 3
2' Muzzle short and broad 5
3   Ear when pressed forward not reaching nose tip . . . . Glossophaginae
3' Ear when pressed forward reaching tip of nose or beyond 4
4   Single papilla on chin or large central papilla on chin surrounded by smaller papillae . . . . Carolliinae
4' Y- or V-shaped shield on chin . . . . Phyllostominae
5   White facial stripes and often median dorsal stripe present . . . . Stenoderminae
5' No white facial stripes 6
6   Face naked, wrinkled, or naked with hornlike structure on nose 8
6' Face not naked, no wrinkles 7
7   Interfemoral membrane absent . . . . Sturnirinae
7' Interfemoral membrane present . . . . Ametrida
8   Face with extreme wrinkling; fold of skin on throat . . . . Centurio
8' Face with folds; hornlike structure on nose; fold of skin on throat . . . . Sphaeronycteris

Natural History

Bats of this subfamily exhibit a wide range of dietary specializations. Smaller forms take a great deal of insect prey in addition to fruit (e.g., *Tonatia*, *Micronycteris*). Some of the larger species are excellent predators feeding on amphibians, reptiles, birds, and small mammals.

Genus *Chrotopterus* Peters, 1865
• *Chrotopterus auritus* (Peters, 1856)
Peters's Woolly False Vampire Bat, Falso Vampiro Orejón

Measurements

	Mean	Min.	Max.	N	Loc.	Source[a]
TL	118.8	110.0	140.0	6	P, A	1
HB	106.4	101.0	109.5	6	Br	2
T	11.7	5.0	18.0	3	P	1
HF	36.8	23.0	52.1	6	P, A	1
E	38.7	23.0	57.0			
	44.7	42.0	48.5	6	Br	2
FA	82.5	78.9	86.0	4	P, A	1
	83.2	81.0	86.5	6	Br	2
Wta	96.3	79.0	120.0	3	A	1

[a](1) Crespo 1982; PCorps, UM; (2) Taddei 1975.

Description

The dental formula is I 2/1, C 1/1, P 2/3, M 3/3. *C. auritus* is known for its robust body and large round eyes. It has a well-developed tail membrane and large oval ears. Almost three-quarters of its forearm is covered with hair, dorsally and ventrally. The bat is chestnut gray to gray dorsally and lighter gray ventrally, with dark wing membranes (Olrog and Lucero 1981; Villa-R. and Villa Cornejo 1969; pers. obs.).

Distribution

Chrotopterus auritus is found from southern Mexico through northern South America and Bolivia

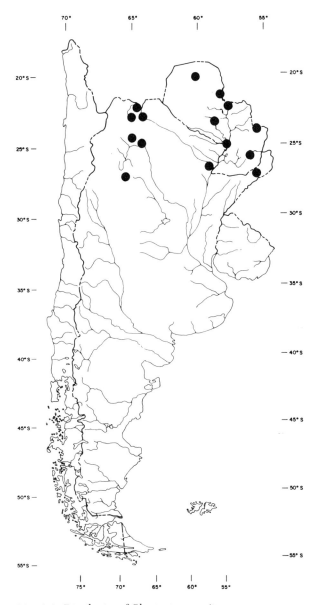

Map 4.4. Distribution of *Chrotopterus auritus*.

to Paraguay, northern Argentina, and southern Brazil. In southern South America it has been collected in Jujuy, Tucumán, Salta, and Misiones provinces, Argentina, and in several localities in Paraguay (Crespo 1982; Honacki, Kinman, and Koeppl 1982; Myers and Wetzel 1983; Ojeda and Mares 1989; Olrog 1973) (map 4.4).

Life History
A single young is born after a gestation period exceeding one hundred days (Taddei 1976).

Ecology
These large predatory bats are found in the warm subtropical forests of Argentina and Paraguay. They are difficult to catch, apparently because their slow flight allows them to avoid mist nets, but they have been caught in dense thickets at between 1 and 2 m, where they apparently search for small mammals and roosting passerines. Stomach contents have been found to contain feathers, the remains of a small passerine, coleopteran and lepidopteran remains, *Marmosa*, and what were tentatively identified as remains of *Ctenomys*. They roost in caves and hollow trees (Crespo 1982; Myers and Wetzel 1983; Olrog 1973, 1979; Taddei 1976; Villa-R. and Villa Cornejo 1969).

Genus *Macrophyllum* Gray, 1838
• *Macrophyllum macrophyllum* (Schinz, 1821)
Long-legged Bat, Falso Vampiro Patilargo

Measurements

	Mean	Min.	Max.	N	Loc.	Source[a]
TL	85.0	80.0	92.0	5	A, Br, P	1
HB	53.4	44.0	85.0			
	46.2	44.0	48.0	8 m	Br	2
T	38.8	35.0	45.0	5	A, Br, P	1
HF	13.0	10.0	15.0	5		
E	18.4	17.5	19.0	8 m	Br	2
FA	36.8	35.5	37.5			
	38.0	36.0	42.0	4	A, Br, P	1
Wta	8.0	7.5	9.0	6		

[a](1) Fornes, Delpietro, and Massoia 1969; Taddei 1975; PCorps; (2) Taddei 1975.

Description
The dental formula is I 2/2, C 1/1, P 2/3, M 3/3. This small, slender phyllostomatid bat is readily distinguishable from others by the peculiar development of the feet. The legs are long, and the feet are strikingly enlarged and equipped with powerful claws. The long tail continues to the distal border of the broad interfemoral membrane, which is a uniform dark brown dorsally, with the venter only slightly lighter and is studded distally with longitudinal rows of projecting dermal denticles (Fornes,

Delpietro, and Massoia 1969; Harrison 1975; pers. obs.).
Chromosome number: $2n = 32$; FN = 56.

Distribution
Macrophyllum macrophyllum is distributed from Yucatán, Mexico, south in scattered localities through South America to northern Argentina and Paraguay (Harrison 1975; Massoia 1980; PCorps) (map 4.5).

Ecology
This bat roosts in caves, houses, and culverts and is insectivorous (Fornes, Delpietro, and Massoia 1969; Harrison 1975).

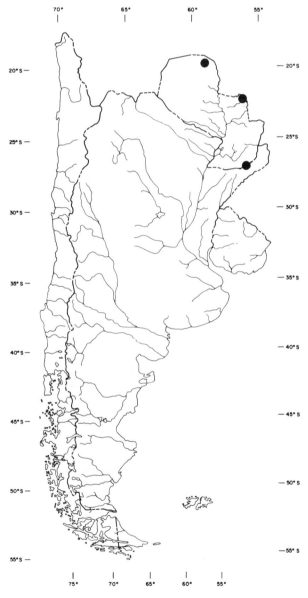

Map 4.5. Distribution of *Macrophyllum macrophyllum*.

Genus *Phyllostomus* Lacépède, 1799
• *Phyllostomus discolor* Wagner, 1843
Pale Spear-nosed Bat, Falso Vampiro Gris

Measurements

	Mean	S.D.	Min.	Max.	N	Loc.	Source[a]
TL	96.3	4.1			10 m	Br	1
	98.3	3.1			16 f		
T	14.9	1.7			10 m		
	15.1	1.3			16 f		
HF	13.2	0.8			10 m		
	13.2	0.7			16 f		
E	21.6	1.8			10 m		
	21.1	1.2			16 f		
FA	61.6	1.8			10 m		
	59.8	1.8			16 f		
Wta	37.5	3.3			10 m		
	40.1	3.3			16 f		
	37.2		34.7	42.7	16 m	Br	2
	34.7		29.5	41.6	13 f		

[a](1) Willig 1983; (2) Taddei 1975.

Description
The dental formula is I 2/2, C 1/1, P 2/2, M 3/3. This stout-bodied bat with short, soft, dark brown to tan fur has a short tail that pierces the tail membrane. The venter is grayish white to yellowish white.

Chromosome number: $2n = 32$; FN = 60.

Distribution
Phyllostomus discolor ranges from Mexico south to southeastern Brazil, Peru, and northern Argentina. In southern South America it has been recorded in eastern Paraguay and in Salta province, Argentina (Honacki, Kinman, and Koeppl 1982; Myers and Wetzel 1983; Olrog 1958) (map 4.6).

Ecology
This species roosts in hollow trees in groups of about twenty-five. In breeding roosts the sex ratio may be one male to twelve females. In Brazil this bat was found to be primarily frugivorous, though it also ate nectar, pollen, flower parts, and insects (Gardner 1977a; Willig 1983).

• *Phyllostomus hastatus* (Pallas, 1767)
Greater Spear-nosed Bat

Measurements

	Mean	S.D.	Min.	Max.	N	Loc.	Source[a]
TL	127.3	4.2			20 m	Br	1
	119.8	5.1			20 f		
T	19.2	1.9			20 m		
	18.5	2.6			20 f		
HF	17.8	1.1			20 m		
	17.1	1.2			20 f		
E	28.9	9.7			20 m		
	27.8	1.5			20 f		
FA	84.3	1.6			20 m		
	81.6	2.2			20 f		
Wta	93.4	8.6			20 m		
	81.7	7.8			20 f		
	93.3		91.1	96.5	4 m	Br	2
	79.5		78.7	80.2	4 f		

[a](1) Willig 1983; (2) Taddei 1975.

Description
The dental formula is I 2/2, C 1/1, P 2/2, M 3/3 (fig. 4.6). The claws are long and strong. Bats of this species from Bolivia are large and robust, with rich, thick chocolate brown fur. The venter is lighter, with an orange tinge. The ears, nose leaf, and wings are all dark.

Chromosome number: $2n = 32$; FN = 58.

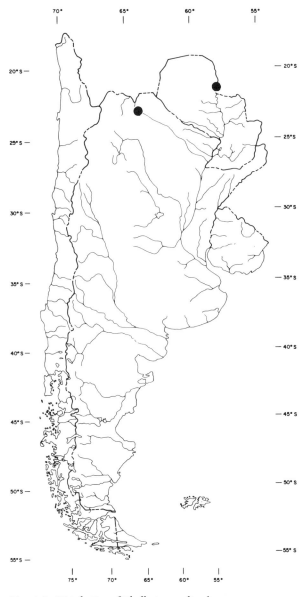

Map 4.6. Distribution of *Phyllostomus discolor*.

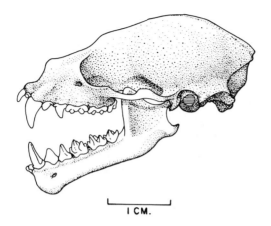

Figure 4.6. Skull of *Phyllostomus hastatus*.

Distribution

Phyllostomus hastatus is distributed from Honduras south to Peru, Bolivia, and Paraguay (Baud 1981; Honacki, Kinman, and Koeppl 1982) (map 4.7).

Ecology

Colonies of ten to one hundred, depending on the size of the roosting location, have been found in hollow trees, termite nests, caves, and palm-thatch roofs. Within a colony males will actively defend groups of females, and temporary harems of thirty females per male may be formed. This bat is an omnivore, eating vertebrates, flowers, and pollen (Carvalho 1961; McCracken and Bradbury 1977, 1981; Tuttle 1970; Willig 1983).

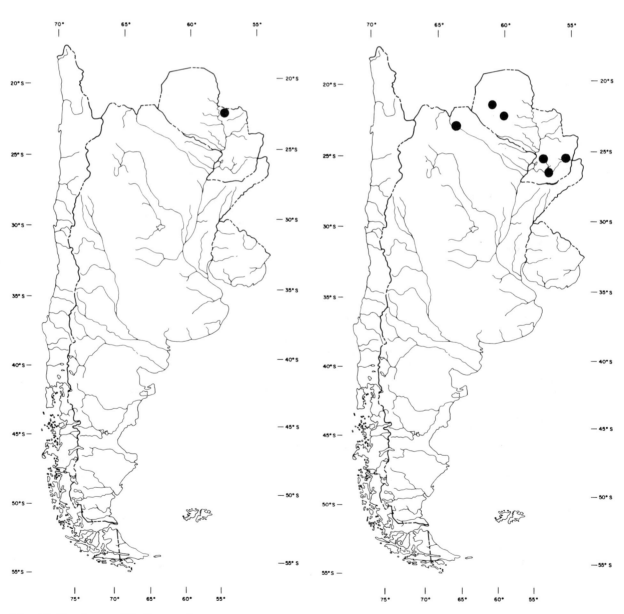

Map 4.7. Distribution of *Phyllostomus hastatus*.

Map 4.8. Distribution of *Tonatia bidens*.

Genus *Tonatia* Gray, 1827
• *Tonatia bidens* (Spix, 1823)
Spix's Round-eared Bat

Measurements

	Mean	Min.	Max.	N	Loc.	Source[a]
TL	61.8	57.0	71.0	5	P	1
	91.7	89.0	95.0	7	Peru	1
T	8.0	0.0	11.0			
	19.0	17.0	21.0	7	Peru	
HF	14.6	14.0	16.0			
E	21.0	18.0	25.0			
	30.6	29.0	32.0	7	Peru	
FA	55.6	54.0	56.8			
	56.0	57.1	54.0	2	P	2
Wta	24.7	23.0	27.0	6	Peru	1

[a](1) Gardner 1976; (2) Myers and Wetzel 1983.

Description
The dental formula is I 2/1, C 1/1, P 2/3, M 3/3. *T. bidens* is a large phyllostomatid with long, broad ears. The proximal half of the forearms is well haired, and there is a well-developed tail membrane. The dorsum ranges from ochraceous tawny to blackish brown, with the venter paler and grayer and washed with pale buff (Goodwin 1942; pers. obs.).
Chromosome number: $2n = 16$; FN = 20 (Genoways and Williams 1984).

Distribution
Tonatia bidens is distributed from Guatemala and Honduras south to Peru and Paraguay. In southern South America it has been collected in Paraguay and northern Argentina (Bárquez 1987; Honacki, Kinman, and Koeppl 1982; Myers and Wetzel 1983) (map 4.8).

Life History
Two females from Peru were found to contain two embryos each (Gardner 1976).

Ecology
In Paraguay *T. bidens* has been collected over isolated ponds in thorn scrub and over a stream in high tropical forest. Specimens have been found roosting with other species in hollow trees. Insect remains and fruit pulp were found in their stomachs (Eisenberg 1989; Myers and Wetzel 1983; Myers, White, and Stallings 1983).

• *Tonatia silvicola* (d'Orbigny, 1836)
d'Orbigny's Round-eared Bat, Falso Vampiro Oreja Redonda

Measurements

	Mean	S.D.	N	Loc.	Source[a]
TL	97.9	3.3	13 m	Br	1
	97.6	3.1	36 f		
T	17.7	2.5	13 m		
	17.7	2.0	36 f		
HF	14.6	0.5	13 m		
	14.7	0.8	36 f		
E	29.6	1.1	13 m		
	28.8	0.8	36 f		
FA	58.2	1.4	13 m		
	57.6	1.0	36 f		
Wta	33.9	3.1	13 m		
	31.8	2.7	36 f		

[a](1) Willig 1983.

Description
The dental formula is I 2/1, C 1/1, P 2/3, M 3/3. The ears are very large. This large bat has ash brown to light brown fur with a V of paler fur across the shoulders. The venter is paler and grayer, with a band or splotch of sparse cream-colored fur across

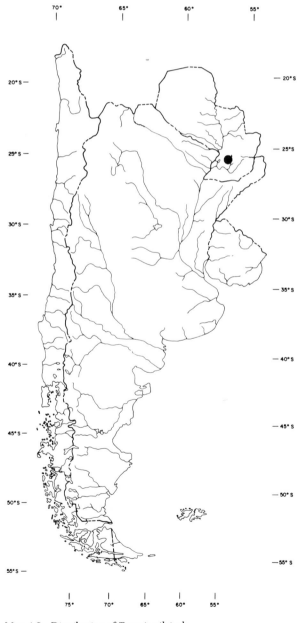

Map 4.9. Distribution of *Tonatia silvicola*.

the chest, and the wing membranes are tan to gray. Chromosome number: $2n = 34$; FN $= 60$ (Genoways and Williams 1984).

Distribution

Tonatia silvicola ranges from Mexico to Bolivia, northern Argentina, Paraguay, and eastern Brazil. In southern South America it is found in eastern Paraguay (see Bárquez 1987; Fornes, Massoia, and Forrest 1967) (map 4.9).

Ecology

In the Brazilian caatinga this species is a foliage-gleaning insectivore. In northern South America it occasionally roosts in termite nests (Eisenberg 1989; Willig 1983).

SUBFAMILY GLOSSOPHAGINAE

Description

These bats are specialized for feeding on pollen and nectar. The tongue is long, and its tip is covered with papillae (fig. 4.3). The loss of the lower incisors is not uncommon (e.g., *Anoura;* fig. 4.7). The rostrum is long, and the ears are short. Some species show great reduction in tail length.

Natural History

These bats feed on pollen and nectar but also supplement their diet with insects. A coevolutionary relationship with certain flowering plants has been proposed for some species.

Genus *Anoura* Gray, 1838
• *Anoura caudifer* (E. Geoffroy, 1818)

Measurements

	Mean	S.D.	N	Loc.	Source[a]
TL	65.8	2.7	12 m	V	1
	65.4	3.5	16 f		
HF	12.1	0.3	12 m		
	12.1	0.4	16 f		

E	14.7	0.5	12 m
	14.4	0.7	16 f
FA	36.2	0.8	19 m
	36.3	0.9	24 f
Wta	10.3	0.9	12 m
	10.0	0.8	15 f

[a](1) Eisenberg 1989.

Description

This is one of the smaller species of *Anoura*. The upperparts are dark brown, and the venter is paler brown.

Chromosome number: $2n = 30$; FN $= 45$ (Eisenberg 1989).

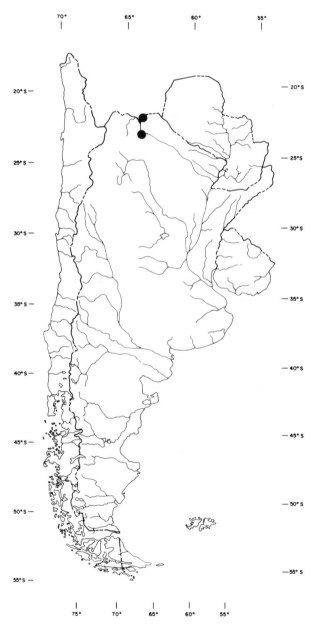

Map 4.10. Distribution of *Anoura caudifer*.

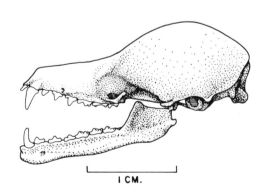

Figure 4.7. Skull of *Anoura geoffroyi*.

Distribution

Anoura caudifer is distributed over northern
South America, south through Bolivia, and into Argentina. In southern South America it has been
caught only in Salta province, Argentina (Honacki,
Kinman, and Koeppl 1982; Olrog and Bárquez 1979)
(map 4.10).

Ecology

This species has been caught in high forest in
Salta province Argentina (Olrog and Bárquez 1979).

Comment

The species was originally described as *A. geoffroyi*, but Bárquez (1987) assigned all specimens to
A. caudifer.

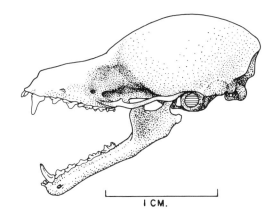

Figure 4.8. Skull of *Glossophaga soricina.*

Genus *Glossophaga* E. Geoffroy, 1818
• *Glossophaga soricina* (Pallas, 1766)
Pallas's Long-tongued Bat, Falso Vampiro Soriciteo

Measurements

	Mean	Min.	Max.	N	Loc.	Source[a]
TL	63.5	50.0	71.0	8	P	1
HB	58.0	50.0	71.0			
	59.5	55.0	63.0	10	B	2
T	5.5	0.0	8.0	8	P	1
	6.0	4.0	8.0	10	B	2
HF	10.4	6.0	12.0	8	P	1
E	15.5	11.0	22.0			
FA	36.0	34.0	37.0	4		
	35.5	31.3	37.2	10	B	2
Wta	15.8	10.5	25.5	3	P	1
	10.8	8.0	14.0	10	B	2

[a](1) PCorps; (2) Ojeda and Bárquez 1978.

Description

The dental formula is I 2/2, C 1/1, P 2/3, M 3/3
(fig. 4.8). This small, long-nosed bat is grayish brown,
slightly lighter on the venter, and has a moderately
elongated muzzle, a short tail, and an unreduced
interfemoral membrane (Phillips 1971).
Chromosome number: $2n = 32$; FN = 60.

Distribution

Glossophaga soricina is distributed from Mexico
south to southeastern Brazil and northern Argentina. In southern South America it has been collected over much of eastern Paraguay and in Salta
and Chaco provinces, Argentina (Fornes and Massoia 1967; Honacki, Kinman, and Koeppl 1982;
Myers and Wetzel 1983; Ojeda and Mares 1989)
(map 4.11).

Life History

The timing of reproduction can be strongly seasonal in habitats with pronounced periodicity in

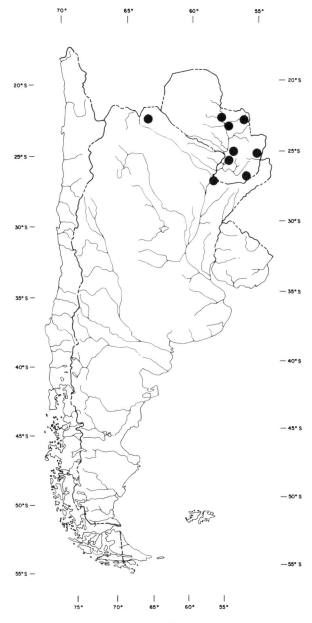

Map 4.11. Distribution of *Glossophaga soricina.*

rainfall, but females are polyestrous and can bear two or three young a year (Eisenberg 1989).

Ecology

Primarily nectarivorous, *G. soricina* is particularly dominant in disturbed and second-growth areas. It can hover in flight while taking nectar from flowers and is also a foliage-gleaning insectivore. This species roosts in abandoned caves and man-made structures in colonies of up to two thousand animals; there is spatial segregation, with females more common in the center of the roost (Eisenberg 1989; Willig 1983).

SUBFAMILY CAROLLIINAE

Diagnosis

The muzzle is long but not as pronounced as in the Glossophaginae, and the naked pad on the chin bears a single large wart. The ears are shorter than in the Phyllostominae. The tail is short or absent, but the uropatagium is present.

Genus *Carollia* Gray, 1838
• *Carollia perspicillata* (Linnaeus, 1758)
Short-tailed Bat, Falso Vampiro Colicorto

Measurements

	Mean	Min.	Max.	N	Loc.	Source[a]
TL	66.5	57.0	75.0	17	P	1
HB	59.3	50.0	72.0			
T	7.7	0.0	15.0	16		
HF	13.4	8.0	23.0	17		
E	19.0	9.0	22.0			
FA	41.2	40.0	43.0	11	P	1
	44.1	37.4	45.8	56	A	2
Wta	16.3	13.0	20.5	17	P	1
	20.4	11.0	24.0	56	A	2

[a](1) PCorps; (2) Ojeda and Bárquez 1978.

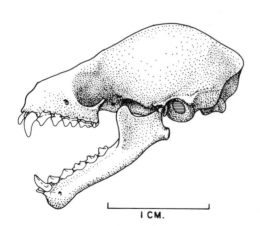

Figure 4.9. Skull of *Carollia perspicillata*.

Description

The dental formula is I 2/2, C 1/1, P 2/2, M 3/3 (see fig. 4.9). The fur color in *C. perspicillata* is highly variable, ranging from almost black through various browns and grays to bright orange. In Paraguay these bats are densely furred—grayish brown dorsally and ventrally, with a tendency for the venter to be lighter gray (Pine 1972).

Distribution

This bat has a wide range, occurring from southern Mexico to Peru, Bolivia, Paraguay, and southeastern Brazil and across northern South America. In southern South America it is found in north-central and

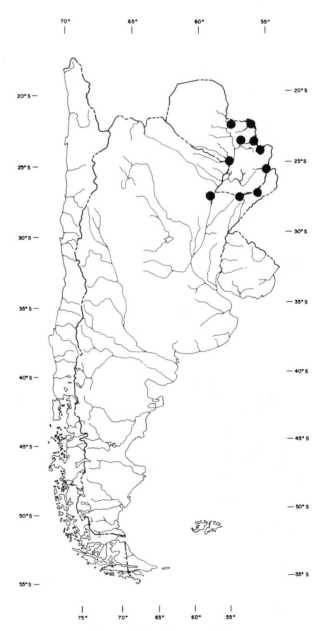

Map 4.12. Distribution of *Carollia perspicillata*.

northeastern Argentina and in Paraguay (Fornes and Massoia 1969; Massoia 1980; Myers and Wetzel 1983) (map 4.12).

Life History

Carollia perspicillata apparently breeds throughout the year in some localities. The female will carry the young with her when she flies, but this may be only while changing roosts. In many parts of their range bats of this species have two birth peaks; thus a female can produce two young annually, with an interbirth interval ranging from 115 to 173 days. Gestation lasts approximately 115–20 days. The young are born in an advanced state, with their eyes open. The newborn remains more or less continuously attached to the mother for the first fourteen days of life. Young weigh approximately 5 g at birth (Fleming 1983; Kleiman and Davis 1979; Pine 1972).

Ecology

The short-tailed bat is characteristic of lower elevations and is found in both tropical evergreen and deciduous forests. There is some indication that in certain parts of its range *C. perspicillata* may migrate seasonally. It will roost almost anywhere that is dark or shaded and that provides some protection, and it can be found singly or in roosts of up to a thousand.

Carollia perspicillata is primarily frugivorous and can harm fruit crops in some areas. In many areas fruit of *Piper* is a major food source, though it also gleans insects from foliage. Males defend small groups of females and thus have a harem breeding system. The definitive volume on the life history and ecology of this bat is Fleming (1988) (see also Fleming 1983; Pine 1972).

SUBFAMILY STURNIRINAE

Diagnosis

These bats are often included within the Stenoderminae. The dental formula is I 2/2, C 1/1, P 2/2, M 3/2–3 (see fig. 4.10). They have a short muzzle and modest ears. They are tailless, and the uropatagium is inconspicuous; a fringe of hairs adorns the remnant. Many species bear patches of stiff yellowish hairs on the shoulders.

Genus *Sturnira* Gray, 1842
• *Sturnira lilium* (E. Geoffroy, 1810)
Yellow-shouldered Bat, Falso Vampiro Flor de Lis

Measurements

	Mean	S.D.	Min.	Max.	N	Loc.	Source[a]
TL	61.9		52.0	69.0	37	P	1
	60.5		53.0	67.0	10 m	A	2
	58.1		48.0	65.0	10 f		
HF	11.9		10.5	14.3	10 m		
	11.8		10.5	17.7	10 f		
	13.2		9.0	17.0	41	P	1
E	16.6		11.0	20.0	38		
	18.2		16.3	19.5	10 m	A	2
	17.9		16.8	18.6	10 f		
FA	42.8		40.3	46.1	10 m		
	43.0		39.4	47.2	10 f		
	42.1		38.0	46.0	30	P	1
Wta	17.2		14.5	23.0	7	P	1
	22.5		15.3	26.0	10 m	A	2
	19.8		14.1	25.0	10 f		
	19.9	0.3			52 m	A	3

[a](1) PCorps; (2) Mares, Ojeda, and Kosco 1981; (3) Crespo 1982.

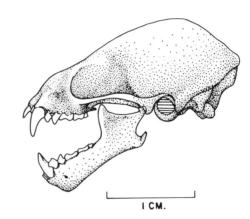

Figure 4.10. Skull of *Sturnira* sp.

Description

The hind legs are furred to the feet. This pretty little bat ranges from a bright orangish brown through grayish brown. It is always lighter across the shoulders, and in all individuals the venter is a duller shade of the dorsal color.

Distribution

The yellow-shouldered bat ranges from southern Mexico to northern Argentina, Uruguay, and eastern Brazil. In southern South America it has been recorded from eastern Paraguay and the Chaco, though in the latter it is probably restricted to riparian forests along the Río Paraguay and the lower reaches of its tributaries. It is reported from several sites in Uruguay, although it may move into this country only seasonally after fruiting trees, and in Argentina it occurs in several of the northern provinces and south into Tucumán (Acosta y Lara 1950; Bárquez 1984a; Honacki, Kinman, and Koeppl 1982; Myers and Wetzel 1983; Olrog and Lucero 1981; Ximénez, Langguth, and Praderi 1972) (map 4.13).

Life History

Sturnira lilium has a litter size of one, and in northeastern Argentina it reproduces from September to April (Crespo 1982).

Ecology

One of the most common bats where it occurs, *S. lilium* seems to be confined to moister habitats, where it roosts in hollow trees and caves. In Jujuy it is common to 2,600 m or where the forest ends, and in Iguazú National Park, Argentina, it was netted in all forest types. This species is exclusively frugivorous and at times will even forage on the ground for fallen fruit; it is occasionally taken on the ground in rodent traps (Crespo 1982; Eisenberg 1989; Mares, Ojeda, and Kosco 1981; Olrog 1979; Villa-R. and Villa Cornejo 1969).

Comment

There may be other species of *Sturnira* present in Argentina as well (M. A. Mares, pers. comm.).

SUBFAMILY STENODERMINAE

Diagnosis

Bats of this subfamily exhibit loss or great reduction of the tail, though the uropatagium is usually well developed. The muzzle is short, and the ears are modest in size. The face frequently bears two pairs of white stripes, and the dorsum may have a white median stripe (see plate 4).

Natural History

These bats are strongly frugivorous and may make local migratory movements.

Genus *Artibeus* Leach, 1821

Description

The dental formula is I 2/2, C 1/1, P 2/2, M 2–3/2–3 (see fig. 4.11). Four white facial stripes may be present, but there is no white line on the dorsum.

Artibeus fimbriatus (Gray, 1838)

Measurements

	Mean	Min.	Max.	N	Loc.	Source[a]
TL	95.2	80	113	22	P	1
T	18.0	13	21			
HF	23.5	21	25			
E	64.6	63	69			
Wta	54.0	45	71	11		

[a](1) UM.

Description

This very large *Artibeus* has long, lax, silky fur. There is a wide median band of dorsal hairs, and the legs and interfemoral membranes are distinctly

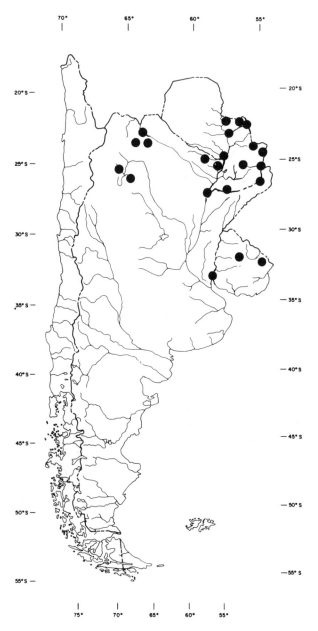

Map 4.13. Distribution of *Sturnia lilium*.

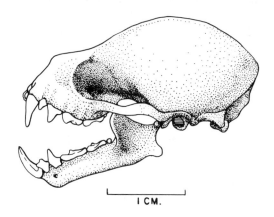

Figure 4.11. Skull of *Artibeus jamaicensis*.

haired. The bat is sooty brown above and a little paler below, with white-tipped hairs giving it a frosted appearance. The face is blackish with facial stripes at least faintly indicated, and the ears are blackish. The wings are usually white tipped (Handley 1989).

Distribution

Artibeus fimbriatus occurs from Rio de Janeiro state, Brazil, south along the coast to eastern Paraguay at altitudes from sea level to 530 m (Handley 1989) (map 4.14).

• *Artibeus jamaicensis* Leach, 1821
Jamaican Fruit-eating Bat, Falso Vampiro Grande

Measurements

	Mean	S.D.	Min.	Max.	N	Loc.	Source[a]
TL	74.9	3.0			20 m	Br	1
	75.9	3.1			20 f		
HB	86.8		83.0	90.0	6	A	2
HF	16.9		15.9	18.1			
E	24.4		22.9	25.6			
	24.6		21.7	28.0	39	B	3
FA	66.6		63.6	68.6	6	A	2
	62.5		57.4	66.8	39	B	3
Wta	53.3		45.0	60.0	6	A	2
	46.4		31.0	62.0	14 m	B	3
	47.5		39.0	66.0	25 f		
	41.0	3.5			20 m	Br	1
	46.4	5.9			20 f		

[a](1) Willig 1983; (2) Mares, Ojeda, and Kosco 1981; (3) Ojeda and Bárquez 1978.

Description

This short-bodied bat is ashy gray to ashy brown above, with the white hair bases showing through. The venter is a lighter gray. Some individuals have faint white eye stripes running above the eye from the nose to the base of the ear.

Distribution

Artibeus jamaicensis is widely distributed from southern Mexico to northern Argentina. Its southernmost limit is in Salta province, Argentina, though there are unconfirmed reports that it occurs in Tucumán province. In Jujuy province it is common at 2,000 m and is found up to 2,600 m (Olrog 1979; Ojeda and Mares 1989; Honacki, Kinman, and Koeppl 1982) (map 4.15).

Life History

Breeding is tied to the maximum abundance of fruit. Breeding colonies of up to twenty-five involve harem defense by the male, and the mating system is strongly polygynous. During lactation females have a daytime roost, and when moving to forage, they deposit their babies in crèches near the feeding tree (Eisenberg 1989; Fleming 1971; Kunz, August, and Burnett 1983).

Ecology

This bat is strongly frugivorous, feeding mainly on figs over much of its range. Apparently *A. jamaicensis* comes into the Paraguayan Chaco only when the trees are in bloom.

It is suspected that owls may be significant predators on this species during the night, and snakes and the bat falcon may be significant predators during the day. When foraging these bats fly in small groups,

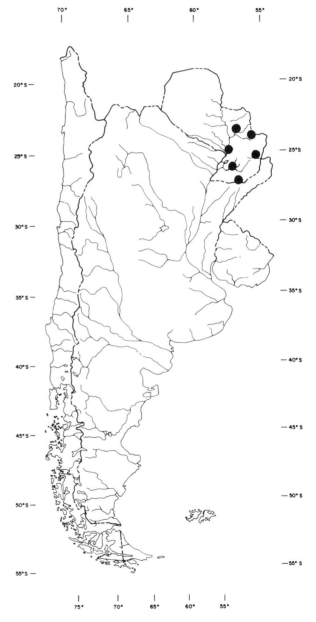

Map 4.14. Distribution of *Artibeus fimbriatus*.

and an individual that is caught will emit distress calls that induce "mobbing behavior" by fellow flock members. This species roosts in hollow trees and well-lighted caves, though it occasionally constructs tents by biting the midrib of large leaves, causing the leaf to fold over (August 1979; Foster and Timm 1976; Goodwin and Greenhall 1961; Morrison 1978; Myers and Wetzel 1983).

Comment

Bárquez (1988) has stated that specimens from northwestern Argentina should be considered *A. planirostris*.

• *Artibeus lituratus* (Olfers, 1818)
Great Fruit-eating Bat, Falso Vampiro Cariblanco

Measurements

	Mean	Min.	Max.	N	Loc.	Source[a]
TL	92.7	81.0	102.0	42	P	1
HF	20.2	16.0	26.0			
E	22.1	10.0	27.0			
FA	72.0	68.0	75.0	34		
	72.2	68.8	74.4	16	B	2
Wta	62.6	50.0	88.0	24	P	1
	66.5			51 m	A	3

[a](1) PCorps; (2) Anderson, Koopman, and Creighton 1982; (3) Crespo 1982.

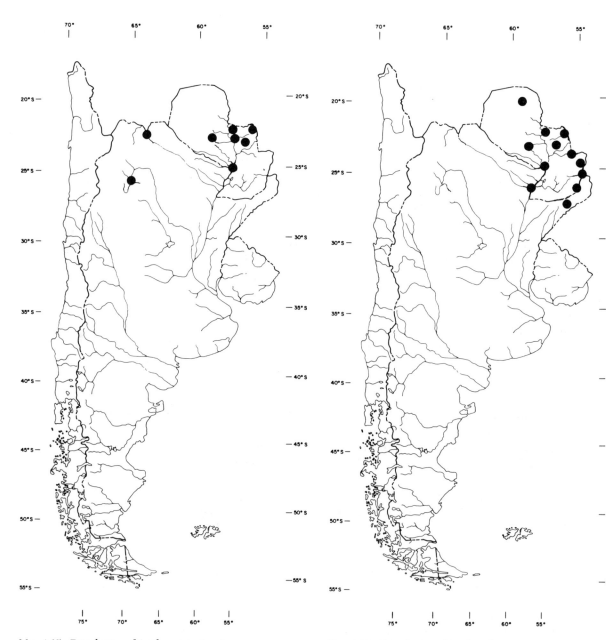

Map 4.15. Distribution of *Artibeus jamaicensis*.

Map 4.16. Distribution of *Artibeus lituratus*.

Description

Artibeus lituratus is distinguishable from other members of *Artibeus* by its large size and clearly defined white eye stripe, which passes from the nose above each eye. Dorsally these bats are brownish gray, and ventrally they are more heavily washed with gray. Some individuals show bleaching around the head and shoulders, giving an almost light tan color (Myers and Wetzel 1983; pers. obs.).

Distribution

Artibeus lituratus is distributed from southern Mexico to southern Brazil, northern Argentina, and Bolivia. In southern South America it has been collected from several localities in Paraguay and northern Argentina (Bárquez 1987; Honacki, Kinman, and Koeppl 1982; Myers and Wetzel 1983; Ojeda and Mares 1989; Villa-R. and Villa Cornejo 1969) (map 4.16).

Life History

In Misiones province, Argentina, *A. lituratus* was found to reproduce from September to February with a litter size of one (Crespo 1982).

Ecology

In eastern Paraguay this bat is abundant and found year-round; in the Chaco it may be migratory, moving in when trees are blooming. In Misiones province, Argentina, it is found in many forest types, in more open areas, and around human-disturbed areas. The diet of *A. lituratus* is pollen and wild and domestic fruit, including palm fruit. In northern South America it roosts in well-lighted caves and palm trees in colonies that can number up to twenty-five (Crespo 1982; Eisenberg 1989; Myers and Wetzel 1983; Villa-R. and Villa Cornejo 1969).

Comment

Bárquez (1988) has stated that specimens from northwestern Argentina should be considered *A. planirostris*.

• *Artibeus planirostris* (Spix, 1823)

Measurements

	Mean	Min.	Max.	N	Loc.	Source[a]
TL	85.0	81.0	90.0	11	A	1
T	17.6	16.0	18.5			
HF	24.5	23.1	26.0			
E	65.6	63.4	67.5			
FA	52.4	47.0	60.0			

[a](1) CM.

Description

This heavy-bodied, medium-sized bat is similar in size to *A. lituratus*. It has ashy brown dorsal col-

oration, with the light gray at the hair bases showing on the shoulders. Dorsally there is faint grizzling, and this is more pronounced on the venter. The wings are dark and the ears lighter (pers. obs.).

Distribution

Artibeus planirostris is found from Colombia south to northern Argentina and eastern Brazil. In southern South America it is found in eastern Paraguay and northwestern Argentina (Bárquez 1987; Honacki, Kinman, and Koeppl 1982; CM) (map 4.17).

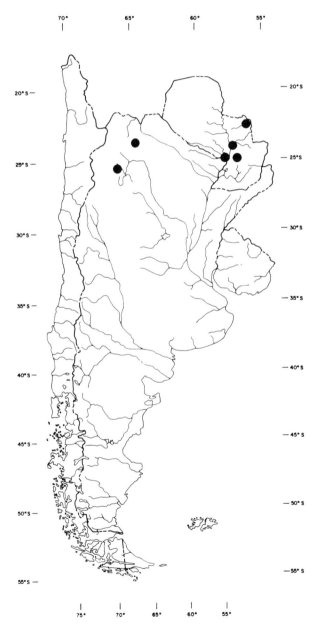

Map 4.17. Distribution of *Artibeus planirostris*.

Comment

Some authors use the genus *Cynomops* for this species.

Genus *Pygoderma* Peters, 1863

• *Pygoderma bilabiatum* (Wagner, 1843)
Ipanema Bat, Falso Vampiro Penacho Blanco

Measurements

	Mean	Min.	Max.	N	Loc.	Source[a]
TL	58.4	53.0	65.0	15	A, P	1
HF	10.5	6.0	15.0	16		
E	16.8	15.0	20.0			
	21.1	19.9	22.5	12	B	2
FA	40.1	36.0	43.2	21	A, P	1
	43.8	41.8	45.8	12	B	2
Wta	21.5	19.0	26.0	— m		
	22.4	21.0	24.5	— f		

[a](1) Bárquez 1984a; Fornes and Delpietro 1969; Olrog 1967; (2) Ojeda and Bárquez 1978.

Description

The dental formula is I 2/2, C 1/1, P 1/1, M 2/2. The upper incisors are unequal in size, with the inner pair larger. In this medium-sized bat the male shows extreme development of a doughnut-shaped mass of glandular tissue surrounding each eye, which is present in the female but much reduced. Both sexes show T-shaped glandular swelling in the ventral surface of the lower jaw and swollen glands lateral to the nose leaf. The nose leaf is well developed, the ears are broadly rounded, and the tragus is reduced. The forearms are heavily furred, but the pelage on the anterior chest and shoulders is sparse, especially in males, leaving a great deal of exposed skin. The uropatagium is semicircular, extends approximately to the knees, and is furred dorsally and ventrally. The dorsal pelage is brown or dark gray, lighter on the shoulders, and there is a lighter venter and a distinctive white epaulet over each shoulder (Fornes and Delpietro 1969; Myers 1981; Owen and Webster 1983; Webster and Owen 1984; pers. obs.).
Chromosome number: $2n = 30-31$; FN = 56.

Distribution

Pygoderma bilabiatum has been found in Suriname, southeastern Brazil, Paraguay, northern Argentina, and Bolivia. In southern South America it has been collected from Paraguay and from two apparently disjunct populations in Argentina, one in the province of Salta and the other in the province of Misiones (Bárquez 1984a, 1987; Myers 1981; Owen and Webster 1983) (map 4.18).

Life History

The litter size in this species is one, and in Paraguay females reproduce in fall and winter (Myers 1981).

Ecology

This species has been collected in mature tropical forest, in subtropical forest, and in second growth bordering forest. It apparently feeds on large fleshy fruit (Carvalho 1961; Myers 1981; Olrog 1967).

Genus *Vampyressa* Thomas, 1900
Yellow-eared Bat

Description

The dental formula is variable: I 2/2, C 1/1, P 2/2, M 2/2, or I 2/1, C 1/1, P 2/2, M 2/2. Head and body length ranges from 43 to 65 mm. The dorsum varies from smoky gray to pale brown or dark brown, and a dorsal stripe is usually not present. The white facial

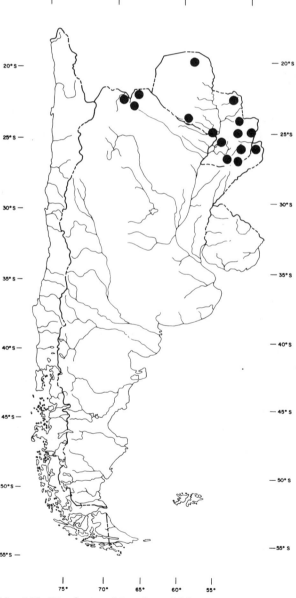

Map 4.18. Distribution of *Pygoderma bilabiatum*.

stripes are prominent in some species but lacking in others. The ears typically have a yellow margin. The genus has been reviewed by Gardner (1977b) and Peterson (1968).

Distribution
The genus occurs from southern Mexico south through the Isthmus to northeastern Brazil and adjacent portions of Amazonian Peru. One species, *V. pusilla*, extends its range to southeastern Brazil and Paraguay but is absent from the Amazonian portion.

Life History
The species of this genus are believed to be largely frugivorous. Reproduction is seasonally timed by the onset of the rains (Gardner 1977a; Wilson 1979).

• *Vampyressa pusilla* (Wagner, 1843)

Measurements

	Mean	Min.	Max.	N	Loc.	Source[a]
TL	49.7	45	56	16	Br, P,	1
T	0.0				Peru	
HF	9.6	9	11			
E	14.8	14	16	17		
FA	33.0	30	35	14		
Wta	8.2			27	CR	2

[a](1) Taddei 1979; FM, PCorps, UM; (2) Lewis and Wilson 1987.

Description
This medium-sized bat has no trace of an external tail; it has a narrow interfemoral membrane and a well-developed nose leaf. There are two pairs of narrow whitish facial stripes. Its dorsal color ranges from gray to reddish brown and, unlike most other species in the genus, it has no dorsal stripe (Goodwin 1963; Lewis and Wilson 1987).
Chromosome number: There are several chromosomal races of this species. In one, from Honduras and Nicaragua, $2n = 18$ and FN $= 20$, while in the other, from Colombia, $2n = 24$ in females and $2n = 23$ in males; FN $= 22$. In Paraguayan specimens $2n = 20$ and FN $= 36$ (Lewis and Wilson 1987).

Distribution
Vampyressa pusilla ranges from Mexico to Peru, Colombia, and Venezuela and south on the western side of the Andes to Peru. There is a disjunct population in eastern Paraguay and southeastern Brazil (Lewis and Wilson 1987; Myers, White, and Stallings 1983) (map 4.19).

Life History
Females probably exhibit bimodal polyestry (Lewis and Wilson 1987).

Ecology
Individuals of this species were captured in Paraguay at the edge of a small clearing in tropical forest.

In Venezuela this bat is strongly associated with moist habitats and multistratal evergreen forest. It is entirely frugivorous and has been described as a common fig specialist (Handley 1976; Lewis and Wilson 1987; Myers, White, and Stallings 1983).

Genus *Vampyrops* Peters, 1865
White-lined Bat

Description
The dental formula is I 2/2, C 1/1, P 2/2, M 3/3. Head and body length ranges from 48 to 98 mm. The basic color of the dorsum is dark brown to almost black, and the white or gray dorsal stripe is prominent, extending from the ears to the tail membrane

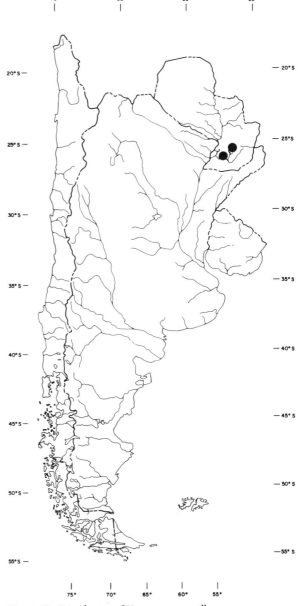

Map 4.19. Distribution of *Vampyressa pusilla*.

(see plate 5). There are prominent white facial stripes.

Chromosome number: All species show $2n = 30$; FN = 56.

Distribution

The genus is distributed from southern Mexico through the Isthmus to northeastern Argentina and Uruguay. Within its range it is divisible into a number of different species.

Life History and Ecology

White-lined bats are basically frugivorous. They roost in groups of three to ten in leafy tangles, tree hollows, or caves. Reproduction usually coincides with the onset of the rainy season and varies locally.

Comment

Some authors refer these species to the genus *Platyrrhinus*.

• *Vampyrops dorsalis* (Thomas, 1900)

Measurements

	Mean	Min.	Max.	N	Loc.	Source[a]
TL	78.0	76.0	80.0	2	B	1
HF	12.5	12.0	13.0			
E	19.0	19.0	19.0			
FA	46.5	44.5	49.1	45		2

[a](1) UM; (2) Anderson, Koopman, and Creighton 1982.

Description

The dorsum is dark gray brown or almost black, with the venter paler. Facial and dorsal stripes are usually present (Sanborn 1955).

Distribution

This species has been recorded in Bolivia and is known from southern South America only through a single record from Paraguay (Anderson, Koopman, and Creighton 1982; PCorps) (map 4.20).

• *Vampyrops lineatus* (E. Geoffroy, 1810)
Falso Vampiro Listado

Measurements

	Mean	Min.	Max.	N	Loc.	Source[a]
TL	69.5	63	75	24	A, P	1
T	0.0					
HF	13.0	8	18	32		
E	19.5	15	24			
FA	46.4	43	49	21		
Wta	26.7	20	36	28		

[a](1) Baud 1981; Fornes and Massoia 1966; PCorps.

Description

This medium-sized phyllostomatid is ash brown to light brown dorsally with a lighter venter and four prominent facial stripes. A pair (only a single stripe on Paraguayan specimens) of white stripes extends from the back of the skull to the base of the uropatagium (pers. obs.).

Chromosome number: $2n = 30$; FN = 56 (Willig and Hollander 1987).

Distribution

Vampyrops lineatus extends from Colombia to Peru, Bolivia, Uruguay, and northern Argentina. In southern South America it has been collected from Paraguay, Uruguay, and northeastern Argentina (Bárquez 1984a; Baud 1981; Honacki, Kinman, and Koeppl 1982; Massoia 1980; Willig and Hollander 1987; Ximénez 1969; UM) (map 4.21).

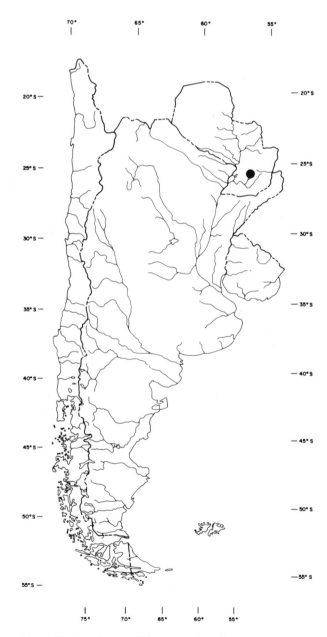

Map 4.20. Distribution of *Vampyrops dorsalis*.

Life History

Females with one and two embryos have been collected. A single neonate weighed 8.3 g. In São Paulo state, Brazil, pregnant females were caught in all months, and in northeastern Brazil there was a bimodal distribution of pregnancy (Fornes and Massoia 1966; González and Vallejo 1980; Willig and Hollander 1987).

Ecology

Vampyrops lineatus may forage in large groups, but in roosts males maintain small harems of seven to fifteen females. They are frugivorous, insectivo-rous, and nectarivorous (González and Vallejo 1980; Sanborn 1955; Willig 1983; Willig and Hollander 1987).

SUBFAMILY DESMODONTINAE (DESMODONTIDAE)

Diagnosis

This subfamily contains three species that range in head and body length from 65 to 90 mm. The dental formula for the three species is highly variable and will be discussed in the species accounts, but the incisors and canines are specialized for cutting, and the cutting edges form a V (fig. 4.12). These teeth are used to make incisions in prey, for these are the true vampire bats. The premolars and molars are greatly reduced, indicating less selection for their retention, since these animals have specialized for their peculiar dietary habits. There is no tail. The nose leaf is reduced to fleshy pads that are U-shaped and surround either nostril (fig. 4.4). The dorsum is generally some shade of brown.

Distribution

Members of this subfamily occur from southern Sonora, Mexico, and southeastern Texas in the United States south across the continent of South America to northern Argentina, Uruguay, and northern Chile.

Natural History

The three species of this subfamily are specialized for feeding on the blood of warm-blooded vertebrates. The saliva contains an enzyme that retards coagulation, and they lap the blood after making an incision. They can walk on their feet and thumbs and

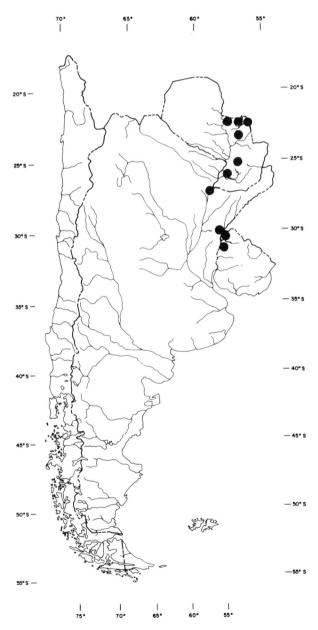

Map 4.21. Distribution of *Vampyrops lineatus*.

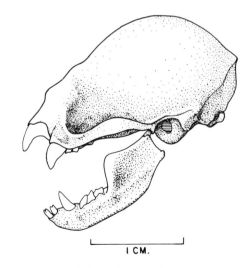

Figure 4.12. Skull of *Desmodus rotundus*.

usually crawl on the body of their prey before feeding. Vampire bats are of medical importance because they have been implicated in transmitting rabies.

Genus *Desmodus* Wied-Neuwied, 1826
• *Desmodus rotundus* (E. Geoffroy, 1810)
Common Vampire Bat, Vampiro de Azara

Measurements

	Mean	Min.	Max.	N	Loc.	Source[a]
TL	81.0	72.0	95.0	24 m	U	1
	86.6	82.0	110.0	9 f		
HF	17.0	14.0	19.0	24 m		
	18.0	16.5	18.0	9 f		
E	16.5	13.0	21.0	24 m		
	18.1	14.0	23.0	9 f		
FA	61.2	59.9	63.7	5 m	P	2
	65.8	64.1	67.6	3 f		
Wta	40.8	33.5	49.0	24 m	U	1
	40.8	35.9	62.5	9 f		

[a](1) Barlow 1965; (2) Myers and Wetzel 1983.

Description
The dental formula is I 1/2, C 1/1, P 2/3, M 0/0 (fig. 4.12). The middle and upper incisors and canines are very large, projecting and sharp edged. Females are usually larger than males. This medium-sized bat has thin, coarse pelage, a small, rounded nose leaf, and no tail. It has small, somewhat rounded, separate ears, a deeply grooved lower lip, a short muzzle, and an unusually strong and elongated thumb, slightly longer than the hindfoot (fig. 4.4). *D. rotundus* is the only Chilean bat to lack a uropatagium. The dorsum is dark gray to gray brown and is strongly demarcated from the silvery gray venter; in Paraguay there is a tendency for some individuals to be lighter on the neck and shoulders (Barlow 1965; Greenhall, Lord, and Massoia 1983; Mann 1978; Myers and Wetzel 1983; Osgood 1943; pers. obs.).

Distribution
The common vampire bat ranges from northern Mexico south to Chile and Argentina. In southern South America it follows the 10° January isotherm, being found as far south as Valparaíso in Chile and Córdoba in Argentina. It is found throughout Uruguay and Paraguay (Barlow 1965; Greenhall, Lord, and Massoia 1983; Myers and Wetzel 1983; Villa-R. and Villa Cornejo 1969) (map 4.22).

Life History
Over much of its range the common vampire bat can apparently breed throughout most of the year. The litter size is usually one, the gestation period is approximately seven months, and neonates weigh 5–7 g. Young are suckled up to 300 days, and there is a gradual switch from milk to regurgitated blood.

After about four months the young accompanies its mother to prey, drinking blood from her feeding site, and after about five months its growth is largely complete. In a Costa Rican population the interbirth interval was ten months, and the infant mortality rate was 54% (Barlow 1965; Crespo 1982; Greenhall, Lord, and Massoia 1983; Langguth and Achaval 1972; Wilkinson 1985a,b).

Ecology
Desmodus rotundus is found in many different habitats and at elevations up to 1,500 m in Jujuy province, Argentina, and 2,000 m in Chile, but it apparently prefers tropical and subtropical woodland and open grasslands for foraging. These bats

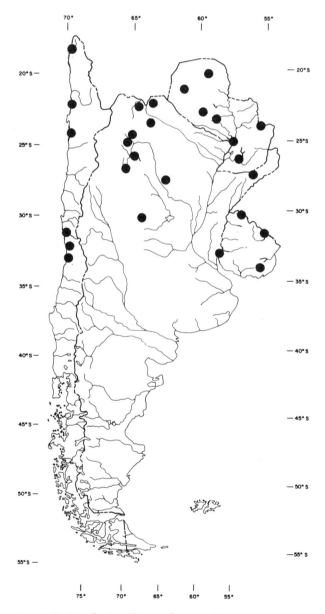

Map 4.22. Distribution of *Desmodus rotundus*.

commonly forage 5–8 km from the roost, but distances of up to 20 km have been recorded.

In different areas, different roost sites are selected. In northern Argentina vampires roost in tree trunks, especially *Chorisia* and *Caesalpinia,* and in Chile they also roost in sea caves. Throughout their range they show a liking for mine shafts, old wells, and abandoned buildings. The colonies are usually small (20–100) but colonies up to about 5,000 have been observed. Vampires commonly share roosts with other species of bats; over their entire range, forty-five different species of bats have been recorded as roosting with vampires.

In a Costa Rican population the social unit is a group of eight to twelve females. Males fight for access to the top of the preferred female roosting site. Males that gain this position mate preferentially with the females in that roost. Within the colony there is extensive sharing of blood by regurgitation, and bats that fail to obtain a blood meal are fed by successful roost mates. Food sharing is based both on degree of relatedness and on opportunity for reciprocation. There is also extensive allogrooming, which takes place preferentially between related bats. Male offspring leave the maternal group and females remain.

At this time the most common prey of vampires is undoubtedly livestock, though there is an interesting observation of vampires feeding on *Otaria* sea lions in Chile. After selecting a "bite site," the vampire bites off a piece of skin using its upper lancetlike teeth; it laps up blood over a period ranging from nine to forty minutes, aided by an anticoagulant in its saliva. Sometimes several bats share one wound.

It seems clear that the spread of humans with their domestic stock has increased the range and abundance of vampire bats. Throughout much of their range these bats represent a considerable economic threat by transmitting rabies (Barlow 1965; Greenhall, Lord, and Massoia 1983; Crespo et al. 1961; González 1973; Langguth and Achaval 1972; Mann 1978; Olrog 1979; Pine, Miller, and Schamberger 1979; Tamayo and Frassinetti 1980; Villa-R. and Villa Cornejo 1969, 1971; Wilkinson 1984, 1985 a,b 1986).

• *Desmodus (Diaemus) youngi* Jentink, 1893
White-winged Vampire Bat

Measurements

	Mean	S.D.	N	Loc.	Source[a]
TL	83.36	5.35	11 m	V	1
	84.00	2.94	4 f		
HB	83.36	5.35	11 m		
	84.00	2.94	4 f		
HF	17.91	1.30	11 m		
	19.50	1.29	4 f		
E	18.18	1.33	11 m		
	18.75	0.50	4 f		
FA	51.65	1.39	11 m		
	53.48	1.24	4 f		
Wta	36.27	2.80	11 m		
	32.20		1 f		

[a](1) USNMNH.

Description

The dental formula is I 1/2, C 1/1, P 1/2, M 2/1. This robust-bodied bat is similar in size to *Desmodus rotundus.* It has thin brown fur, paling to white on the shoulders. The wings are pale brown and their tips are white, which separates it from *D. ro-*

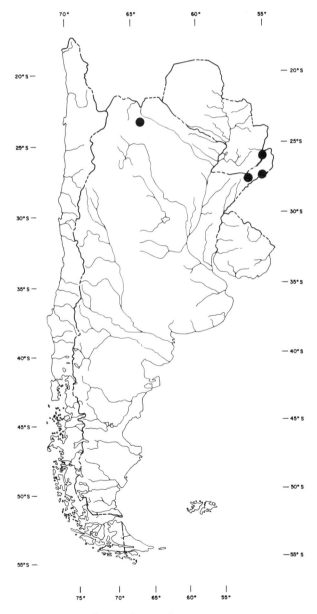

Map 4.23. Distribution of *Desmodus youngi.*

tundus. White is also found in blotches along the posterior wing edge (pers. obs.).

Chromosome number: $2n = 32$; FN = 60 (Baker et al. 1982).

Distribution

Desmodus youngi is distributed from Mexico to northern Argentina and eastern Brazil. Until recently this bat was not known from southern South America, the first individual being caught in Misiones province, Argentina, a range extension of over 450 km. It is now known from Misiones and Jujuy provinces, Argentina (Bárquez 1984b; Delpietro et al. 1973; Honacki, Kinman, and Koeppl 1982; Massoia 1980) (map 4.23).

Ecology

Like *D. rotundus*, this species feeds on the blood of higher vertebrates. Glands in the mouths of males produce a sharp odor, but their exact function is unknown. The bats roost in hollow trees, and colony size can range up to thirty (Goodwin and Greenhall 1961).

SUPERFAMILY VESPERTILIONOIDEA

These bats appear to be the most specialized for flight, based on the anatomy of the pectoral girdle. The greater tuberosity of the humerus is enlarged, and on the upstroke of the wingbeat it articulates with the scapula, thus mechanically terminating the upswing (Miller 1907).

FAMILY FURIPTERIDAE

Smoky Bats

Diagnosis

The thumb is extremely reduced or absent. The claw, if present, is minute and functionless (see fig. 4.4). There are two genera, *Furipterus* and *Amorphochilus*.

Distribution

This family is confined to the New World tropics, ranging from Costa Rica across northern South America to eastern Brazil.

Genus *Amorphochilus* Peters, 1877
Amorphochilus schnablii Peters, 1877

Measurements

	Mean	Min.	Max.	N	Loc.	Source[a]
TL	76.1	66	98	13	Peru	1
HB	48.9	39	67			
T	27.2	23	31			
HF	10.0	10	10	8		
E	14.7	13	15	10		
FA	36.8	36	39	12		

[a](1) FM.

Description

The dental formula is I 2/3, C 1/1, P 2/3, M 3/3. This small bat has long, light-gray fur, lightly washed with brown. The tail membrane is light brown, and the wings and ears are brownish gray (pers. obs.).

Distribution

This poorly known bat is found from Isla Puno in Ecuador south along the coast to northern Chile (Mann 1978; Tamayo and Frassinetti 1980) (map 4.24).

Ecology

In Chile, *A. schnablii* is found along the coast near the mouths of rivers. It feeds on small dipter-

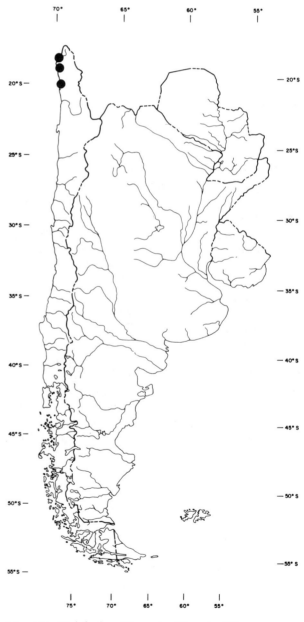

Map 4.24. Distribution of *Amorphochilus schnablii*.

ans and lepidopterans, and it roosts in cracks in rocks along the arid northern coast (Mann 1978).

FAMILY VESPERTILIONIDAE

Diagnosis

These bats are highly specialized for flight. The trochiter of the humerus is much larger than the trochin and has a surface of articulation with the scapula more than half as large as the glenoid fossa. The tail is rather long and extends to the edge of a wide interfemoral membrane. These bats are of medium to small size, with forearm lengths ranging from 90 to 24 mm. The nostrils and lips are relatively unmodified, and there are no decorative facial folds. The dental formula is highly variable and will be discussed under the accounts for the genera. A key to the genera is given in table 4.3.

Table 4.3   Key to the Genera of the Vespertilionidae of Southern South America

1 Dorsal surface of interfemoral membrane furred at least halfway to distal edge . . . . *Lasiurus*	
1′ Dorsal surface of interfemoral membrane naked 2	
2 Only one premolar per side 3	
2′ Two or more premolars per side . . . . *Myotis*	
3 Ears 25% of head and body length . . . . *Histiotus*	
3′ Ears 12% or less of head and body length . . . . *Eptesicus*	

Distribution

This family is worldwide in its distribution except for Antarctica and New Zealand.

Natural History

Vespertilionid bats are usually specialized as aerial insectivores, although some show specializations for feeding on fish (e.g., *Pizonyx*). Bats of this family roost in caves, hollow trees, or other sheltered areas. These are the common bats of the North Temperate Zone, but they also occupy the tropics. In the Temperate Zone they are specialized for winter hibernation. Echolocation in the capture of flying insects has been well studied in certain species of this family (Griffin 1958; Gould 1977).

Genus *Eptesicus* Rafinesque, 1820
Big Brown Bat, Murciélago con Orejas de Ratón

Description

The dental formula is I 2/2, C 1/1, P 1/1, M 3/3. The genus is distinguishable from *Myotis* by the reduction in premolar number (see figs. 4.13 and 4.14). Head and body length ranges from 35 to 75 mm and forearm length from 28 to 55 mm. In spite of the common name, there is a considerable size range

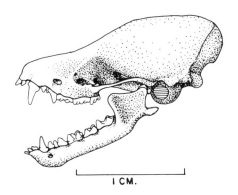

Figure 4.13. Skull of *Myotis nigricans*.

among the species of the genus. The dorsum is usually dark brown to almost black, and the venter is paler.

Distribution

This genus contains approximately thirty species and is worldwide in its distribution except for the Arctic and Antarctic regions.

• *Eptesicus brasiliensis* (Desmarest, 1819)
Brazilian Brown Bat, Murciélago Pardo

Measurements

	Mean	Min.	Max.	N	Loc.	Source[a]
TL	101.3	89.0	110.0	20 m	U	1
	103.6	92.0	115.0	45 f		
	85.1	78.0	94.0	18	A, U	2
T	43.0	30.0	50.0	20 m	U	1
	40.7	35.0	50.0	45 f		
	28.1	25.0	30.0	18	A, U	2
HF	10.3	7.5	12.0	20 m	U	1
	10.0	7.5	12.0	45 f		
E	14.6	11.5	16.0	20 m		
	14.4	11.5	16.0	45 m		
FA	39.7	37.5	42.0	21	A, U	2
Wta	10.2	8.6	15.0	20 m	U	1
	11.2	9.3	15.3	45 f		

[a](1) Barlow 1965; (2) Acosta y Lara 1950; Olrog 1959.

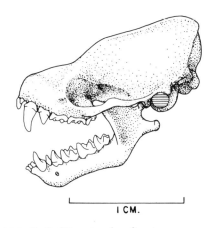

Figure 4.14. Skull of *Eptesicus brasiliensis*.

Description

Eptesicus brasiliensis is a medium-sized vespertilionid bat that in the southern part of its distribution ranges from dark grayish with a yellowish cast dorsally and grayish white on the venter to dark brown with a redish cast dorsally and lighter with a yellow wash on the venter. The flight membranes are brownish. No single character can distinguish between *E. brasiliensis*, *E. diminutus*, and *E. furinalis*, but in any given area they should differ in size. In identifying species, they must be properly sexed (Barlow 1965; Williams 1978; pers. obs.).

Distribution

The Brazilian brown bat is widely distributed from Mexico south to northern Argentina and Uruguay. In southern South America it is found throughout Uruguay, in Paraguay, and in Argentina south to about 35° S, though southernmost specimens may be *E. furinalis*, according to Bárquez (Bárquez 1987; Barlow 1965; Honacki, Kinman, and Koeppl 1982; Olrog and Lucero 1981; UM) (map 4.25).

Ecology

In eastern Paraguay this bat has been caught over streams in tropical forest; in Uruguay it is ubiquitous but seems to show a preference for foraging in edge situations (Barlow 1965; Myers, White, and Stallings 1983).

• *Eptesicus diminutus* Osgood, 1915
Murciélago Dorado

Measurements

	Mean	Min.	Max.	N	Loc.	Source[a]
TL	87.8	81.7	91.0	7	A, P, U	1
HB	55.4	53.0	59.0			
T	33.5	32.0	35.0			
HF	7.0	5.9	9.0			
E	13.6	11.0	15.0			
FA	34.4					1
Wta	5.6	4.9	6.3			

[a](1) Acosta y Lara 1951; Barlow 1965; Williams 1978; CM, UM.

Description

Eptesicus diminutus is the smallest of the South American *Eptesicus* and approaches the size of *Myotis*. In Uruguay this species may be confused with *Myotis albescens* but can be distinguished by the configuration of the external border of the pinna: in *M. albescens* it is essentially straight, whereas in *E. diminutus* it is convex. Its pelage ranges from dark chestnut brown with an orangish tint dorsally and paler and more yellowish ventrally to dark grayish with pale yellowish tips to the hairs dorsally and dark gray tinted with brown and whitish ventrally. The flight membranes are dark brown (Acosta y Lara 1951; Barlow 1965; Williams 1978; pers. obs.).

Distribution

Eptesicus diminutus ranges from Venezuela south to Paraguay, Uruguay, and northern Argentina. In southern South America it has been recorded from eastern Paraguay, western Uruguay, and the northern part of Argentina (Barlow 1965; Davis 1966; Myers, White, and Stallings 1983; Olrog and Lucero 1981) (map 4.26).

Ecology

Eptesicus diminutus, *E. furinalis*, and *E. brasiliensis* can all be caught in the same net in the rain forests along the Río Paraná in eastern Paraguay. *E. diminutus* has been caught in houses and in trees (Acosta y Lara 1951; Myers, White, and Stallings 1983; Romaña and Ábalos 1950).

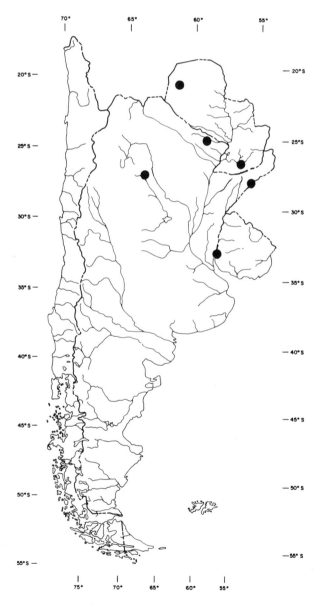

Map 4.25. Distribution of *Eptesicus brasiliensis*.

Comment

This species includes *E. fidelis* and *E. dorianus* (Honacki, Kinman, and Koeppl 1982).

• *Eptesicus furinalis* (d'Orbigny, 1847)
Murciélago Pardusco

Measurements

	Mean	Min.	Max.	N	Loc.	Source[a]
TL	92.8	88.0	96.0	20 m	P	1
	97.0	86.0	106.0	29 f		
T	36.8	32.0	41.0	20 m		
	38.6	29.0	45.0	29 f		
HF	9.4	8.0	10.0	20 m		
	9.7	8.0	11.0	29 f		
E	16.4	13.0	18.0	20 m		
	16.2	14.0	18.0	29 f		
FA	38.0	35.2	39.9	20 m		
	38.4	36.7	40.5	29 f		
	39.9	37.4	40.8	5	A	2
Wta	9.3	8.0	11.0			

[a](1) Myers and Wetzel 1983; (2) Mares, Ojeda, and Kosco 1981.

Description

Specimens of *E. furinalis* from eastern Paraguay tend to be blackish dorsally and paler on the venter with a yellowish tent, whereas those from the Chaco are lighter, highly variable in dorsal color, and smaller (Myers and Wetzel 1983; pers. obs.).

Distribution

Distributed from Mexico to northern Argentina, in southern South America *E. furinalis* is found throughout Paraguay and across northern Argentina

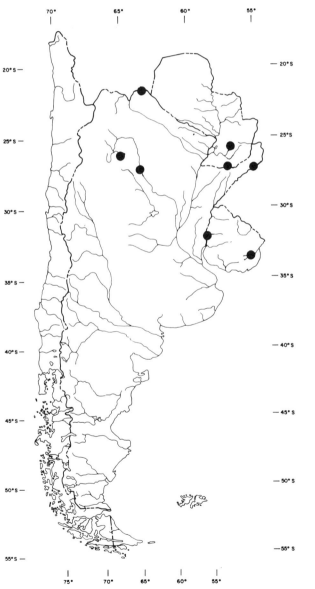

Map 4.26. Distribution of *Eptesicus diminutus*.

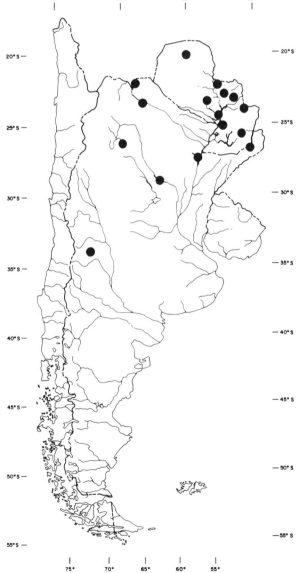

Map 4.27. Distribution of *Eptesicus furinalis*.

south to Mendoza province (Davis 1966; Honacki, Kinman, and Koeppl 1982; Myers and Wetzel 1983; R. A. Ojeda, pers. comm.; Romaña and Ábalos 1950; Williams 1978) (map 4.27).

Life History

In Paraguay the first pregnant females of *E. furinalis* appear in late July and August, and there are probably two periods of breeding. The average number of embryos for the first breeding period is 1.9, and for the second breeding period it is always 1. Gestation lasts slightly more than three months; copulation follows soon after parturition, and the females can store sperm. The age of first reproduction for both sexes is less than one year (Myers 1977).

Ecology

Bats of the species probably do not hibernate but undergo daily torpor (Myers 1977).

Eptesicus innoxius (Gervais, 1841)
Murciélago Pigmeo

Description

Two specimens measured TL 90–91; T 36–38; HF 8–9; E 10–13; FA 38.9–39.1. This small chocolate brown bat has a gray dorsum and faint white lines on the venter. The wings are dark (Villa-R. and Villa Cornejo 1969; pers. obs.).

Distribution

Eptesicus innoxius is found in Ecuador, Peru, and northwestern Argentina (Jujuy province). It probably also occurs in northern Chile (Honacki, Kinman, and Koeppl 1982; Villa-R. and Villa Cornejo 1969) (map 4.28).

Ecology

The specimens from Argentina were caught in early evening from a net across a small stream in mixed forest (Villa-R. and Villa Cornejo 1969).

Comment

Bárquez (1987) suggests that the Argentine specimens of *E. innoxius* should be referred to *E. furinalis*.

Genus *Histiotus* Gervais, 1856
Murciélago Orejudo

Description

Head and body length ranges from 54 to 70 mm and the forearm from 42 to 52 mm. This genus resembles *Eptesicus* in dental formula and shape of the skull (fig. 4.15), but it has extremely large ears compared with *Eptesicus*. For example, *E. fuscus* with a total length of 117 mm will have an ear ap-

proximately 17 mm long, whereas a comparable-sized *Histiotus* will have an ear 31 mm long. The upperparts of the body are generally light brown or grayish brown; the underparts are slightly paler.

Distribution

This genus is confined to South America, generally in the southern parts.

Life History and Ecology

These aerial insectivores are found in small (three to seventeen) colonies inside natural cavities or buildings.

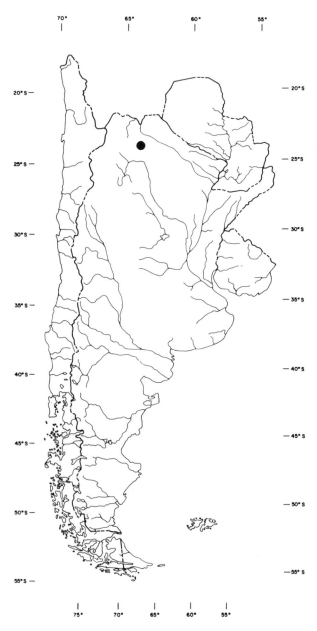

Map 4.28. Distribution of *Eptesicus innoxius*.

Histiotus macrotus (Poeppis, 1835)
Murciélago Orejón Grande

Measurements

	Mean	Min.	Max.	N	Loc.	Source[a]
TL	105.4	94	120.0	9	A, C	1
HB	57.1	48	70.0			
T	48.3	42	55.0			
HF	9.3	5	12.0			
E	31.7	25	37.0			
FA	48.7	44	51.2			
Wta	12.0	11	13.0	3		

[a](1) Olrog 1959; Osgood 1943; Santiago; CM.

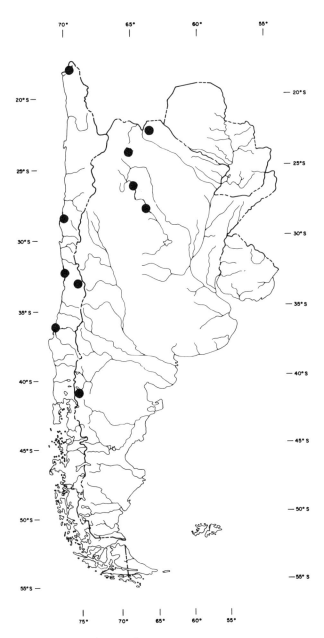

Map 4.29. Distribution of *Histiotus macrotus*.

Description

This medium-sized vespertilionid has very large ears connected at the base by a membrane. When folded down, the ears pass the nose, a character that separates this species from the shorter-eared *H. montanus*. The dorsum is light brown, and the venter is whitish gray (Mann 1978; Osgood 1943; pers obs.).

Distribution

Histiotis macrotus is found in Chile, southern Bolivia, western Argentina, and southern Peru (Honacki, Kinman, and Koeppl 1982; Olrog 1959; Tamayo and Frassinetti 1980) (map 4.29).

Life History

In southern Argentina reproduction seems to be well synchronized. All females examined bore fetuses in only the right uterine horn (Pearson and Pearson 1989).

Ecology

This species has been found roosting in roofs and in cracks in deep mines (Mann 1978; Pearson and Pearson 1989).

Comment

This species was identified as *H. montanus* by Mares, Ojeda, and Kosco (1981) (see also Ojeda and Mares 1989).

• *Histiotus montanus* (Philippi and Landbeck, 1861)
Murciélago Orejón Chico

Measurements

	Mean	Min.	Max.	N	Loc.	Source[a]
HB	61.7	53.0	65.0		U	1
T	47.3	42.0	50.0	6		
HF	8.4	8.0	9.5			
E	21.4	20.0	22.0			
FA	45.0	44.0	48.0			

[a](1) Acosta y Lara 1950.

Description

This medium-sized bat has very large ears not connected by a membrane. It is the only large-eared bat in Uruguay. The dorsum is light grayish brown to brownish tan, and the venter is slightly lighter (Barlow 1965; Osgood 1943; pers. obs.).
Chromosome number: $2n = 50$ (Williams and Mares 1978).

Distribution

Histiotus montanus is distributed from Venezuela south to Chile, Argentina, and Uruguay. In southern South America it is found in Chile south to the Strait of Magellan, in much of Uruguay, and in Ar-

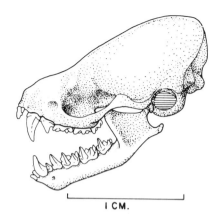

Figure 4.15. Skull of *Histiotus montanus*.

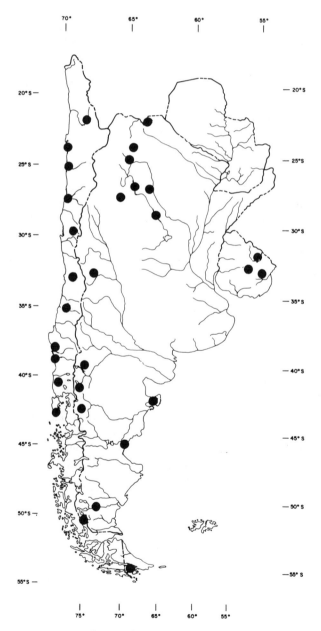

Map 4.30. Distribution of *Histiotus montanus*.

gentina south to at least Santa Cruz province. It has not been recorded in Paraguay but will probably be found to occur there (Acosta y Lara 1950; Barlow 1965; Daciuk 1977; Tamayo and Frassinetti 1980) (map 4.30).

Life History

In Argentine Patagonia pregnant bats are found between August and November. Unlike that of the sympatric *Myotis chiloensis*, their breeding is not tightly synchronized. Implantation is always in the right horn of the uterus, litter size is one, and age of first reproduction is approximately one year (Mann 1978; Pearson and Pearson 1989).

Ecology

This wide-ranging bat is most commonly found in Uruguay in dense woodlands, but it also inhabits zones with few trees. In Chile it is principally a species of savannas and woodlands and demonstrates remarkable tolerance for cold temperatures. *H. montanus* is generally found in small colonies in mines, caves, roofs, and tree holes (Barlow 1965; Greer 1966; Mann 1978; Tamayo and Frassinetti 1980; Villa-R. and Villa Cornejo 1969).

Histiotus velatus (I. Geoffroy, 1824)
Murciélago Orejón Tropical

Measurements

	Mean	Min.	Max.	N	Loc.	Source[a]
HB	62.7	60.0	65	3	U	1
T	42.5	39.0	46			
HF	8.8	8.5	9	2		
E	23.2	22.5	23	3		
Wta	13.0	12.0	14	2	Br	2

[a](1) AMNH; (2) Peracchi 1968.

Description

This bat's large ears are not connected at the base. It is grayish to tannish dorsally, with the gray at the base of the hairs prominent. The venter is appreciably lighter than the dorsum, and in some specimens it is white (pers. obs.).

Distribution

Histiotus velatus is found in southern Brazil, Uruguay, Paraguay, and northern Argentina (Honacki, Kinman, and Koeppl 1982; Massoia 1980; Thomas 1898) (map 4.31).

Life History

In southern Brazil this species begins to reproduce in September (Peracchi 1968).

Ecology

Histiotus velatus in southern Brazil roosts in roofs in groups averaging twenty-three (range, 12–30 $n = 6$); there are always more males than females,

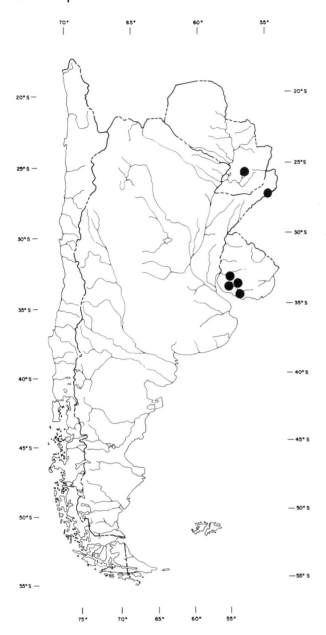

Map 4.31. Distribution of *Histiotus velatus*.

but the composition of sexes at peak reproduction is not recorded (Peracchi 1968).

Genus *Lasiurus* Gray, 1831

Description

The dental formula is usually I 1/3, C 1/1, P 2/2, M 3/3 (fig. 4.16). Species of this genus have a considerable range in size. Head and body length ranges from 50 to 90 mm, tail length from 40 to 75 mm, and the forearm from 37 to 57 mm. Weight ranges from 3 to 6 g. The distinguishing feature of this group is that the dorsal surface of the interfemoral membrane is

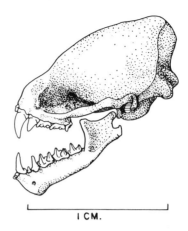

Figure 4.16. Skull of *Lasiurus borealis*.

well haired for at least half its length and usually for its entire length (see fig. 4.4).

Distribution

These bats are confined to the New World and have colonized the Galápagos and Hawaiian Islands. They are distributed from Canada to Argentina.

Life History and Ecology

Bats of the genus *Lasiurus* have been studied in North America, but little is known about their habits in South America. In the north they make seasonal migrations and are aerial insectivores. *L. borealis* has two or three young, thus departing from the general rule of one young characteristic for most of the Chiroptera.

• *Lasiurus borealis* (Müller, 1776)
Red Bat, Murciélago Peludo Rojizo

Measurements

	Mean	Min.	Max.	N	Loc.	Source[a]
TL	96.5	76.0	110.0	27	A, Br, C, P, U	1
HB	51.3	42.0	61.0	28		
T	45.1	30.0	54.0	27		
HF	8.0	7.0	10.0	28		
E	9.4	6.0	14.7	27		
FA	39.1	35.0	42.0			
Wta	10.4	8.5	12.4			

[a](1) Acosta y Lara 1950; Barlow 1965; Crespo 1974; Greer 1966; Mares, Ojeda, and Kosco 1981; Myers and Wetzel 1983; Olrog 1959; Peracchi 1968.

Description

Lasiurus borealis is a moderate-sized lasiurine with long, pointed wings, low, broad ears, and a triangular tragus. The tail is entirely contained within the uropatagium, which is well developed and densely covered with hair on its dorsal surface. This bat is easily recognized because of its bright rufous color, lightly frosted with gray in some individuals.

The venter is lighter than the dorsum. Males are often more brightly colored than females (Barlow 1965; Osgood 1943; Shump and Shump 1982a).

Distribution

Lasiurus borealis has an extremely large range, being found from southern Canada to southern Chile. In southern South America it has been recorded from Uruguay, throughout Paraguay, in Argentina south to about 40°, and in Chile to at least 52° S. This species migrates seasonally and may extend its range farther south during warmer years (Acosta y Lara 1950; Crespo 1974; Myers and Wetzel 1983; Osgood 1943; Rau and Yáñez 1979; Tamayo and Frassinetti 1980) (map 4.32).

Life History

The red bat has a gestation period of eighty to ninety days, a lactation period of about thirty-eight days, and a litter size that ranges from one to five but averages two or three. The young are born at about 0.5 g and open their eyes three to four days after birth (Greer 1966; Mann 1978; Shump and Shump 1982a).

Ecology

Lasiurus borealis is a migratory species, traveling in small groups. There is evidence that the sexes have different summer ranges. These bats roost individually or in small clusters among leaves of shrubs or trees, but they can also be found roosting in rocky areas. In the Northern Hemisphere they are very good hibernators. These insectivorous bats forage primarily in open woodland or edge situations, and in Chile they were found to eat primarily nocturnal lepidopterans (Barlow 1965; Greer 1966; Mann 1978; Schneider 1946; Shump and Shump 1982a).

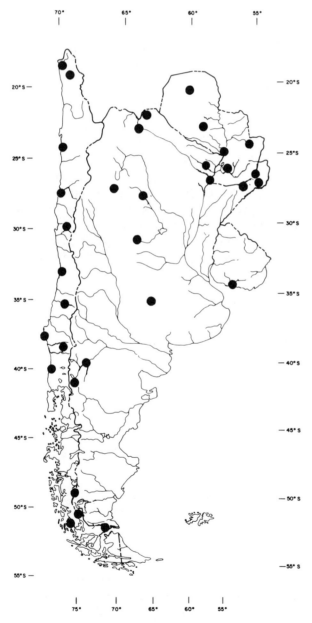

Map 4.32. Distribution of *Lasiurus borealis*.

• *Lasiurus cinereus* (Beauvois, 1796)
Hoary Bat, Murciélago Blanquizco

Measurements

	Mean	Min.	Max.	N	Loc.	Source[a]
TL	126.9	109.0	150.0	11	A, P	1
HB	70.4	62.0	95.5			
T	55.7	43.0	63.0			
HF	10.7	9.4	12.6	12		
E	12.4	10.0	17.1	11		
FA	53.3	46.0	55.0	12		
Wta	18.8	17.0	20.5	2		

[a](1) Mares, Ojeda, and Kosco 1981; Myers and Wetzel 1983; Olrog 1959; PCorps.

Description

Lasiurus cinereus is distinct from all other lasiurines because of its large size. Like other lasiurines, it has a heavily furred interfemoral membrane. Other members of the genus have yellowish to reddish pelage, whereas this species is mixed dark brownish, reddish brown, or grayish tinged with white to produce a frosty or hoary appearance. There are whitish shoulder and wrist patches and a yellowish throat patch (Barlow 1965; Osgood 1943; Shump and Shump 1982b).

Distribution

Like other lasiurines, *L. cinereus* is very widely distributed from southern Canada and Hawaii to Chile and Argentina. In southern South America it is found in Uruguay, Paraguay, and much of Chile and Argentina. Also like other lasiurines, this species moves seasonally, expanding its range south during the warmer months (Acosta y Lara 1950; Barlow 1965; Myers and Wetzel 1983; Sanborn and Crespo 1957; Shump and Shump 1982b; Tamayo and Frassinetti 1980) (map 4.33).

Life History

The hoary bat has a litter size of one to four, with an average of two. In the North Temperate Zone it exhibits delayed implantation. After birth the young cling to the female by day and are left hanging at the roost site during the night. Purposeful flight is first

seen at day thirty-three. In Argentina the young are born from November to December (Sanborn and Crespo 1957; Shump and Shump 1982b).

Ecology

Little is known of the seasonal movements of *L. cinereus*, but it appears that the sexes may forage in different areas except during mating. In Chile it has been speculated that this species may migrate altitudinally. In Salta province, Argentina, these bats are common in gallery forests along permanent streams, and in Malleco province, Chile, they were taken in forested regions and orchards.

Lasiurus cinereus roosts solitarily in trees, apparently preferring conifers and fruit trees in Uruguay. They are strong, swift fliers and appear to show a strong preference for moths (Acosta y Lara 1950; Greer 1966; Mares 1973; Sanborn and Crespo 1957; Shump and Shump 1982b).

• *Lasiurus ega* (Gervais, 1856)
Southern Yellow Bat, Murciélago Leonado

Measurements

	Mean	Min.	Max.	N	Loc.	Source[a]
TL	118.3	111.0	126.0	31 m	P	1
	126.1	117.0	132.0	32 f		
T	50.1	42.0	58.0	31 m		
	51.7	45.0	55.0	32 f		
HF	9.8	8.0	11.0	31 m		
	10.5	9.0	13.0	32 f		
E	18.7	16.0	20.0	31 m		
	19.0	17.0	21.0	32 f		
FA	45.1	42.9	46.9	31 m		
	47.6	46.3	48.9	32 f		
Wta	12.3	10.0	15.0	10		2

[a](1) Myers and Wetzel 1983; (2) UM.

Description

This species is easily separated from the other southern South American species of *Lasiurus* by its pale tan color with a grayish wash both ventrally and dorsally. It has pale ears, a pale tail membrane, and dark wing membranes.

Distribution

Lasiurus ega is distributed from the southern United States to Argentina and Uruguay. In southern South America it is commonly found in Paraguay and Uruguay and in Argentina to about 40° S. There are occasional unsubstantiated records of its appearance in Chile (Barlow 1965; Honacki, Kinman, and Koeppl 1982; Myers and Wetzel 1983; Tamayo and Frassinetti 1980) (map 4.34).

Life History

In western Paraguay breeding begins in the fall, and the young are born in the spring after a gestation

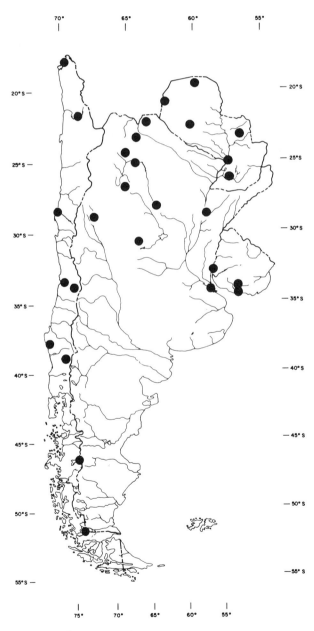

Map 4.33. Distribution of *Lasiurus cinereus*.

period of 3 to 3.5 months. The number of embryos averages 2.9 (range 2–4; $n = 17$). The young are capable of breeding in their first year, and the females can store sperm (Myers 1977).

Ecology

This species often roosts in palms and apparently does not hibernate, though it does undergo daily torpor (Acosta y Lara 1950; Barlow 1965; Myers 1977).

Comment

Some authors refer *L. ega* to the genus *Dasypterus*.

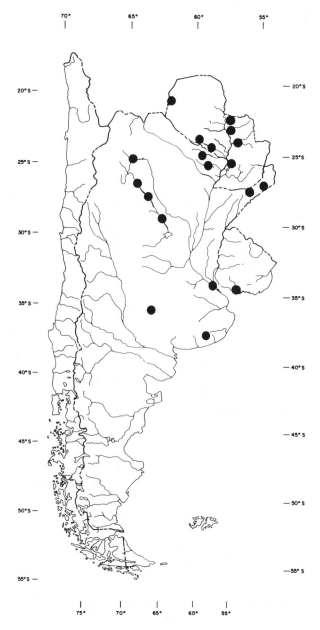

Map 4.34. Distribution of *Lasiurus ega*.

Genus *Myotis* Kaup, 1829
Little Brown Bat

Description

The dental formula is usually I 2/3, C 1/1, P 3/3, M 3/3 (fig. 4.13). Head and body length ranges from 35 to 80 mm, tail length from 40 to 60 mm, and forearm length from 29 to 68 mm. The dorsum is usually some shade of brown, and the underparts are somewhat paler. The Neotropical species were reviewed by LaVal (1973).

Distribution

The genus is worldwide in its distribution except for the Arctic and Antarctic regions.

• *Myotis albescens* (E. Geoffroy, 1806)
Paraguayan Myotis, Murciélago Blancuzo

Measurements

	Mean	Min.	Max.	N	Loc.	Source[a]
TL	86.4	79.0	96.0	133	P	1
	80.5	75.0	90.0	21	A, P, U	2
T	36.6	27.0	40.0	133	P	1
	34.6	30.0	42.0	21	A, P, U	2
HF	9.1	8.0	11.0	133	P	1
E	15.2	12.0	18.0			
FA	34.6	32.5	37.3	133		
	33.9	32.0	37.0	20	A, P, U	2
Wta	5.7	3.0	7.7			
	7.3	6.4	8.2	8 m	U	3
	7.6	7.5	7.7	2 f		

[a](1) Myers and Wetzel 1983; (2) Acosta y Lara 1950; Mares, Ojeda, and Kosco 1981; PCorps; (3) Barlow 1965.

Description

This medium-sized *Myotis* has a short, thin snout. The interfemoral membranes are sparsely furred, with the fur not reaching the knees on the dorsal surface. *M. albescens* is one of the most distinctive of all Neotropical *Myotis* species because of its frosted appearance; its dark brown or black fur is tipped with golden yellow or silvery white. In Uruguay this species has dark gray pelage with yellowish tips on the longest dorsal hairs, though some individuals have a dark brown dorsum. The venter is always whitish shading to ochraceous on the underside of the neck. The wings are dark or gray, and the wing membranes are light brown to black. In Paraguay the venter is quite light, and the fur at the base of the tail membrane is pure white (Acosta y Lara 1950; Barlow 1965; LaVal 1973).

Distribution

Myotis albescens ranges from southern Mexico to Uruguay and northern Argentina. In southern South America it is found throughout Paraguay and Uruguay, and in Argentina it occurs as far south as Mendoza province, although this point may be another species of *Myotis* (Barlow 1965; Myers and Wetzel 1983; Roig 1965) (map 4.35).

Life History

In Paraguay the first births occur in October, followed by copulation and a second pregnancy. Some females may even breed a third time. The gestation period is three months, though it may be slightly less for the second and third pregnancies. The litter size is always one, and lactation lasts about one month. Males probably breed in their first year, and females can store sperm (Myers 1977).

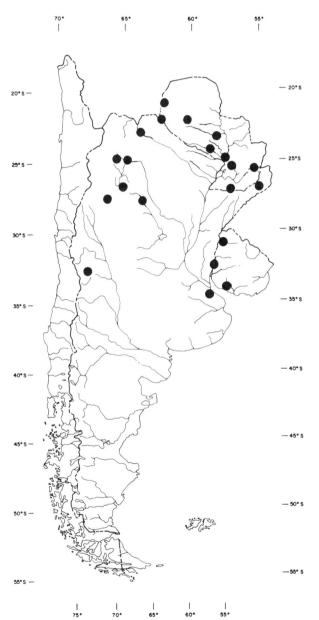

Map 4.35. Distribution of *Myotis albescens*.

Ecology

Although a few specimens of this species are known from above 1,500 m, most have been taken below 500 m. In Uruguay *M. albescens* is found in open woodlands and forages in open grasslands. It roosts in inhabited buildings, walls, crevices in rocks, and trees, typically near fast-moving streams. In Paraguay it is apparently dependent of human habitations, where it typically roosts in large groups. Most likely this bat does not hibernate but does undergo daily torpor (Barlow 1965; González 1973; LaVal 1973; Myers 1977).

Myotis atacamensis (Lataste, 1892)

Measurements

	Mean	Min.	Max.	N	Loc.	Source[a]
TL	88.8	82	94	10	C	1
HB	49.3	47	52			
T	39.4	31	44			
HF	7.8	6	10			
FA	37.8	35	40			

[a](1) Miller and Allen 1928.

Description

This bat is among the smallest of the Neotropical *Myotis*. It has long blond fur and a tiny skull. Sparse fur extends distally on the dorsum of the uropatagium to a point halfway from the knee to the ankle and fur is sparse or absent on the rest of the membranes, which are pale brown (LaVal 1973; Miller and Allen 1928).

Distribution

Myotis atacamensis is known only from southern Peru and northern Chile. In Chile it extends south to about 33° (Tamayo and Frassinetti 1980) (map 4.36).

Ecology

This bat is found from the coastal deserts at sea level to an altitude of 2,400 m (Tamayo and Frassinetti 1980).

Myotis chiloensis (Waterhouse, 1840)

Measurements

	Mean	S.D.	Min.	Max.	N	Loc.	Source[a]
TL	86.3		72	97	15	C	1
T	38.4		28	46			
HF	8.0		6	10	16		
E	12.3		11	14	7		
FA	37.3		33	42			
Wta	6.96	0.57			67 f	A	2
	6.58	0.50			17 m		

[a](1) FM, Santiago; (2) Pearson and Pearson 1989.

Description

This medium-sized bat is the only *Myotis* over the southern two-thirds of Chile but may overlap with *M. atacamensis* in the north. The sparse fur on the dorsal surface on the uropatagium rarely extends past the knees; the rest of the membranes are very sparsely haired or bare. The color varies from blond in the north to dark brown in the south, and the wing membranes vary from light brown to black (LaVal 1973; Osgood 1943).

Distribution

Myotis chiloensis is found in Chile and recently has been recorded in Argentine Patagonia. It ranges from at least 30° S to Tierra del Fuego (Koopman 1967; LaVal 1973; Pearson and Pearson 1989) (map 4.37).

Life History

In Argentine Patagonia this species exhibits highly synchronous breeding with all recorded parturitions in December. In a pattern very similar to that of hibernating vespertilionids in North America, spermatogenesis takes place during summer when females copulate, and they retain sperm over the winter. Implantation is always in the right horn of the uterus, and after a gestation period of approxi-

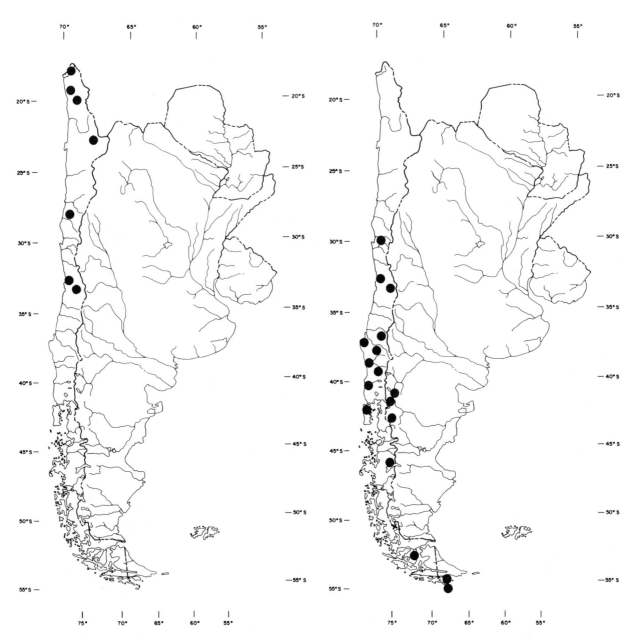

Map 4.36. Distribution of *Myotis atacamensis*.

Map 4.37. Distribution of *Myotis chiloensis*.

mately fifty-three days a single young weighing 2.3 g ($n = 1$) is born. Most young females are not pregnant, but a few conceive in their first year (Pearson and Pearson 1989).

Ecology

Distributed over a large latitudinal range. *M. chiloensis* has been collected in matorral, savanna, and moist forest. It appears to be uncommon in all areas but is generally found near open fresh water, where it forages for insects. It roosts in colonies in roofs, caves, and crevices, and females form nursery colonies. Bats of this species store fat (some males attained a proportion of 6%–13% fat by weight) and may overwinter singly or in small groups, but apparently not in large groups. They pass the winter in hibernation interrupted by occasional bouts of feeding during mild weather (Greer 1966; Mann 1978; Pearson and Pearson 1989; Tamayo and Frassinetti 1980).

Myotis levis (I. Geoffroy, 1824)
Murciélago Común

Measurements

	Mean	Min.	Max.	N	Loc.	Source[a]
TL	88.3	77	97.0	13	A, P	1
HB	51.8	44	58.0			
T	36.5	30	42.0			
HF	7.8	7	9.3			
E	14.1	11	16.0			
Wta	4.6	4	5.4			

[a](1) CM, PCorps.

Description

This pretty little bat is mottled chestnut to tan. There is a dark cast to the dorsum because the dark hair bases show through. The belly is gray washed with white (pers. obs.).

Distribution

Myotis levis is found in northern and central Argentina, southeastern Brazil, and Uruguay (Honacki, Kinman, and Koeppl 1982) (map 4.38).

Ecology

This bat is locally common in Uruguay, where it forages primarily over open country, in open woodland, or over streams. It characteristically roosts in colonies numbering in the thousands (Barlow 1965).

• *Myotis nigricans* (Schinz, 1821)
Black Myotis, Murciélago Castaño

Measurements

	Mean	Min.	Max.	N	Loc.	Source[a]
TL	81.8	71.0	91.0	110	P	1
	83.3	78.0	89.0	14 m	U	2
	84.4	79.0	93.0	17 f		

T	33.7	25.0	40.0	110	P	1
	38.1	35.0	40.0	14 m	U	2
	38.4	34.0	45.0	17 f		
HF	8.1	6.0	10.0	110	P	1
E	14.4	11.0	18.0			
FA	32.7	29.8	35.8			
	38.4	33.0	42.0	7	P	3
Wta	5.2	4.8	6.0	14 m	U	2
	5.6	4.8	6.4	14 f		

[a](1) Myers and Wetzel 1983; (2) Barlow 1965; (3) PCorps.

Description

This small vespertilionid has silky, rarely woolly, fur that is highly variable in color, ranging from

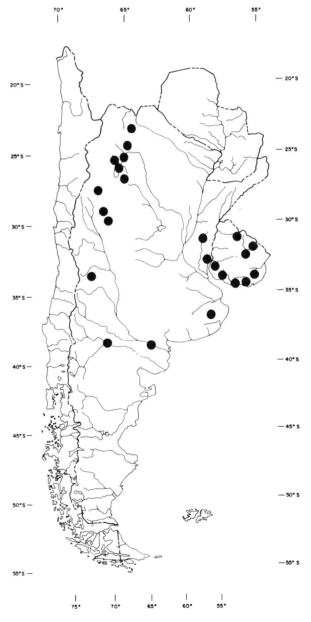

Map 4.38. Distribution of *Myotis levis*.

brown to black, although reddish molts are known. In general the underparts are slightly lighter, and the flight membranes do not contrast sharply with the pelage color (Barlow 1965; LaVal 1973).

Distribution

Myotis nigricans is found from southern Mexico to northern Argentina. In southern South America its range includes all of Paraguay and the northern provinces of Argentina (Bárquez 1987; Myers and Wetzel 1983; Wilson and LaVal 1974) (map 4.39).

Life History

In Paraguay breeding appears to occur during most of the year, though most females copulate in May to December, with the first birth peak coming in late October. The second peak occurs in late December and early January. There is the possibility of a third peak for some females. The gestation period is slightly less than three months, but there may be a delay in implantation of the blastocyst. Litter size is one, and the lactation lasts about one month. Males can breed year round, females store sperm, and both males and females can breed at about four months of age.

Young remain attached to their mothers for the first two or three days and then are left behind in large groups or "crèches" when the mothers go to fly and feed. Adult weight is reached by week two, and

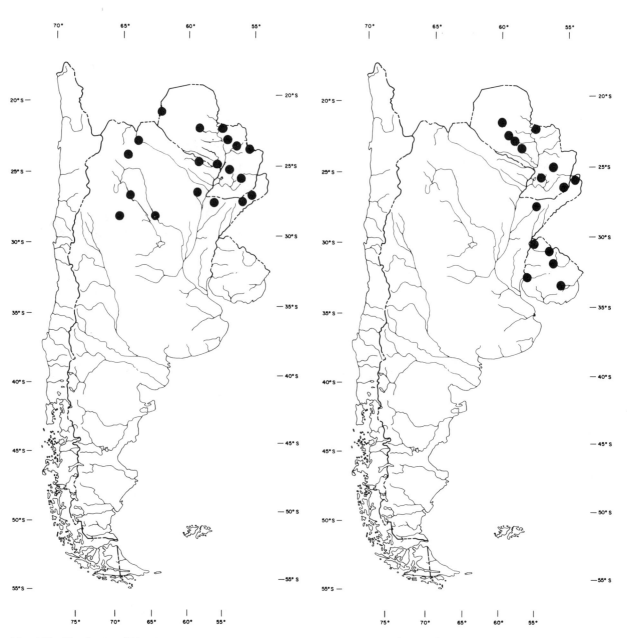

Map 4.39. Distribution of *Myotis nigricans*.

Map 4.40. Distribution of *Myotis riparius*.

flight begins in week three. Some individuals are known to reach seven years of age in the wild (Myers 1977; Wilson and LaVal 1974).

Ecology

Myotis nigricans occurs in all the vegetation zones within its large range, but it appears to be uncommon over 1,200 m. In Uruguay it shows a distinct preference for subtropical woodland. In Paraguay this species depends on human habitation for roosts and forages above man-made cattle ponds. Males are often found roosting singly outside the breeding season, whereas females tend to occur in groups. In one study there was an indication of

stable groups with the possibility of harems (Barlow 1965; LaVal 1973; Myers 1977).

• *Myotis riparius* Handley, 1960
Murciélago Ocráceo

Measurements

	Mean	Min.	Max.	N	Loc.	Source[a]
TL	82.0			3 m	P	1
	87.0	86.0	88	2 f		
T	35.0			3 m		
	38.0	38.0	38	2 f		
HF	8.0		9	3 m		
	9.0	8.0		2 f		
E	15.0			3 m		
	15.0	14.0	16	2 f		
FA	33.4	32.8	34	3 m		
	33.9	33.8	39	2 f		
Wta	4.6	45.0	72			

[a](1) Myers and Wetzel 1983; (2) UM.

Description

This small to medium-sized *Myotis* has short, woolly fur, varying from dark gray to bright cinnamon dorsally and slightly lighter ventrally. This species can be confused with only one other *Myotis*, *M. simus*. *M. riparius* has longer fur, which can be bicolored, whereas *M. simus* is always monocolored (LaVal 1973).

Distribution

Myotis riparius is found from Honduras to Uruguay, northern Argentina, and Paraguay, where it is widely distributed (Honacki, Kinman, and Koeppl 1982; LaVal 1973; Myers and Wetzel 1983; Ximénez, Langguth, and Praderi 1972) (map 4.40).

Ecology

This appears to be primarily a lowland species of *Myotis* (LaVal 1973).

Myotis ruber (E. Geoffroy, 1806)
Murciélago Ruber

Measurements

	Mean	Min.	Max.	N	Loc.	Source[a]
TL	91.2	88.0	95.2	11	Br, P, U	1
HB	49.8	48.0	53.0			
T	41.4	38.0	46.6			
HF	8.8	7.8	10.0			
E	15.0	13.0	17.0	4		
FA	40.2	39.0	40.8	8		
Wta	7.3	6.0	8.0			

[a](1) Acosta y Lara 1950; Miller and Allen 1928; UM.

Description

M. ruber is a Neotropical *Myotis* with reddish, monocolored fur of medium length, thick and silky in texture. Fur is thick on the basal third of the dorsal surface of the uropatagium and extends one-third to one-half the distance from the knee to the ankle.

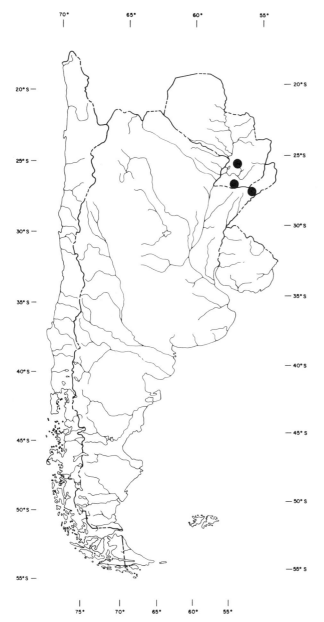

Map 4.41 Distribution of *Myotis ruber.*

Within its known range this species is larger than *M. nigricans*, *M. albescens*, and *M. riparius*. Of these, only *M. riparius* is similar in color, but it is much smaller (LaVal 1973).

Distribution

This species has a fairly limited range, being known only from southeastern Brazil, Paraguay, and northeastern Argentina (Honacki, Kinman, and Koeppl 1982) (map 4.41).

• *Myotis simus* Thomas, 1901

Measurements

	Mean	Min.	Max.	N	Loc.	Source[a]
TL	97.3	96.0	100.0	4 f	P	1
	81.5	78.0	85.0	2	A	2
T	38.8	37.0	41.0	4 f	P	1
	35.0	30.0	40.0	2	A	2
HF	10.0	9.0	11.0	4 f	P	1
E	14.3	13.0	15.0			
FA	38.7	38.4	39.1			
	37.0	33.0	41.0	2	A	2
Wta	8.0			1		3

[a](1) Myers and Wetzel 1983; (2) Fornes 1972; (3) UM.

Description

This medium-sized *Myotis* has extremely short, woolly fur with the dorsum usually bright orange or somewhat duller and the venter slightly lighter. The fur barely extends onto the uropatagium, and the other membranes are bare; the membrane color ranges from brown to black, and the ears are light (LaVal 1973).

Distribution

Myotis simus is distributed over much of the Amazon basin and south into Bolivia and Paraguay and northern Argentina. In Paraguay it may be restricted to the basin of the Río Paraguay and the lower Chaco (Fornes 1972b; Honacki, Kinman, and Koeppl 1982; Myers and Wetzel 1983) (map 4.42).

Ecology

Individuals of this species have been caught in a tree roost with *Noctilio albiventris* (Fornes 1972b).

FAMILY MOLOSSIDAE

Free-tailed Bats, Mastiff Bats, Murciélagos de Cola de Ratón

Diagnosis

The family contains ten genera and about eighty species. Head and body length range from 40 to 130 mm, tail from 14 to 80 mm, and forearm from 27 to 85 mm. The fibula is well developed, supporting the lower leg. In these bats the tail extends beyond the edge of the interfemoral membrane, and this single character is diagnostic for the entire family; hence the common name "free-tailed bat" (see fig. 4.4).

Distribution

This family is worldwide in its distribution but is mainly confined to tropical regions.

Natural History

These aerial insectivores are swift, high fliers, and their wings are long and narrow. Freeman (1981) has made a thorough morphometric analysis of the family. Jaw size covaries positively with body size, suggesting that there is great specialization for certain size classes of prey. Thus when several species co-occur they tend to present an array of sizes that

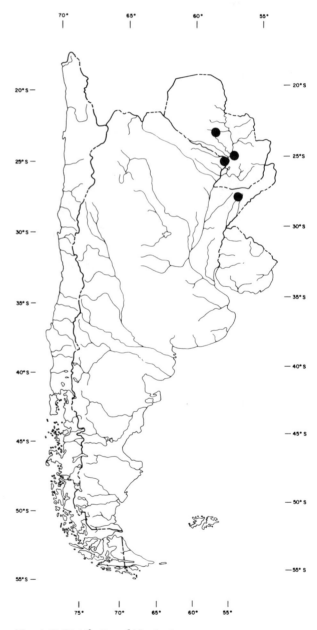

Map 4.42 Distribution of *Myotis simus*.

probably reflects this specialization. Beetles, Coleoptera or moths, and Lepidoptera seem to predominate in their diet. Most species nest in caves, tunnels, or hollow trees, and some are strongly colonial. In the northern parts of their range these bats may be seasonally migratory over moderate distances (see *Tadarida brasiliensis*).

Genus *Eumops* Miller, 1906
Bonneted Bat, Mastiff Bat

Description
The dental formula is I 1/2, C 1/1, P 2/2, M 3/3 (see fig. 4.17). There are eight species, showing an extreme range in size. Head and body length may vary between 40 and 130 mm, and the tail ranges from 35 to 80 mm. The large ears are rounded and usually connected across the head at the base. Males of some species have throat sacs that produce glandular secretions when in breeding condition. The genus has been treated in a monograph by Eger (1977).

Distribution
Species of this genus are distributed from southern California and Florida in the United States south through Central America to southern South America as far as northern Argentina.

• *Eumops auripendulus* (Shaw, 1800)
Slouch-eared Bat, Moloso Alilargo

Measurements

	Mean	Min.	Max.	N	Loc.	Source[a]
TL	136.8	128	146	10	A, P	1
HB	83.3	68	93			
T	53.0	45	60			
HF	11.3	9	14			
E	21.1	12	24	9		
FA	64.1	62	66	6		

[a](1) Fornes 1964; Fornes and Massoia 1968; Massoia 1976; PCorps.

Description
This medium-sized molossid is blackish brown to dark chestnut brown, slightly lighter on the venter,

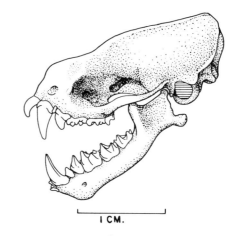

Figure 4.17. Skull of *Eumops glaucinus*.

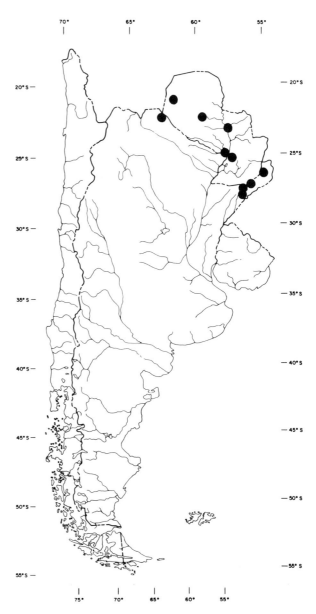

Map 4.43. Distribution of *Eumops auripendulus*.

Table 4.4   Key to the Common Genera of the Molossidae of Southern South America

1 One upper premolar per side 2	
1' Two upper premolars per side 3	
2 One lower incisor per side .... *Molossus*	
2' Two lower incisors per side; ears separate .... *Molossops*	
3 Upper lip grooved and separable into halves ..... *Tadarida*	
3' Upper lip not divided 4	
4 Ears almost reaching snout tip .... *Eumops*	
4' Ears falling far short of snout tip .... *Promops*	

Note: Tadarida includes *Nyctinomops.*

with dark wing membranes. It has long hairs on the feet extending past the tips of the claws (Eger 1977; Massoia 1976).

Distribution

Eumops auripendulus is found from southern Mexico to northern Argentina and Paraguay. In southern South America it has been recorded from Paraguay and the Argentine provinces of Salta and Misiones (Eger 1977; Fornes 1964; Massoia 1976; Myers and Wetzel 1983) (map 4.43).

Ecology

In Argentina this species inhabits subtropical forest (Massoia 1976).

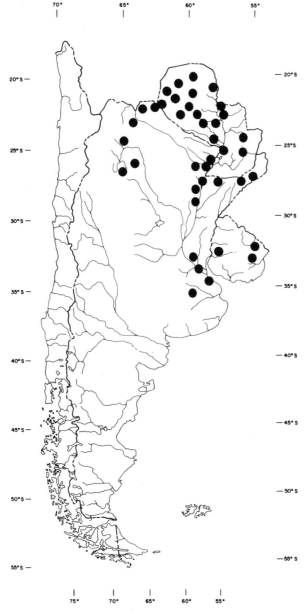

Map 4.44. Distribution of *Eumops bonariensis*.

• *Eumops bonariensis* (Peters, 1874)
Peter's Mastiff Bat, Moloso Orejiancho

Measurements

	Mean	Min.	Max.	N	Loc.	Source[a]
TL	101.2	92.0	114.0	27 m	P	1
	99.0	92.0	110.0	40 f		
T	35.2	30.0	40.0	27 m		
	33.8	27.0	39.0	40 f		
HF	9.7	8.0	11.0	27 m		
	9.3	7.0	10.0	40 f		
E	19.9	18.0	23.0	27 m		
	19.2	18.0	22.0	40 f		
FA	44.2	41.9	46.2	27 m		
	43.6	40.2	45.7	40 f		
Wta	12.6	11.0	19.0	10	A, P	2

[a](1) Myers and Wetzel 1983; (2) Mares, Ojeda, and Kosco 1981; PCorps.

Description

This is the smallest species of *Eumops*. The dorsum is dark chocolate brown, with a paler venter and dark brown membranes. The color varies geographically in Paraguay; bats from the northwestern Chaco are lighter than those from the eastern Chaco and eastern Paraguay (Barlow 1965; Eger 1977; Myers and Wetzel 1983).

Distribution

Eumops bonariensis ranges from southern Mexico to central Argentina. In southern South America it is found throughout Paraguay, from only certain locations in Uruguay, and in Argentina as far south as Buenos Aires province (Barlow 1965; Crespo 1958; Eger 1977; Myers and Wetzel 1983) (map 4.44).

Ecology

This is one of the common house bats of the Paraguayan Chaco, roosting in roofs. In Uruguay it is commonly netted in open thorn woodlands, and it emits, both in flight and at rest, a series of distinctive high-pitched peeps audible to humans (Barlow 1965; Myers and Wetzel 1983).

• *Eumops dabbenei* Thomas, 1914
Moloso Grande

Measurements

	Mean	Min.	Max.	N	Loc.	Source[a]
TL	189.7	188.0	192.0	3 m	P	1
	187.5	184.0	191.0	3 f		
T	63.3	62.0	66.0	3 m		
	62.5	59.0	66.0	3 f		
HF	18.0	18.0	18.0	3 m		
	18.0	18.0	18.0	3 f		
E	31.7	31.0	32.0	3 m		
	29.0	28.0	30.0	3 f		
FA	78.1	78.0	78.1	3 m		
	78.1	78.7	78.1	3 f		
Wta	74.3	50.0	100.0	3		2

[a](1) Myers and Wetzel 1983; (2) Ibañez 1979.

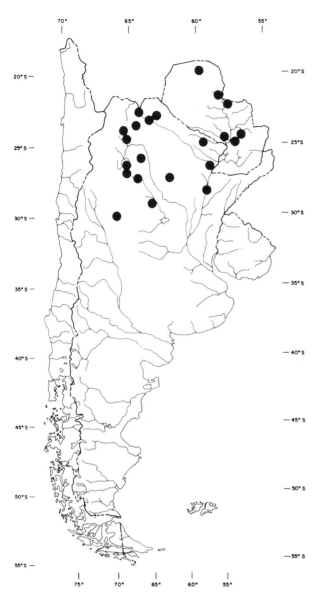

Map 4.45. Distribution of *Eumops dabbenei*.

Description

This large molossid can easily be distinguished from *E. perotis* by its shorter ears, pointed tragus, and more massive skull. Males have greatly enlarged gular glands. It is light brown to orangish brown dorsally and ventrally, with dark ears and wing membranes (Harrison, Pendleton, and Harrison 1979; Myers and Wetzel 1983).

Distribution

Eumops dabbenei is known from Colombia, Venezuela, Paraguay, and Argentina. In Paraguay it is found in the Chaco, and in Argentina it has been collected from Chaco, Tucumán, and Santa Fe provinces (Bárquez 1987; Harrison, Pendleton, and Harrison 1979; Honacki, Kinman, and Koeppl 1982; Myers and Wetzel 1983) (map 4.45).

Ecology

In Paraguay this species roosts in hollow trees (Myers and Wetzel 1983).

Comment

Eumops underwoodi is included in this species (Ibañez 1979).

• *Eumops glaucinus* (Wagner, 1843)
Wagner's Mastiff Bat, Moloso Negro

Measurements

	Mean	Min.	Max.	N	Loc.	Source[a]
TL	149.0	142.0	153.0	3 m	P	1
	144.0	142.0	146.0	2 f		
T	51.3	49.0	53.0	3 m		
	51.0	50.0	52.0	2 f		
HF	14.0	14.0	14.0	3 m		
	13.5	13.0	14.0	2 f		
E	28.0	28.0	29.0	3 m		
	27.0	26.0	28.0	2 f		
FA	61.6	60.8	62.7	3 m		
	60.1	59.8	60.4	2 f		
Wta	11.0	10.0	12.0	2	P	2

[a](1) Myers and Wetzel 1983; (2) PCorps.

Description

The dorsum of this species is light brown to chestnut, and the venter is much lighter. The contrast between the dorsum and the venter is much more pronounced than in the similar *E. auripendulus* (Massoia 1976).

Distribution

Eumops glaucinus is distributed from southern Florida to northern Argentina. In southern South America it is known from the Paraguayan Chaco and northwestern Argentina (Eger 1977; Massoia 1976; Myers and Wetzel 1983) (map 4.46).

Ecology

This bat is apparently a typical inhabitant of subtropical forests (Massoia 1976).

• *Eumops perotis* (Schinz, 1821)
Greater Mastiff Bat, Moloso Gigante

Measurements

	Mean	Min.	Max.	N	Loc.	Source[a]
TL	169.2	150.0	188.0	16	A, P	1
	178.0	171.0	185.0	3 m	P	2
T	56.4	50.8	62.3	16	A, P	1
	59.5	59.0	60.0	3 m	P	2
HF	17.3	14.2	20.0	16	A, P	1
	17.0	16.0	18.0	3 m	P	2
E	40.1	34.5	44.0	16	A, P	1
FA	77.9	74.0	81.0	17		
	81.6	80.5	83.2	3 m	P	2
Wta	64.0	45.5	73.0	4	B, P	3

[a](1) Crespo 1958; Massoia 1976; Myers and Wetzel 1983; Villa-R. and Villa Cornejo 1969; PCorps; (2) Myers and Wetzel 1983; (3) Massoia 1976; UM.

Description

Eumops perotis has large ears connected at the bases by a membrane, and its tail protrudes above and beyond the tail membrane. It is light chocolate brown with white at the bases of the hairs showing through both ventrally and dorsally. The venter is often ashy brown, and the wings, tail, and ears are dark.

Distribution

Eumops perotis is distributed from the United States to northern Argentina and eastern Brazil. In southern South America it is found in Paraguay and broadly distributed over northern and north-central Argentina (Baud 1981; Eger 1977) (map 4.47).

Genus *Molossops* Peters, 1865
Malaga's Free-tailed Bat

Description

The dental formula is I 1/2, C 1/1, P 1/2, M 3/3 (but see *M. neglectus*). Head and body length ranges from 40 to 95 mm, and the tail shows a corresponding range from 14 to 30 mm. The forearm averages 28–51 mm. The dorsum tends to be a yellow brown to chocolate brown, and the underparts usually contrast, being gray or slate colored.

Distribution

The seven species within this genus are distributed from southern Mexico through Central America to northern Argentina.

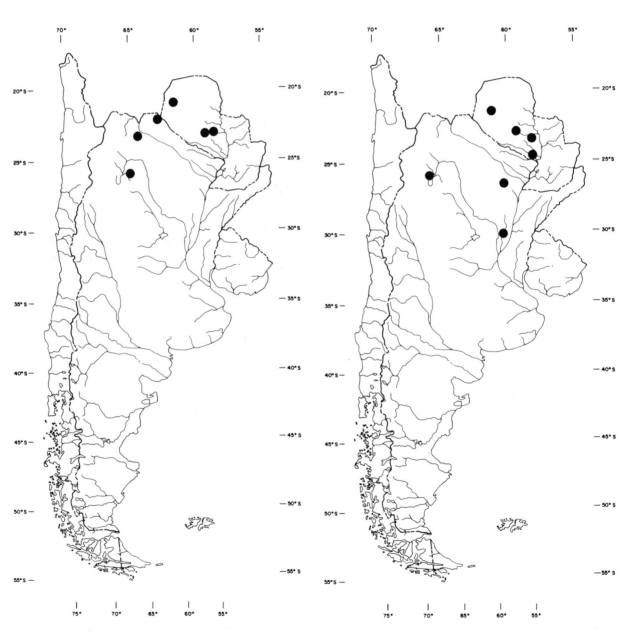

Map 4.46. Distribution of *Eumops glaucinus*.

Map 4.47. Distribution of *Eumops perotis*.

Life History and Ecology

Species of *Molossops* are specialized for feeding on insects. Some are highly gregarious and may live in colonies of up to seventy-five. In forested habitats these bats tend to roost in hollow trees.

• *Molossops abrasus* (Temminck, 1827)
Moloso Chico

Measurements

	Mean	Min.	Max.	N	Loc.	Source[a]
TL	121.1	113.0	130	14	P	1
HB	82.9	73.0	89			
T	37.6	33.0	42			
HF	11.2	9.0	13			
E	19.5	18.0	23			
FA	48.1	48.1		1		
Wta	33.0	27.0	42	12		

[a](1) Myers and Wetzel 1983; UM.

Description

This bat is reddish brown dorsally and ventrally, with a faint yellowish wash on some individuals. The ears and wing membranes are dark and contrast with the dorsum (pers. obs.).

Distribution

Molossops abrasus is found from Venezuela south to northern Argentina. In southern South America it has been collected in Paraguay and in Misiones province, Argentina (Honacki, Kinman, and Koeppl 1982; Massoia 1980; Myers and Wetzel 1983) (map 4.48).

Ecology

In Paraguay, *M. abrasus* has been caught in riparian forest over streams as well as in thorn forest (UM).

Comment

Some authors refer *Molossops abrasus* to *M. brachymeles*.

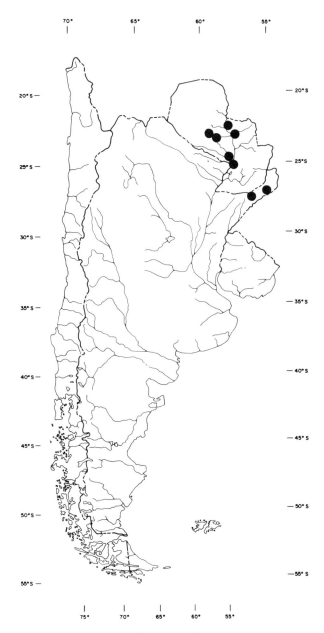

Map 4.48. Distribution of *Molossops abrasus*.

• *Molossops planirostris* (Peters, 1865)
Dog-faced Bat, Moloso Hocico Aplanado

Measurements

	Mean	Min.	Max.	N	Loc.	Source[a]
HB	55.7	51.0	59.0	15 m	Br	1
	52.9	50.0	54.5	15 f		
T	25.2	21.5	28.0	15 m		
	25.6	23.5	28.0	15 f		
E	13.3	12.5	14.0	15 m		
	12.7	11.5	14.0	15 f		
FA	31.6	30.0	34.0	15 m		
	30.7	29.5	31.5	15 f		
Wta	8.9	5.9	10.5	13 m		
	6.9	5.5	8.2	2 f		

[a](1) Vizotto and Taddei 1976.

Description

This very small molossid is pale brown dorsally with the white at the bases of the hairs showing through. Ventrally it is white on the chin and chest, with the white continuing in a medial line to the base of the tail. The rest of the venter is colored like the dorsum.

Distribution

Molossops planirostris ranges from Panama to northern Argentina. In southern South America it is found in Paraguay and in the Argentine province of Salta (Honacki, Kinman, and Koeppl 1982; Myers and Wetzel 1983; Olrog and Bárquez 1979) (map 4.49).

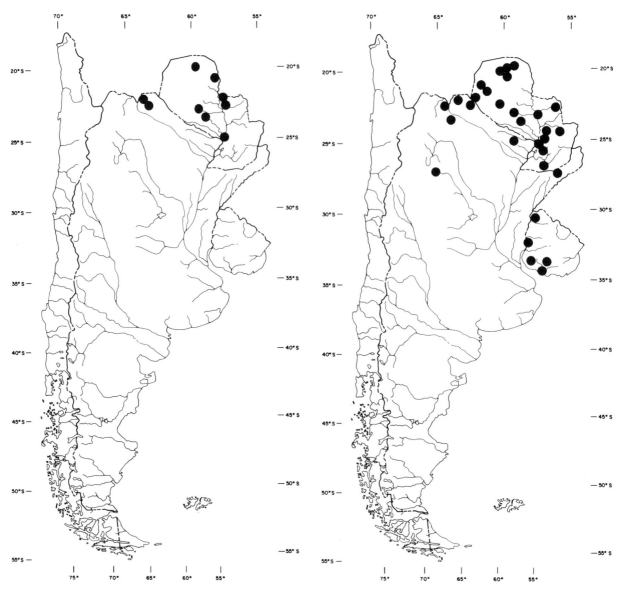

Map 4.49. Distribution of *Molossops planirostris*.

Map 4.50. Distribution of *Molossops temminckii*.

Life History

This species has a litter size of one (Vizotto and Taddei 1976).

Ecology

Molossops planirostris is a fast-flying aerial insectivore that roosts in cavities in trees and posts in colonies of up to eight. It never seems to roost with other species (Vizotto and Taddei 1976; Willig 1983).

• *Molossops temminckii* (Burmeister, 1854)
Moloso Pigmeo

Measurements

	Mean	Min.	Max.	N	Loc.	Source[a]
TL	73.8	66.0	79.0	48 m	P	1
	72.3	61.0	85.0	50 f		
	69.3	61.0	85.0	42	A, P, U	2
T	26.0	20.0	33.0	48 m	P	1
	22.9	14.0	29.0	50 f		
	21.8	17.0	30.0	42	A, P, U	2
HF	7.6	5.0	9.0	48 m	P	1
	7.5	6.0	9.0	50 f		
E	13.0	10.0	15.0	48 m		
	13.1	10.0	15.0	50 f		
FA	30.0	27.8	32.8	48 m		
	29.7	28.5	31.0	50 f		
	29.8	28.0	32.0	35	A, P, U	2
Wta	5.6	5.2	6.5	5 m	A	3
	5.4	5.0	5.6	5 f	A	3

[a](1) Myers and Wetzel 1983; (2) Barlow 1965; Villa-R. and Villa Cornejo 1969; Ximénez 1969; PCorps; (3) Mares, Ojeda, and Kosco 1981.

Description

Molossops temminckii is distinguished from the similar *M. planirostris* by shorter, wider wings and ears set farther apart (4 mm vs. 2 mm). This small molossid exhibits striking variation in Paraguay; individuals in the northwestern Chaco are light brown dorsally and tan ventrally and are smaller, whereas those in the eastern Chaco are blackish dorsally and ventrally and are 10% larger. In Uruguay they are brown above and paler ventrally, with pale brown ears and wing membranes (Barlow 1965; Myers and Wetzel 1983; Vizotto and Taddei 1976).

Distribution

This species ranges from Colombia to northern Argentina and Uruguay. In southern South America it is found throughout most of Paraguay, in several locations in Uruguay, and across the northern portion of Argentina (Honacki, Kinman, and Koeppl 1982; Myers and Wetzel 1983; Olrog and Lucero 1981) (map 4.50).

Life History

Molossops temminckii has a litter size of one (Mares, Ojeda, and Kosco 1981; Vizotto and Taddei 1976).

Ecology

In Paraguay most individuals were caught in thorn scrub over ponds or at the edge of clearings, and in Salta province, Argentina, this is the most common bat in Chaco vegetation when free water is present. It may be seen flying before sunset, and it roosts in roofs, tree hollows, and fence posts in colonies with a maximum of three individuals of both sexes (Mares, Ojeda, and Kosco 1981; Myers and Wetzel 1983; Vizotto and Taddei 1976).

Genus *Molossus* E. Geoffroy, 1805

Description

The dental formula is I 1/1, C 1/1, P 1/2, M 3/3. Head and body length ranges from 50 to 95 mm, tail length from 20 to 70 mm, and the forearm from 33 to 60 mm. The general dorsal coloration is reddish brown to dark chestnut brown. It has been remarked that several species have two color phases, dark brown to almost black, and a lighter brown phase.

Distribution

This New World genus is distributed from northern Mexico south to northern Argentina and Uruguay.

• *Molossus ater* E. Geoffroy, 1805
Red Mastiff Bat, Moloso Coludo

Measurements

	Mean	Min.	Max.	N	Loc.	Source[a]
TL	129.8	121.0	140.0	18	P	1
HB	85.4	75.0	95.0			
T	44.4	40.0	48.0			
HF	13.0	10.0	15.0			
E	16.7	11.4	18.0			
FA	49.7	48.0	51.5	4 m		
	47.8	45.2	50.0	4 f		
	50.0	46.6	48.7	4	A	
Wta	30.3	25.0	38.0	14	P	

[a](1) Myers and Wetzel 1983; Villa-R. and Villa Cornejo 1969; UM.

Description

This medium-sized dark mollosid is dark brown ventrally and dorsally, with the tan at the bases of the hairs showing through.

Distribution

Molossus ater ranges from Mexico south to northern Argentina. In southern South America it has been recorded in Paraguay and in the northern-

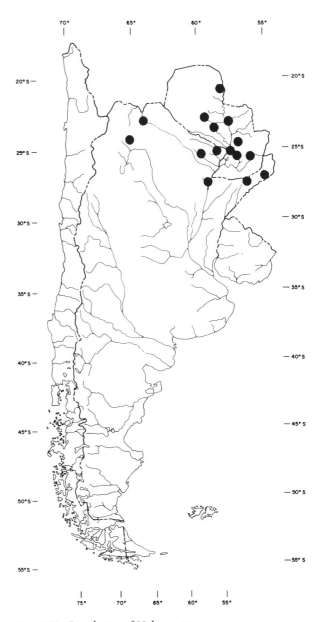

Map 4.51. Distribution of *Molossus ater*.

most provinces of Argentina (Honacki, Kinman, and Keoppl 1982; Massoia 1970, 1980; Myers and Wetzel 1983; Romaña and Ábalos 1950; Villa-R. and Villa Cornejo 1969) (map 4.51).

Life History

In the Brazilian Amazon females are reproductively active throughout most of the year, and pregnancy lasts two to three months (Marques 1986).

Ecology

This molossid roosts in roofs and small hollows in trees (Myers and Wetzel 1983; Villa-R. and Villa Cornejo 1969).

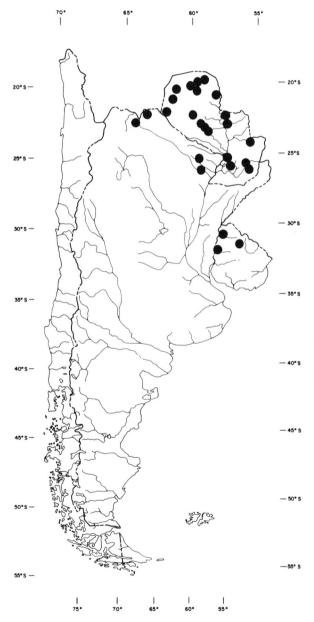

Map 4.52. Distribution of *Molossus molossus*.

Comment

Some authors refer *Molossus ater* to *M. rufus*.

• *Molossus molossus* (Pallas, 1766)
Pallas's Mastiff Bat

Measurements

	Mean	Min.	Max.	N	Loc.	Source[a]
TL	101.8	96.0	110.0	24 m	P	1
	97.5	90.0	106.0	47 f		
	115.0	108.0	121.0	9 m	U	2
	118.0	110.0	120.0	10 f		
T	34.7	29.0	41.0	24 m	P	1
	35.4	31.0	39.0	47 f		
	41.8	36.0	47.0	9 m	U	2
	39.8	33.0	45.0	10 f		
HF	9.3	7.0	10.0	24 m	P	1
	9.0	7.0	11.0	47 f		
	10.9	9.0	12.0	9 m	U	2
	11.0	10.0	12.0	10 f		
E	13.0	11.0	14.0	24 m	P	1
	12.8	12.0	14.0	47 f		
FA	39.1	36.2	40.1	24 m		
	38.7	34.3	42.1	47 f		
	41.6	40.7	42.5	14 m	B	3
	40.9	39.3	42.7	43 f		
Wta	19.7	17.5	21.7	9 m	U	2
	18.8	16.1	21.0	10 f		

[a](1) Myers and Wetzel 1983; (2) Barlow 1965; (3) Bárquez 1983b.

Description

This fairly small molossid is grayish brown to dark chocolate brown on the dorsum and wing membranes and slightly lighter on the venter (Barlow 1965).

Distribution

Molossus molossus ranges from southern Mexico to northern Argentina and Uruguay. In southern South America it is found in Uruguay, throughout most of Paraguay, and in northern Argentina (Barlow 1965; Mares, Ojeda, and Kosco 1981; Massoia 1970; Myers and Wetzel 1983) (map 4.52).

Life History

Young bats emit cries that are recognizable by their own mothers. The young nurse for sixty-five days, achieve adult forearm length in sixty days, and begin thermoregulation at twenty days of age (Haussler, Moller, and Schmidt 1981).

Ecology

In northern South America reproduction takes place at the onset of the rainy season, and females form nursery colonies for rearing the young. Although all the young hang as a compact group while the mothers forage, each female can identify her own upon returning.

In Uruguay this bat inhabits subtropical woodlands and edge situations. It has been found roosting in buildings, in small groups and singly. Bats from

Bolivia were found to be entirely insectivorous, eating mostly beetles (Barlow 1965; Bárquez 1983b; Eisenberg 1989; Myers and Wetzel 1983).

Comment

In the past some specimens have been referred to *M. major*.

Genus *Mormopterus* Peters, 1865
Mormopterus kalinowskii (Thomas, 1893)

Comment

This molossid has a small range, being described from Peru and northern Chile (Tamayo and Frassinetti 1980) (map 4.53).

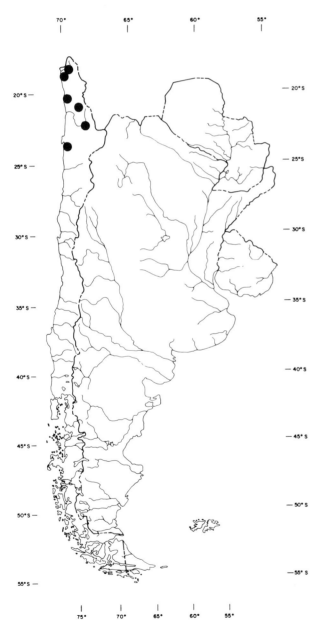

Map 4.53. Distribution of *Mormopterus kalinowskii*.

Genus *Tadarida* Rafinesque, 1814
Free-tailed Bat

Description

The dental formula is I 1/2, C 1/1, P 2/2, M 3/3. This genus contains approximately thirty-five species, with considerable variation in body size. Head and body length ranges from 45 to 100 mm, the tail from 20 to 60 mm, and the forearm from 27 to 65 mm. Males have a small glandular throat sac. The dorsum is generally reddish brown to black, and the venter tends to be somewhat paler.

Distribution

This genus is distributed in the subtropical and tropical portions of both the Old World and the New World.

• *Tadarida brasiliensis* (I. Geoffroy, 1824)
Brazilian Free-tailed Bat, Moloso Común

Measurements

	Mean	Min.	Max.	N	Loc.	Source[a]
TL	104.9	98.0	110.0	14	A	1
T	37.0	34.0	41.0			
HF	8.0	7.5	8.5			
E	17.2	15.0	18.5			
FA	43.5	41.0	47.0			
Wta	13.2	10.0	15.5			

[a](1) Fornes and Massoia 1967.

Description

The dental formula and skull are distinctive (fig. 4.18). This medium-sized molossid is dark brown dorsally and slightly lighter ventrally and has thick, leathery ears (Barlow 1965; Osgood 1943).

Distribution

Tadarida brasiliensis is found from the United States to southern Argentina and south-central Chile. In southern South America it is found throughout Uruguay, in eastern and probably also

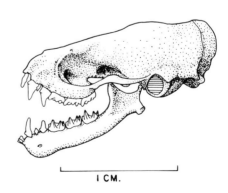

Figure 4.18. Skull of *Tadarida brasiliensis*.

western Paraguay, in Chile south to Valdivia, and in Argentina south to at least the province of Chubut (Barlow 1965; Mann 1978; Myers and Wetzel 1983; Romaña and Ábalos 1950; Tamayo and Frassinetti 1980) (map 4.54).

Life History

In North America a single young is born and is left in a crèche while the mother feeds. Young females may reproduce in the year following their birth. In Chile females give birth to one young between August and November (Davis, Herreid, and Short 1962; Mann 1978).

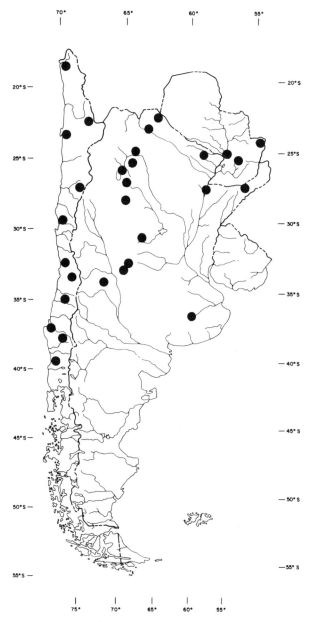

Map 4.54. Distribution of *Tadarida brasiliensis*.

Ecology

Tadarida brasiliensis is one of the best-studied bats in the United States, where it migrates several hundred kilometers seasonally. In the state of Texas these bats form large colonies in caves. While rearing the young, females form nursery colonies spatially segregated from male groups. Populations of this species may exhibit seasonal torpor, but hibernation is not as profound as in other vespertilionids.

These bats are found in many different habitats, from the northern deserts to the southern rain forests of Chile, and they apparently move seasonally. Their roosts have been found in buildings and in natural cavities. They commonly catch nocturnal lepidopterans and larger beetles as well as occasionally taking prey from the ground. In Uruguay *T. brasiliensis* is commonly found in *Tyto alba* pellets (Acosta y Lara 1950; Barlow 1965; Davis, Herreid, and Short 1962; Greer 1966; Mann 1978; Villa-R. and Villa Cornejo 1969).

Genus *Nyctinomops* Miller, 1902

Description

Use of *Nyctinomops* as a genus follows Honacki, Kinman, and Koeppl (1982). In species of this group the ears meet at the midline, though the point of intersection may be deeply notched, whereas in *T. brasiliensis* there is a slight gap in the fold of skin between the ears. This genus or subgenus includes members distributed from the southwestern United States south across most of South America to northern Argentina.

• *Nyctinomops laticaudatus* (E. Geoffroy, 1805)
Broad-tailed Bat, Moloso Colilargo

Measurements

	Mean	Min.	Max.	N	Loc.	Source[a]
TL	110.1	100.0	116.0	12	A, P	1
HB	67.8	63.0	75.0			
T	42.3	35.0	53.0			
HF	10.4	9.8	11.0	11		
E	20.5	19.0	21.0			
FA	44.6	42.0	46.7			
Wta	14.5	13.0	16.0	10		

[a](1) Bárquez and Ojeda 1975; Mares, Ojeda, and Kosco 1981; Myers and Wetzel 1983; PCorps.

Description

This species is a dark chocolate brown dorsally with a lighter venter and has large, dark ears. Individuals from the Paraguayan Chaco are smaller and lighter than those from eastern Paraguay (Myers and Wetzel 1983).

Distribution

Nyctinomops laticaudatus ranges from Mexico to northern Argentina. Few specimens have been reported from southern South America, but it is known from Paraguay and northern Argentina (Bárquez and Ojeda 1975; Honacki, Kinman, and Koeppl 1982; Mares, Ojeda, and Kosco 1981; Myers and Wetzel 1983) (map 4.55).

Ecology

In the Argentine province of Salta, individuals were caught in a lush area of mixed forest and orchards (Mares, Ojeda, and Kosco 1981).

• *Nyctinomops macrotis* (Gray, 1839)
Big Free-tailed Bat, Moloso Castaño

Measurements

	Mean	Min.	Max.	N	Loc.	Source[a]
TL	131.0	130.0	133	4	A, U	1
HB	78.3	75.0	80			
T	51.2	48.0	55	6		
HF	11.0			1		
E	22.9	15.0	28	7		
FA	55.9	48.1	61	5		
Wta	16.0			1	A	2

[a](1) Acosta y Lara 1950; Crespo 1958; Fornes 1964; Gonzáles 1977; (2) CM.

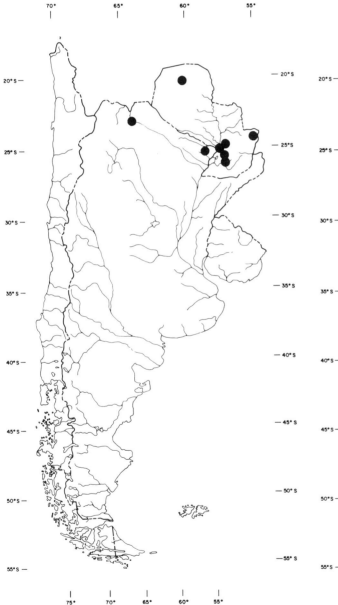

Map 4.55. Distribution of *Nyctinomops laticaudatus*.

Map 4.56. Distribution of *Nyctinomops macrotis*.

Description

This species has soft, ash brown fur, longer than that of *N. laticaudata*. The venter is only slightly lighter than the dorsum. This bat has much larger ears than *N. laticaudata*.

Distribution

Nyctinomops macrotis is broadly distributed from Canada south to Argentina. In southern South America it has been recorded at one location in Paraguay, through northwestern Argentina, and in Uruguay, though it does not seem to be a resident there (Crespo 1958; Fornes 1964; González 1977; Honacki, Kinman, and Koeppl 1982; Mares 1973; Myers and Wetzel 1983; Ximénez, Langguth, and Praderi 1972) (map 4.56).

Ecology

This species was caught in transitional forest and montane Chaco in Salta province, Argentina (Ojeda and Mares 1989).

Genus *Promops* Gervais, 1855

Description

The dental formula is I 1/2, C 1/1, P 2/2, M 3/3. In these intermediate-sized bats, head and body length ranges from 60 to 90 mm, tail length from 45 to 75 mm, and forearm from 43 to 63 mm. The dorsum is drab brown to glossy black, with the venter paler.

Distribution

Species of this genus are distributed from Central America south to Paraguay. There are two species within the range covered by this volume.

• *Promops centralis* Thomas, 1915
Thomas's Mastiff Bat, Moloso Rufo

Measurements

	Mean	Min.	Max.	N	Loc.	Source[a]
TL	132.7	124.0	140	10	A, P	1
HB	79.6	73.0	85			
T	53.1	47.0	56			
HF	12.2	10.0	13			
E	17.3	16.0	18			
FA	53.7	51.7	55	8		
Wta	24.1	20.0	30	9		

[a](1) Baud 1981; Massoia 1976; Myers and Wetzel 1983; UM.

Description

This molossid is very dark chocolate brown above and a lighter gray below. The ears and wing membranes are dark.

Distribution

Promops centralis ranges from Mexico to northern Argentina. In southern Argentina it has been collected in Paraguay and the Argentine province of Formosa (Honacki, Kinman, and Koeppl 1982; Massoia 1976; Myers and Wetzel 1983) (map 4.57).

Ecology

In northern South America this species has been found under palm fronds in colonies of up to six (Goodwin and Greenhall 1961).

Comment

Promops centralis includes *P. occultus*.

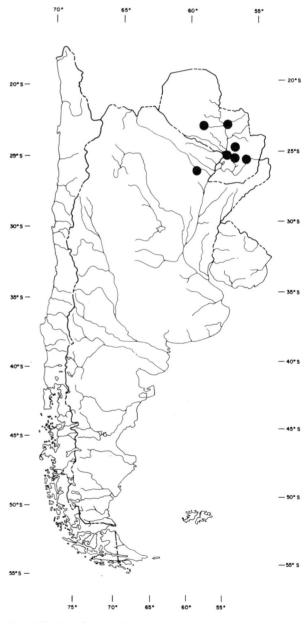

Map 4.57. Distribution of *Promops centralis*.

• *Promops nasutus* (Spix, 1823)
Moloso Moreno

Measurements

	Mean	Min.	Max.	N	Loc.	Source[a]
TL	118.5	112.0	126.0	10	A, P	1
HB	71.0	63.0	86.0			
T	49.0	41.0	51.0	9		
HF	7.7	61.0	110.0			
E	14.2	13.0	17.0			
FA	47.7	46.0	49.3	9		
Wta	12.3	11.5	14.0	6		

[a](1) Myers and Wetzel 1983; BA, UConn.

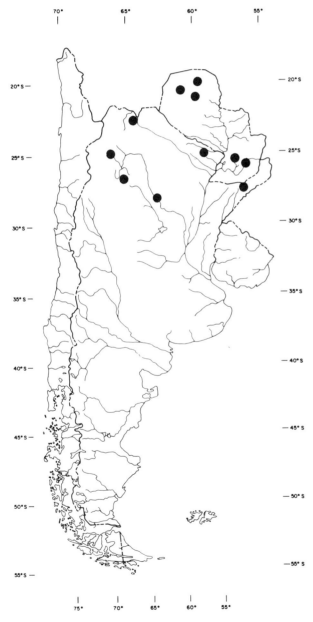

Map 4.58. Distribution of *Promops nasutus*.

Description

This species is smaller and lighter colored than
P. centralis. The dorsum is brown with the tan at
the bases of the hairs showing through; the venter is
lighter and more ash brown (pers. obs.).

Distribution

Promops nasutus ranges from Venezuela to northern Argentina. In southern South America it is found
in Paraguay and in northern Argentina south into
Catamarca province (Bárquez and Lougheed 1990;
Crespo 1958; Honacki, Kinman, and Koeppl 1982;
Lucero 1983; Myers and Wetzel 1983; BA, UConn)
(map 4.58).

References

Acosta y Lara, E. F. 1950. Quirópteros del Uruguay.
Comun. Zool. Mus. Hist. Nat. Montevideo 3(58):
1–71.

———. 1951. Un nuevo quiróptero para el Uruguay. *Comun. Zool. Mus. Hist. Nat. Montevideo*
3(64): 1–4.

Anderson, S., K. F. Koopman, and G. K. Creighton. 1982. Bats of Bolivia: An annotated checklist.
Amer. Mus. Novitat. 2750:1–24.

Arata, A. A., and J. B. Vaughan. 1970. Analyses of
the relative abundance and reproductive activity
of bats in southwestern Colombia. *Caldasia* 10:
517–28.

August, P. V. 1979. Distress calls in *Artibeus jamaicensis*. In *Vertebrate ecology in the northern
Neotropics*, ed. J. F. Eisenberg, 151–59. Washington, D.C.: Smithsonian Institution Press.

Ayala, S. C., and A. D'Alessandro. 1973. Insect
feeding of some Colombian fruit-eating bats. *J.
Mammal.* 54:266–67.

Baker, R. J., M. W. Haiduk, L. W. Robbins, A. Cadena, and B.F. Koop. 1982. Chromosomal studies of bats and their implications. In *Mammalian
biology in South America*, ed. M. A. Mares and
H. H. Genoways, 303–44. Pymatuning Symposia
in Ecology 6. Special Publications Series. Pittsburgh: Pymatuning Laboratory of Ecology, University of Pittsburgh.

Baker, R. J., J. Knox Jones, Jr., and D. C. Carter,
eds. 1976. *Biology of bats of the New World family Phyllostomatidae, part 1.* Special Publications
of the Museum 10. Lubbock: Texas Tech University Press.

———. 1977. *Biology of bats of the New World family Phyllostomatidae, part 2.* Special Publications
of the Museum 13. Lubbock: Texas Tech University Press.

————. 1979. *Biology of bats of the New World family Phyllostomatidae, part 3.* Special Publications of the Museum 16. Lubbock: Texas Tech University Press.

Barlow, J. C. 1965. Land mammals from Uruguay: Ecology and zoogeography. Ph.D. diss., University of Kansas.

Bárquez, R. M. 1983a. Una nueva localidad para la distribución de *Peropteryx macrotis macrotis* (Wagner) (Chiroptera-Emballonuridae). *Hist. Nat.* 3(21): 185–86.

————. 1983b. Breves comentarios sobre *Molossus molossus* (Chiroptera-Molossidae) de Bolivia. *Hist. Nat.* 3(18): 169–73.

————. 1984a. Morfometría y comentarios sobre la colección de murciélagos de la Fundación Miguel Lillo, familias Emballonuridae, Noctilionidae, Mormoopidae, Phyllostomatidae, Furipteridae, Thyropteridae (Mammalia, Chiroptera). *Hist. Nat.* 3(25): 213–23.

————. 1984b. Significativa extensión del rango de distribución de *Diaemus youngii* (Yentink, 1893) (Mammalia, Chiroptera, Phyllostomidae). *Hist. Nat.* 4(7): 67–68.

————. 1987. Los murciélagos de Argentina. Ph.D. diss., Facultad de Ciencias Naturales e Instituto Miguel Lillo. Universidad Nacional de Tucumán.

————. 1988. Notes on identity, distribution, and ecology of some Argentine bats. *J. Mammal.* 69: 873–76.

Bárquez, R. M., and S. C. Lougheed. 1990. New distributional records of some Argentine bat species. *J. Mammal.* 71: 261–63.

Bárquez, R. M., and R. A. Ojeda. 1975. *Tadarida laticaudata*, un nuevo molósido para la fauna argentina (Chiroptera, Molossidae). *Neotrópica* 21(66): 137–38.

Baud, F.-J. 1981. Expédition du Muséum de Genève au Paraguay: Chiroptères. *Rev. Suisse Zool.* 88(2): 567–81.

Bonaccorso, F. J. 1979. Foraging and reproductive ecology in a Panamanian bat community. *Bull. Florida State Mus., Biol. Sci.* 24(4): 359–408.

Bradbury, J. W., and L. Emmons. 1974. Social organization of some Trinidad bats. 1. Emballonuridae. *Z. Tierpsychol.* 36: 137–83.

Bradbury, J. W., and S. L. Vehrencamp. 1976a. Social organization and foraging in emballonurid bats. 1. Field studies. *Behav. Ecol. Sociobiol.* 1: 337–81.

————. 1976b. Social organization and foraging in emballonurid bats. 2. A model for the determination of group size. *Behav. Ecol. Sociobiol.* 1: 383–404.

————. 1977a. Social organization and foraging in emballonurid bats. 3. Mating systems. *Behav. Ecol. Sociobiol.* 2: 1–17.

————. 1977b. Social organization and foraging in emballonurid bats. 4. Parental investment patterns. *Behav. Ecol. Sociobiol.* 2: 19–30.

Brown, J. H. 1968. Activity patterns of some Neotropical bats. *J. Mammal.* 49: 754–57.

Brown, P. E., T. W. Brown, and A. D. Grinnell. 1983. Echolocation, development, and vocal communication in the lesser bulldog bat, *Noctilio albiventris. Behav. Ecol. Sociobiol.* 13: 287–98.

Carvalho, C. T. 1961. Sobre os hábitos alimentares de phillostomídeos (Mammalia, Chiroptera). *Rev. Biol. Trop.* 9(1): 53–60.

Crespo, J. A. 1958. Nuevas especies y localidades de quirópteros para Argentina (Mammalia, Chiroptera). *Neotrópica* 4(13): 27–32.

————. 1974. Comentarios sobre nuevas localidades para mamíferos de Argentina y de Bolivia. *Rev. Mus. Argent. Cienc. Nat. "Bernardino Rivadavia," Zool.* 11(1): 1–31.

————. 1982. Ecología de la comunidad de mamíferos del Parque Nacional Iguazú, Misiones. *Rev. Mus. Argent. Cienc. Nat. "Bernardino Rivadavia," Ecol.* 3(2): 45–162.

Crespo, J. A., J. M. Vanella, B. D. Blood, and J. M. de Carlo. 1961. Observaciones ecológicas del vampiro *Desmodus r. rotundus* (Geoffroy) en el norte de Córdoba. *Rev. Mus. Argent. Cienc. Nat. "Bernardino Rivadavia," Zool.* 6(4): 131–60.

Daciuk, J. 1977. Notas faunísticas y biocecológicas de Península Valdés y Patagonia. 20. Presencia de *Histiotus montanus montanus* (Philippi y Landbeck), 1861 en la Península Valdés (Chiroptera, Vespertilionidae). *Neotrópica* 23(69): 45–46.

Davis, R. B., C. F. Herreid, and H. Short. 1962. Mexican free-tailed bats in Texas. *Ecol. Monogr.* 32: 311–46.

Davis, W. B. 1966. Review of South American bats of the genus *Eptesicus. Southwest. Nat.* 11(2): 245–74.

Delpietro, H., R. D. Lord, L. Lázaro, and R. García. 1973. Extensión de la distribución del vampiro de alas blancas (*Diaemus youngi*). *Physis*, sec. C, 32(84): 224.

Eger, J. L. 1977. Systematics of the genus *Eumops* (Chiroptera: Molossidae). *Royal Ontario Mus. Life Sci. Contrib.* 110: 1–69.

Eisenberg, J. F. 1981. *The mammalian radiations: An analysis of trends in evolution, adaptation, and behavior.* Chicago: University of Chicago Press.

————. 1989. *Mammals of the Neotropics.* Vol. 1.

Mammals of the northern Neotropics: Panama, Colombia, Venezuela, Guyana, Suriname, French Guiana. Chicago: University of Chicago Press.

Eisenberg, J. F., and D. E. Wilson. 1978. Relative brain size and feeding strategies in the Chiroptera. *Evolution* 32(4): 740–51.

Erkert, H. G. 1982. Ecological aspects of bat activity rhythms. In *Ecology of bats*, ed. T. Kunz, 201–42. New York: Plenum.

Fenton, M. B. 1982. Echolocation, insect hearing and feeding ecology of insectivorous bats. In *Ecology of bats*, ed. T. Kunz, 261–86. New York: Plenum.

Findley, J. S, and D. E. Wilson. 1982. Ecological significance of chiropteran morphology. In *Ecology of bats*, ed. T. Kunz, 243–60. New York: Plenum.

Fleming, T. H. 1971. *Artibeus jamaicensis:* Delayed embryonic development in a Neotropical bat. *Science* 171:402–4.

———. 1982. Foraging strategies of plant visiting bats. In *Ecology of bats*, ed. T. Kunz, 287–326. New York: Plenum.

———. 1983. *Carollia perspicillata* (murciélago candelaro, lesser short-tailed fruit bat). In *Costa Rican natural history*, ed. D. H. Janzen, 457–58. Chicago: University of Chicago Press.

———. 1988. *The short-tailed fruit bat*. Chicago: University of Chicago Press.

Fleming, T. H., E. T. Hooper, and D. E. Wilson. 1972. Three Central American bat communities: Structure, reproductive cycles, and movement patterns. *Ecology* 53(4): 555–69.

Fornes, A. 1964. Consideraciones sobre *Eumops abrasus* y *Tadarida molossa* (Mammalia, Chiroptera, Molossidae). *Acta Zool. Lilloana* 20: 171–75.

———. 1972a. *Anoura geoffroyi geoffroyi* Gray, nuevo género para la república Argentina (Chiroptera, Phyllostomidae, Glossophaginae). *Physis* 31(82): 51–53.

———. 1972b. *Myotis (Hesperomyotis) simus* Thomas, nueva especie para la Argentina (Chiroptera, Vespertilionidae). *Neotrópica* 18(56): 87–89.

Fornes, A., and H. Delpietro. 1969. Sobre *Pygoderma bilabiatum* (Wagner) en la república Argentina (Chiroptera, Phyllostomidae, Stenodermatinae). *Physis* 29(78): 141–44.

Fornes, A., H. Delpietro, and E. Massoia. 1969. *Macrophyllum macrophyllum* (Wied) nuevo género y especie para la república Argentina (Chiroptera, Phyllostomidae, Phyllostominae). *Physis* 28(77): 323–26.

Fornes, A., and E. Massoia. 1966. *Vampyrops lineatus* (E. Geoffroy) nuevo género y especie para la república Argentina (Chiroptera, Phyllostomidae). *Physis* 26(71): 181–84.

———. 1967. Procedencias argentinas nuevas o poco conocidas para murciélagos (Noctilionidae, Phyllostomidae, Vespertilionidae y Molossidae). *Segundo Jornal Entomoepidimiologia Argentina 1965*, 1:133–45. Buenos Aires.

———. 1968. Nuevas procedencias argentinas para *Noctilio labialis, Sturnira lilium, Molossops temminckii* y *Eumops abrasus* (Mammalia, Chiroptera). *Physis* 28(76): 37–38.

———. 1969. La presencia de *Carollia perspicillata perspicillata* (L.) en la república Argentina (Chiroptera, Phyllostomidae, Carolliinae). *Physis* 28(77): 322.

Fornes, A., E. Massoia, and G. E. Forrest. 1967. *Tonatia sylvicola* (d'Orbigny) nuevo género y especie para la república Argentina (Chiroptera, Phyllostomidae). *Physis* 27(74): 149–52.

Foster, M. S., and R. M. Timm. 1976. Tent making by *Artibeus jamaicensis* with comments on plants used by bats for tents. *Biotropica* 8:265–69.

Freeman, P. W. 1979. Specialized insectivory: Beetle-eating and moth-eating molossid bats. *J. Mammal.* 69:467–79.

———. 1981. A multivariate study of the family Molossidae: Morphology, ecology, evolution. *Fieldiana: Zool.* 7.

Gardner, A. L. 1976. The distributional status of some Peruvian mammals. *Occas. Pap. Mus. Zool. Louisiana State Univ.* 48:1–18.

———. 1977a. Feeding habits. In *Biology of bats of the New World family Phyllostomatidae, part 2*, ed. R. J. Baker, J. Knox Jones, and D. C. Carter, 293–350. Special Publications of the Museum 13. Lubbock: Texas Tech University Press.

———. 1977b. Chromosomal variation in *Vampyressa* and a review of chromosomal evolution in the Phyllostomidae (Chiroptera). *Syst. Zool.* 2: 300–318.

Genoways, H. H., and S. L. Williams. 1984. Results of the Alcoa Foundation–Suriname expeditions. 9. Bats of the genus *Tonatia* (Mammalia: Chiroptera) in Suriname. *Ann. Carnegie Mus.* 53(1): 327–46.

González, J. C. 1973. Observaciones sobre algunos mamíferos de Bopicuá (Dpto. de Río Negro, Uruguay). *Comun. Mus. Mun. Hist. Nat. Río Negro, Uruguay* 1(1): 1–14.

———. 1977. Sobre la presencia de *Tadarida molossus* Pallas (Chiroptera, Molossidae) en el Uruguay. *Rev. Biol. Uruguay* 5(1): 27–30.

González, J. C., and S. Vallejo. 1980. Notas sobre *Vampyrops lineatus* (Geoffroy), del Uruguay (Phyllostomidae, Chiroptera). *Comun. Zool. Mus. Hist. Nat. Montevideo* 10(144): 1–8.

Goodwin, G. G. 1942. A summary of recognizable species of *Tonatia*, with descriptions of two new species. *J. Mammal.* 23:204–9.

————. 1963. American bats of the genus *Vampyressa* with the description of a new species. *Amer. Mus. Novitat.* 2125:1–24.

Goodwin, G. G., and A. M. Greenhall. 1961. A review of the bats of Trinidad and Tobago. *Bull. Amer. Mus. Nat. Hist.* 122(3): 187–302.

Gould, E. 1977. Echolocation and communication. In *Biology of bats of the New World family Phyllostomatidae, part 2*, ed. R. J. Baker, J. Knox Jones, Jr., and D. C. Carter, 247–80. Special Publications of the Museum 13. Lubbock: Texas Tech University Press.

Greenhall, A. M., G. Joermann, and U. Schmidt. 1983. *Desmodus rotundus. Mammal. Species* 202:1–6.

Greenhall, A. M., R. D. Lord, and E. Massoia. 1983. *Key to the bats of Argentina*. Special Publication 5. Buenos Aires: Pan American Zoonoses Center.

Greer, J. K. 1966. Mammals of Malleco province, Chile. *Publ. Mus., Michigan State Univ., Biol. Ser.* 3(2): 49–152.

Griffin, D. R. 1958. *Listening in the dark*. New Haven: Yale University Press.

Griffin, D. R., F. A. Webster, and C. R. Michael. 1960. The echolocation of flying insects by bats. *Anim. Behav.* 8:141–54.

Handley, C. O., Jr. 1976. Mammals of the Smithsonian Venezuelan project. *Brigham Young Univ. Sci. Bull., Biol. Ser.* 20(5): 1–90.

————. 1989. The *Artibeus* of Gray, 1838. In *Advances in Neotropical mammalogy*, ed. K. H. Redford and J. F. Eisenberg, 443–69. Gainesville, Fla.: Sandhill Crane Press.

Harrison, D. L. 1975. *Macrophyllum macrophyllum. Mammal. Species* 62:1–3.

Harrison, D. L., N. G. E. Pendleton, and G. C. D. Harrison. 1979. *Eumops dabbenei* Thomas, 1914 (Chiroptera: Molossidae), a free-tailed bat new to the fauna of Paraguay. *Mammalia* 43:251–52.

Haussler, U., E. Moller, and U. Schmidt. 1981. Zur Haltung und Jugendentwicklung von *Molossus molossus*. Z. *Säugetierk.* 46:337–51.

Honacki, J. H., K. E. Kinman, and J. W. Koeppl, eds. 1982. *Mammal species of the world*. Lawrence, Kans.: Allen Press and Association of Systematics Collections.

Hood, C. S., and J. Knox Jones, Jr. 1984. *Noctilio leporinus. Mammal. Species* 216:1–7.

Hood, C. S., and J. Pitocchelli. 1983. *Noctilio albiventris. Mammal. Species* 197:1–5.

Hooper, E. T., and J. H. Brown. 1968. Foraging and breeding in two sympatric species of Neotropical bats, genus *Noctilio. J. Mammal.* 49: 310–12.

Howell, D. J. 1974. Bats and pollen: Physiological aspects of chiropterophily. *Comp. Biochem. Physiol.* 48A:263–76.

Humphrey, S. R., and F. J. Bonaccorso. 1979. Population and community ecology. In *Biology of bats of the New World family Phyllostomatidae, part 3*, ed. R. J. Baker, J. Knox Jones, Jr., and D. C. Carter, 409–41. Special Publications of the Museum 16. Lubbock: Texas Tech University Press.

Humphrey, S. R., F. J. Bonaccorso, and T. L. Zinn. 1983. Guild structure of surface-gleaning bats in Panama. *Ecology* 64(2): 284–94.

Husson, A. M. 1962. *The bats of Suriname*. Zoologische Verhandelingen 58. Leiden: E. J. Brill.

————. 1978. *The mammals of Suriname*. Leiden: E. J. Brill.

Ibañez, C. 1979. Nuevos datos sobre *Eumops dabbenei* Thomas, 1914 (Chiroptera, Molossidae). *Doñana, Acta Vert.* 6(2): 248–52.

Jepsen, G. L. 1970. Bat origins and evolution. In *Biology of bats*, vol. 1, ed. W. A. Wimsatt, 1–64. New York: Academic Press.

Kleiman, D. G., and T. M. Davis. 1979. Ontogeny and maternal care. In *Biology of bats of the New World family Phyllostomatidae, part 3*, ed. R. J. Baker, J. Knox Jones, Jr., and D. C. Carter, 387–402. Special Publications of the Museum 16. Lubbock: Texas Tech University Press.

Koopman, K. F. 1967. The southernmost bats. *J. Mammal.* 48:487–88.

————. 1982. Biogeography of the bats of South America. In *Mammalian biology in South America*, ed. M. A. Mares and H. H. Genoways, 273–302. Pymatuning Symposia in Ecology 6. Special Publication Series. Pittsburgh: Pymatuning Laboratory of Ecology, University of Pittsburgh.

Kunz, T. H., ed. 1982. *Ecology of bats*. New York: Plenum.

Kunz, T. H., P. V. August, and C. D. Burnett. 1983. Harem social organization in cave roosting *Artibeus jamaicensis* (Chiroptera: Phyllostomidae). *Biotropica* 15(2): 133–38.

Langguth, A., and F. Achaval. 1972. Notas ecológicas sobre el vampiro *Desmodus rotundus rotundus* (Geoffroy) en el Uruguay. *Neotrópica* 18(55): 45–53.

LaVal, R. K. 1973. A revision of the Neotropical bats

of the genus *Myotis*. *Bull. Nat. Hist. Mus., Los Angeles County* 15:1–54.

Lewis, S. E., and D. E. Wilson. 1987. *Vampyressa pussilla*. *Mammal. Species* 292:1–5.

Lucero, M. M. 1983. Lista y distribución de aves y mamíferos de la provincia de Tucumán. Ministerio de Cultura y Educación, Fundación Miguel Lillo, *Miscelánea* 75:5–53.

McCracken, G. F., and J. W. Bradbury. 1977. Paternity and genetic heterogeneity in the polygynous bat *Phyllostomus hastatus*. *Science* 198:303–6.

———. 1981. Social organization and kinship in the polygynous bat *Phyllostomus hastatus*. *Behav. Ecol. Sociobiol.* 8:11–34.

McNab, B. K. 1969. The economics of temperature regulation in Neotropical bats. *Comp. Biochem. Physiol.* 31:227–68.

———. 1982. Evolutionary alternatives in the physiological ecology of bats. In *Ecology of bats*, ed. T. H. Kunz, 151–200. New York: Plenum.

Mann, G. 1978. *Los pequeños mamíferos de Chile*. Guyana: Zoología 40. Santiago: Universidad de Concepción.

Mares, M. A. 1973. Climates, mammalian communities and desert rodent adaptations: An investigation into evolutionary convergence. Ph.D. diss., University of Texas at Austin.

Mares, M. A., R. A. Ojeda, and M. P. Kosco. 1981. Observations on the distribution and ecology of the mammals of Salta province, Argentina. *Ann. Carnegie Mus.* 50(6): 151–206.

Marques, S. A. 1986. Activity cycle, feeding and reproduction of *Molossus ater* (Chiroptera: Molossidae) in Brazil. *Bol. Mus. Paraense Emilio Goeldi, Zool.* 2(2): 159–79.

Massoia, E. 1970. Contribución al conocimiento de los mamíferos de Formosa con noticias de los que habitan zonas viñaleras. *IDIA* 276:55–63.

———. 1976. Cuatro notas sobre murciélagos de la república Argentina (Mollosidae y Vespertilionidae). *Physis*, sec. C, 35(91): 257–65.

———. 1980. Mammalia de Argentina. 1. Los mamíferos silvestres de la provincia de Misiones. *Iguazú* 1(1): 15–43.

Miller, G. S., Jr. 1907. The families and genera of bats. *U.S. Nat. Mus. Bull.* 57:1–282.

Miller, G. S., Jr., and G. M. Allen. 1928. The American bats of the genera *Myotis* and *Pizonyx*. *U.S. Nat. Mus. Bull.* 144:175–213.

Morrison, D. W. 1978. Lunar phobia in a Neotropical fruit bat, *Artibeus jamaicensis* (Chiroptera: Phyllostomidae). *Anim. Behav.* 26:852–55.

Myers, P. 1977. Patterns of reproduction of four species of vespertilionid bats in Paraguay. *Univ. Calif. Publ. Zool.* 107:1–41.

———. 1978. Sexual dimorphism in size of vespertilionid bats. *Amer. Nat.* 112(986): 701–11.

———. 1981. Observations on *Pygoderma bilabiatum* (Wagner). *Z. Säugetierk.* 46:146–51.

Myers, P., and R. M. Wetzel. 1983. Systematics and zoogeography of the bats of the Chaco Boreal. *Misc. Publ. Mus. Zool. Univ. Michigan* 165:1–59.

Myers, P., R. White, and J. Stallings. 1983. Additional records of bats from Paraguay. *J. Mammal.* 64:143–45.

Novick, A., and B. A. Dale. 1971. Foraging behavior in fishing bats and their insectivorous relatives. *J. Mammal.* 52:817–18.

Ojeda, R. A., and R. M. Bárquez. 1978. Contribución al conocimiento de los quirópteros de Bolivia. *Neotrópica* 24(71): 33–38.

Ojeda, R. A., and M. A. Mares. 1989. *Mammals of Salta*. Lubbock: Texas Tech University Press.

Olrog, C. C. 1958. Notas mastozoológicas sobre la colección del Instituto Miguel Lillo (Tucumán). *Acta Zool. Lilloana* 16:91–95.

———. 1959. Notas mastozoológicas. 2. Sobre la colección del Instituto Miguel Lillo. *Acta Zool. Lilloana* 17:403–19.

———. 1967. *Pygoderma bilabiatum*, un murciélago nuevo para la fauna argentina (Mammalia, Chiroptera, Phyllostomidae). *Neotrópica* 13:104.

———. 1973. Alimentación del falso vampiro *Chrotopterus auritus* (Mammalia, Phyllostomidae). *Acta Zool. Lilloana* 30:5–6.

———. 1979. Los mamíferos de la selva húmeda, Cerro Calilegua, Jujuy. *Acta Zool. Lilloana* 33:9–14.

Olrog, C. C., and R. M. Bárquez. 1979. Dos quirópteros nuevos para la fauna argentina. *Neotrópica* 25(74): 185–86.

Olrog, C. C., and M. M. Lucero. 1981. *Guía de los mamíferos argentinos*. Tucumán, Argentina: Ministerio de Cultura y Educación, Fundación Miguel Lillo.

Osgood, W. H. 1943. The mammals of Chile. *Field Mus. Nat. Hist. Zool. Ser.* 30:1–268.

Owen, R. D., and W. D. Webster. 1983. Morphological variation in the Ipanema bat, *Pygoderma bilabiatum*, with description of a new subspecies. *J. Mammal.* 64:146–49.

Pearson, O. P., and A. K. Pearson. 1989. Reproduction of bats in southern Argentina. In *Advances in Neotropical mammalogy*, ed. K. H. Redford and J. F. Eisenberg, 549–66. Gainesville, Fla.: Sandhill Crane Press.

Peracchi, A. L. 1968. Sobre os hábitos de *Histiotus*

velatus (Geoffroy, 1824) (Chiroptera, Vespertilionidae). *Rev. Brasil. Biol.* 28(4): 469–73.

Peterson, L. 1968. A new bat of the genus *Vampyressa* from Guyana, South America, with a brief systematic of the genus. *Roy. Ontario Mus., Life Sci. Contrib.* 73:1–17.

Phillips, C. J. 1971. *The dentition of glossophagine bats: Development, morphological characteristics, variation, pathology, and evolution.* Miscellaneous Publications 54. Lawrence: Museum of Natural History, University of Kansas.

Pine, R. H. 1972. *The bats of the genus* Carollia. Technical Monograph 8. College Station: Texas Agricultural Experiment Station, Texas A&M University.

Pine, R. H., S. D. Miller, and M. L. Schamberger. 1979. Contributions to the mammalogy of Chile. *Mammalia* 43:339–76.

Ralls, K. 1976. Mammals in which females are larger than males. *Quart. Rev. Biol.* 51:245–76.

Rau, J. R., and J. Yáñez. 1979. Nuevos registros de *Lasiurus borealis* en Magallanes. *Not. Mens. Mus. Nac. Hist. Nat.* (Santiago) 23(274–75): 13–14.

Roig, V. G. 1965. Elenco sistemático de los mamíferos y aves de la provincia de Mendoza y notas sobre su distribución geográfica. *Bol. Est. Geogr.* 12(49): 175–222.

Romaña, C., and J. W. Ábalos. 1950. Lista de los quirópteros de la colección del Instituto de Medicina Regional, y sus parásitos. *Anal. Inst. Med. Reg.* 3:111–17.

Sanborn, C. C. 1937. American bats of the subfamily Emballonurinae. *Field Mus. Nat. Hist. Zool. Ser.* 20(24): 321–54.

———. 1955. Remarks on the bats of the genus *Vampyrops*. *Fieldiana: Zool.* 37:403–13.

Sanborn, C. C., and J. A. Crespo. 1957. El murciélago blanquizco (*Lasiurus cinereus*) y sus subespecies. *Bol. Mus. Argent. Cienc. Nat. "Bernardino Rivadavia," Zool.* 4:1–13.

Sazima, I. 1976. Observations on the feeding habits of phyllostomatid bats (*Carollia, Anoura,* and *Vampyrops*) in southeastern Brazil. *J. Mammal.* 57:381–82.

Schneider, C. O. 1946. Catálogo de los mamíferos de la provincia de Concepción. *Bol. Soc. Biol. Concepción* 21:67–83.

Shump, K. A., Jr., and A. U. Shump. 1982a. *Lasiurus borealis. Mammal. Species* 183:1–6.

———. 1982b. *Lasiurus cinereus. Mammal. Species* 185:1–5.

Taddei, W. A. 1975. Phyllostomidae (Chiroptera) do norte-ocidental do estado de São Paulo. 2. Glosso-phaginae; Carolliinae; Sturnirinae. *Cienc. Cult.* 27(7): 723–34.

———. 1976. The reproduction of some Phyllostomidae (Chiroptera) from the northwestern region of the state of São Paulo. *Bol. Zool. Univ. São Paulo* 1:313–30.

———. 1979. Phyllostomidae (Chiroptera) do norte-ocidental do estado de São Paulo. 3. Stenodermatinae. *Cienc. Cult.* 31(8): 900–914.

Tamayo, M., and D. Frassinetti. 1980. Catálogo de los mamíferos fósiles y vivientes de Chile. *Mus. Nac. Hist. Nat., Chile* 37:323–99.

Thomas, O. 1898. On the small mammals collected by Dr. Borelli in Bolivia and northern Argentina. *Boll. Mus. Zool. Anat. Comp.* 13(315): 1–4.

Tuttle, M. D. 1970. Distribution and zoogeography of Peruvian bats, with comments on natural history. *Univ. Kansas Sci. Bull.* 49(2): 45–86.

Tuttle, M. D., and D. Stevenson. 1982. Growth and survival of bats. In *Ecology of bats,* ed. T. H. Kunz, 105–50. New York: Plenum.

Vaughan, T. 1959. *Functional morphology of three bats:* Eumops, Myotis, *and* Macrotus. Miscellaneous Publications 12. Lawrence: Museum of Natural History, University of Kansas.

Villa-R., B., and M. Villa Cornejo. 1969. Algunos murciélagos del norte de Argentina. In *Contributions in mammalogy,* ed. J. Knox Jones, Jr., 409–29. Miscellaneous Publications 51. Lawrence: Museum of Natural History, University of Kansas.

———. 1971. Observaciones acerca de algunos murciélagos del norte de Argentina, especialmente de la biología del vampiro *Desmodus r. rotundus. An. Inst. Biol. Univ. Nal. Auton. Mexico, Ser. Zool.* 41(1): 107–48.

Vizotto, L. D., and V. A. Taddei. 1976. Notas sobre *Molossops temminckii temminckii* e *Molossops planirostris* (Chiroptera-Molossidae). *Naturalia* 2:47–59.

Webster, W. D., and R. D. Owen. 1984. *Pygoderma bilabiatum. Mammal. Species* 178:1–3.

Wilkinson, G. S. 1984. Reciprocal food sharing in the vampire bat. *Nature* 308(5955): 181–84.

———. 1985a. The social organization of the common vampire bat. 1. Pattern and cause of association. *Behav. Ecol. Sociobiol.* 17:111–21.

———. 1985b. The social organization of the common vampire bat. 2. Mating system, genetic structure, and relatedness. *Behav. Ecol. Sociobiol.* 17:123–34.

———. 1986. Social grooming in the common vampire bat, *Desmodus rotundus. Anim. Behav.* 34: 1880–89.

Williams, D. F. 1978. Taxonomic and karyologic comments on small brown bats, genus *Eptesicus* from South America. *Ann. Carnegie Mus.* 47(16): 361–83.

Williams, D. F., and M. A. Mares. 1978. Karyologic affinities of the South American big-eared bat, *Histiotus montanus* (Chiroptera, Vespertilionidae). *J. Mammal.* 59:844–46.

Willig, M. R. 1983. Composition, microgeographic variation, and sexual dimorphism in caatinga and cerrado bat communities from northeast Brazil. *Bull. Carnegie Mus. Nat. Hist.* 23:1–131.

Willig, M. R., and R. R. Hollander. 1987. *Vampyrops lineatus. Mammal. Species* 275:1–4.

Wilson, D. E. 1973a. Reproduction in Neotropical bats. *Period. Biol.* 75:215–17.

———. 1973b. Bat faunas: A trophic comparison. *Syst. Zool.* 22:14–29.

———. 1979. Reproductive patterns. In *Biology of bats of the New World family Phyllostomatidae,* part 3, ed. R. J. Baker, J. Knox Jones, Jr., and D. C. Carter, 317–78. Special Publications of the Museum 16. Lubbock: Texas Tech University Press.

Wilson, D. E., and R. K. LaVal. 1974. *Myotis nigricans. Mammal. Species* 39:1–3.

Wimsatt, W. A. 1970a. *Biology of bats.* Vol. 1. New York: Academic Press.

———. 1970b. *Biology of bats.* Vol. 2. New York: Academic Press.

———. 1977. *Biology of bats.* Vol. 3. New York: Academic Press.

Ximénez, A. 1969. Dos nuevos géneros de quirópteros para el Uruguay (Phyllostomidae-Molossidae). *Comun. Zool. Mus. Hist. Nat. Montevideo* 10(125): 1–8.

Ximénez, A., A. Langguth, and R. Praderi. 1972. Lista sistemática de los mamíferos del Uruguay. *Anal. Mus. Nac. Hist. Nat. Montevideo* 7(5): 1–45.

5 Order Primates

Diagnosis

If we exclude the fossil Plesiadapiformes, the order Primates consists of a group of species that exhibit rather unspecialized physical characteristics. Except for the extremely terrestrially adapted forms, most have flexible digits and retain five fingers and five toes. Only in some brachiating forms is the thumb lost. The shoulder joint is freely movable, and the radius and ulna remain unfused. The orbits of the skull are directed forward, and a strong postorbital bar separates the eye socket from the temporal fossa. The braincase is relatively large, and in the evolutionary history of this order one can see a trend toward progressive enlargement of the cerebral hemispheres and cerebellum (Hill 1957, 1960).

Distribution

The Recent distribution of this order, exclusive of man (*Homo sapiens*), is in both the Old World and New World tropics except for Australasia. In the Old World some members of the genus *Macaca* extend their distributions into the Temperate Zone, but in the Western Hemisphere distributions are confined more or less between 23° north and 24° south latitude.

History and Classification

The earliest representatives of the order Primates can be distinguished in the Paleocene, when the first fossils appear in North America. By the time of the Eocene, lemurlike primates are recognizable from North America and Europe. In some manner the early prosimians of the Eocene made their way to Asia and Africa, with one or two stocks transiting to Madagascar, where they became isolated and underwent an adaptive radiation. The New World primates, the Ceboidea, had an independent origin from the Old World primates and became established on what was then the island continent of South America, where they underwent an extensive adaptive radiation. The earliest ceboid primates have been found in the Oligocene strata of South America. The Cercopithecoidea and the Hominoidea had an Old World origin and underwent an adaptive radiation roughly parallel to the radiations in South America. In the New World tropics there are currently only three families extant, the Callithricidae, the Cebidae, and the recently immigrant family Hominidae (Szalay and Delson 1979). Hill (1957, 1960, 1962) summarizes anatomical data for the New World primates.

FAMILY CALLITHRICIDAE (CALLITRICHIDAE)
Marmosets and Tamarins, Titis

Diagnosis

These small primates are highly adapted for an arboreal life, with nonprehensile tails. Rather than bearing nails at the ends of the digits, they have secondarily acquired clawlike nails. Head and body length ranges from 150 to 370 mm. The tail can be shorter or longer than the head and body. There are five Recent genera, of which one occurs in the area dealt with in this volume: *Callithrix*. The dental formula for the family is I 2/2, C 1/1, P 3/3, M 2/2 (see fig. 5.1), except in *Callimico*, which retains the third molar for a total of 36 teeth. Based on this characteristic and other skull features, the genus *Callimico* has often been assigned to its own family (Hill 1957; Hershkovitz 1977). The standard reference for the family is Hershkovitz (1977).

Distribution

Species of this family do not occur north of Panama, but they range south to Paraguay. The species

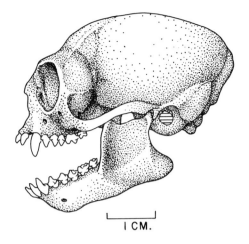

Figure 5.1. Skull of *Callithrix argentata*.

are mainly confined to tropical forests and do not reach extremely high elevations.

Natural History

The marmosets are diurnal and strongly arboreal. They feed on fruits and insects and live in small groups. The female typically produces twins at birth; *Callimico* is exceptional in being uniparous. During a rearing cycle, the male participates by carrying the young and offering them food. The ecology and behavior of the family are reviewed by Kleiman (1977).

Genus *Callithrix* Erxleben, 1777
Callithrix argentata (Linnaeus, 1776)
Black-tailed Silvery Marmoset, Cai Poshy

Measurements

	Mean	Min.	Max.	N	Loc.	Source[a]
TL	545.8	491	604	13	Br, P	1
HB	233.5	191	281			
T	312.2	240	351			
HF	67.2	60	70			
E	27.2	21	32	6		
Wta	440.0	380	500	2		

[a](1) Allen 1916; Schaller 1983; Vieira 1945; PCorps.

Description

Callithrix argentata melanura has fully exposed, bare ears, is brown on the forehead, crown, back, and outer sides of limbs, and has distinctive whitish hip and thigh patches (plate 5) (Hershkovitz 1977).

Distribution

Callithrix argentata is distributed from northern Brazil south to Paraguay. The subspecies *C. argentata melanura* occurs in central and western Brazil, in adjacent parts of Bolivia, and into northern Paraguay. It is the only species of *Callithrix* in the south to range outside Brazil and is confined to the northeastern Chaco of Paraguay (Stallings 1984; Stallings and Mittermeier 1983) (map 5.1).

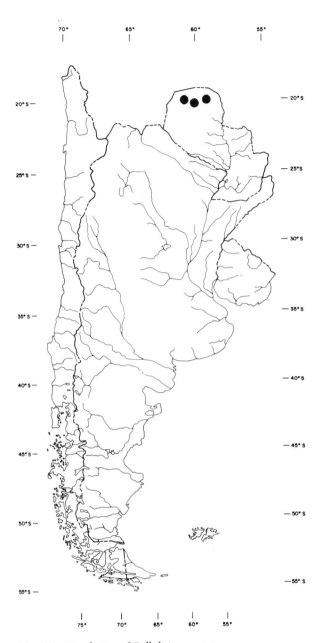

Map 5.1. Distribution of *Callithrix argentata*.

Ecology

In Paraguay this marmoset was found in two types of forest: an intermediate formation somewhat higher than the typical scrub forest, with a canopy of 5–10 m, and more commonly in a forest with a canopy of 15–20 m. In the high forest it shelters at night in holes in trees and also among the branches of tree cacti. In the Mato Grosso area of Brazil it was observed to come to the ground and may routinely cross open areas or forage for insects on the ground. In Brazil one stomach was full of beetle remains, and in Paraguay five stomachs held buds and tiny black seeds.

Group size in the Chaco ranged from 5 to 14 animals, with a mean of 4.6 in one area and 7.8 in another. Ecological density estimates in these areas ranged from 7.36/km² to 27.45/km². In Mato Grosso state, Brazil, group size was calculated at 5.6 (range 3–12; *n* = 7) (Miller 1930; Schaller 1983; Stallings 1984; Stallings and Mittermeier 1983).

FAMILY CEBIDAE

Diagnosis

The dental formula is I 1/1, C 2/2, P 3/3, M 3/3 (see figs. 5.2 and 5.3). The posterior molar may be missing in a small proportion of individuals within a larger population (15% for *Ateles*). There are eleven Recent genera, including some twenty-nine species. Five genera (*Cebus, Ateles, Alouatta, Brachyteles,* and *Lagothrix*) have prehensile tails. All species are small to medium in size.

Distribution

The species of this family currently occur from southern Veracruz, Mexico, to the gallery forests on the upper reaches of the Río Paraná in Argentina. All are adapted to multistratal tropical evergreen forests, but the genera *Alouatta* and *Cebus* have adapted to the semideciduous forests of the more xeric portions of northern and southern South America.

Natural History

All members of this family are diurnal except *Aotus*, which is the only nocturnal New World primate. Most species are adapted to a diet of fruit, seeds, small vertebrates, and invertebrates. *Alouatta* is exceptional in that it generally includes 50%

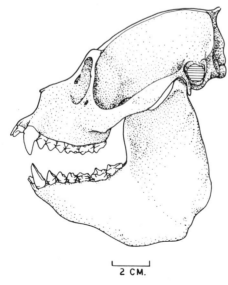

Figure 5.2. Skull of *Alouatta* sp.

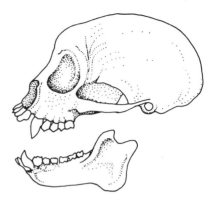

Figure 5.3. Skull of *Cebus apella* (modified from Mares, Ojeda, and Barquez 1989) (typical skull CB = 90 mm).

foliage in its diet. The species of *Aotus, Callicebus,* and *Pithecia* show monogamous tendencies. All females of this family typically bear a single young.

Genus Alouatta Lacépède, 1799
Howler Monkey, Aullador, Arguato

Description

Six species are recognized within this genus, and two occur in the region covered by this volume: *A. caraya* and *A. fusca*. Members of this genus are among the largest of the New World primates. Head and body length ranges from 559 to 915 mm, and the prehensile tail is from 585 to 920 mm. The weight of adult males ranges from 6.5 to 7.8 kg, and that of females from 4.5 to 6.6 kg. Color patterns are variable from species to species. In *A. caraya* the males are black and the females and juveniles are buff, one of the few cases of color dimorphism within the New World primates.

Adult females of *A. caraya* are about 69% of the average adult male's weight. The hyoid bone is extremely enlarged and ossified in both the male and the female (but more so in the male) and no doubt influences the low fundamental frequency of the long call (Thorington, Rudran, and Mack 1979).

Distribution

The genus is the most widely distributed of the New World primates, with a range extending from Veracruz, Mexico, to northern Argentina. The species tolerate a range of habitat types varying from semideciduous tropical forests to multistratal tropical evergreen forests (Eisenberg 1979; Crockett and Eisenberg 1987).

Life History and Ecology

Howler monkeys are essentially mixed feeders; they include up to 50% leaves in their diet, the rest being fruit. The species of the genus exhibit variations in group size but are basically polygynous; the males are larger and tend to dominate a matriline of females. Long-term studies of howler monkeys have for the most part been carried out in Panama and Venezuela (Crockett and Eisenberg 1987). The function of the loud, long calls has been analyzed by Sekulic (1981) and Whitehead (1989).

Alouatta caraya (Humboldt, 1812)
Black Howler, Mono Aullador Negro, Carayá

Measurements

	Mean	Min.	Max.	N	Loc.	Source[a]
TL	1,711.5	1,075.0	1,260.0	12 m	A, Br,	1
	1,043.3	991.0	1,100.0	8 f	P	
HB	549.8	500.0	610.0	12 m		
	483.8	460.0	510.0	8 f		
T	591.2	540.0	650.0	12 m		
	559.5	526.0	600.0	8 f		
HF	142.5	140.0	145.0	2 m		
	140.0			1 f		
E	41.0			1 m		
Wta	6.87 kg	4.6 kg	9.8 kg	14 m		
	4.84 kg	3.1 kg	7.8 kg	10 f		

[a](1) Allen 1916; Thorington, Ruiz, and Eisenberg 1984; Krieg 1948; Kühlhorn 1954; Schaller 1983; Pope 1966; PCorps.

Description
Like other members of the genus *Alouatta, A. caraya* demonstrates extreme sexual dimorphism, in color as well as in size. Males are all black and weigh about 6.5 kg, whereas females are yellowish brown and weigh about 4.5 kg and infants are yellow to reddish buff. In captivity young males change pelage color at between thirty-two and forty months; in the wild they change at about fifty-four months. In Argentina females never exceeded 6.5 kg, but some males weighed 9.5 kg (Jones 1983; Pope 1966; Thorington, Ruiz, and Eisenberg 1984).

Distribution
Alouatta caraya is confined to southern Brazil, Paraguay, Bolivia, and northern Argentina. In Argentina it is found in Salta, Formosa, Chaco, Santa Fe, and Misiones provinces and along gallery forests to southern Corrientes. In eastern Paraguay this howler monkey commonly occurs in forested habitats, and in the Paraguayan Chaco it generally occupies the gallery forests along the Río Paraguay and Río Pilcomayo as well as the small rivers that drain the Chaco. *A. caraya* can also be found sporadically throughout the more xeric areas of the contiguous forested Chaco (Cabrera 1939; Olrog 1984; Stallings 1984; Stallings and Mittermeier 1983; UM) (map 5.2).

Life History
Females give birth to a single young, rarely twins, after a gestation period of about 187 days. The birth-weight of a single young born in captivity was 113 g. In northeastern Argentina and southwestern Brazil young are born between August and October, and in north-central Argentina the birth peak is at the onset of the dry season. Females stop growing at sexual maturity, though males apparently continue growing for several years. Males do not show adult coloration until about 4.5 years of age and about 5 kg weight. In the wild animals may live at least twenty years (Crespo 1982; Miller 1930; Pope 1966; Shoemaker 1979; Thorington, Ruiz, and Eisenberg 1984).

Ecology
Alouatta caraya occurs in small, multiple-male troops with the males age graded. In northern Ar-

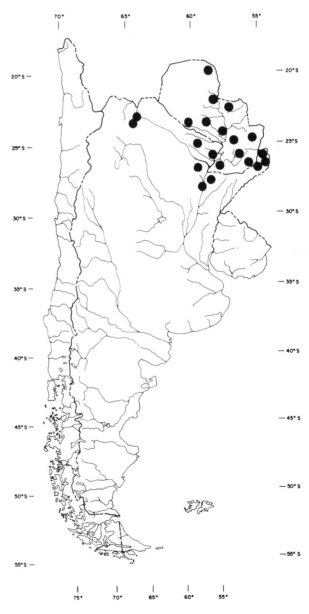

Map 5.2. Distribution of *Alouatta caraya*.

gentina near Puerto Bermejo the adult sex ratio was 0.56, with infants making up 6%–14% of the troop and juveniles 16%–21%. The mean troop size ranged from 7.2 to 8.9 adults (range 3–19). The animals were found at an ecological density of 130/km². This agrees with another study in northern Argentina that found nine troops to average 7.2 individuals (range 3–15). In an island population in Corrientes, group size varied seasonally from 2.5 to 10, with single males also recorded. In the Paraguayan Chaco the largest group counted was 10, and the average troop contained 7 (*n* = 2). In a study conducted in Mato Grosso state, Brazil, the average group size was 7.2, and an average group had 1.8 adult males, 0.6 subadult males, 2.4 females, 0.8 yearlings, and 1.6 young.

Within the past seventy-five years yellow fever swept through the range of this species, and it is not known how it may have affected its distribution and abundance (Crespo 1982; Piantanida et al. 1984; Rathbun and Gache 1980; Schaller 1983; Stallings 1984; Thorington, Ruiz, and Eisenberg 1984; Rumiz 1990).

Alouatta fusca (E. Geoffroy, 1812)
Mono Aullador Rufo

Measurements

	Mean	Min.	Max.	N	Loc.	Source[a]
TL	1,001.0	880	1,150	10 m	Br	1
	956.5	870	1,055	10 f		
HB	550.3	500	641	10 m		
	494.3	420	540	10 f		
T	451.7	370	590	10 m		
	458.7	400	555	10 f		
HF	141.0	140	143	3 m		
	131.7	120	140	3 f		

[a](1) R. W. Thorington (pers. comm.), from Museu de Zoologia, São Paulo.

Description

In southern South America the subspecies *A. fusca clamitans* exhibits color dimorphism: adult males are bright reddish with a golden tint, and most adult females are brown, as are immature males (plate 5) (Cordeiro da Silva 1981; Kinzey 1982).

Distribution

Alouatta fusca is found only in southeastern and south-central Brazil and in the Argentine province of Misiones. Both *A. caraya* and *A. fusca* are reported from the department of Montecarlo in Misiones province. Another population of *A. fusca* has been reported from Bolivia, but it is not clear whether this is based on accurate identification of specimens (Crespo 1954; Honacki, Kinman, and Koeppl 1982) (map 5.3).

Ecology

This howler seems to prefer moister forests than *A. caraya* and is found in the Brazilian Atlantic forest, the southern Araucaria forest of Brazil, and into the tropical forest of Misiones province, Argentina. In the Brazilian state of São Paulo group size ranged from 2 to 22, with an average of 5.76 (*n* = 25), and the mean population density was 80.9/km² (±32.5). The average group consisted of 1.76 adult males and 2.3 adult females. In this Araucaria forest 70.6% of the feeding observations were on leaves and 29.4% on fruit. In a forest in Rio Grande do Sul *A. fusca*

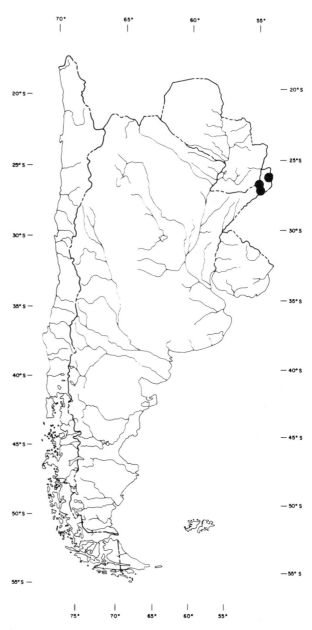

Map 5.3. Distribution of *Alouatta fusca.*

was found eating figs (*Ficus*), leaves, *Inga* fruit, and *Cecropia* fruit. Like *A. caraya*, this species was adversely affected by yellow fever (Chitolina and Sander 1981; Cordeiro da Silva 1981; Crespo 1974).

Comment

Some authors list this species as *Alouatta guariba*.

Genus *Aotus* (Illiger, 1811)
Aotus azarae (Humboldt, 1812)
Night Monkey, Mono de Noche, Cai Pyhare, Mbirikina

Measurements

	Mean	Min.	Max.	N	Loc.	Source[a]
TL	734.2	650	800	27	A, Br,	1
HB	330.4	285	380		P	
T	403.9	355	450			
HF	94.4	85	105			
E	32.1	22	41			
Wta	962.9	600	1,360	7		

[a](1) Allen 1916; Crespo 1974; PCorps, UConn.

Description

The upperparts and sides are grayish or pale buffy agouti, and the upper surface of the proximal portion of the tail is pale orange or straw colored. The facial markings are distinct. The eyeshine is bright pink (plate 5) (Hershkovitz 1983; Rathbun and Gache 1980).
Chromosome number: In northern Argentina $2n = 50$ in females and $2n = 49$ in males (Mudry de Pargament, Colillas, and Brieux de Salum 1984).

Distribution

Aotus azarae is distributed throughout the Chaco Boreal, where it ranges north of the Río Pilcomayo to the northern Paraguayan border with Bolivia and west of the Río Paraguay to about 62° S. In Argentina this species has been recorded from Formosa and Chaco provinces (Crespo 1974; Hershkovitz 1983; Rathbun and Gache 1980; Stallings 1984) (map 5.4).

Life History

Both in captivity and in the wild there appears to be no pronounced birth season, but in the Paraguayan Chaco females were seen with infants only in August, September, and October. In captivity the interbirth interval is eight to nine months (Dixson et al. 1979; Rathbun and Gache 1980; J. R. Stallings, unpubl. data).

Ecology

In Formosa province, Argentina, *Aotus* habitat is highly disjunct, being a mosaic of high and low forest islands surrounded by brushlands and seasonally flooded savanna. Night monkeys are restricted to both low and high forests that have relatively dense canopies and highly diverse plant species. In the northern Paraguayan Chaco the night monkey is common both in 5–10 m canopy low scrub forest and in 15–20 m canopy higher forest. *Aotus* in the southern Paraguayan Chaco occurs in habitat similar to that of Argentine animals.

In Argentina the mean group size was 2.9 animals ($n = 25$), with a range of 1–4 and a density of 17–24 km². Similarly, in Paraguay the mean group size in two locations was 2.7 and 3.1 (range 1–5; $n = 27$).

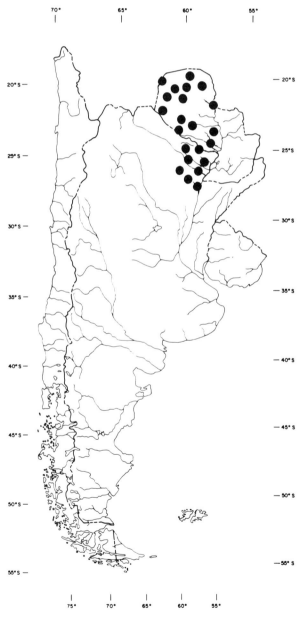

Map 5.4. Distribution of *Aotus azarae*.

In more xeric areas of the Chaco, group size is smaller than in more mesic areas. Ecological density estimates for two locations were 8.85 and 14.36 km². It is likely that the groups consisted of an adult male, an adult female, and one young.

Aotus is usually active during the night but can be seen during overcast, cool, damp days. In El Beni, Bolivia, activity started with low-intensity vocalizations about fifteen minutes after sunset. The monkeys returned to sleeping sites ten to twenty minutes before dawn. Feeding was done mostly during the early hours of the night. In the Paraguayan Chaco, during the day *Aotus* pairs and family groups can be seen huddled together among the branches of large tree cacti (Rathbun and Gache 1980; Stallings 1984; Stallings and Mittermeier 1983; Stallings et al. 1989).

Genus *Callicebus* Thomas, 1903
• *Callicebus moloch* (Hoffmannsegg, 1807)
Dusky Titi Monkey, Cai Ygau

Measurements

	Mean	Min.	Max.	N	Loc.	Source[a]
TL	687.5	580	740	6	Br, P	1
HB	282.5	225	313			
T	405.0	355	440			
HF	84.8	79	91			
E	31.2	30	32			
Wta	752.0	690	800	5		

[a](1) Schaller 1983; PCorps.

Description
The dorsum is gray, reddish, or brown and buff; the venter is similar with occasional lateral areas of orange, red, or buff. The feet are gray, red, dark brown, or black. The tail is either the same color as the dorsum or dark gray and frequently has a white tip (plate 5) (Jones and Anderson 1978; J. R. Stallings, unpubl. data).

Distribution
Callicebus moloch is found in eastern Colombia and in Ecuador, Peru, Bolivia, Paraguay, and western Brazil. In Paraguay it is restricted to the Chaco from the northern border with Bolivia south to about 23° S and from the Río Paraguay west to about 61° W (Honacki, Kinman, and Koeppl 1982; Stallings 1984) (map 5.5).

Life History
Callicebus moloch forms family groups consisting of a monogamous pair and one or two young. In the Paraguayan Chaco neonates have been observed only during October and November (J. R. Stallings, unpubl. data).

Ecology
In the Paraguayan Chaco the dusky titi monkey is common both in 5–10 m canopy low scrub forest and 15–20 m canopy higher forest. In Mato Grosso state, Brazil, it is found in deciduous forests, in bamboo forests with emergent trees, and in scrub forest bordering floodplains.

These highly arboreal monkeys usually inhabit small home ranges. They have a rather rich vocal repertoire. A bonded male and female will produce a duet calling sequence, which has several elements including bellows and pants. Countercalling be-

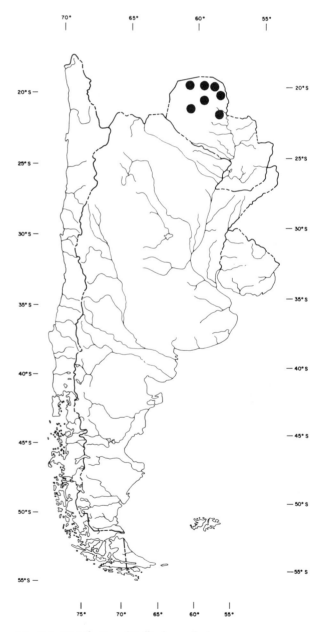

Map 5.5. Distribution of *Callicebus moloch*.

tween adjacent social groups is involved in territorial defense.

The basic unit of social organization is the pair and their various-aged offspring. Therefore group size is usually small: in Paraguay the average was 2.8 (range 1–6; $n = 26$); in Mato Grosso it was 3.3 (range 2–5; $n = 17$). In the Paraguayan Chaco *C. moloch* was found at an ecological density of 17.7 km². The diet is largely fruit, though leaves and insects are included. One stomach from a Brazilian animal contained 90% seed pulp and 10% *Atta* ants.

Like *Aotus, Callicebus* groups can be seen on

cool, damp days huddled on branches and will seek refuge in tree cacti (Jones and Anderson 1978; Robinson 1977, 1979a,b; Schaller 1983; Stallings 1984, unpubl. data; Stallings and Mittermeier 1983; Stallings et al. 1989).

Genus *Cebus* Erxleben, 1777
• *Cebus apella* (Linnaeus, 1758)
Tufted Capuchin, Cai Común, Mono

Measurements

	Mean	Min.	Max.	N	Loc.	Source[a]
TL	833.5	780.0	970.0	15	A, Br,	1
HB	402.9	363.0	480.0			
T	430.6	394.0	490.0			
HF	117.7	110.0	130.0	14		
E	35.8	28.0	41.0	10		
Wta	3.34 kg	2.5 kg	4.0 kg	50 m	P	2
	2.45 kg	2.0 kg	4.5 kg	50 f		

[a](1) Cabrera 1939; Crespo 1950, 1974; PCorps, UM; (2) K. Hill, unpubl. data (weights used only for males >2.5 kg and females >2.0 kg).

Description
This robust species has a brownish general body color. The cap on top of the head is composed of short, erect dark brown hairs that in males form ridges on either side of the crown. It contrasts sharply with the light brown body color (plate 5). Males tend to be almost 1 kg heavier than females.

Distribution
Cebus apella ranges all the way from Colombia south to northern Argentina. In southern South America it is found throughout eastern Paraguay and may make its way into the neighboring Chaco and into Argentina, at least in the northern forests from Jujuy and Salta provinces discontinuously east to the forests of Misiones province. The most southerly recorded location for *C. apella* is Parque Nacional El Rey in central Salta (Brown, Chalukian, and Malmierca 1984; Cabrera 1939; Stallings 1984; BA, UM) (map 5.6).

Life History
A single young is born after a gestation period of approximately 160 days. Young males apparently do not become sexually mature until at least seven years of age. Females may conceive in their fourth year. In Paraguay young are seen only from September to December, whereas in Misiones province, Argentina, births take place from the end of winter throughout the spring (Clark 1983; Crespo 1982; Eisenberg 1989).

Ecology
In Peru the diet of this strictly diurnal species includes fruit, palm nuts, seeds, and a considerable

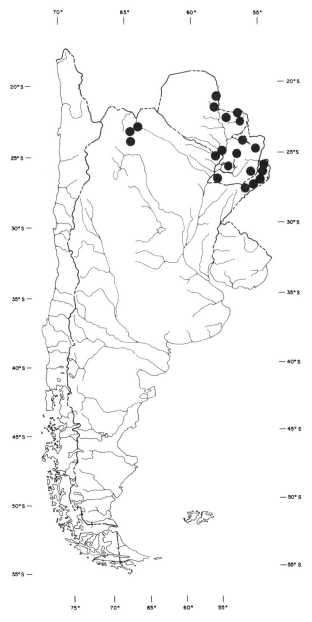

Map 5.6. Distribution of *Cebus apella*.

quantity of insects. The mating system is polygamous; generally a single adult male is dominant over all the males in the troop. Communication patterns have been described for captives.

In eastern Paraguay *C. apella* is most often seen in high and intermediate forests. There groups ranged from one to ten, with most groups having seven (*n* = 74), and troop home ranges varied from 0.25 to 0.40 ha. In Misiones province, Argentina, the average troop size was 10.8 (range 3–15), and in Mato Grosso state, Brazil, it was 8 (range 3–20). In Paraguay the capuchin was found at an ecological density of twenty-eight individuals per square kilometer. In central Salta province, Argentina, *C. apella* has very large home ranges, and bromeliads compose a large proportion of the diet (Brown, Chalukian, and Malmierca 1984; Clark 1983, Crespo 1982; Eisenberg 1989; Schaller 1983; Stallings 1984; Terborgh 1983).

Comment
Cebus apella paraguayanus has a karyotype of 54. This species includes *C. paraguayensis* (Matayoshi et al. 1986).

References

Allen, J. A. 1916. Mammals collected on the Roosevelt Brazilian expedition, with field notes by Leo E. Miller. *Bull. Amer. Mus. Nat. Hist.* 34: 559–610.

Brown, A., S. Chalukian, and L. Malmierca. 1984. Habitat y alimentación de *Cebus apella* en el N.O. argentino y la disponibilidad de frutos en el dosel arbóreo. *Rev. Mus. Argent. Cienc. Nat. "Bernardino Rivadavia," Zool.* 13(1–60): 273–80.

Cabrera, A. 1939. Los monos de la Argentina. *Physis* 16: 3–29.

Chitolina, O. P., and M. Sander. 1981. Contribuição ao conhecimento da alimentação de *Alouatta guariba clamitans* Cabrera, 1940 em habitat natural no Rio Grande do Sul (Cebidae, Alouattinae). *Iheringia, Ser. Zoo.* (Porto Alegre) 59: 37–44.

Clark, D. 1983. Observaciones de *Cebus apella* en el Parque Nacional Ybycui. Unpubl. MS. in authors' files.

Cordeiro da Silva, E. 1981. A preliminary survey of brown howler monkeys (*Alouatta fusca*) at the Cantareira Reserve (São Paulo, Brazil). *Rev. Brasil. Biol.* 41(4): 897–909.

Crespo, J. A. 1950. Nota sobre mamíferos de Misiones nuevos para Argentina. *Comun. Inst. Nac. Invest. Cienc. Nat. Mus. Argent. Cienc. Nat. "Bernardino Rivadavia," Zool.* 1(14): 1–14.

———. 1954. Presence of the reddish howling monkey (*Alouatta guariba clamitans* Cabrera) in Argentina. *J. Mammal.* 35: 117–18.

———. 1974. Comentarios sobre nuevas localidades para mamíferos de Argentina y de Bolivia. *Rev. Mus. Argent. Cienc. Nat. "Bernardino Rivadavia," Zool.* 11(1): 1–31.

———. 1982. Ecología de la comunidad de mamíferos del Parque Nacional Iguazú, Misiones. *Rev. Mus. Argent. Cienc. Nat. "Bernardino Rivadavia," Ecol.* 3(2): 45–162.

Crockett, C. M., and J. F. Eisenberg. 1987. Howlers: Variations in group size and demography. In *Primate societies*, ed. B. B. Smuts, D. L. Cheney, R. M. Seyfarth, R. W. Wrangham, and T. T. Struhsaker, 54–68. Chicago: University of Chicago Press.

Dixson, A. F., R. D. Martin, R. C. Bonney, and D. Fleming. 1979. Reproductive biology of the owl monkey (*Aotus trivirgatus griseimambra*). Unpublished paper presented at the seventh International Congress on Primatology, Bangalore, India.

Eisenberg, J. F. 1979. Habitat, economy and society: Some correlations and hypotheses for the Neotropical primates. In *Primate ecology and human origins*, ed. I. S. Bernstein and E. O. Smith, 215–62. New York: Garland Press.

———. 1989. *Mammals of the Neotropics.* Vol. 1. *Mammals of the northern Neotropics: Panama, Colombia, Venezuela, Guyana, Suriname, French Guiana.* Chicago: University of Chicago Press.

Hershkovitz, P. 1977. *Living New World monkeys (Platyrrhini).* Vol. 1. Chicago: University of Chicago Press.

———. 1983. Two new species of night monkeys, genus *Aotus* (Cebidae, Platyrrhini): A preliminary report on *Aotus* taxonomy. *Amer. J. Primatol.* 4: 209–43.

Hill, W. C. O. 1957. *Primates.* Vol. 3. Edinburgh: Edinburgh University Press.

———. 1960. *Primates.* Vol. 4. Edinburgh: Edinburgh University Press.

———. 1962. *Primates.* Vol. 5. Edinburgh: Edinburgh University Press.

Honacki, J. H., K. E. Kinman, and J. W. Koeppl, eds. 1982. *Mammal species of the world.* Lawrence, Kans.: Allen Press and Association of Systematics Collections.

Jones, C., and S. Anderson. 1978. *Callicebus moloch. Mammal. Species* 112: 1–5.

Jones, C. B. 1983. Social organization of captive black howler monkeys (*Alouatta caraya*): "Social competition" and the use of nondamaging behavior. *Primates* 24: 25–39.

Kinzey, W. G. 1982. Distribution of primates and forest refuges. In *Biological diversification in the tropics*, ed. G. T. Prance, 155–82. Proceedings of the Fifth International Symposium of the Association for Tropical Biology. New York: Columbia University Press.

Kleiman, D. G., ed. 1977. *The biology and conservation of the Callitrichidae*. Washington, D.C: Smithsonian Institution Press.

Krieg, H. 1948. *Zwischen Anden und Atlantik*. Munich: Carl Hanser.

Kühlhorn, F. 1954. Säugetierkundliche Studien aus Süd-Mattogrosso. 2. Edentata, Rodentia. *Säugetierk. Mitt.* 2:66–72.

Mares, M. A., R. A. Ojeda, and R. M. Barquez. 1989. *Guide to the mammals of Salta province, Argentina*. Norman: University of Oklahoma Press.

Matayoshi, T., E. Howlin, N. Nasazzi, C. Nagle, E. Gadow, and H. N. Seuanez. 1986. Chromosome studies of *Cebus apella*: The standard karyotype of *Cebus apella paraguayanus* Fischer, 1829. *Amer. J. Primatol.* 10:185–93.

Miller, F. W. 1930. Notes on some mammals of southern Mato Grosso, Brazil. *J. Mammal.* 11:10–22.

Mudry de Pargament, M. D, O. J. Colillas, and S. Brieux de Salum. 1984. The *Aotus* from northern Argentina. *Primates* 25(4): 530–37.

Olrog, C. C. 1984. El mono caraya. In *Mamíferos fauna argentina*, vol. 1. Buenos Aires: Centro Editor de America Latina.

Piantanida, M., S. Puig, N. Nani, F. Rossi, L. Cavanna, S. Mazzucchelli, and A. Gil. 1984. Introducción al estudio de la ecología y etología del mono aullador (*Alouatta caraya*) en condiciones naturales. *Rev. Mus. Argent. Cienc. Nat. "Bernardino Rivadavia," Ecol.* 3(3): 163–92.

Pope, B. 1966. The population characteristics of howler monkeys (*Alouatta caraya*) in northern Argentina. *Amer. J. Phys. Anthropol.* 24:361–70.

Rathbun, G. B., and M. Gache. 1980. Ecological survey of the night monkey, *Aotus trivirgatus*, in Formosa province, Argentina. *Primates* 21:211–19.

Robinson, J. 1977. The vocal regulation of spacing in the titi monkey, *Callicebus moloch*. Ph.D. diss., University of North Carolina.

———. 1979a. An analysis of the organization of vocal communication in the titi monkey *Callicebus moloch*. *Z. Tierpsychol.* 49:381–405.

———. 1979b. Vocal regulation of use of space by groups of titi monkeys *Callicebus moloch*. *Behav. Ecol. Sociobiol.* 5:1–15.

Rumiz, D. 1990. *Alouatta caraya*: Population density and demography. *Amer. J. Primatol.* 21:279–94.

Schaller, G. B. 1983. Mammals and their biomass on a Brazilian ranch. *Arq. Zool. São Paulo* 31(1):1–36.

Sekulic, R. 1981. The significance of howling in the red howler monkeys *Alouatta seniculus*. Ph.D. diss., University of Maryland.

———. 1982. Daily and seasonal patterns of roaring and spacing in four red howler *Alouatta seniculus* troops. *Folia Primatol.* 39:22–48.

Shoemaker, A. H. 1979. Reproduction and development of the black howler monkey *Alouatta caraya* at Columbia Zoo. *Int. Zoo Yearb.* 19:150–55.

Stallings, J. R. 1984. Status and conservation of Paraguayan primates. M.A. thesis, University of Florida.

Stallings, J. R., and R. A. Mittermeier. 1983. The black-tailed marmoset (*Callithrix argentata melanura*) record from Paraguay. *Amer. J. Primatol.* 4:159–63.

Stallings, J. R., L. West, W. Hahn, and I. Gamarra. 1989. Primates and their relation to habitat in the Paraguayan Chaco. In *Advances in Neotropical mammalogy*, ed. K. H. Redford and J. F. Eisenberg, 425–42. Gainesville, Fla.: Sandhill Crane Press.

Szalay, F. S., and E. Delson. 1979. *The evolutionary history of primates*. New York: Academic Press.

Terborgh, J. 1983. *Five New World primates: A study in comparative ecology*. Princeton: Princeton University Press.

Thorington, R. W., Jr., R. Rudran, and D. Mack. 1979. Sexual dimorphism of *Alouatta seniculus* and observation on capture techniques. In *Vertebrate ecology in the northern Neotropics*, ed. J. F. Eisenberg, 97–106. Washington, D.C.: Smithsonian Institution Press.

Thorington, R. W., J. C. Ruiz, and J. F. Eisenberg. 1984. A study of a black howling monkey (*Alouatta caraya*) population in northern Argentina. *Amer. J. Primatol.* 6:357–66.

Vieira, C. 1945. Sôbre uma coleção de mamíferos de Mato Grosso. *Arqu. Zool. São Paulo* 4:395–430.

Whitehead, J. M. 1989. The effect of the location of a simulated intruder on responses to long-distance vocalizations of mantled howling monkeys, *Alouatta palliata palliata*. *Behaviour* 108 (1–2):73–103.

6   Order Carnivora (Fissipedia)

Diagnosis

The order Carnivora includes most extant terrestrial mammals specialized for predation on other vertebrates. Many members are omnivores, and most are terrestrial or scansorial, except the semiaquatic otters. Dental formulas are somewhat variable, with reduction in tooth number pronounced in the Felidae, and will be presented under the family accounts. The canine is conical and prominent. The modern Carnivora are usually defined based on the major carnassial shearing effect of the molariform teeth; the upper fourth premolar shears against the lower first molar. The digestive system shows no extreme modifications, and the cecum is usually small. The brain is relatively large and the braincase somewhat inflated, with the tympanic bullae enlarged and hemispheric. Of the eight extant families, five occur in South America.

We treat the Pinnipedia separately, since we consider them monophyletic. Although clearly derived from the Carnivora, they are sufficiently distinct to warrant ordinal status.

Distribution

The Recent distribution includes all continents except Antarctica and Australia; however, the dog (*Canis familiaris*) was introduced to Australia by aboriginal humans about 4,000–7,000 B.P., possibly later than its introduction to the Western Hemisphere. Members of this order tolerate a variety of habitats, having extended to the extreme climatic conditions of the Arctic and the alpine zones of high mountains.

History and Classification

In Paleocene times, the suborder Creodonta first appeared and radiated into the Eocene. Some creodonts persisted until the early Pliocene, but in general the original suborder was replaced by the suborder Fissipedia, which first appeared in the middle Paleocene as the family Miacidae, known from North America, Europe, and Asia. The modern families of carnivores are recognizable in the early Oligocene. Members of the order Carnivora were originally absent from South America and Australia; the carnivore niches there were filled by a parallel radiation within the order Marsupialia, resulting in the Dasyuridae (Australia) and the Borhyaenidae and Didelphidae (South America). The first true carnivores to enter South America appeared in the Miocene as an early raccoonlike form (Linares 1981). The remaining carnivores currently found there entered the continent via the Panamanian land bridge during the late Pliocene. Standard references for the biology of the Carnivora are Ewer (1973) and Gittleman (1989).

Note: In addition to the species discussed in this volume, an additional bear, *Tremarctos ornatus*, has been found in department of Tarija, Bolivia, and may have been found or may eventually be found in Salta province, Argentina (Brown and Rumiz, n.d.). A second, much more problematic species is the "Andean wolf," *Dasycyon hagenbecki*, described from a skin obtained from a dealer in Argentina (Krumbiegel 1953). Its status remains in doubt (Dieterlen 1954). A third species of carnivore, the Falkland Island wolf, was driven to extinction by Europeans by the end of the 1800s (Allen 1942).

FAMILY CANIDAE

Diagnosis

The dental formula for New World species is I 3/3, C 1/1, P 4/4, M 2/3, but some variation is shown in the number of molars. In the genus *Speothos*, molar number varies 1–2/2 (figs. 6.1 and 6.2). The

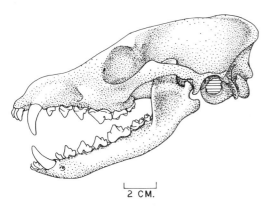

Figure 6.1. Skull of *Dusicyon culpaeus.*

canines are long and prominent. The molars have small crushing surfaces, and the lower first molar and upper third premolar provide the carnassial shearing surfaces. The rostrum of the skull is long. Members of this family are adapted for a cursorial gait; they walk on their toes, and some reduction in toe number is demonstrable. There are generally five toes on the forefoot and four on the hind foot, but on the forefoot one toe is reduced to a dewclaw. The claws are nonretractile.

Distribution

Species of this family occur on all continents except Antarctica. They were introduced to Australia by early humans.

Genus *Cerdocyon* Hamilton-Smith, 1839
• *Cerdocyon thous* (Linnaeus, 1766)
Zorro de Monte

Measurements

	Mean	Min.	Max.	N	Loc.	Source[a]
TL	952.7	840.0	1,060.0	25	A, P	1
	981.1	825.0	1,070.0	13	U	2
HB	640.2	545.0	750.0	25	A, P	1
T	312.5	275.0	382.0			
	316.6	220.0	375.0	13	U	2
HF	136.5	125.0	155.0	21	A, P	1
	148.8	138.0	164.0	13	U	2
E	71.1	65.0	85.0	24	A, P	1
	73.6	65.0	86.0	13	U	2
Wta	5.6 kg	3.5 kg	8.2 kg	17	A, P	1
	6.6 kg	5.7 kg	8.9 kg	13	U	2

[a](1) Crespo 1982; BA, PCorps; (2) Barlow 1965.

Description

This fox is nearly the same size as *Dusicyon gymnocercus* but is slightly more robust, with shorter and coarser fur (plate 6) and a slightly shorter snout. The dorsum is grizzled grayish brown, with black conspicuous along the middorsum, and there may be black over the shoulders and in varying amounts on the hind legs and forearms. In some specimens

there is a reddish cast on the cheeks. The lips, ears, feet, and top of the tail are all black; the venter is grayish white. There is a good deal of individual variation, with some individuals much darker than others (Barlow 1965; Berta 1982; Stains 1975; pers. obs.).

Distribution
Cerdocyon thous is distributed from Colombia and Venezuela south to Uruguay and Paraguay, and in northern Argentina it ranges south to the province of Entre Ríos (Crespo 1984; Honacki, Kinman, and Koeppl 1982; BA) (map 6.1).

Life History
In captivity animals mature sexually at about nine months and breed twice a year. Litter size ranges from three to six (*n* = 4), young are born weighing 120 to 160 g, gestation averages fifty-six days (range 52–59), and pups open their eyes at fourteen days. During the first thirty days they take only milk; until ninety days they take both solid food and milk; and after ninety days the pups are fully weaned. At about five months the family group breaks up. In Venezuela pregnant females are found throughout the year, but pups are most commonly seen in February and March (Berta 1982; Brady 1978).

Ecology
This fox is found in a variety of habitats, from savanna to woodland, and up to at least 2,000 m in northwestern Argentina. In Uruguay it prefers forest and edge areas, separating itself from *Dusicyon gymnocercus*, which is found in more open country, but, in northeastern Argentina it is found in all habitat types.

In the Venezuelan llanos *C. thous* is nocturnal and pair bonded, and individuals were usually seen in pairs. Larger groups always consisted of adults with their young. Adults traveled extensively within

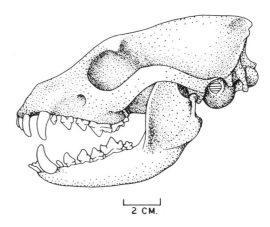

Figure 6.2. Skull of *Speothos venaticus.*

relatively small home ranges, from 0.6 to 0.9 km². Partners may locate one another by a characteristic high-pitched whistle.

Cerdocyon thous is omnivorous: in Venezuela 104 stomachs were found to contain 26% small mammals, 24% fruits, 13% amphibians, 11% insects (mostly orthopterans), 10% reptiles, and 9% birds. In the low llanos, crustaceans, principally crabs, are a very common food during the wet season. In Uruguay one stomach contained palm fruit, one skink, and two grasshoppers; in Paraguay one stomach had four snails, one frog, and two crabs; in Misiones province, Argentina, one stomach had 40% plant material, in-

cluding mushrooms, one *Dasypus* sp., insects, and frogs; and in southern Brazil two stomachs contained grasshoppers plus *Astrocarpus* and *Syzygium* seeds. Four stomachs from Pantanal foxes contained frogs, fruit, crabs, beetles, birds, and fish (Barlow 1965; Berta 1982; Bisbal and Ojasti 1980; Brady 1979; Coimbra-Filho 1966; Crespo 1982; Langguth 1975; Montgomery and Lubin 1978; Olrog 1979; Schaller 1983; Sunquist, Sunquist, and Daneke 1989; UConn).

Genus *Chrysocyon* Smith, 1839
Chrysocyon brachyurus (Illiger, 1815)
Maned Wolf, Aguará Guazú, Lobo de Crin

Measurements

	Mean	Min.	Max.	N	Loc.	Source[a]
TL	1,491.0	1,470.0	1,512.0	2	Peru, P	1
HB	1,030.0	950.0	1,070.0	9	Br	2
	1,093.5	1,067.0	1,120.0	2	Peru, P	1
T	445.0	380.0	490.0	9	Br	2
	397.5	350.0	445.0	2	Peru, P	1
HF	295.0	275.0	320.0	9	Br	2
E	168.0	150.0	180.0			
Wta	23.3 kg	20.5 kg	25.8 kg			
	24.2 kg	23.5 kg	25.0 kg	2	Peru, P	1

[a](1) Dietz 1981; (2) Hofmann, Ponce del Prado, and Otte 1975–76; UConn.

Description

The maned wolf is unmistakable, with its long, thick buff red pelage, long legs, and large ears (plate 6). It is the largest canid in South America, standing about 900 mm at the shoulder. Its legs are black, as is part of the erectile mane, and there is white on the pinnae, under the chin, and on the tip of the tail (Dietz 1981; Kleiman 1972).

Distribution

Chrysocyon brachyurus is found from the lowlands of Bolivia south through central Brazil into Paraguay, Argentina, and northeastern Brazil. This species is poorly represented in collections (Honacki, Kinman, and Koeppl 1982) (map 6.2).

Life History

The maned wolf is monoestrous, with matings in the Southern Hemisphere occurring between April and June, and most litters are born from June to September. The gestation period is sixty-two to sixty-six days, and the litter size averages 2.2 (range 1–5; n = 25). The pups are born weighing about 350 g and open their eyes at between eight and nine days (Brady and Ditton 1979; Dietz 1981, 1984).

Ecology

Chrysocyon frequents grassland and scrub habitats. In a comprehensive field study in central Brazil, maned wolves spent 34% of their time in open grass-

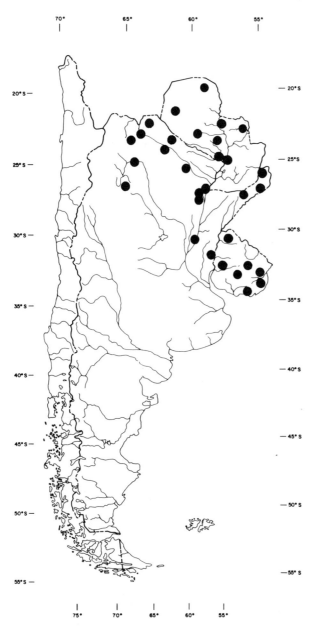

Map 6.1. Distribution of *Cerdocyon thous*.

land, 43% in cerrado, and 24% in forest. In northern Argentina this species inhabits grasslands, often rather marshy, preferring those that are interrupted by thickets, gallery forests, and palm stands. *Chrysocyon* is mostly nocturnal and spends the day resting in thick cover.

Individual adults are strictly territorial against same-sex adults. They are facultatively monogomous, with a long-term pair-bond between the male and the female, though animals are usually seen alone. There is a good deal of scent marking, and pair mates often communicate over long distances using a high amplitude roar/bark. In the wild males do not care for the young but are primarily territorial defenders. Females regurgitate food for the young, and in captivity males do so as well.

Maned wolves are omnivores, eating small vertebrates and invertebrates and large quantities of fruit, particularly that of *Solanum lycocarpum*. In central Brazil small rodents were found in 77.9% of the scats examined ($n = 68$), birds were found in 17.6%, armadillos in 8.8%, *Solanum* fruit in 82.4%, and grass in 2.9%. In central Brazil, out of 740 scats examined, 28% of the volume was small mammals, 2.3% was birds, and 57.6% was *Solanum lycocarpum* fruit (Brady and Ditton 1979; Dietz 1981, 1984; Rasmussen and Tilson 1984; Schaller 1983; G. B. Schaller and A. Tarak, unpublished data; UConn).

Genus *Dusicyon* Hamilton-Smith, 1839
South American Fox, Zorro

Description

There has been much controversy concerning the taxonomy of the South American canids, but in this volume we recognize *Dusicyon* as a valid genus, in conformity with Langguth (1975). Species of *Dusicyon* are medium-sized canids ranging from the small *D. sechurae* (HB 445 mm) to the large, robust *D. culpaeus* (HB may exceed 1,000 mm). The dorsum exhibits some pattern of brown, often speckled with black. The venter is usually some shade of cream to white (see species accounts and plate 6).

Distribution

Depending on one's taxonomic preferences, the genus can have a broad or a restricted range. The members as so defined in this volume are South American and are to be found in almost all major habitat categories from puna to cerrado.

Dusicyon culpaeus (Molina, 1792)
Culpeo, Zorro Colorado

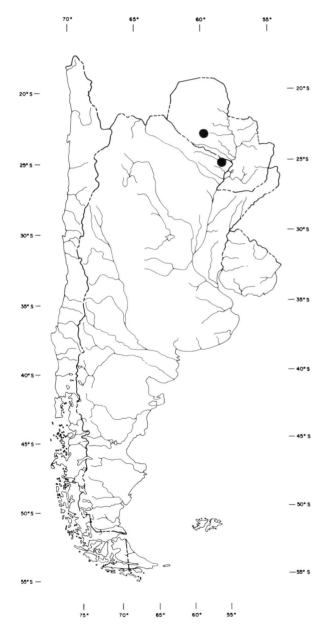

Map 6.2. Distribution of *Chrysocyon brachyurus*.

Measurements

	Mean	Min.	Max.	N	Loc.	Source[a]
TL	1,150.0	820.0	1,520	150 m	A	1
	1,102.0	900.0	1,240	107 f		
	1,172.5	1,110.0	1,240	6	C	2
T	412.0	300.0	510	130 m	A	1
	395.0	315.0	455	108 f		
	439.5	422.0	465	6	C	
HF	163.0	140.0	181	143 m	A	1
	152.0	130.0	160	106 f		
	176.3	165.0	190	6	C	2
E	88.0	80.0	95	144 m	A	1
	84.0	76.0	90	104 f		
	98.9	95.0	103	6	C	2
Wta	8.05 kg	4.00 kg	13 kg	150 m	A	1
	6.69 kg	3.65 kg	11 kg	106 f		

[a](1) Crespo and Carlo 1963; (2) Greer 1966.

Description

After *Chrysocyon*, this is the largest of the South American canids. There is much color and size variation, but the overall impression is of a large, broadheaded fox with a wide muzzle and considerable orange on its coat. In Chile it ranges from 700 to 900 mm in body length over its range from 33° to 54° south latitude, being smallest in the north and larger toward the south (the opposite pattern from *D. griseus*). In Chile foxes of this species usually are reddish on the sides, with the middorsum showing a brownish stripe. Foxes from Argentina are also highly variable in color. The black and gray dorsum is almost spotted, producing an agouti effect. The

amount of black on the tail is highly variable. There is a distinct reddish wash on the face, cheeks, limbs, and sometimes the tail, and the venter is cream to almost pale orange. There is a mutant form with a much more pronounced orange color to its coat (Crespo 1962; Fuentes and Jaksić 1979).

Distribution

Dusicyon culpaeus is distributed along the Andes from Colombia to southernmost South America. It is found throughout Chile and in Argentina along the Andes from Jujuy province south to Patagonia. From Río Negro province south it is found across the width of the country (Crespo 1984; Honacki, Kinman, and Koeppl 1982) (map 6.3).

Life History

In central Argentina males produce sperm from June to the middle of October, and the females are monoestrous, ovulating from early August to October. Females are fertile before one year of age. The gestation period is fifty-five to sixty days, the number of embryos per female averages 5.2 ($n = 6$), and litters are produced from October to December. Lactation lasts about two months. In Chile births were recorded in November, and the litter size ranged from three to five (Crespo and Carlo 1963; Housse 1953; Rabinovich et al. 1987).

Ecology

Dusicyon culpaeus is found in many kinds of habitats up to at least 4,500 m. Most of these contemporary habitats are arid or semiarid, but it does penetrate the dense subantarctic forests of the Patagonian Andes. In Neuquén province, Argentina, this species is found in all available habitats but is most common in areas with abundant cover, where it occurs at densities of about 0.72 km².

Dusicyon culpaeus is largely nocturnal and in some areas shows seasonal movements seeking hares and sheep. In Peru two stomachs contained lizards, birds, and small mammals, and animals were found to be active both day and night. In Chile this species eats lizards, small birds, mice, and the mouse opossums (*Marmosa*). Its food habits have been well studied and vary seasonally and geographically. In most areas rodents make up a major portion of the diet; principal prey species include *Octodon degus*, *Abrocoma bennetti*, *Reithrodon*, sp. and *Akodon olivaceus*. In areas where livestock is common, carrion is a frequent component of the diet, and in some areas the bird *Chloephaga picta* is a major food item. The introduced hare (*Lepus*) is commonly taken, but most of those eaten (83.3%) were juveniles, suggesting that nest predation is the main way these foxes catch hares. *D. culpaeus* also takes

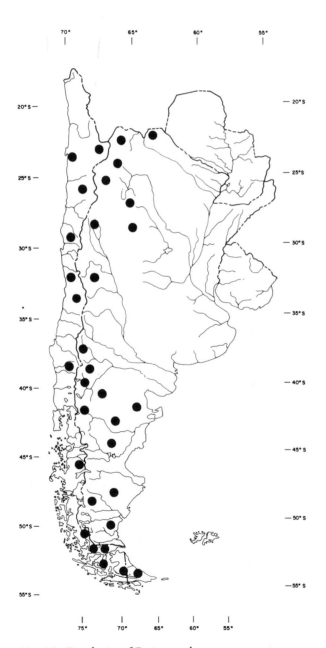

Map 6.3. Distribution of *Dusicyon culpaeus*.

rabbits out of snares. Plant material, especially fruit, was commonly eaten in some areas.

In Argentina these foxes became much more common with widespread sheep ranching and the introduction of the hare (*Lepus*). *D. culpaeus* in Argentina has a narrower diet than *D. griseus*, being exclusively carnivorous. In an extensively studied population in Neuquén province ninety-six stomachs where examined: rodents and hares occurred in 61.4%, domestic stock (probably mostly sheep carrion) in 27.4%, and birds in 6.1%. Hares were the most important prey throughout the year. This species is important as a furbearer, though less so than smaller species of *Dusicyon*: between 1976 and 1979, 32,000 skins were exported from Argentina (Crespo 1975; Crespo and Carlo 1963; Jaksić, Yáñez, and Rau 1980; Jaksić and Yáñez 1983; Jaksić, Schlatter, and Yáñez 1983; Langguth 1975; Lucero 1983; Mann 1945; Mares and Ojeda 1984; Meserve, Shadrick, and Kelt 1987; Pearson 1951; Rabinovich et al. 1987; Simonetti 1986; Yáñez and Jaksić 1978; Yáñez and Rau 1980).

Dusicyon griseus (Gray, 1837)
Zorro Gris, Chilla

Measurements

	Mean	Min.	Max.	N	Loc.	Source[a]
TL	868.2	735	1,030	18	A, C	1
HB	556.9	446	660	14		
T	314.4	202	427			
HF	122.2	87	147	13		
E	72.0	60	85	6		
Wta	3.99 kg	2.5 kg	5.45 kg	4		

[a](1) Daciuk 1974; Greer 1966; Osgood 1943; Pine, Miller, and Schamberger 1979; CM, FM, Santiago.

Description

In Chile this species ranges in body length from 520 to 670 mm, with the largest animals in the north (34°) and the smallest in the south (54°). This is the opposite of the pattern seen in *D. culpaeus*. *D. griseus* is a small gray fox with a blackish middorsal stripe and a black tail tip, and Chilean specimens have a well-marked black chin. The venter is cream, the underside of the tail is mixed pale tawny and black (plate 6), and some specimens have a faint reddish cast. On Chiloé Island, Chile, foxes of this species are almost black (Fuentes and Jaksić 1979; Osgood 1943).

Distribution

Dusicyon griseus is found in central Chile and in Argentina along the Andes, from at least Salta province south to Río Negro province and then across the whole country south to Tierra del Fuego (Crespo 1984; Honacki, Kinman, and Koeppl 1982; Olrog and Lucero 1981) (map 6.4).

Life History

In Chile most births appear to be in November, with litter sizes ranging from two to five. The gestation period is fifty-three to fifty-eight days (Housse 1953; Rabinovich et al. 1987).

Ecology

In Chile this fox is found in a broad range of habitats from forests to open grassland, though in Argentina it typically occurs in arid and semiarid temperate areas of Patagonia and the Andes. It is both diurnal and nocturnal. In Argentina it ranges from densities of 4.35 km² in bushy littoral to 0.95 km² in steppe habitat.

The food habits of *D. griseus* have been well

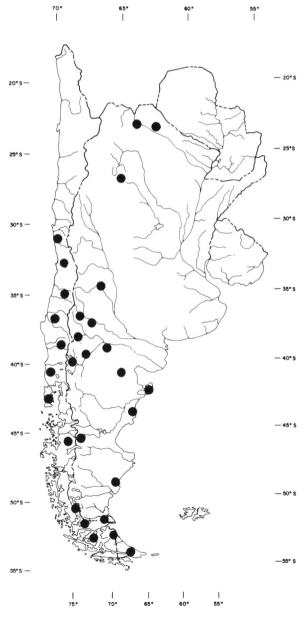

Map 6.4. Distribution of *Dusicyon griseus*.

studied. In northern Argentina stomachs have been found to contain plant material, insects, and *Ctenomys*. In Chile this species is reported to prefer birds. Of fifty stomachs from Tierra del Fuego, 87% contained plant material, mostly grass; 45% contained insects, mostly coleopterans; 27.1% contained rodents; and sheep carrion was commonly encountered. Other studies have shown a greater frequency of small rodents, particularly *Ctenomys magellanicus*, *Octodon degus*, and *Abrocoma bennetti*. Birds are taken, as are lizards. Most of the samples examined contained insects (especially lepidopteran larvae and adult beetles) and plant material (especially grass, leaves, and fruit). *D. griseus* was introduced onto some southern islands to control feral, introduced rabbits (*Oryctolagus*) but apparently does not kill many. The few that are killed are mostly juveniles (Atalah, Siefield, and Venegas 1980; Jacksić and Yáñez 1983; Jaksić, Schlatter, and Yáñez 1980; Jaksić, Yáñez, and Rau 1983; Mares 1973; Pine, Miller, and Schamberger 1979; Rabinovich et al. 1987; Yáñez and Jaksić 1978).

Comment
Dusicyon griseus includes the darker form *D. fulvipes* from Chiloé Island, Chile.

Dusicyon gymnocercus (G. Fischer, 1914)
Zorro Pampa

Measurements

	Mean	Min.	Max.	N	Loc.	Source[a]
TL	920.4	830.0	998.0	23	A, P	1
	959.6	920.0	1,078.0	26	U	2
	1,010.0	965.0	1,060.0	10 m	A	3
	940.0	860.0	1,020.0	16 f		
HB	590.9	520.0	722.0	23	A, P	1
T	329.5	226.0	380.0			
	344.8	300.0	380.0	26	U	2
	352.0	320.0	365.0	10 m	A	3
	319.0	270.0	356.0	16 f		
HF	126.7	112.0	145.0	22	A, P	1
	151.6	125.0	173.0	26	U	2
	140.0	135.0	155.0	10 m	A	3
	128.0	115.0	145.0	16 f		
E	79.7	71.0	90.0	23	A, P	1
	87.0	80.0	95.0	26	U	2
	86.0	80.0	90.0	10 m	A	3
	84.0	80.0	90.0	16 f		
Wta	3.97 kg	2.4 kg	5.0 kg	18	A, P	1
	5.95 kg	4.5 kg	7.9 kg	26	U	2
	4.63 kg			116 m	A	3
	4.21 kg			163 f		

[a](1) Massoia 1982; PCorps, UConn; (2) Barlow 1965; (3) Crespo 1971.

Description
This species is grayish on the dorsum and grayish mixed with yellowish on the sides. The ears and parts of the neck are reddish, as are the lower parts of legs, and the underparts are markedly white. The full tail is a mixture of grayish, reddish, and black. The dorsum is similar to that of *C. thous*, but the black of the dorsal midline is broken by white longitudinal speckles (Barlow 1965; pers. obs.).

Distribution
Dusicyon gymnocercus is found in eastern Bolivia, southern Brazil, western Paraguay, and the eastern provinces of Argentina north of Río Negro province (Crespo 1984; Honacki, Kinman, and Koeppl 1982) (map 6.5).

Life History
In central Argentina this species mates between August and October, and pups are born between

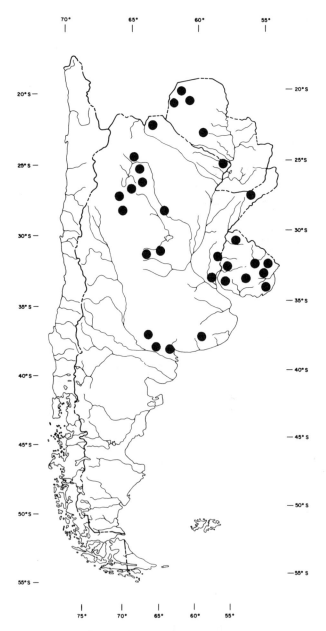

Map 6.5. Distribution of *Dusicyon gymnocercus*.

September and October. Females are monoestrous, the gestation period is approximately fifty-eight to sixty days, the average number of embryos is 3.4 ($n = 72$), and females can breed by eight to twelve months. Most of the breeding is done by the youngest females, and few individuals live longer than three years (Crespo 1971).

Ecology

Dusicyon gymnocercus is an inhabitant of the grasslands, pampas, and open woodlands of Uruguay, the Paraguayan Chaco, and eastern Argentina. In the arid Patagonian steppe to the south it is replaced by *D. griseus*. *D. gymnocercus* occurs with *Cerdocyon thous* over a good portion of its range but seems to prefer more open habitats. *D. gymnocercus* is nocturnal where it is hunted. Like other species of *Dusicyon*, it is omnivorous, and plant food, especially fruit, makes up one-quarter of the total diet ($n = 230$ stomachs). In La Pampa province, Argentina, the most important food is mammals, particularly the young of the hare, *Lepus*, which was found in 33% of the stomachs. Next most important was the mouse *Graomys griseoflavus*. Sheep carrion was found in fewer than 20% of the samples. Birds, particularly tinamous, of the family Tinamidae, were found in a third of the stomachs. Insects and reptiles were less important (Barlow 1965; Crespo 1971, 1975, 1982; Langguth 1975).

Genus *Speothos* Lund, 1839
• *Speothos venaticus* (Lund, 1842)
Bush Dog, Zorro Vinagre

Description

The dental formula differs from the basic canid pattern, being I 3/3, C 1/1, P 4/4, M 1–2/2. One specimen from Argentina measured TL 810; HB 700, T 110; HF 110; E 45 (Crespo 1974); three adults from Paraguay weighed 5.4 to 5.6 kg (K. Hill, unpubl. data). This canid is unmistakable because of its short ears, very short legs, and uniformly dark pelage (plate 6). It is stocky, with a short tail, small, rounded ears, and a thick muzzle. The fur is dark brown, grading to a more golden color on the neck, with no distinctive markings. The venter is darker than the dorsum (Kleiman 1972; pers. obs.).

Distribution

Speothos venaticus is found from Panama south through Paraguay to northern Argentina (Crespo 1974b; Hill and Hawkes 1983; Honacki, Kinman, and Koeppl 1982) (map 6.6).

Life History

In captivity the gestation period is approximately sixty-five days, and three to five is the usual litter

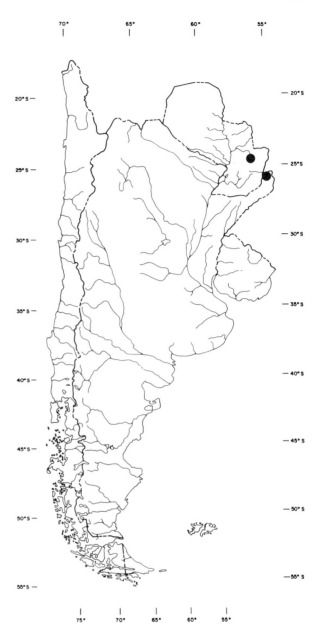

Map 6.6 Distribution of *Speothos venaticus*.

size. Pups open their eyes at eight to seventeen days and eat solid food at thirty-eight to seventy-one days. The male provisions the female and young throughout the rearing phase. Pups share food and show a much higher social tolerance than many other canids. The vocal repertoire is very rich, and when foraging as a group animals use a short whine as a contact call (Biben 1982a,b, 1983; Brady 1981; Drüwa 1977).

Ecology

Bush dogs in the wild are very poorly studied. They are apparently found only in moist forest, although this species has been seen in central Brazil where the only forest was gallery forest.

All evidence from captivity indicates a strong pair-bond between an adult male and an adult female (Drüwa 1977). Young animals that remain with their parents after attaining maturity do not reproduce. Apparently the dominant female supresses estrus in her daughters. Unlike most other canids, this species uses short-distance contact calls, and this combined with the high tolerance in pups suggests that bush dogs may be social. This is corroborated by two observations in Argentina and Paraguay, where a male and a female were trapped together and a male and three females were collected from the same burrow. In both cases the animals were adults. The animals urine mark in a unique fashion, standing on the forelegs and micturating on a spot higher than would have been possible had they remained on all fours.

It has been suggested that bush dogs, with their short legs, specialize on large caviomorph rodents such as the paca (*Agouti paca*), which they chase into burrows. In Mato Grosso state, Brazil, a bush dog was seen to capture a paca out in the open (Crespo 1974b; Eisenberg 1989; K. Hill, unpubl. data; Kleiman 1972; Deutsch 1983).

FAMILY PROCYONIDAE

Diagnosis

The molars of this family tend to be broad, without the shearing edge so characteristic of canids and felids. In this feature Procyonidae resemble the bear family (Ursidae). The dental formula is usually I 3/3, C 1/1, P 4/4, M 2/2 (fig. 6.3), except in the genus *Potos*, where the number of premolars is 3/4. There is no toe reduction, and members of the family are either plantigrade or semiplantigrade. They are medium to small, with a total length ranging from 600 to 1,350 mm.

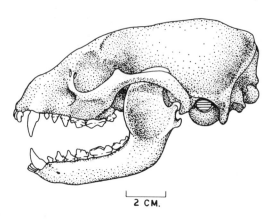

Figure 6.3. Skull of *Procyon cancrivorus*.

Distribution

If one excludes the Asiatic lesser panda, *Ailurus*, and the giant panda, *Ailuropoda*, from the procyonids, then the current distribution of the family is confined to the New World. Raccoons and their allies are distributed from Canada to Argentina. They are adapted to a variety of habitats, but they generally occur only where there is tree cover.

Genus *Nasua* Storr, 1780
• *Nasua nasua* (Linnaeus, 1766)
Coati, Coatí, Pisote

Measurements

	Mean	Min.	Max.	N	Loc.	Source[a]
TL	999.4	665.0	1,280.0	30	A, Br, P	1
HB	550.0	345.0	890.0	29		
T	464.3	320.0	560.0			
HE	92.8	80.0	108.0	22		
E	44.9	37.0	50.0	20	P	
Wta	3.18 kg	1.00 kg	5.50 kg	11		
	4.89 kg	2.25 kg	7.75 kg	16 m		2
	3.63 kg	2.00 kg	5.90 kg	22 f		

[a](1) Schaller 1983; Vieira 1945; BA, PCorps, UConn; (2) K. Hill, unpubl. data.

Description

With its long snout and its ringed tail, usually held perpendicular to the body, the coati is unmistakable. The short ears are heavily furred, and each has a white spot. The feet are dark and equipped with long, strong claws. Individuals from eastern Paraguay are grizzled orange and black dorsally and orange tan on the sides, chin, and venter. The tail has pronounced but ill-defined black and orange tan bands. There is white on the face. Individuals from other populations may have more black and less orange, with the face less marked with white (pers. obs.).

Distribution

Nasua nasua is found from southern Venezuela south throughout Paraguay to northern Argentina, where it occurs at least to Santa Fe Province. It has also been introduced onto Robinson Crusoe Island off the coast of Chile (Honacki, Kinman, and Koeppl 1982; Pine, Miller, and Schamberger 1979; BA) (map 6.7).

Life History

Young are born after a seventy-seven-day gestation period, in an arboreal nest constructed by the female. Litter size ranges from one to six. During the early part of life young remain in the nest, but when they are two to three weeks old they begin to accompany their mother, who rejoins other females with young. In northeastern Argentina coatis reproduce from October to February, and the litter size is three to six (Crespo 1982; Eisenberg 1989).

Ecology

These versatile animals are found in many vegetation types ranging from thorn scrub to moist tropical forest. They occur in groups ranging from two to as many as twenty. In Panama they have been well studied. Several females form permanent bands and forage together with their young. The males join the female bands during the breeding season, but at other times of the year they forage alone.

Coatis are diurnal and scansorial, foraging arboreally and also terrestrially, by probing their sensitive snouts into the forest litter. They are omnivorous, eating a great deal of fruit at certain seasons, and are also opportunistic predators on vertebrates. Two stomachs from Brazilian animals contained fish, snakes, crabs, millipedes, sowbugs, spiders, slugs, cicadas, beetles, ants, and termites, and they also feed on palm nuts and figs.

In 1935 two pregnant females escaped from captivity on Robinson Crusoe Island in the Juan Fernández group. In 1972 the population of coatis on this small island was estimated at four thousand (Crespo 1982; Eisenberg 1989; Pine, Miller, and Schamberger 1979; Kaufmann 1962; Russell 1981, 1983; Schaller 1983).

Comment

Nasua narica may well be a valid species (J. Kaufmann, pers. comm.). The late Ralph Wetzel had partially described a third species before his untimely death (R. Wetzel, pers. comm.).

Genus *Procyon* Storr, 1780

• *Procyon cancrivorus* (F. Cuvier, 1798)
Crab-eating Raccoon, Osito Lavador, Mayuato

Measurements

	Mean	Min.	Max.	N	Loc.	Source[a]
TL	1,002.6	885.0	1,258.0	14	A, B, Br, P	1
HB	678.2	553.0	880.0	13		
T	323.1	260.0	420.0	14		
HF	142.1	113.0	170.0			
E	53.5	40.0	65.0	14		
Wta	8.81 kg	7.5 kg	10.2 kg	7		

[a](1) Crespo 1974a, 1982; Schaller 1983; BA, FM, UConn, UM.

Description

This unmistakable procyonid resembles its northern conspecific but has longer legs and shorter fur. It is agouti black and yellow dorsally, yellowing on the sides and venter. The tail is well furred and striped, and the face has a prominent dark mask across the eyes (plate 7).

Distribution

Procyon cancrivorus is found from Costa Rica south through eastern and western Paraguay, Uruguay, and into the northern Argentine provinces at least as far south as Santa Fe province (Barlow, 1965; Honacki, Kinman, and Koeppl 1982; BA, UConn, UM) (map 6.8).

Life History

In Paraguay a female was collected with six embryos, and in northeastern Argentina the crab-eating raccoon reproduces from May to July, with a litter size from two to four. The developmental schedule for the young is similar to that of *P. lotor* (Crespo 1982; Löhmer 1976; UConn).

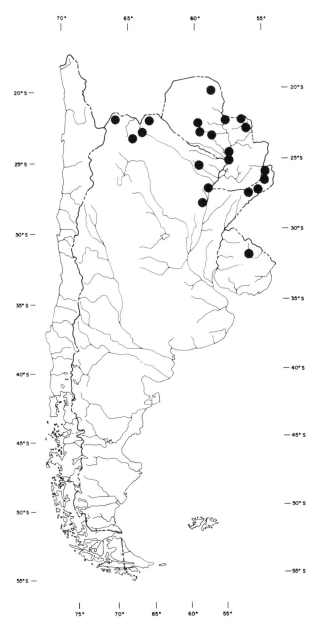

Map 6.7. Distribution of *Nasua nasua*.

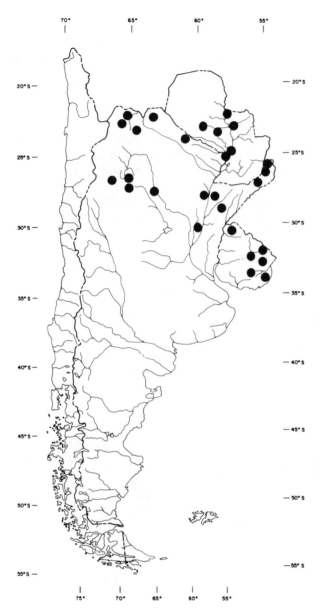

Map 6.8. Distribution of *Procyon cancrivorus*.

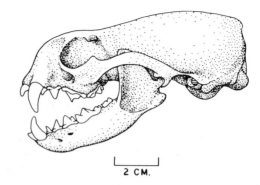

Figure 6.4. Skull of *Lutra* sp.

formula for most South American forms is I 3/3, C 1/1, P 3/3, M 1/2. The otters (*Lutra*) do not show the reduction in premolar number, and their formula is 4/3 (see figs. 6.4 and 6.5). The body tends to be elongated, with relatively short legs and usually a long tail. The family is classically divided into five subfamilies, three of which occur in South America: Mustelinae (the weasels), Mephitinae (the skunks), and Lutrinae (the otters).

Distribution

Species of the weasel family occur on all continents except Antarctica and Australia. The family had a northern origin, since mustelids first appear in the Oligocene in North America and Europe. Species of the Mustelidae entered South America in the late Pliocene and rapidly occupied the small-carnivore ecological niches.

Genus *Conepatus* Gray, 1837
Hog-nosed Skunk, Zorrino

Description

Head and body length ranges from 300 to 490 mm, and the tail is from 160 to 210 mm; weight may

Ecology

Procyon cancrivorus is found in many different habitats, from xeric Chaco vegetation to moist forest, but apparently always near water. It is basically nocturnal, spending the day asleep in trees. One stomach examined was filled with fruit pulp (Crespo 1982; Lucero 1983; Schaller 1983).

FAMILY MUSTELIDAE
Weasels and Their Allies

Diagnosis

These small to medium-sized carnivores generally show a reduction in molar number. The dental

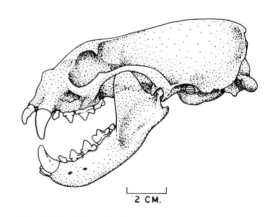

Figure 6.5. Skull of *Eira barbara*.

reach 4 kg. The muzzle is bare, and the nose pad is large and flat. The claws of the forepaws are well developed. Hog-nosed skunks have a typical pelage pattern, black on the sides and venter with either a strongly contrasting white dorsum or two white dorsal stripes (see plate 7).

Distribution

The genus *Conepatus* is distributed from the southwestern United States to northern Argentina. It is broadly tolerant of many habitat types. The distribution of the species within the genus was reviewed by Kipp (1965).

Conepatus chinga (Molina, 1792)
Zorrino Común, Zorrino Andino

Measurements

	Mean	Min.	Max.	N	Loc.	Source[a]
TL	517.9	410.0	685	8	P	1
HB	288.0	220.0	325.0			
	171.8	146.0	212.0	29 m	U	2
	170.0	133.0	200.0	29 f		
T	229.9	180.0	386.0	8	P	1
	377.4	325.0	420.0	16 m	U	2
	337.7	290.0	396.0	29 f		
HF	58.8	50.0	84.0	8	P	1
	71.1	61.0	78.0	16 m	U	2
	64.1	55.0	71.0	29 m		
E	24.0	13.0	32.0	8	P	1
	28.0	22.0	35.0	16 m	U	2
	25.8	21.0	29.0	24 f		
Wta	1.50 kg			1	P	1
	2.24 kg	1.59 kg	2.95 kg	16 m	U	2
	1.53 kg	1.12 kg	2.04 kg	24 f		

[a](1) PCorps, UConn; (2) Van Gelder 1968.

Description

Conepatus chinga is blackish to brownish black with a varying amount of white hair. The white pattern varies from stripes that extend the length of the body and onto the base of the tail to no stripes at all and only a few white hairs on the head. In Uruguay the male is larger than the female and much heavier (Van Gelder 1968; pers. obs.).

Distribution

Conepatus chinga is found from Bolivia south through Uruguay and western Paraguay into Argentina at least as far south as Neuquén province and central Chile (Honacki, Kinman, and Koeppl 1982; Mann 1945; Osgood 1943) (map 6.9).

Life History

In Uruguay this species is apparently monoestrous with a protracted breeding season (Barlow 1965).

Ecology

This skunk is found in many habitats from the Paraguayan Chaco into the precordillerean steppe. In Tucumán province, Argentina, it occurs to at least 3,000 m. In Uruguay *C. chinga* prefers open country and has apparently become more abundant as woodland has been replaced by grassland. It digs its own burrow as well as using burrows dug by other animals. It is nocturnal and eats mostly beetles, orthopterans, and spiders. The stomach of one individual collected in Paraguay contained ten frogs (Barlow 1965; Lucero 1983; UM).

Comment
Considered by some authors as *Conepatus rex*.

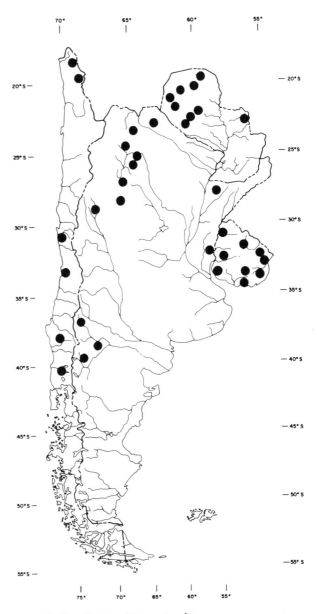

Map 6.9. Distribution of *Conepatus chinga*.

Conepatus humboldtii Gray, 1837
Zorrino Patagónico, Zorrino Chico

Measurements

	Mean	Min.	Max.	N	Loc.	Source[a]
TL	422.6	295	570	7	A, C	1
HB	252.4	123	370			
T	170.1	110	200			
HF	52.1	45	60			
E	19.3	10	25			

[a](1) Allen 1905; Daciuk 1974; Osgood 1943; Wolffsohn 1923; BA.

Description
This skunk is smaller than *C. chinga*. In Chile it has rather soft, silky fur ranging from blackish brown to reddish brown. Most animals have two narrow white stripes that stretch from the head to the tail, but there is considerable variation in their size and location (Allen 1905; Osgood 1943; pers. obs.).

Distribution
Conepatus humboldtii is found in southern Argentina and adjacent parts of Chile from the Strait of Magellan north to Chubut and Río Negro provinces (map 6.10).

Ecology
In Chilean Patagonia these skunks were nocturnal during the austral summer. Daytime resting sites, most often in shrub or forest cover, consisted of earthen tunnels, rock piles, and brush piles and were usually changed every day. When active, skunks were most often found foraging in grass habitats. Skunks were solitary with overlapping home ranges; the home range of one female was 16.4 ha, and the home ranges of six juveniles ranged from 7.4 to 11.6 ha (Fuller et al. 1987).

Comment
Considered by some authors as *Conepatus castaneus*.

Genus *Eira* H. Smith, 1842
• *Eira barbara* (Linnaeus, 1758)
Tayra, Hurón Mayor

Measurements

	Mean	Min.	Max.	N	Loc.	Source[a]
TL	959.4	860	1,040.0	8	A, B, P	1
HB	613.8	526	680.0			
T	333.6	290	373.0	7		
HF	90.1	70	103.0	8		
E	35.9	30	40.0	7		
Wta	3.94 kg	2 kg	5.2 kg	5		

[a](1) Allen 1916; BA, FM, PCorps, UConn.

Description
This large mustelid is unmistakable (plate 7). It has a slender body, long legs equipped with strong claws, and a long tail. The coat color is variable; in general it is brown to black with varying amounts of white or tan on the head and throat, extending onto the chest. In Argentina animals have very dark bodies, with the head and shoulders much lighter brown, and a conspicuous cream-colored blaze on the chest. The tayra's black eyes shine blue-green at night (Kaufmann and Kaufmann 1965; pers. obs.).

Distribution
The tayra is found from southern Mexico south to Paraguay and northern Argentina, where it occurs at least as far south as Tucumán province (Honacki, Kinman, and Koeppl 1982; Lucero 1983) (map 6.11).

Life History
In captivity a female comes into estrus six months after giving birth. After a gestation period of sixty-three to seventy days a litter of two is produced.

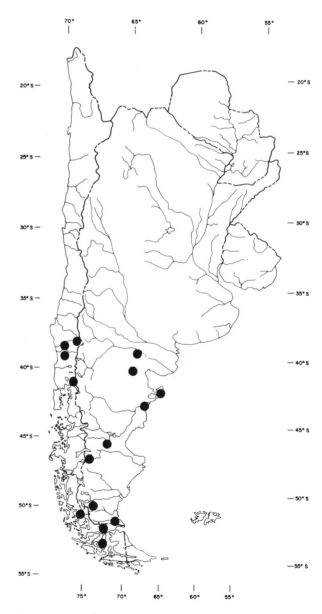

Map 6.10. Distribution of *Conepatus humboldtii*.

Young weigh about 100 g and open their eyes at thirty-eight to forty-one days. In captivity a female reproduced for the first time at twenty-three months (Poglayen-Neuwall 1978; Vaughn 1974).

Ecology

Tayras are active both during the day and at night but are primarily diurnal. They are usually encountered as solitary individuals or family groups. In the Venezuelan llanos a female had a home range of 9 km² through which she traveled extensively.

Tayras are found predominantly in forest, though they will range away from trees. They are excellent climbers and are often seen high in trees. Although often termed a frugivore, this large mustelid can be an active predator; individuals have been seen killing

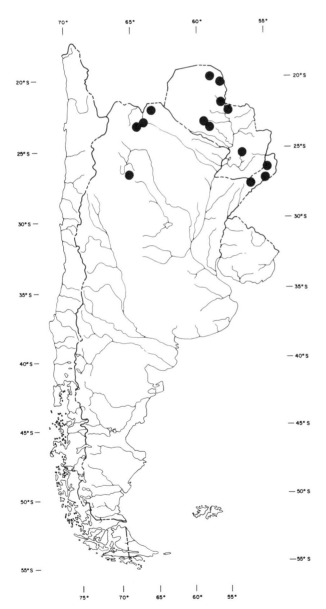

Map 6.11. Distribution of *Eira barbara*.

iguanas and chasing agoutis and monkeys (*Saimiri* and *Saguinus midas*). In Belize scats contained (by occurrence; n = 31) 68% fruit and 58% arthropods; small vertebrates were also found. In Venezuela, fruit and vertebrates were taken with equal frequency, and the arboreal spiny rat was the vertebrate most commonly taken (Galef, Mittermeier, and Bailey 1976; Kaufmann and Kaufmann 1965; Konecny 1989; Sunquist, Sunquist, and Daneke 1989).

Genus *Galictis* Bell, 1826
Grison, Hurón

Description

This genus contains three species (*G. cuja*, *G. allamandi*, and *G. vittata*) distributed from southern Veracruz, Mexico, to Argentina. Most authorities consider *G. allamandi* to be a junior synonym of *G. vittata* (Honacki, Kinman, and Koeppl 1982). The genus *Grison* is a synonym for *Galictis*. Head and body length ranges from 475 to 550 mm; the tail is short, usually less than 150 mm. Animals may weigh up to 3.2 kg. The limbs, throat, and venter are black, and the black extends to the face. A white stripe from the forehead to the shoulders demarcates the ventral color from the dorsum, which has a grizzled salt-and-pepper pattern (plate 7).

Distribution

Galictis vittata occurs from southern Veracruz to Brazil, where it is replaced by *G. cuja* (Krumbiegel 1942). It usually occurs below 1,200 m but is broadly tolerant of the vegetation cover, being found in both dry deciduous forests and multistratal tropical rain forests.

Life History and Ecology

In Venezuela the grison is active in the early morning and late afternoon and at night. One female had a home range of 4.2 km². These active predators feed on reptiles, small birds, and rodents. They shelter in burrows and will occupy the burrows of armadillos (Sunquist, Sunquist, and Daneke 1989).

Galictis cuja (Molina, 1782)
Hurón Menor

Measurements

	Mean	Min.	Max.	N	Loc.	Source[a]
TL	542.7	425.0	667.0	23	A, C, P, Peru, U	1
HB	388.6	280.0	508.0			
T	154.1	120.0	193.0			
HF	56.1	32.0	75.0	22		
E	22.8	13.0	30.0	20		
Wta	1.58 kg	1.02 kg	2.45 kg	5		

[a](1) Barlow 1965; Daciuk 1974; Greer 1966; Pearson 1957; Texera 1974b; Wolffsohn 1923; BA, PCorps, Santiago, UConn.

Description

Galictis cuja is smaller than *G. vittata;* it resembles a large weasel in its proportions but is more robust and has a striking black-and-white pattern. The face, sides, and underparts are black and sharply delineated from the yellowish gray dorsum. On the head a white stripe extends across the forehead and down the sides of the neck. There is much variation in the amount of lighter dorsal coloration as well as in the size and placement of the white stripes (Barlow 1965; Osgood 1943; pers. obs.).

Distribution

Galictis cuja is found from southern Peru throughout Uruguay, in Paraguay, in central Chile, and in Argentina south to Chubut province. A record from Tierra del Fuego may be a misidentification (Honacki, Kinman, and Koeppl 1982; J. A. Iriarte, pers. comm.; Texera 1974b; BA, Santiago) (map 6.12).

Ecology

The grison is found in a great range of habitats that have water and good cover, to at least 3,500 m. They are also found in the xeric Chaco of Paraguay. In Peru one specimen was collected in the afternoon with a stomach full of mice and a lizard. In Chile this species was trained to drive chinchillas out of rock piles. In Patagonia (Valdés Peninsula) grisons were seen in a group of three and a group of four, possibly mother and young (Barlow 1965; Lucero 1983; Osgood 1943; Pearson 1957; A. Taber, pers. comm.).

Genus *Lutra* Brunnich, 1771
Freshwater Otter, Perro da Agua

Description

The dental formula is I 3/3, C 1/1, P 4/3, M 1/2 (see fig. 6.4.) A cylindrical body form, very short ears, webbed feet, and stout vibrissae characterize this aquatic carnivore. The tail is thick at the base, tapering to a point, and the body contours are smooth (plate 8). There are approximately twelve species worldwide.

Distribution

The genus is distributed in freshwater and coastal areas of North and South America, most of Africa, Europe, Asia, and parts of Southeast Asia. Three species occur in southern South America.

Lutra felina (Molina, 1792)
Lobito Marino, Gatuna

Measurements

	Mean	Min.	Max.	N	Loc.	Source[a]
TL	910.0			9	C	1
	870.0			3		2
HB						
T	340.0			9		1
	340.0			3		2
HF	97.0			6		1
	90.0			3		2
E	15.0					
Wta	4.5 kg	3.2 kg	5.8 kg			3

[a](1) Osgood in Cabello 1983; (2) Cabello 1983; (3) Duplaix 1982.

Description

The top edge of the rhinarium is flat, in contrast to the biconcave edge of *L. provocax* (see plate 8). This small otter is a uniform dark coffee brown on the dorsum, and the venter is sometimes lighter. Juveniles are darker than adults (Cabello 1978; Osgood 1943; Sielfeld 1983).

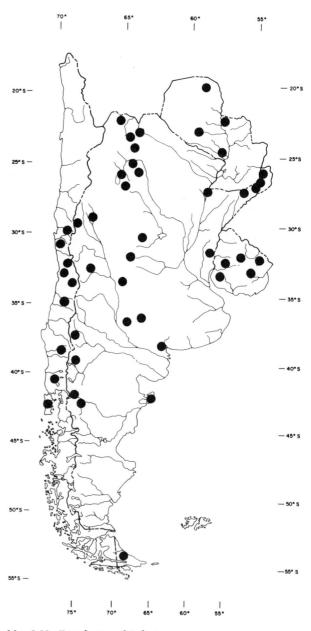

Map 6.12. Distribution of *Galictis cuja.*

Distribution

Lutra felina is found from northern Peru south along the whole Chilean coast in isolated populations to the Strait of Magellan, where it is found in Argentina as well. The original range has decreased because of hunting (Massoia 1976; Melquist 1984; Olrog 1950; Torres, Yáñez, and Cattan 1979) (map 6.13).

Life History

In the southern part of the range there seems to be a birth peak between September and October. The gestation period is between sixty and sixty-five days, and the litter size is two. Young are born in earthen dens or rock crevices and stay with the female for approximately ten months (Cabello 1978, 1983; Sielfeld 1983).

Ecology

Lutra felina is virtually confined to marine waters, though there are reports of this species' swimming up rivers. Apparently limited to places on the shore with rocky outcroppings, *L. felina* is therefore found in disjunct populations all down the Chilean coast. In the northern part of its range *L. felina* occurs in small, scarce populations because of a paucity of suitable habitat. In areas where it occurs, densities vary from 0.04 to 10 per kilometer of coastline. The areas where *L. felina* is found are characterized by heavy seas and strong winds. It stays within about 500 m of the coast and is mostly solitary, although groups of three have been seen.

These otters usually sleep ashore and give birth in dens in the rocks. They will shelter in caves open at water level. The afternoon is the time when most of the aquatic activity takes place. Food dives last fifteen to forty-five seconds, and the prey is usually eaten in the water. On Chiloé Island 73% of the feeding observations were on shellfish and 27% on fish. Analysis of feces showed that the "shellfish" category consisted of 72% crustaceans and 28% mollusks. Some fruit was also eaten. Another study showed that *L. felina* ate primarily carnivorous fish, crustaceans, and gastropods, with some predation on birds. They have also been reported as eating freshwater prawns, crabs, cuttlefish, and some algae.

Darwin reported this species to be very abundant in the last century. Since that time heavy hunting for fur has greatly reduced its numbers (Allen 1905; Brownell 1978; Cabello 1978, 1983; Castilla 1982; Castilla and Bahamondes 1979; Melquist 1984; Sielfeld 1983; Van Zyll de Jong 1972).

• *Lutra longicaudis* (Olfers, 1818)
Lobito Común, Guairao

Measurements

	Mean	Min.	Max.	N	Loc.	Source[a]
TL	1,053.0	890.0	1,200.0	3	A, U	1
HB	513.0	360.0	660.0			
T	540.0	370.0	840.0			
HF	120.0	94.0	144.0			
E	19.3	18.0	22.0			
Wta	5.8 kg	4.5 kg	7.1 kg	2		

[a](1) Barlow 1965; Massoia 1976; BA.

Description

This otter is a lustrous grayish brown dorsally, slightly lighter on the venter. The tip of the muzzle and the mandible are usually blazed with yellowish white. The nose pad distinguishes this species from *L. provocax* (see plate 8) (Barlow 1965).

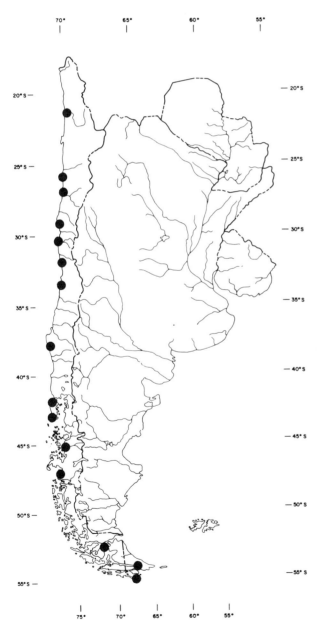

Map 6.13. Distribution of *Lutra felina*.

Distribution

Lutra longicaudis is broadly distributed from Mexico south to Uruguay, probably Paraguay, and Argentina across the northern part of the country to Buenos Aires province (Honacki, Kinman, and Koeppl 1982; Massoia 1976; Melquist 1984) (map 6.14).

Life History

In northeastern Argentina *L. longicaudis* reproduces in the spring with a litter size of two or three. In central Brazil a female had three embryos (Crespo 1982; AMNH).

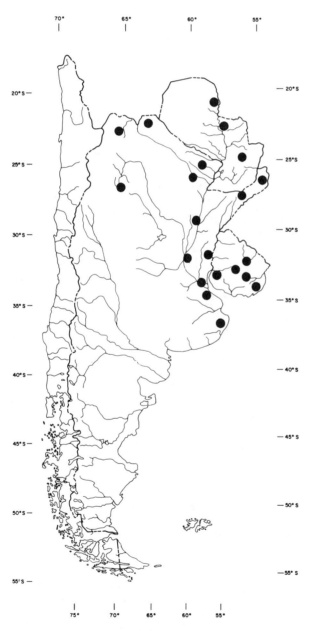

Map 6.14. Distribution of *Lutra longicaudis*.

Ecology

Lutra longicaudis is found in areas that meet its habitat requirements: ample riparian vegetation along permanent streams and lakes and a readily available year-round food source. The animals shelter in terrestrial burrows that they dig themselves. In Jujuy province, Argentina, they are found at least to 3,000 m elevation. Hunting has considerably reduced their abundance and distribution. These otters eat fish as well as mollusks and crustaceans (Crespo 1982; Griva 1978; Massoia 1976; Melquist 1984; Olrog 1979).

Comment

Considered by some authors as *Lutra platensis*.

Lutra provocax Thomas, 1908
Lobito Patagónico, Huillín

Measurements

	Mean	Min.	Max.	N	Loc.	Source[a]
TL	980.0	920	1,010	4	A, C	1
HB	597.5	570	610			
T	382.5	350	400			
HF	119.5	108	125			
E	261.0					

[a](1) Massoia 1976; Osgood 1943; Wolffsohn 1923.

Description

This otter is larger than *L. felina* and is further distinguished by possessing a biconcave upper edge to its rhinarium, in contrast to the straight edge of *L. felina* (plate 8). Its dorsum is a rich dark brown, and its venter is a silvery whitish, contrasting quite strongly with the dorsal color and with the darker venter of *L. felina* (Osgood 1943; Pine, Miller, and Schamberger 1979).

Distribution

Lutra provocax is found only in central and southern Chile and adjacent parts of Argentina. In much of its original Chilean range it has been exterminated by hunting. In Argentina it is found along the Andes from the southern part of Neuquén province all the way to Tierra del Fuego (Cabello 1983; Melquist 1984; García-Mata 1978; Massoia 1976; Miller et al. 1983) (map 6.15).

Life History

Females have four nipples and produce litters of one or two and perhaps up to four (Sielfeld 1983).

Ecology

Lutra provocax is found in both marine and fresh waters. In Argentina it occurs primarily in freshwater lakes and streams of the lake district, but in southern Chile glacial ice may prevent its occupying many of the potential bodies of fresh water. In the

Patagonian archipelago it lives on the rocky coasts and in protected canals in areas where there are few waves. It is also found in these areas only in places characterized by a coastal strip of *Drimys, Nothofagus betuloides*, and *Maytenus*. This vegetation is not found in the open coastal areas where *L. felina* occurs.

In the southern part of Chile an examination of 281 feces showed that 75% had fish and 63% had crustaceans, although this otter also eats mollusks and birds. In a site in Argentina 99% of the scats examined had crustaceans and only 2% had fish. There appears to be much overlap in diet between *L. fel-*

ina and *L. provocax*, but they are apparently never found sympatrically.

In the lake district of Argentina the diet is mostly crustaceans and mollusks, with only a few fish. The introduction of salmonids may have affected the diet of this otter, because these fish have outcompeted the native fish and are themselves too fast for the otters to catch (Chehebar 1983, 1985; Chehebar et al. 1984; Melquist 1984; Sielfeld 1983).

Genus *Pteronura* Gray, 1837
• *Pteronura brasiliensis* (Gmelin, 1788)
Giant River Otter, Lontra Gigante, Arirai

Measurements

	Mean	Min.	Max.	N	Loc.	Source[a]
TL	1,625.0	1,450	1,800.0	—	—	1
T	550.0	450	650.0			
Wta	29.10 kg	24 kg	34.2 kg			
	24.25 kg	24 kg	24.5 kg	2	A, Br	2

[a](1) Duplaix 1982; (2) Autuori and Deutsch 1977; Crespo 1982.

Description
This otter is unmistakable because of its size. Its long tail becomes dorsoventrally flattened in its distal half, and the broad head is flattened. The eyes are large, the ears are small and round, and the feet are large and well webbed. The fur, thick, glossy, and short, is shiny chocolate black when wet, drying to brown. The lip, chin, throat, and chest may be spotted with creamy white that can be almost absent or expanded to form a bib (plate 8) (Duplaix 1980).

Distribution
Pteronura brasiliensis is found in the major river systems of South America east of the Andes as far south as northern Argentina, where it is found in the province of Misiones. It is also found in Paraguay and once occurred in Uruguay but is probably now very rare there or extinct. Its range has been greatly decreased by hunting (Barlow 1965; Honacki, Kinman, and Koeppl 1982; Massoia 1976; Melquist 1984) (map 6.16).

Life History
In captivity the gestation period was recorded as sixty-five to seventy days, and litters ranged from two to five. Otters were born at about 200 g and opened their eyes at thirty-one days. In northeastern Argentina this species reproduces in spring and summer, and litter sizes of two and three have been noted. In Suriname litters of one to three are born at the beginning of the dry season (Autuori and Deutsch 1977; Crespo 1982; Duplaix 1980).

Ecology
These large, diurnal otters seem to prefer slow-moving creeks and rivers. In Suriname, where it has

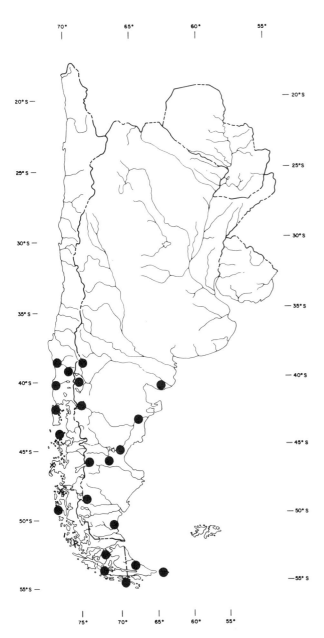

Map 6.15. Distribution of *Lutra provocax*.

been well studied, *Pteronura* moves seasonally as
water levels and fish populations change. Otters occur in groups of up to twenty, composed of a pair-
bonded male and female and related individuals.
This group tends to occupy one stretch of a river and
to scent mark certain areas, as well as using common
latrines. This species is highly vocal and is easily recognized by its "barks." A large portion of time is
spent ashore grooming. Cubs are born in a den at-
tended by both parents. Otters travel and hunt in
groups, pursuing fish by sight. In Suriname chara-
coid fish were preferred (Duplaix 1980).

Genus *Lyncodon* Gervais, 1845
Lyncodon patagonicus (Blainville, 1842)
Huroncito

Description
Two individuals from Argentina and Chile ex-
hibited the following measurements: TL 430–45;
HB 332–55; T 90–98; HF 351 (Koslowsky 1904;
Wolffsohn 1923). The ears are very small, concealed
by the surrounding fur, and the short tail is bushy.
This small, slender weasel has coloration similar to
Galictis but is smaller, and the top of the head is

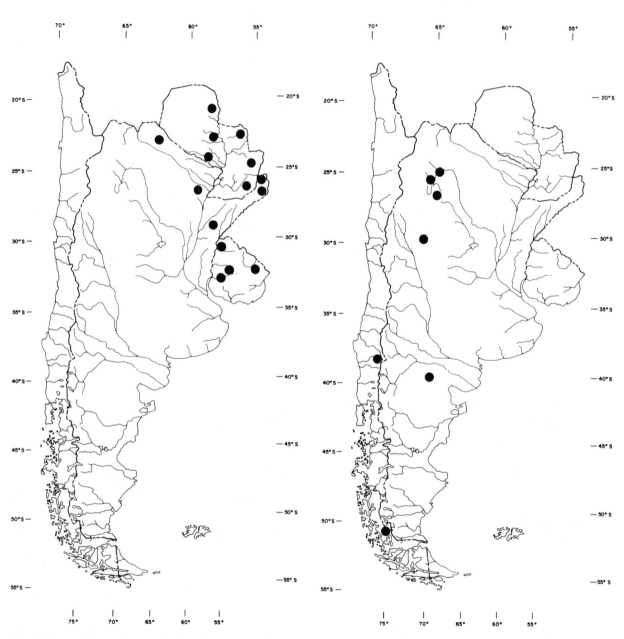

Map 6.16. Distribution of *Pteronura brasiliensis*.

Map 6.17. Distribution of *Lyncodon patagonicus*.

creamy or white, extending as a broad stripe to the shoulders. It also has long white hairs on the dorsum, unlike *Galictis* (plate 7). The throat and venter are dark brown, like the sides of the body (Allen 1905; Osgood 1943; pers. obs.).

Distribution

Lyncodon patagonicus is found in Argentina from Salta province south along the western part of the country to Santa Cruz province and then into Chile along the southern Argentine border. This species is poorly represented in collections (Allen 1905; Olrog 1979; Osgood 1943; Peña 1966) (map 6.17).

Ecology

Very little is known of this interesting carnivore. It is reported to be nocturnal or crepuscular and to enter the burrows of *Ctenomys* and *Microcavia* in pursuit of prey. The long pelage on the neck is erected when the animal is cornered (Koslowsky 1904; Mares 1973).

FAMILY FELIDAE
Cats

Diagnosis

The tooth formula is extremely modified in this family; most notably, the number of molars is reduced. The dental formula typically is I 3/3, C 1/1, P 2–3/2, M 1/1 (figs. 6.6 and 6.7). The auditory bulla is inflated and subdivided into two chambers. Five toes are retained on the forefeet and four on the hind feet; the claws are retractile except in the African cheetah (*Acinonyx*). The eyes are directed forward, and the rostrum is rather short.

Distribution

The Recent distribution includes all continents except Australia and Antarctica. Members of the

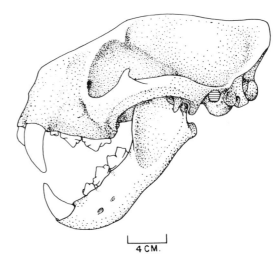

Figure 6.7. Skull of *Panthera onca*.

family Felidae are distinguishable in the Oligocene. The early cats appear in North America, Eurasia, and later in Africa and are placed in the subfamily Nimravinae. Subsequently one could recognize two divergent groups, the true cats and the saber-toothed cats; the latter are frequently placed in their own subfamily, the Machairodontinae. Members of the Felidae entered South America during the Pliocene and rapidly replaced the large carnivorous birds (Phorusrhachidae) that then occupied the medium-to-large cursorial carnivore niche (Vuilleumier 1985).

Genus *Felis* Linnaeus, 1758
Felis colocolo Molina, 1810
Gato de Pajonal

Measurements

	Mean	Min.	Max.	N	Loc.	Source[a]
TL	854.7	710	960.0	14	A, C, Peru, U	1
HB	570.6	435	660.0			
T	278.7	220	322.0			
HF	123.4	110	139.0	10		
E	53.3	36	72.2	13		
Wta	2.95 kg			1		

[a](1) Barlow 1965; Cabrera 1961; Daciuk 1974; Pearson 1951; Wolffsohn 1908, 1923; BA.

Description

This account includes *F. pajeros*. This robust bobcat-sized (*Lynx rufus*) cat varies widely in color. In Chile the dorsum and sides are mottled reddish brown and gray with faint banding. The venter is white with black stripes, and the forelimbs have bold black stripes separated with white. In Argentina the striping on the forelimbs and hind limbs is

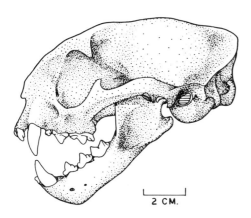

Figure 6.6. Skull of *Felis pardalis*.

the only constant character. The dorsum can be almost a uniform gray brown or a basic gray to tan, broken by brown rosettes and black lines in the mid-dorsum. In quite a few individuals, the hair is considerably longer than on other South American felids. The tail is full and well haired, usually banded with black (plate 9) (Barlow 1965; Osgood 1943; pers. obs.).

Distribution

Felis colocolo is distributed from Peru south through Uruguay and in western Paraguay into Chile, from approximately Coquimbo to Chillan. *F.*

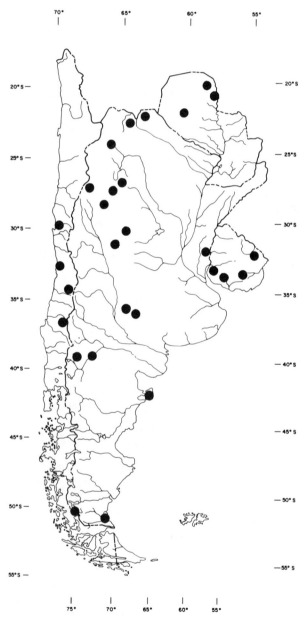

Map 6.18. Distribution of *Felis colocolo.*

colocolo is found in the northwestern and central portion of Argentina south through most of the country (Tamayo and Frassinetti 1980; Ximénez 1961; BA, UM) (map 6.18).

Life History

The litter size is two or three, and one female in captivity gave birth at twenty-four months of age (Eaton 1984; Rabinovich et al. 1987).

Ecology

This cat occupies a greater range of habitats than any other southern South American felid. It is found above 5,000 m in the Andes and in the Paraguayan Chaco. In Uruguay it prefers low areas in or near swamps and marshes with tall grass, and in Chile it has been taken in cloud forest. It is reported to be nocturnal and highly arboreal.

Felis colocolo is an important species in the fur trade, and over 78,000 skins were exported from Argentina between 1976 and 1979 (Barlow 1965; Mares and Ojeda 1984; Pearson 1951; Schamberger and Fulk 1974; Wolffsohn 1908).

Felis geoffroyi d'Orbigny and Gervais, 1844
Geoffroy's Cat, Gato Montés

Measurements

	Mean	Min.	Max.	N	Loc.	Source[a]
TL	781.2	708.0	940.0	20	A, P	1
	956.4	892.0	1,004.0	5 m	U	2
HB	502.9	422.0	665.0	20	A, P	1
T	277.2	240.0	332.0	21		
	353.2	341.0	365.0	5 m	U	2
HF	100.3	92.0	125.0	19	A, P	1
	135.2	128.0	147.0	5 m	U	2
E	49.5	35.0	65.0	13	A, P	1
	55.6	52.0	57.0	5 m	U	2
Wta	2.36 kg	2.0 kg	2.8 kg	7	A, P	1
	5.20 kg	3.6 kg	6.0 kg	5 m	U	2
	3.70 kg	3.2 kg	4.1 kg	5 m		3
	3.10 kg	2.6 kg	3.4 kg	5 f		3

[a](1) Berrie 1978; Cabrera 1961; Daciuk 1974; PCorps, UConn; (2) Barlow 1965; (3) Ximénez 1973.

Description

Felis geoffroyi is a distinctive small, lithely built, spotted cat (plate 9). Its basic body color is gray to reddish brown with black spots, not rosettes, as the predominant pattern. The ground color lightens on the sides. On the sides, neck, and legs these spots can join to form indistinct parallel stripes. It is equal in size to a very large domestic cat, but with a proportionally shorter tail and longer, more flattened head. The tail is the same basic color as the dorsum and is narrowly banded with black stripes. Melanistic individuals are not unusual (Barlow 1965; Ximénez 1975; pers. obs.)

Distribution

Geoffroy's cat is found from Bolivia across southern Brazil and into the Paraguayan Chaco, throughout Uruguay and in Argentina south to the southernmost part of the continent. From there it goes into Chile (Honacki, Kinman, and Koeppl 1982; Melquist 1984; Texera 1974a; Ximénez 1973; BA, UConn) (map 6.19).

Life History

In captivity *F. geoffroyi* has a gestation period of seventy-four to seventy-six days and a litter size of one to three. Kittens are born weighing 65 g and open their eyes at about twelve days. In Uruguay this cat has one litter a year of two or three kittens (Scheffel and Hemmer 1975; Ximénez 1975).

Ecology

Felis geoffroyi primarily frequents subtropical and temperate regions of South America. It ranges from sea level to 3,300 m in Bolivia. In Uruguay its preferred habitats are open woodland, brushy areas, open savannas, and marshes. It apparently does not occupy subtropical rain forest or the southern coniferous forests, being replaced in the latter habitat by *F. guigna*. This primarily nocturnal hunter preys on small rodents and birds. Stomachs have been found to contain *Cavia, Ctenomys, Phyllotis, Lepus,* and the birds *Myiopsitta* and *Nothura.* One stomach had five *Oryzomys,* two *Nothura,* and an unidentified passeriform.

This is the most important species of spotted cat in the skin trade of the southern cone. Over 340,000 skins were exported from Argentina between 1976 and 1979 (Barlow 1965; Gibson 1899; Melquist 1984; Ojeda and Mares 1984; Ximénez 1973, 1975; UConn).

Felis guigna Molina, 1782
Kokod, Gato Guigna

Measurements

	Mean	Min.	Max.	N	Loc.	Source[a]
TL	701.1	665.0	722.0	7	A, C	1
	649.7	605.0	700.0		C	2
HB	473.9	424.0	510.0		A, C	
T	227.3	210.0	241.0			2
	222.1	195.0	250.0		C	1
HF	101.8	86.0	116.0	6	A, C	2
	94.6	90.0	100.0	7	C	1
E	42.2	35.0	49.0	6	A, C	2
	41.9	38.0	46.0	7	C	1
Wta	2.23 kg	2.08 kg	2.5 kg	3		

[a](1) Greer 1966; Koslowsky 1904; Osgood 1943; Wolffsohn 1923; (2) Allen 1919.

Description

This housecat-size spotted felid has a brown background color covered with black spots. The dorsum can vary from light to rather dark brown. There is very little tendency for the spots to form rosettes, though some individuals have streaks on the head and shoulders. The venter is spotted as well, and the brown tail is narrowly ringed with blackish bands (plate 9). Melanistic individuals are not uncommon (Osgood 1943; pers. obs.).

Distribution

The kokod is found in south-central Chile and adjacent Argentina in the Andean areas (Cabrera 1961;

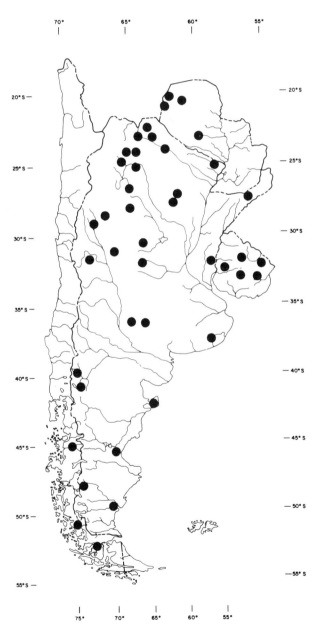

Map 6.19. Distribution of *Felis geoffroyi.*

Honacki, Kinman, and Koeppl 1982; Melquist 1984; FM) (map 6.20).

Ecology

This cat inhabits the moist coniferous forests of the southern Andes. In Malleco province, Chile, it lives in wooded and semiopen areas of the central valley as well as in the forests of the Andes. It is reported to be locally abundant, but very little fieldwork has been attempted. *F. guigna* is an excellent climber, and stomachs have been found to contain small rodents, the introduced *Rattus*, and birds (Greer 1966; Koslowsky 1904; Melquist 1984).

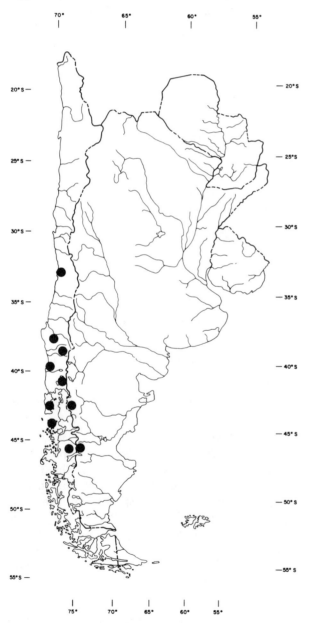

Map 6.20. Distribution of *Felis guigna*.

Felis jacobita Cornalia, 1865
Andean Cat, Gato Andino

Measurements

	Mean	Min.	Max.	N	Loc.	Source[a]
TL	1,046.7	990	1,120	3	A, Peru	1
HB	605.7	577	640			
T	441.0	413	480			
HF	124.0	115	133	2		
E	631.0					
Wta	4.0 kg			1		

[a](1) Cabrera 1961; Pearson 1957.

Description

This large cat resembles a small snow leopard (*Panthera unica*) with its long, luxurious fur. It is larger than *F. colocolo* and *F. guigna* and lacks the spinal crest of long hair seen on many specimens of *F. colocolo*. The dorsum is pale gray, spotted, and transversely striped with blackish or brownish, and the underparts are white. There are two faint middorsal stripes on the neck and three lines on the forehead. The feet are well furred except for the pads. The tail is long and banded with about seven rings, though some may be very faint (plate 9). The legs are banded with two to three bars of black (Osgood 1943; Pearson 1957; pers. obs.)

Distribution

Felis jacobita is found in the Andes from southwestern Bolivia to northern Chile and northern Argentina (Honacki, Kinman, and Koeppl 1982; Melquist 1984) (map 6.21).

Ecology

This rare felid inhabits the high Andes, where it seems to be confined to treeless, rocky, semiarid, and arid areas above about 3,000 m. In Peru an individual was collected at 5,100 m in a barren region of rocks and bare ground with scattered clumps of bunchgrass and small bushes. Another individual was observed in similar habitat in the Argentine province of Tucumán at 4,250 m (Cabrera 1961; Melquist 1984; Pearson 1957; Scrocchi and Halloy 1986).

• *Felis pardalis* Linnaeus, 1758
Ocelot, Gato Onza, Ocelote

Measurements

	Mean	Min.	Max.	N	Loc.	Source[a]
TL	1,097.4	900.0	1,277.0	7	A, P	1
HB	742.0	560.0	867.0			
	822.0	710.0	850.0	4	A	2
T	355.4	301.0	410.0	7	A, P	1
	375.0	345.0	410.0	4	A	2
HF	160.0	155.0	164.0			

E	55.3	40.0	75.0	7	A, P	1
	54.0	50.0	61.0	4	A	2
Wta	7.0 kg	5.5 kg	9.0 kg	3	A, Br	1
	8.8 kg	7.0 kg	10.8 kg	5	V	3

[a](1) Schaller 1983; BA, UConn; (2) Cabrera 1961; (3) Mondolfi 1986.

Description

The ocelot is the largest of the small spotted cats. The upperparts are grayish to cinnamon colored, with black markings forming streaks on the neck or elongated spots on the body. The tail exhibits incomplete banding. Spotting and banding extend to the dorsal surfaces of the limbs and onto the venter, whose basic color is white with occasional black spots (plate 9).

Distribution

Felis pardalis is found from the southern United States to Paraguay and northern Argentina. In Argentina it has never been known south of Entre Ríos province (Honacki, Kinman, and Koeppl 1982; Melquist 1984; BA) (map 6.22).

Life History

In captivity females reproduce from 2.5 to at least 12 years of age. In Venezuela the litter size averages

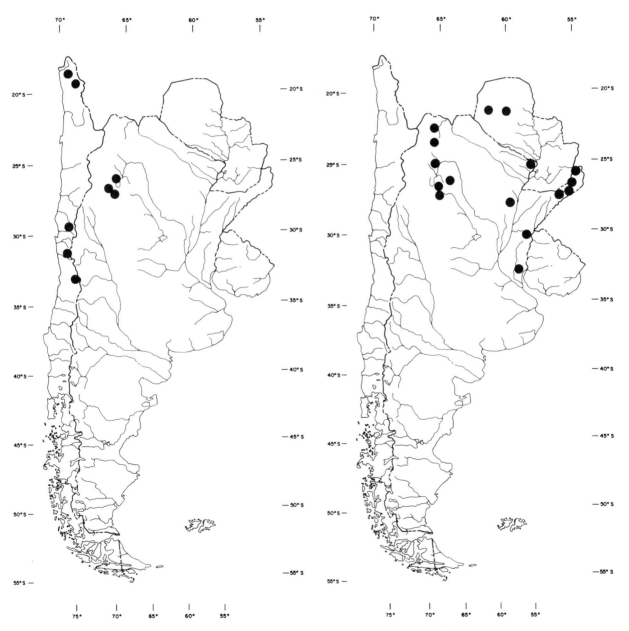

Map 6.21. Distribution of *Felis jacobita*.

Map 6.22. Distribution of *Felis pardalis*.

1.6 (range 1–3). In northeastern Argentina ocelots reproduce from October to January, and litter size is two or three (Crespo 1982; Eaton 1984; Mondolfi 1986).

Ecology

Ocelots are found in a wide range of habitats, from thorn forest to tropical moist forest. Animals are solitary except when the female is being courted or has cubs. In Venezuela movements of radio-tagged ocelots were variable; adult females traveled about half as far during a night as adult males. The larger home ranges of males encompassed the smaller ranges of adult females. Animals were usually nocturnal, though they were sometimes active during the day. In Belize ocelots were most active in the early morning and late evening. There, home ranges were almost entirely within areas of second-growth vegetation.

Ocelots seem to do most of their hunting at night on the ground. In Venezuela an examination of fifteen ocelot stomachs showed a predominance of rodents (*Proechimys*, *Dasyprocta*, and others), but also armadillos, a sloth, marsupials, an iguana, and snakes and frogs. In the llanos of Venezuela, where the cane rat *Zygodontomys* is seasonally abundant, the ocelot will eat this species. In the Brazilian Pantanal ocelots can be active twenty-four hours a day and have been found to eat fish and howler monkeys (*Alouatta*). In Belize an examination of forty-nine scats showed *Didelphis marsupialis* and *Philander opossum* to be the prey most commonly taken. In Amazonian Peru larger rodents and marsupials predominate in the diet. *Mazama americana*, *Tamandua*, and *Dasypus* were also taken (Emmons 1988; Konecny 1989; Mondolfi 1986; Schaller 1983; Schaller, Quiggley, and Cranshaw 1984; Ludlow and Sunquist 1987; Sunquist, Sunquist, and Daneke 1989).

• *Felis tigrina* Schreber, 1775
Little Spotted Cat, Gato Tigre, Chivi

Measurements

	Mean	Min.	Max.	N	Loc.	Source[a]
TL	752.9	676	815.0	13	A, Br, Co, Cr, Peru	1
HB	483.8	431	539.0			
T	269.2	245	298.0			
HF	103.5	96	110.0			
E	43.3	30	50.0	11		
Wta	2.22 kg	2 kg	2.45 kg	3	Br, Peru, V	2

[a](1) Cabrera 1961; Gardner 1971; BA, FM; (2) Mondolfi 1986; FM.

Description

Excluding *Felis guigna*, this is the smallest of the spotted cats. Compared with *F. wiedii* it is more gracile, the eyes and ears are relatively larger, and the snout is narrower. The ground color is dark tawny. On the neck are several stripes, extending as irregular broken bands down the center of the back, and along the sides are elongate rosettes, with the ground color darker within the rosettes. The venter is whitish between dark spots (plate 9) (Cabrera 1961; pers. obs.).

Distribution

The little spotted cat is found from Costa Rica south into northern Argentina (Honacki, Kinman, and Koeppl 1982; Melquist 1984) (map 6.23).

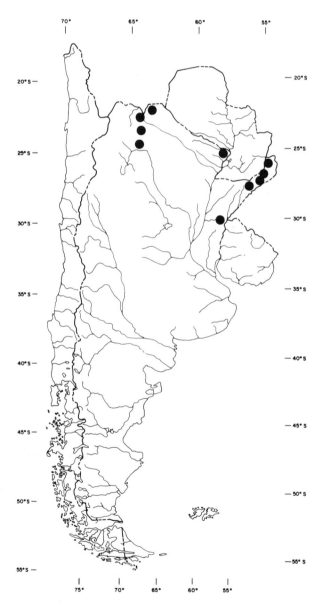

Map 6.23. Distribution of *Felis tigrina*.

Life History

In captivity the gestation period is sixty-two to seventy-four days, and the litter size is two to four (Leyhausen and Falkena 1966).

Ecology

In Venezuela this species is strictly a forest dweller and shows a preference for humid evergreen forests up to 3,000 m. It is generally terrestrial but climbs with ease. In Costa Rica it was taken at 3,500 m. Stomachs have been found to contain small rodents, a shrew, and a small passerine bird (Gardner 1971; Mondolfi 1986).

• *Felis wiedii* Schinz, 1821
Margay, Gato Pintado, Gato Brasileiro

Measurements

	Mean	Min.	Max.	N	Loc.	Source[a]
TL	893.8	820.0	938.7	6	A, U	1
HB	529.5	465.0	584.0	6	A, U	1
T	364.3	330.0	400.0			
HF	114.9	105.0	125.0	5	A, U	1
E	42.3	33.0	54.0	3		
Wta	3.22 kg	2.6 kg	3.9 kg	4	U, V	1, 2

[a](1) Barlow 1965; Cabrera 1961; Ximénez and Palerm 1971; BA; (2) Mondolfi 1986.

Description

Felis wiedii is smaller than *F. geoffroyi*, with softer fur and a much longer tail. The fur is longer than in many of the other spotted cats, thick and very soft, and the tail is proportionately longer than in the ocelot. This feature, together with the margay's smaller size, distinguishes the species (see plate 9). Young margays, however, are very similar in appearance to young ocelots. The margay has the same basic dorsal ground color, from grayish to cinnamon, strongly marked with bands and spots on the dorsum and incomplete black stripes in a transverse pattern on the tail. The venter is white (Barlow 1965; Cabrera 1961; Eisenberg 1989; Osgood 1914).

Distribution

The margay is found from the southern United States to Uruguay and northern Argentina to about 34° S (Cabrera 1961; Honacki, Kinman, and Koeppl 1982; Melquist 1984; Ojeda and Mares 1989; BA) (map 6.24).

Life History

In captivity margay kittens are born weighing about 165 g, open their eyes at eleven to sixteen days, and are weaned at about fifty days of age. In northeastern Argentina they reproduce in July and August and have a litter size of one (Crespo 1982; Petersen and Petersen 1978).

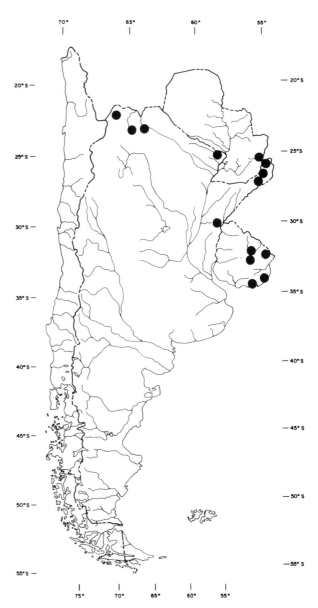

Map 6.24. Distribution of *Felis wiedii*.

Ecology

Felis weidii is an animal of tropical and subtropical forests. In Venezuela the margay is more specific in its habitat requirements than the ocelot, being found in humid lowland tropical forests as well as in premontane moist forests and cloud forest. It is much more arboreal than the ocelot, and unlike other cats the margay can pronate and supinate its hind foot. Thus, when the animal is descending a tree, the hind foot can rotate around the ankle so that the cat can hang vertically during descent like a squirrel. This arboreal ability is reflected in that margays in Belize always rested aboveground, escaped through the trees, and took arboreal rodents as their most frequent prey (n = 27 scats). Stomachs have also been found to contain squirrels. In Panama the most frequent prey was *Dasyprocta* and *Proechimys* (Barlow 1965; Enders 1935; Konency 1989; Leyhausen 1963; Mondolfi 1986).

• *Felis yagouaroundi* E. Geoffroy, 1803
Jaguarundi, Gato Eyra, Gato Moro

Measurements

	Mean	Min.	Max.	N	Loc.	Source[a]
TL	1,021.0	850.0	1,225.0	13	A, B, P	1
HB	640.5	510.0	779.0			
T	380.9	280.0	450.0			
HF	111.7	95.0	121.0	7		
E	36.5	28.0	45.0	11		
Wta	2.95 kg	2.5 kg	3.9 kg	7	A, P, V	1, 2

(1) Cabrera 1961; Crespo 1974a; Zapata 1982; BA, UConn;
(2) Mondolfi 1986.

Description
The jaguarundi is distinctive for its lithe, mustelid-shaped body and its lack of spots. A medium-sized cat, *F. yagouaroundi* has a flattened head, relatively small ears, and rather short legs. Its pelt has no spots or stripes at any age and comes in a variety of colors from reddish through gray to black (plate 9). Different-colored animals can be found in the same population (Cabrera 1961; Mondolfi 1986; Zapata 1982; pers. obs.).

Distribution
Felis yagouaroundi ranges from southern Texas through much of Argentina. It has been collected in both eastern and western Paraguay and in Argentina as far south as Río Negro province (Honacki, Kinman, and Koeppl 1982; BA, UM) (Map 6.25).

Life History
In captivity the gestation period ranges from seventy-two to seventy-five days. Litter sizes range from one to four, and the age of first reproduction is about two years (Hulley, 1976).

Ecology
In Venezuela the jaguarundi is found in a great variety of habitats, including semiarid thorn forest, deciduous forest, swampy grassland, and moist tropical forest. In southern South America it appears to inhabit a similar range of habitats.

Felis yagouaroundi is reported as being active both during the day and at night, but in Belize it is essentially diurnal. It is primarily terrestrial and eats a large variety of small vertebrates. An examination of sixteen stomachs from Venezuela revealed a predominance of rodents, birds, and lizards, with an armadillo and several rabbits as well. In Belize small mammals were the major diet, principally *Sigmodon* taken from old fields. Other prey items included *Didelphis*, small birds, arthropods, and some fruit (Konecny 1989; Mondolfi 1986; Zapata 1982).

Genus *Panthera* Oken, 1816
• *Panthera onca* (Linnaeus, 1758)
Jaguar, Yaguar, Yaguareté, Tigre Americano, El Tigre

Measurements

	Mean	Min.	Max.	N	Loc.	Source[a]
TL	2,072.0	1,970.0	2,120	5 m	Br	1
	1,882.0	1,670.0	2,030	5 f		
HB	1,472.0	1,360.0	1,570	5 m		
	1,336.7	1,270.0	1,420	3 f		

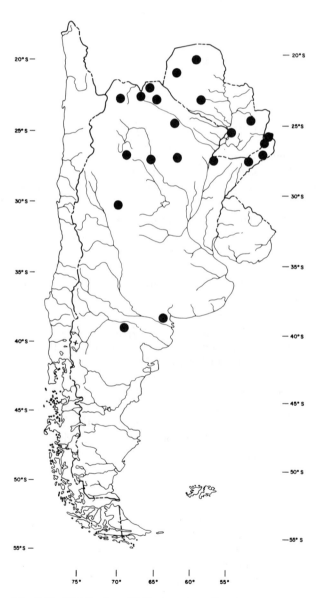

Map 6.25. Distribution of *Felis yagouaroundi*.

T	600.0	550.0	640	5 m	
	600.0	590.0	610	3 f	
Wta	95.3 kg	79.5 kg	119 kg	5 m	
	73.2 kg	60.5 kg	85 kg	5 f	
	99.5 kg			24 m	2
	75.5 kg			17 f	

[a](1) Schaller and Vasconcelos 1978; Schaller and Crawshaw 1980; (2) Almeida, n.d.

Description

The jaguar is unmistakable because of its large size, powerful build, and glossy spotted coat (plate 8). This is the largest of the American cats; it has a robust build and a relatively short tail, short, rounded ears, and large feet. The female is almost always smaller than the male. Jaguars from the Pantanal of Brazil are reportedly some of the largest in South America, so the measurements reported in this account may be larger than for jaguars in other parts of their range. The ground color of the dorsum ranges from light ochraceous buff to golden tawny; the ventral surfaces and the inner surfaces of the legs have a white ground. There is a great deal of variation in the rosette spotting (Almeida 1976; Cabrera 1961; Nelson and Goldman 1933).

Distribution

The jaguar was formerly distributed from the southwestern United States to northern Argentina, but it has been exterminated from much of its range. It occurs in the Paraguayan Chaco and was found in Argentina south probably to the Río Santa Cruz or Río Negro (Arra 1974; Carman 1984; Honacki, Kinman, and Koeppl 1982; Ojeda and Mares 1982) (map 6.26).

Life History

In captivity jaguars can come into estrus every month. The gestation period ranges from 101 to 105 days, the litter size is two, and cubs are born weighing 970 g. The kittens open their eyes within 3 days of birth, eat solid food at 75 days, and cease nursing at about 157 days of age. In Venezuela females are aseasonally polyestrous and have litters averaging two kittens (range 1–4). In northeastern Argentina jaguars reproduce between March and July and have litters of two or three kittens. Cubs may stay with their mothers for a year and a half before dispersing (Almeida, n.d.; Crespo 1982; Mondolfi and Hoogesteijn 1986; Stehlik 1971).

Ecology

Jaguars occur in a wide variety of habitats but seem to require abundant cover, water, and sufficient prey. They are found in the xeric Chaco as well as in tropical forest. In the Brazilian Pantanal, where jaguars have been studied, female jaguars range over at least 25–38 km² and males over an area more than twice as large. The ranges of females overlap, and the ranges of males include those of several females. However, it seems that the land tenure system of jaguars may vary with prey density. Jaguars are solitary except when females are in heat or have cubs.

Jaguars are most active after dusk and before dawn, though they may be active throughout the day. The loud, repeated, hoarse cough (like someone sawing wood) seems to be an important form of long-distance communication. An animal may stay at a kill for several days, and kills are sometimes covered with vegetation.

In different areas jaguars take different prey items

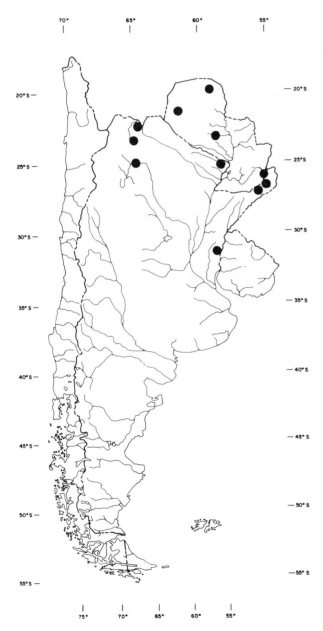

Map 6.26. Distribution of *Panthera onca*.

in different frequencies. They are remarkably catholic in their diet. In Venezuela jaguars eat a wide range of prey, from caiman eggs to tortoises, porcupines, armadillos, collared peccaries, and brocket deer. Where cattle are common, calves can become a major part of the diet. In the Pantanal, of thirty-six stomachs examined, twelve contained cattle, five capybaras, four white-lipped peccaries, four collared peccaries, and two caimans. Other prey included *Tamandua* anteaters, howler monkeys, coatis, tapirs, giant anteaters, marsh deer, jabiru storks, rheas, fish, and land tortoises. There appears to be variation between areas in favored foods, which may be due to preferences passed from female to offspring. In Belize armadillos, pacas, and brocket deer accounted for 70% of identified prey.

The range and density of this species have both been negatively affected by human hunting (Almeida, n.d., 1976; Mondolfi and Hoogesteijn 1986; Rabinowitz and Nottingham 1986; Schaller 1983; Schaller and Vasconcelos 1978; Schaller and Crawshaw 1980; Schaller, Quiggley, and Crawshaw 1984).

Genus *Puma* Jardine, 1834
• *Puma concolor* (Linnaeus, 1771)
Puma, León Americano

Measurements

	Mean	Min.	Max.	N	Loc.	Source[a]
TL	1,699.2	1,403	2,285.0	20	A, Br, C, P	1
	1,554.0			4 f	C	2
HB	1,080.2	900	1,470.0	20	A, Br, C, P	1
T	619.0	469	815.0			
	571.0			4 f	C	2
HF	235.7	200	263.0	14	A, Br, C, P	1
	220.0			4 f	C	2
E	84.6	80	101.0	13	A, Br, C, P	1
	82.0			4 f	C	2
Wta	35.4 kg	26 kg	54.5 kg	4	Br, C, P	1
	23.6 kg			4 f	C	2

[a](1) Allen 1905; Cabrera 1961; Courtin, Pacheco, and Eldridge 1980; Greer 1966; Schaller 1983; BA, UConn; (2) Sanborn 1954.

Description

The puma is unmistakable because of its large size, monocolored coat, and lithe build (plate 8). The average size may vary greatly over its geographic range. There are several color phases, including reddish, dark brownish red, gray, and tawny. The tail is long and often dark tipped, and the ears are short and rounded. The lateral muzzle and backs of the ears are also often black, while the chin, median muzzle, and ventral area are creamy white. Young are spot-

ted and striped (Currier 1983; Sanborn 1954; pers. obs.).

Distribution

The puma ranges from northern Canada south to southern Chile and Argentina, though it is extinct or very rare over much of its former range. In southern South America it is found in Paraguay, Uruguay, Argentina south to Tierra del Fuego, and all of Chile (Cabrera 1961; Currier 1983; Lahille 1899; Ximénez 1972; UConn) (map 6.27).

Life History

The gestation period is eighty-two to ninety-six days, litter size varies from one to six, young are

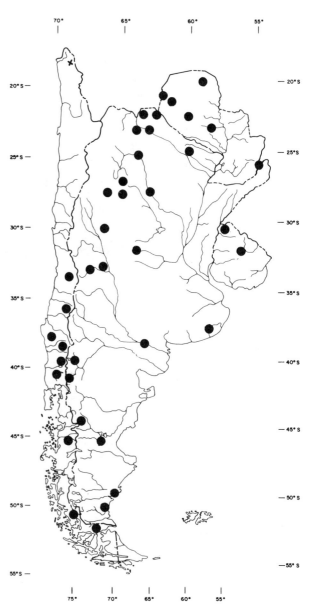

Map 6.27. Distribution of *Puma concolor*.

born weighing about 400 g, and adult weight is attained at about two to four years of age. In northern Argentina pumas reproduce from May to July and have a litter of two or three cubs (Crespo 1982; Currier 1983).

Ecology

Puma concolor is found in a wide range of habitats from the moist tropical forests of northeastern Argentina to above the tree line in the Andes. Tracks have been recorded above 5,800 m in southern Peru. In North America *P. concolor* is polygamous, and individuals have stable home ranges. Animals are solitary except for females with cubs. In the Brazilian Pantanal pumas were active throughout the day and night and used the same areas as jaguars, but seem to avoid each other. In this area the home range of one male was 32 km².

In Chile, near Osorno, a puma avoided the densest brush but used more open brushy areas and most other vegetation types. This same Chilean puma killed fifteen pudus in 249 days, hunting both by day and at night. In Panama pumas preyed on collared peccaries, *Mazama* and *Odocoileus* deer, pacas, agoutis, *Proechimys*, iguanas, and snakes. In southwestern Brazil they have been recorded as killing *Ozotoceros* and *Mazama* deer, a *Tamandua* anteater, and a rhea. In the Paraguayan Chaco one puma stomach contained a peccary, a rabbit, and a *Tolypeutes* armadillo. An examination of twenty-one scats from the same area showed *Dasypus* armadillos in seven, collared peccary in three, *Mazama gouazoubira* deer in four, rabbits in two, coatis in one, and *Tamandua* anteaters in four.

In Patagonia pumas prey on European hares (*Lepus*) and guanacos (*Lama*), and young guanacos are especially vulnerable. Pumas hunt from elevated, concealed positions and kill prey with a throat bite that crushes the trachea. In San Juan province, Argentina, both vicunas and guanacos are killed by pumas, with no preference shown for certain age or sex classes. Guanacos were taken more frequently than vicunas. Puma scats also contained eggshells, feathers, beetles, insect larvae, and rodent hair. In an extensive study of 409 puma scats from Torres del Paine, Chile, 92% of all prey items were mammalian—mostly European hares. Guanacos represented the next most frequent prey item. Yearling and juvenile guanacos were taken more frequently than expected (Cajal and López 1987; Courtin, Pacheco, and Eldridge 1980; Currier 1983; Enders 1935; Iriarte 1988; Miller 1930; Pearson 1951; Schaller and Crawshaw 1980; Schaller, Quiggley, and Crawshaw 1984; J. Stallings, unpubl. data; Wilson 1984; UConn).

Comment

Puma tracks are usually smaller than jaguar tracks, but those of a very large puma can be distinguished from those of a small jaguar by the toes, which radiate like spokes from a wheel around a smaller pad instead of pointing forward from a larger pad (Almeida 1976). *Puma concolor* is listed by some authors as *Felis concolor*.

References

Allen, G. M. 1942. *Extinct and vanishing mammals of the Southern Hemisphere.* Special Publication 11. Washington, D.C.: American Committee for International Wild-Life Protection.

Allen, J. A. 1905. Mammalia of southern Patagonia. In *Reports of the Princeton University expeditions to Patagonia, 1896–1899,* vol. 3, *Zoology.* Stuttgart: Schweizerbart'sche Verlagshandlung.

———. 1916. New mammals collected on the Roosevelt Brazilian expedition. *Bull. Amer. Mus. Nat. Hist.* 35:523–30.

———. 1919. Notes on the synonymy and nomenclature of the smaller spotted cats of tropical America. *Bull. Amer. Mus. Nat. Hist.* 41:341–419.

Almeida, A. de. 1976. *Jaguar hunting in the Mato Grosso.* England: Stanwill Press.

———. n.d. Some feeding and other habits, measurements and weights of *Panthera onca pallustris,* the jaguar of the "Pantanal" region of Mato-Grosso and Bolivia. Photocopy.

Arra, M. A. 1974. Distribución de *Leo onca* (L) en Argentina (Carnivora, Felidae). *Neotrópica* 20(63): 156–58.

Atalah G., A., W. Sielfeld K., and C. Venegas C. 1980. Antecedentes sobre el nicho trófico de *Canis g. griseus* Gray 1836 en Tierra del Fuego. *Anal. Inst. Pat. Punta Arenas* (Chile) 11:259–71.

Autuori, M. P., and L. A. Deutsch. 1977. Contribution to the knowledge of the giant Brazilian otter, *Pteronura brasiliensis* (Gmelin 1788), Carnivora, Mustelidae. *Zool. Gart.,* n.s. 47:1–8.

Barlow, J. C. 1965. Land mammals from Uruguay: Ecology and zoogeography. Ph.D. diss., University of Kansas.

Berrie, P. M. 1978. Home range of young female Geoffroy's cat in Paraguay. *Carnivore* 1(1): 132–33.

Berta, A. 1982. *Cerdocyon thous. Mammal. Species* 186:1–4.

Biben, M. 1982a. Object play and social treatment of prey in bush dogs and crab-eating foxes. *Behaviour* 79:201–11.

————. 1982b. Ontogeny of social behavior related to feeding in the Crab-eating fox (*Cerdocyon thous*) and the bush dog (*Speothos venaticus*). *J. Zool.* (London) 196:207–16.

————. 1983. Comparative ontogeny of social behaviour in three South American canids, the maned wolf, crab-eating fox and bush dog: Implications for sociality. *Anim. Behav.* 31:814–26.

Bisbal, F., and J. Ojasti. 1980. Nicho trófico del zorro *Cerdocyon thous* (Mammalia, Carnivora). *Acta Biol. Venez.* 10(4): 469–96.

Brady, C. A. 1978. Reproduction, growth and parental care in crab-eating foxes, *Cerdocyon thous*, at the National Zoological Park, Washington. *Int. Zoo Yearb.* 18:130–34.

————. 1979. Observations on the behavior and ecology of the crab-eating fox, *Cerdocyon thous*. In *Vertebrate ecology in the northern Neotropics*, ed. J. F. Eisenberg, 161–71. Washington, D.C.: Smithsonian Institution Press.

————. 1981. Vocal repertoires of the bush dog *Speothos venaticus*, crab-eating fox *Cerdocyon thous*, and maned wolf *Chrysocyon brachyurus*. *Anim. Behav.* 29:649–69.

Brady, C. A., and M. K. Ditton. 1979. Management and breeding of maned wolves *Chrysocyon brachyurus* at the National Zoological Park, Washington. *Int. Zoo Yearb.* 19:171–76.

Brown, A. D., and D. I. Rumiz. n.d. Habitat and distribution of the spectacled bear (*Tremarctos ornatus*) in the southern limit of its range. In *Proceedings of the First International Symposium on the Spectacled Bear*. Chicago: Lincoln Park Zoo. In press.

Brownell, R. L. 1978. Ecology and conservation of the marine otter *L. felina*. In *Otters*, ed. N. Duplaix, 104–6. Proceedings of the First Working Meeting of the Otter Specialist Group. Gland, Switzerland: IUCN.

Cabello, C. C. 1978. La nutria de mar *L. felina* en la Isla de Chiloé. In *Otters*, ed. N. Duplaix, 108–18. Proceedings of the First Working Meeting of the Otter Specialist Group. Gland, Switzerland: IUCN.

————. 1983. *La nutria de mar en la Isla de Chiloé*. Corporación Nacional Forestal, Ministero de Agricultura, Boletín Técnico 6. Santiago: República de Chile.

Cabrera, A. 1961. Los félidos vivientes de la república Argentina. *Rev. Mus. Argent. Cienc. Nat. "Bernardino Rivadavia," Zool.* 6(5): 161–247.

Cajal, J. L., and N. E. López. 1987. El puma como depredador de camélidos silvestres en la Reserva San Guillermo, San Juan, Argentina. *Rev. Chil. Hist. Nat.* 60:87–91.

Carman, R. L. 1984. Límite austral de la distribución del tigre o yaguareté (*Leo onca*) en los siglos XVIII y XIX. *Rev. Mus. Argent. Cienc. Nat. "Bernardino Rivadavia," Zool.* 13(1–60): 293–96.

Castilla, J. C. 1982. Nuevas observaciones sobre conducta, ecología y densidad de *Lutra felina* (Molina 1782) (Carnivora: Mustelidae) en Chile. *Mus. Nac. Hist. Nat. Publ. Ocas.* 38:197–206.

Castilla, J. C., and I. Bahamondes. 1979. Observaciones conductuales y ecológicas sobre *Lutra felina* (Molina) 1782 (Carnivora: Mustelidae) en las zonas central y centro-norte de Chile. *Arch. Biol. Med. Exp.* 12:119–32.

Chehebar, C. E. 1983. Relevamiento del Huillín, *Lutra provocax*, en el Parque Nacional Nahuel Huapi. Unpubl. report, Ministerio de Economía, Secretaría de Agricultura y Ganadería, Administración de Parques Nacionales, Argentina.

————. 1985. A survey of the southern river otter *Lutra provocax* Thomas in Nahuel Huapi National Park, Argentina. *Biol. Conserv.* 32:299–307.

Chehebar, C., A. Gallur, G. Giannico, M. D. Gottelli, and P. Yorio. 1984. Relevamiento del Huillín, *Lutra provocax*, en los Parques Nacionales Lanín, Puelo y los Alerces y evaluación de su estado de conservación en Argentina. Unpubl. report, Ministerio de Economía, Secretaría de Agricultura y Ganadería, Administración de Parques Nacionales, Argentina.

Coimbra-Filho, A. F. 1966. Notes on the reproduction and diet of Azara's fox *Cerdocyon thous azarae* and the hoary fox *Dusicyon vetulus* at Rio de Janeiro Zoo. *Int. Zoo Yearb.* 6:168–69.

Contreras, A. O. 1985. Algunos comentarios acerca del aguara guazu *Chrysocyon brachyurus* (Illiger, 1815), en la provincia de Corrientes, Argentina (Mammalia: Carnivora: Canidae). *Hist. Nat.* 5(14): 119–20.

Courtin, S. L., N. V. Pacheco, and W. D. Eldridge. 1980. Observaciones de alimentación, movimientos y preferencias de habitat del puma, en el islote rupanco. *Medio Ambiente* (Valdivia, Chile) 4(2): 50–55.

Crespo, J. A. 1962. Una mutación de pelaje en el zorro colorado, *Dusicyon culpaeus* (Molina) (Mammalia: Carnivora). *Neotrópica* 8(27): 115–16.

————. 1971. Ecología del zorro gris *Dusicyon gymnocercus antiquus* (Ameghino) en la provincia de la Pampa. *Rev. Mus. Argent. Cienc. Nat. "Bernardino Rivadavia," Ecol.* 1(5): 147–205.

————. 1974a. Comentarios sobre nuevas localidades para mamíferos de Argentina y de Bolivia.

Rev. Mus. Argent. Cienc. Nat. "Bernardino Rivadavia," Zool. 11(1): 1–31.

———. 1974b. Incorporación de un género de canidos a la fauna de Argentina (Fam. Canidae: *Speothos venaticus* (Lund) 1943). *Comun. Mus. Argent. Cienc. Nat. "Bernardino Rivadavia," Zool.* 4(6): 37–39.

———. 1975. Ecology of the pampas gray fox and the large fox (culpeo). In *The wild canids*, ed. M. W. Fox, 179–91. New York: Van Nostrand Reinhold.

———. 1982. Ecología de la comunidad de mamíferos del Parque Nacional Iguazú, Misiones. *Rev. Mus. Argent. Cienc. Nat. "Bernardino Rivadavia," Ecol.* 3(2): 45–162.

———. 1984. Los zorros. *Fauna Argent.* 52:1–32.

Crespo, J. A., and J. M. de Carlo. 1963. Estudio ecológico de una población de zorros colorados *Dusicyon culpaeus culpaeus* (Molina) en el oeste de la provincia de Neuquén. *Rev. Mus. Argent. Cienc. Nat. "Bernardino Rivadavia," Ecol.* 1(1): 1–55.

Currier, M. J. P. 1983. *Felis concolor. Mammal. Species* 200:1–7.

Daciuk, J. 1974. Notas faunísticas y bioecológicas de Península Valdés y Patagonia. 12. Mamíferos colectados y observados en la Península Valdés y zona litoral de los Golfos San José y Nuevo (provincia de Chubut, república Argentina). *Physis*, sec. C, 33(86): 23–39.

Deutsch, L. 1983. An encounter between bush dog (*Speothos venaticus*) and paca (*Agouti paca*). *J. Mammal.* 64:532–33.

Dieterlen, F. 1954. Über den Haarbau des Andenwolfes, *Dasycyon hagenbecki* (Krumbiegel, 1949). *Säugetierk. Mitt.* 2:26–31.

Dietz, J. M. 1981. Ecology and social organization of the maned wolf (*Chrysocyon brachyurus*). Ph.D. diss., Michigan State University.

———. 1984. *Ecology and social organization of the maned wolf* (Chrysocyon brachyurus). Smithsonian Contributions to Zoology 392. Washington, D.C.: Smithsonian Institution Press.

Drüwa, P. 1977. Beobachtungen zur Geburt und natüralichen Aufzucht von Waldhunden (*Speothos venaticus*) in der Gefangenschaft. *Zool. Gart.* n.s., 47:109–37.

Duplaix, N. 1980. Observations on the ecology and behavior of the giant river otter *Pteronura brasiliensis* in Suriname. *Rev. Ecol. (Terre et Vie)* 34:496–620.

———. 1982. Contribution à l'écologie et à l'ethologie de *Pteronura brasiliensis* Gmelin 1788 (Carnivora, Lutrinae); Implications évolutives. Ph.D. diss., Université de Paris–Sud Centre d'Orsay.

Eaton, R. L. 1984. Survey of smaller felid breeding. *Zool. Gart.*, n.s., 54(1/2): 101–20.

Eisenberg, J. F. 1989. *Mammals of the Neotropics.* Vol. 1. *Mammals of the northern Neotropics: Panama, Colombia, Venezuela, Guyana, Suriname, French Guiana.* Chicago: University of Chicago Press.

Emmons, L. H. 1988. A field study of ocelots (*Felis pardalis*) in Peru. *Rev. Ecol. (Terre et Vie)* 43: 133–57.

Enders, R. K. 1935. Mammalian life histories from Barro Colorado Island, Panama. *Bull. Mus. Comp. Zool. (Harvard)* 78(4): 385–502.

Ewer, R. F. 1973. *The carnivores.* Ithaca, N.Y.: Cornell University Press.

Fuentes, E. R., and F. M. Jaksić. 1979. Latitudinal size variation of Chilean foxes: Tests of alternative hypotheses. *Ecology* 60:43–47.

Fuller, T. K., W. E. Johnson, W. L. Franklin, and K. A. Johnson. 1987. Notes on the Patagonian hognosed skunk (*Conepatus humboldtii*) in southern Chile. *J. Mammal.* 68:864–67.

Galef, B. G., Jr., R. A. Mittermeier, and R. C. Bailey. 1976. Predation by the tayra (*Eira barbara*). *J. Mammal.* 57:760–61.

García-Mata, R. 1978. Nota sobre el status de *L. provoca* en la Argentina. In *Otters*, ed. N. Duplaix, 68–74. Proceedings of the First Working Meeting of the Otter Specialist Group. Gland, Switzerland: IUCN.

Gardner, A. L. 1971. Notes on the little spotted cat, *Felis tigrina oncilla* Thomas, in Costa Rica. *J. Mammal.* 52:464–65.

Gibson, E. 1899. Field-notes on the wood-cat of Argentina (*Felis geoffroyi*). *Proc. Zool. Soc. London* 1899:928–29.

Gittleman, J. L., ed. 1989. *Carnivore behavior, ecology, and evolution.* Ithaca, N.Y.: Cornell University Press.

Greer, J. K. 1966. Mammals of Malleco province, Chile. *Publ. Mus., Michigan State Univ. (Biol. Ser.)* 3(2): 49–152.

Griva, E. E. 1978. El programa de cría y preservación de *L. patensis* en Argentina. In *Otters*, ed. N. Duplaix, 86–103. Proceedings of the First Working Meeting of the Otter Specialist Group. Gland, Switzerland: IUCN.

Hill, K., and K. Hawkes. 1983. Neotropical hunting among the Ache of eastern Paraguay. In *Adaptive responses of native Amazonians*, ed. R. B. Hames and W. T. Vickers, 139–88. New York: Academic Press.

Hofmann, R. K., C. F. Ponce del Prado, and K. C. Otte. 1975–76. Registro de dos nuevas especies de mamíferos para el Perú, *Odocoileus dichoto-*

mus (Illiger–1811) y *Chrysocyon brachyurus* (Illiger–1811) con notas sobre su habitat. *Rev. Forestal Perú* 6:61–81.

Honacki, J. H., K. E. Kinman, and J. W. Koeppl, eds. 1982. *Mammal species of the world.* Lawrence, Kans.: Allen Press and Association of Systematics Collections.

Housse, R. P. R. 1948. Las zorras de Chile o chacales americanos. *Rev. Univ.* 34:33–56.

———. 1953. *Animales salvajes de Chile en su clasificación moderna.* Santiago: Ediciones de la Universidad de Chile.

Hulley, J. T. 1976. Maintenance and breeding of captive jaguarundis (*Felis yagouaroundi*) at Chester Zoo and Toronto. *Int. Zoo Yearb.* 16:120–22.

Iriarte, J. A. 1988. Feeding ecology of the Patagonian puma (*Felis concolor*) in Torres del Paine National Park, Chile. M.A. thesis, University of Florida.

Jaksić, F. M., R. P. Schlatter, and J. L. Yáñez. 1980. Feeding ecology of central Chilean foxes, *Dusicyon culpaeus* and *Dusicyon griseus. J. Mammal.* 61:254–60.

Jaksić, F. M., and J. L. Yáñez. 1983. Rabbit and fox introductions in Tierra del Fuego: History and assessment of the attempts at biological control of the rabbit infestation. *Biol. Conserv.* 26:367–74.

Jaksić, F. M., J. L. Yáñez, and J. R. Rau. 1983. Trophic relations of the southernmost populations of *Dusicyon* in Chile. *J. Mammal.* 64:693–97.

Kaufmann, J. H. 1962. Ecology and social behavior of the coati, *Nasua narica* on Barro Colorado Island, Panama. *Univ. Calif. Public. Zool.* 60: ·95–222.

Kaufmann, J. F., and A. Kaufmann. 1965. Observations of the behavior of tayras and grisons. *Z. Säugetierk.* 30:146–55.

Kipp, H. 1965. Beitrag zur Kenntnis der Gattung *Conepatus* Molina, 1782. *Z. Säugertierk.* 30: 193–256.

Kleiman, D. G. 1972. Social behavior of the maned wolf (*Chrysocyon brachyurus*) and bush dog (*Speothos venaticus*): A study in contrast. *J. Mammal.* 53:791–806.

Konecny, M. J. 1989. Movement patterns and food habits of four sympatric carnivore species in Belize, Central America. In *Advances in Neotropical mammalogy,* ed. K. H. Redford and J. F. Eisenberg, 243–64. Gainesville, Fla.: Sandhill Crane Press.

Koslowsky, J. 1904. Dos mamíferos de Patagonia, cazados en el Valle del Lago Blanco (territorio del Chubut). *Rev. Mus. La Plata* 11:129–32.

Krumbiegel, I. 1942. Die Säugetiere der Südamerikaexpeditionen Prof. Dr. Kriegs. 17. Hyrare und Grisons (*Tayra* und *Grison*). *Zool. Anz.* 139: 81–108.

———. 1953. Der Andenwolf, *Dasycyon hagenbecki* (Krumbiegel, 1949). *Säugetierk. Mitt.* 1: 97–104.

Lahille, F. 1899. Ensayo sobre la distribución geográfica de los mamíferos en la república Argentina. In *Primera reunión del Congreso Científico Latin Americo,* 3:165–206. Buenos Aires: Ed. Cía. Sudamericana de Billetes de Banco.

Langguth, A. 1975. Ecology and evolution in the South American canids. In *The wild canids,* ed. M. W. Fox, 192–206. New York: Van Nostrand Reinhold.

Leyhausen, P. 1963. Über südamerikanische Pardelkatzen. *Z. Tierpsychol.* 20:627–40.

Leyhausen, P., and M. Falkena. 1966. Breeding the Brazilian ocelot-cat *Leopardus tigrinus* in captivity. *Int. Zoo Yearb.* 6:176–82.

Linares, O. J. 1981. Tres nuevos carnívoros prociónidos fósiles de Mioceno de Norte y Sudamérica. *Ameghiniana* 18:113–21.

Löhmer, R. 1976. Zur Verhaltensontogenese bei *Procyon cancrivorus* (Procyonidae). *Z. Säugetierk.* 41:42–58.

Lucero, M. M. 1983. Lista y distribución de aves y mamíferos de la provincia de Tucumán. Ministerio de Cultura y Educación, Fundación Miguel Lillo. *Miscelánea* 75:5–53.

Ludlow, M. E., and M. E. Sunquist. 1987. Ecology and behavior of ocelots in Venezuela. *Nat. Geog. Res.* 3(4): 447–61.

Mann, G. 1945. Mamíferos de Tarapacá: Observaciones realizadas durante una expedición al alto norte de Chile. *Biológica* 2:23–141.

Mares, M. A. 1973. Climates, mammalian communities and desert rodent adaptations: An investigation into evolutionary convergence. Ph.D. diss., University of Texas at Austin.

Mares, M. A., and R. A. Ojeda. 1984. Faunal commercialization and conservation in South America. *BioScience* 34(9): 580–84.

Massoia, E. 1976. Mammalia. In *Fauna de agua dulce de la república Argentina.* vol. 44, *Mammalia.* Buenos Aires: Fundación para la Educación, la Ciencia y la Cultura.

———. 1982. *Dusicyon gymnocercus lordi,* una nueva subespecie del "zorro gris grande" (Mammalia Carnivora Canidae). *Neotrópica* 28(80): 147–52.

Melquist, W. E. 1984. Status survey of otters (Lutrinae) and spotted cats (Felidae) in Latin America. Completion report to IUCN (contract 9006).

Meserve, P. L., E. J. Shadrick, and D. A. Kelt. 1987. Diets and selectivity of two Chilean predators in the northern semi-arid zone. *Rev. Chil. Hist. Nat.* 50:93–99.

Miller, F. W. 1930. Notes on some mammals of southern Mato Grosso, Brazil. *J. Mammal.* 11: 10–22.

Miller, S. D., J. Rottman, K. J. Raedeke, and R. D. Taber. 1983. Endangered mammals of Chile: Status and conservation. *Biol. Conserv.* 25: 335–52.

Mondolfi, E. 1986. Notes on the biology and status of the small wild cats in Venezuela. In *Cats of the world: Biology, conservation and management,* ed. S. Douglas Miller and D. D. Everett, 125–46. Washington, D.C.: National Wildlife Federation.

Mondolfi, E., and R. Hoogesteijn. 1986. Notes on the biology and status of the jaguar in Venezuela. In *Cats of the world: Biology, conservation, and management,* ed. S. Douglas Miller and D. D. Everett, 85–124. Washington, D.C. National Wildlife Federation.

Montgomery, G. G., and Y. D. Lubin. 1978. Social structure and food habits of crab-eating fox (*Cerdocyon thous*) in Venezuelan llanos. *Acta Cient. Venez.* 29:382–83.

Nelson, E. W., and E. A. Goldman. 1933. Revision of the jaguars. *J. Mammal.* 14:221–40.

Ojeda, R. A., and M. A. Mares. 1982. Conservation of South American mammals: Argentina as a paradigm. In *Mammalian biology in South America,* ed. M. A. Mares and H. H. Genoways, 505–22. Pymatuning Symposia in Ecology 6. Special Publication Series. Pittsburgh: Pymatuning Laboratory of Ecology, University of Pittsburgh.

———. 1984. La degradación de los recursos naturales y la fauna silvestre en Argentina. *Incerciencia* 9(1): 21–26.

———. 1989. *A biogeographic analysis of the mammals of Salta province, Argentina.* Special Publications of the Museum 27. Lubbock: Texas Tech University Press.

Olrog, C. C. 1950. Notas sobre mamíferos y aves del archipiélago de Cabo de Hornos. *Acta Zool. Lilloana* 9:509–11.

———. 1979. Los mamíferos de la selva húmeda, Cerro Calilegua, Jujuy. *Acta Zool. Lilloana* 33: 9–14.

Olrog, C. C., and M. M. Lucero. 1981. *Guía de los mamíferos argentinos.* Tucumán, Argentina: Ministerio de Cultura y Educación, Fundación Miguel Lillo.

Osgood, W. H. 1914. Mammals of an expedition across northern Peru. *Field Mus. Nat. Hist., Zool. Ser.* 10(12): 143–85.

———. 1943. The mammals of Chile. *Field Mus. Nat. Hist., Zool. Ser.* 30:1–268.

Pearson, O. P. 1951. Mammals in the high-lands of southern Peru. *Bull. Mus. Comp. Zool.* 106(3): 117–74.

———. 1957. Additions to the mammalian fauna of Peru and notes on some other Peruvian mammals. *Breviora* 73:1–7.

Peña, L. E. 1966. Dos especies raras de mamíferos. *Not. Mens. Mus. Nac. Hist. Nat.* (Santiago) 11(123): 7–8.

Petersen, M. K., and M. K. Petersen. 1978. Growth rates and other post-natal developmental changes in margays. *Carnivore* 1(1): 87–92.

Pine, R. H., S. D. Miller, and M. L. Schamberger. 1979. Contributions to the mammalogy of Chile. *Mammalia* 43:339–76.

Poglayen-Neuwall, I. 1978. Breeding, rearing and notes on the behaviour of tayras *Eira barbara* in captivity. *Int. Zoo Yearb.* 18:134–40.

Rabinovich, J., A. Capurro, P. Folgarait, T. Kitzberger, G. Kramer, A. Novaro, M. Puppo, and A. Travaini. 1987. Estado del conocimiento de 12 especies de la fauna silvestre Argentina e valor comercial. Documento presentado, para su estudio y discusión, al 2° taller de trabajo: "Elaboración de propuestas de investigación orientada al manejo de la fauna silvestre de valor comercial," Buenos Aires.

Rabinowitz, A. R., and B. G. Nottingham, Jr. 1986. Ecology and behaviour of the jaguar (*Panthera onca*) in Belize, Central America. *J. Zoo.* (London), 210:149–59.

Rasmussen, J. L., and R. L. Tilson. 1984. Food provisioning by adult maned wolves (*Chrysocyon brachyurus*). *Z. Tierpsychol.* 65:346–52.

Russell, J. K. 1981. Exclusion of adult male coatis from social groups: Protection from predation. *J. Mammal.* 62(1): 206–8.

———. 1983. Altruism in coati bands: Nepotism or reciprocity? In *Social behavior of female vertebrates,* ed. S. K. Wasser, 263–90. New York: John Wiley.

Sanborn, C. C. 1954. Weights, measurements, and color of the Chilean forest puma. *J. Mammal.* 35:126–28.

Schaller, G. B. 1983. Mammals and their biomass on a Brazilian ranch. *Arq. Zool. São Paulo* 31(1): 1–36.

Schaller, G. B., and P. G. Crawshaw, Jr. 1980. Movement patterns of jaguar. *Biotropica* 12: 161–68.

Schaller, G. B., H. B. Quiggley, and P. G. Crawshaw. 1984. Biological investigations in the Pantanal, Mato Grosso, Brazil. *Nat. Geogr. Soc. Res. Repts.* 17:777–92.

Schaller, G. B., and J. M. C. Vasconcelos. 1978. Jaguar predation on capybara. *Z. Säugetierk.* 43: 296–310.

Schamberger, M., and G. Fulk. 1974. Mamíferos del Parque Nacional Fray Jorge. *Idesia* (Chile) 3: 167–79.

Scheffel, W., and H. Hemmer. 1975. Breeding Geoffroy's cat *Leopardus geoffroyi salinarum* in captivity. *Int. Zoo Yearb.* 15:152–54.

Scrocchi, G. J., and S. P. Halloy. 1986. Notas sistemáticas, ecológicas, etológicas y biogeográficas sobre el gato andino *Felis jacobita* Cornalia (Felidae, Carnivora). *Acta Zool. Lilloana* 38(2): 157–70.

Sielfeld, W. 1983. *Mamíferos marinos de Chile.* Santiago: Ediciones de la Universidad de Chile.

Simonetti, J. A. 1986. Human-induced dietary shift in *Dusicyon culpaeus. Mammalia* 50(3): 406–8.

Stains, H. J. 1975. Distribution and taxonomy of the Canidae. In *The wild canids,* ed. M. W. Fox, 3–26. New York: Van Nostrand Reinhold.

Stehlik, J. 1971. Breeding jaguars *Panthera onca* at Ostrava Zoo. *Int. Zoo Yearb.* 11:116–18.

Sunquist, M. E., F. Sunquist, and D. E. Daneke. 1989. Ecological separation in a Venezuelan llanos carnivore community. In *Advances in Neotropical mammalogy,* ed. K. H. Redford and J. F. Eisenberg, 197–232. Gainesville, Fla.: Sandhill Crane Press.

Tamayo, M., and D. Frassinetti. 1980. Catálogo de los mamíferos fósiles y vivientes de Chile. *Mus. Nac. Hist. Nat. Chile* 37:323–99.

Texera, W. A. 1974a. Nuevos antecedentes sobre mamíferos de Magallanes. 1. La distribución del gato montes (*Felis geoffroyi leucobapta*) (Mammalia: Felidae) en la region de Magallanes. *Anal. Inst. Patagonia, Punta Arenas* (Chile) 5(1–2): 189–92.

———. 1974b. Nuevos antecedentes sobre mamíferos de Magallanes. 3. El quique (*Galictis cuja cuja*) (Mammalia: Mustelidae), una nueva adición a la fauna mamal de Magallanes, Chile. *Anal. Inst. Patagonia, Punta Arenas* (Chile) 5(1–2): 195–98.

Torres, D., J. Yáñez, and P. Cattan. 1979. Mamíferos marinos de Chile: Antecedentes y situación actual. *Biol. Pesquera, Chile* 11:49–81.

Van Gelder, R. G. 1968. The genus *Conepatus* (Mammalia, Mustelidae): Variation within a population. *Amer. Mus. Novitat.* 2322:1–37.

Van Zyll de Jong, C. G. 1972. *A systematic review of the Nearctic and Neotropical river otters (genus* Lutra, *Mustelidae, Carnivora).* Life Sciences Contribution 80. Toronto: Royal Ontario Museum.

Vaughn, R. 1974. Breeding the tayra *Eira barbara* at Antelope Zoo, Lincoln. *Int. Zoo Yearb.* 14: 120–22.

Vieira, C. 1945. Sôbre uma coleção de mamíferos de Mato Grosso. *Arqu. Zool. São Paulo* 4:395–430.

Vuilleumier, F. 1985. Fossil and Recent avifaunas and the interamerican interchange. In *The great American biotic interchange,* ed. G. Stehli and S. David Webb, 387–424. New York: Plenum.

Wilson, P. 1984. Puma predation on guanacos in Torres del Paine National Park, Chile. *Mammalia* 48:515–22.

Wolffsohn, J. A. 1908. Contribuciones a la mamalogía chilena: 1. Sobre el *Felis colocolo,* Mol. *Rev. Chil. Hist. Nat:* 12:165–72.

———. 1923. Medidas máximas y mínimas de algunos mamíferos chilenos colectados entre los años 1896 y 1917. *Rev. Chil. Hist. Nat.* 27:159–67.

Ximénez, A. 1961. Nueva subespecie del gato pajero en el Uruguay. *Comun. Zool. Mus. Hist. Nat. Montevideo* 5(88): 1–8.

———. 1972. Notas sobre felidos neotropicales. 4. *Puma concolor* spp. en el Uruguay. *Neotrópica* 18(55): 37–39.

———. 1973. Notas sobre félidos neotropicales. 3. Contribución al conocimiento de *Felis geoffroyi* d'Orbigny y Gervais, 1844 y sus formas geográficas (Mammalia, Felidae). *Pap. Avulsos Zool. São Paulo* 27(3): 31–43.

———. 1975. *Felis geoffroyi. Mammal. Species* 54: 1–4.

Ximénez, A., and E. Palerm. 1971. Confirmación de la presencia de *Felis wiedii wiedii* Schinz (Carnivora, Felidae), en el Uruguay. *Bol. Soc. Zool. Uruguay* 1:7–10.

Yáñez, J., and F. Jaksić. 1978. Rol ecologico de los zorros (*Dusicyon*) en Chile central. *Anal. Mus. Hist. Nat. Valparaíso* (Chile), 11:105–12.

Yáñez, J., and J. Rau. 1980. Dieta estacional de *Dusicyon culpaeus* (Canidae) en Magallanes. *Anal. Mus. Hist. Nat. Valparaíso* 13:189–91.

Zapata, A. R. P. 1982. Sobre el yaguarundi o gato eira, *Felis yagouaroundi ameghinoi* Holmberg y su presencia en la provincia de Buenos Aires, Argentina. *Neotrópica* 28(80): 165–70.

7   Order Pinnipedia
(Seals, Sea Lions, and Walruses)

Diagnosis

The body form of pinnipeds is strongly modified for swimming. The hind feet and forefeet are in the form of flippers, with the digits completely enclosed in the integument. The tail is short, and the body is spindle shaped. The teeth are usually simplified in structure, often peglike or conical (fig. 7.1), but the feeding specialization of such forms as *Lobodon* yields a modified form of the molars and premolars (fig. 7.2).

Distribution

Seals and their relatives are found worldwide in the oceans close to continental shorelines or polar ice. A few species are landlocked in subarctic lakes. They reach their greatest species diversity in the temperate to Arctic and Antarctic latitudes and are generally absent from tropical waters except for the monk seals (King 1956).

History and Classification

The order Pinnipedia is sometimes considered a suborder of the Carnivora (Honacki, Kinman, and Koeppl 1982). The Pinnipedia are first detected in the middle Miocene, already extremely specialized for an aquatic existence. The present-day Pinnipedia may have had a dual origin and thus be an artificial assemblage, since the ancestors of the true seals (Phocidae) may have derived from mustelid precursors while the eared seals (Otariidae) may have derived from an ancestral form that also gave rise to modern canids and ursids. In contrast, evidence presented by Berta, Ray, and Wyss (1989) and by Sarich (1969) supports a monophyletic origin for the group. The extant pinnipeds are usually grouped into three families: the Otariidae, or eared seals, including the fur seals and sea lions; the Odobenidae,

including the walruses; and the Phocidae, or "earless" seals. This last group is characterized by loss of the external ear (pinna) and by hind limbs incapable of rotating forward, thus limiting movement on land. The biology of the Pinnipedia is summarized by King (1964).

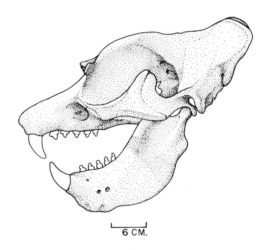

Figure 7.1. Skull of *Mirounga leonina*.

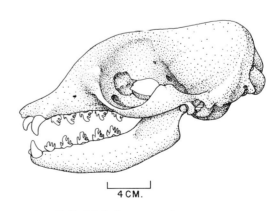

Figure 7.2. Skull of *Lobodon carcinophagus*.

FAMILY OTARIIDAE

Diagnosis

The dental formula is I 3/2, C 1/1, P 4/4, M 1–2/1. There is a wide range of size within the family, from 60 kg to over 1,000 kg. Total length ranges from 1.5 to 3.5 m. Species of this family are strongly size dimorphic, with the males much larger than the females. The body is fusiform, highly adapted for aquatic life, and the limbs are modified into flippers, with the nails well developed. The hind limbs can be brought under the body for partial support. The external ear (pinna) is present but small.

Distribution

Distribution is coastal or insular. The family is broadly distributed from the northeast coast of Asia southward along the Pacific coast of North America and South America and extends on the Atlantic coast of South America to approximately 15° south latitude. Populations also occur off southern Australia and New Zealand and the major islands in the south Atlantic, near the coast of southern Africa, and in the extreme south Indian Ocean.

Natural History

Fur seals and sea lions spend most of their lives at sea, returning to land to breed and to rear their young. Sites for breeding and rearing are usually on isolated offshore islands or inaccessible rocky coasts. These animals are primarily fish feeders. A single young is produced after a gestation period of approximately eleven months. The time spent ashore varies widely. Polar populations spend only a short time on land (usually less than a month), while populations in the lower latitudes may spend several months on land and have a more protracted rearing phase. The males are strongly polygynous and usually stake out a favorable beach area and defend it against other males. Females usually arrive ashore later than males and take up residence in a male's territory. (See species accounts.)

Genus *Arctocephalus* E. Geoffroy and F. Cuvier, 1826

Description

Arctocephalus is strongly sexually dimorphic in size; males may be 2.7 m in head and body length and weigh over 700 kg, while females are more gracile, rarely exceeding 1.9 m with a maximum weight of 120 kg. The basic dorsal color is a reddish to dark brown. Breeding males have thick fur on the neck, and its white tips yield a frosted appearance (plate 10).

Distribution

This genus ranges in the Southern Hemisphere following the cold Humboldt current to the Galápagos Islands in South America.

Life History and Ecology

The Southern Hemisphere fur seals are at sea much of their lives but return to isolated insular beaches to breed and give birth. After giving birth the female usually nurses the young for about a week, then she must return to the sea to feed. The female may suckle the young for four to twelve months after birth, but she alternates feeding bouts with time ashore nursing. A similar rearing pattern is shown by the sea lion. (See species accounts.)

Arctocephalus australis (Zimmermann, 1783)
Lobo de los Pelos

Description

Adult males are much larger than females; they reach 1.9 to 3 m in total length and weigh more than 159 kg, while females reach 1.4 to 2 m and 48.5 kg. The snout is pointed and is larger than that of *A. flavescens*. The rhinarium is globular with nasal openings pointed forward, distinguishing it from *A. phillipi*, which has the openings pointed down. This large marine mammal has small external ears and short, thick pelage composed of relatively stiff outer hairs and soft, dense underfur. Dorsally it is dark coffee brown with gray tints, and ventrally it has cinnamon tints. The pups are born black and gain adult coloration after about three months (Sielfeld 1983; Vaz-Ferreira 1979a).

Distribution

Arctocephalus australis is found on the coasts of South America from Rio de Janeiro (Brazil) around the tip of the continent and up the west coast to Lima, Peru. It is also found on the Juan Fernández Islands, Galápagos Islands, and Malvinas Islands. It does not breed throughout this range. *A. australis* has been eliminated from most of its former distribution in Chile by hunting of pups (Daciuk 1974; Honacki, Kinman, and Koeppl 1982; Miller et al. 1983; Scheffer 1958) (map 7.1).

Life History

The birth season is in November and December, with mating postpartum. One pup is born, weighing 3–5 kg, and lactation lasts six to twelve months. The age of first reproduction is probably four to five years; pups swim before two months of age (Housse 1952; Sielfeld 1983; Trillmich and Majluf 1981; Vaz-Ferreira 1979a).

Ecology

Males of *A. australis* are polygamous, setting up territories in November, and in some areas they defend an average of fifteen females and do not eat during the entire reproductive period. There are also groups of nonreproductive bachelor males. Breeding groups are normally established in rocky places. The sex ratio (male: female) in breeding rookeries in some areas varies from 1:1 to 1:13, averaging 1:6.5. In Peru this species breeds in huge sea caves. Females coming in from the ocean call to attract their pups, then they smell them to confirm identity.

In some parts of its range *A. australis* is a krill (*Euphausia*) feeder, but it is reported to take a broad range of other food items including fish and squid. It feeds over wide areas of the continental shelf and beyond. No definite migration is shown, but seasonally there is wide seaward dispersion (King 1983; Sielfeld 1983; Trillmich and Majluf 1981; Vaz-Ferreira 1979a).

Arctocephalus gazella (Peters, 1875)
Lobo Marino Grácil

Description

This species is physically similar to *A. australis* but smaller and with a shorter, more pug-nosed snout. *A. gazella* can be separated from other *Arctocephalus* by its slender, widely spaced postcanine

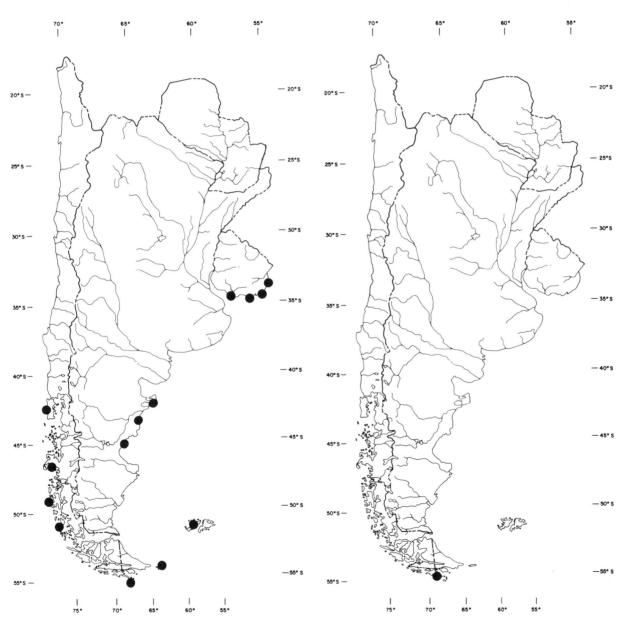

Map 7.1. Distribution of *Arctocephalus australis*.

Map 7.2. Distribution of *Arctocephalus gazella*.

teeth. Males reach 2.2 m in total length and weigh 125–200 kg, while females measure 1.4 m and weigh 25–50 kg. The back and sides are gray to brownish, depending on how long the seal has been ashore. The throat and breast are creamy, and the venter is a dark ginger. Unlike *A. tropicalis*, this species does not have a definite yellow face and chest. In adult males there is a heavy mane around the neck and shoulders that has a grizzled appearance (Bonner 1979; King 1983; Scheffer 1958; Sielfeld 1983).

Distribution

Arctocephalus gazella is found on the islands south of the Antarctic convergence and occasionally on the islands at the extreme tip of the continent (Sielfeld 1983; Texera 1974) (map 7.2).

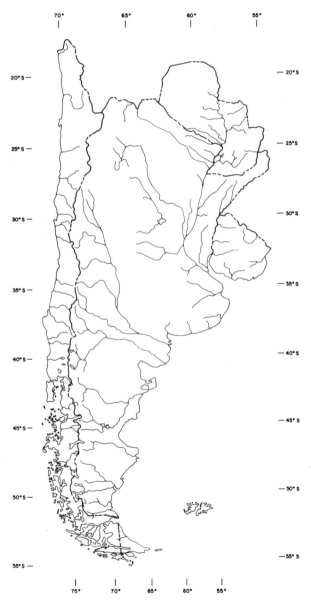

Map 7.3. Distribution of *Arctocephalus philippii*.

Life History

On South Georgia Island mature bulls start to establish territories in the middle of November. The dominant bulls have territories at the water's edge. Cows come to shore toward the end of November and give birth within a day or two of arriving. They come into estrus about eight days postpartum. Lactation lasts 110–15 days, and by the end of April most pups have gone to sea (King 1983).

Ecology

During the nineteenth century this species was hunted almost to extinction. In one breeding population on the South Shetland Islands the male:female ratio was 1:3.5. A pup tagged on South Georgia Islands was found at Bahía Orange, Isla Hoste, Chile. These fur seals are primarily krill feeders (Bonner 1979; Cattan et al. 1982; Sielfeld 1983; Texera 1974).

Arctocephalus philippii (Peters, 1866)

Description

Adult males reach approximately 2 to 3 m in length and weigh 140 kg, while adult females are much smaller at 1.4 to 2 m and 50 kg. The dorsum is gray black, and the venter is coffee black. The head and neck have yellow tints, and the area around the lips and vibrissae is yellow (Aguayo 1979; Sielfeld 1983).

Distribution

Arctocephalus philippii is restricted to the Juan Fernández Islands and to San Félix and San Ambrosio Islands off the coast of central Chile (Sielfeld 1983) (map 7.3).

Life History

The reproductive system of this species is similar to that of *A. australis* except that this species uses caves, some with underwater entrances (Sielfeld 1983).

Ecology

This species was thought to have become extinct because of extensive hunting for pelts. Several million seals were taken off the islands in the sixteenth and seventeenth centuries for sale in China. *A. philippii* hauls out chiefly on solid lava rock at the base of cliffs, on ledges, and in caves. The stomach of one individual contained fish and cephalopods (Aguayo 1979; Hubbs and Norris 1971; Sielfeld 1983).

Genus *Otaria* Peron, 1816
Otaria byronia (Blainville, 1820)
Lobo de un Pelo

Description

This large marine mammal has a large head, a short, broad snout, small external ears, and coarse pelage with no underfur (plate 10). Adult males can

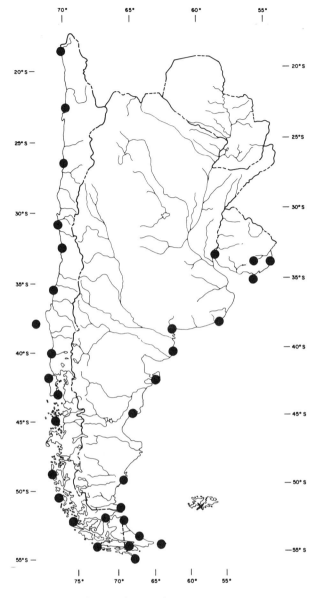

Map 7.4. Distribution of *Otaria byronia*.

reach 2.56 to 3.5 m and at least 300 to 340 kg, while the much smaller females reach 2 to 2.5 m and at least 144 kg. The males are much more robust than the females, which lack a mane. The color is very variable: often adult males are dark brown to orange with a mane of a paler shade and the belly shading to a lighter yellow. The females are dark to light brown, and the pups are black (Osgood 1943; Sielfeld 1983; Vaz-Ferreira 1979b).

Distribution

Otaria byronia is widely distributed in South America, from northern Peru down the length of Chile, up the coast of Argentina and Uruguay to southeastern Brazil. It is also found on offshore islands like the Malvinas (Majluf and Trillmich 1981; Scheffer 1958) (map 7.4).

Life History

Newborn male pups are 82 cm long and weigh 14.2 kg, while females measure 79 cm and weigh 11.5 kg. In Uruguay most births are from the end of December to the middle of January. Pups suckle six to twelve months, and a female may nurse both a pup and a yearling. Females mate postpartum. In Chile births are from September to March, depending on the latitude; lactation lasts about ten months, and young enter the water at about two months (Sielfeld 1983; Vaz-Ferreira 1979b).

Ecology

Males are polygynous, and in Uruguay they establish territories from the end of November to the end of February, on sand or gravel beaches where the male:female ratio may be as high as 1:15. In Chile males set up breeding territories in September on beaches or in caves, and females come late in September. Outside the reproductive period these seals live in the pelagic ocean, visiting land only to sun. Individuals have also been found a long way up freshwater rivers and in lakes in southern Chile. *O. byronia* is an opportunistic feeder, taking small schooling fish, bottom fish, squid, and occasional seabirds.

This species was the object of an intense fur trade that reached its height between 1920 and 1954, declining because of overkill. There has been a reduction in the range of this species because of commercial harvesting and killing by fishermen (Brandenburg 1938; Housse 1952; Iriarte and Jaksić 1986; Sielfeld 1983; Vaz-Ferreira 1979b).

FAMILY PHOCIDAE

Diagnosis

The dental formula is I 2–3/1–2, C 1/1, P 4/4, M 0–2/0–2. Species of this family are highly adapted to aquatic life, spending as little time as possible on land or ice. They come ashore only for reproduction. The hind flippers cannot be rotated under the body and are not used in terrestrial locomotion. No external ear is present. Size varies widely; weights range from 80 to 450 kg for most species, but male elephant seals (*Mirounga*) may reach 3,600 kg. The seals of the Southern Hemisphere show interesting variations on size dimorphism. In the Antarctic seals, males are slightly smaller than females, whereas in the elephant seal males are much larger than females.

Distribution

The family is distributed worldwide off the coasts of all continents, but it reaches its greatest species richness in polar regions. Only one genus, *Monachus*, is tropical in its distribution.

Natural History

These animals are highly adapted for aquatic life. They come ashore to give birth and to breed. Landing sites range from rocky coasts to pack ice. Many species eat fish, but others are specialized for feeding on squid, octopus, and even macroplankton. Most species of the Phocidae show shortened lactation; the basic strategy is for the female to feed heavily during pregnancy, accumulate fat reserves, and then wean the young after an intensive but short period of nursing. (See species accounts; see also Oftedal, Boness, and Tedman 1987.)

Genus *Hydrurga* Gistel, 1848
Hydrurga leptonyx (Blainville, 1820)
Leopard Seal, Leopardo Marino

Description

This largest Antarctic seal is easily identified by its disproportionally large, reptilelike head and pronounced neck, its graceful, streamlined body, and its long, tapered foreflippers. Adult males average 2.79 m in total length (range 2.5–3.2) and weigh an average of 324 kg (range 200–455), while the females average 2.91 m (range 2.41–3.38) and 367 kg (range 225–591).

The dorsum is dark gray, and the lower flanks and venter are a deep silvery white with dark and light stripes or blotched. When disturbed on land the seal alertly lifts its head, exposing the white-and-black blotched throat (Erickson and Hofman, in Brown et al. 1974; Hofman 1979; Scheffer 1958; Sielfeld 1983).

Distribution

Hydrurga leptonyx has a circumpolar distribution in the Southern Hemisphere, having been recorded in the oceans of Antarctica, southern South America, and southern Australia. It has been recorded in the Patagonian canals and off the southern coasts of Chile and Argentina. It used to be plentiful off the Malvinas Islands but has not been seen there in recent years. Waifs have been found as far north as Buenos Aires province, Argentina (Castello and Rumboll 1978; Honacki, Kinman, and Koeppl 1982; Schneider 1946; Sielfeld 1978, 1983) (map 7.5).

Life History

The leopard seal reproduces from September to December on the pack ice. Lactation probably lasts about four weeks (Hofman 1979; King 1983; Sielfeld 1983).

Ecology

Hydrurga leptonyx migrates northward to ice-free islands in the winter from the pack ice where it breeds in the summer. Outside the breeding season it is found singly. The leopard seal earns its name by being the only seal that regularly feeds on warm-blooded animals. It is known to take penguins, seals, and carrion as well as fish, squid, and krill (Erickson and Hofman, in Brown et al. 1974; Hofman 1979; Scheffer 1958; Sielfeld 1983).

Genus *Leptonychotes* Gill, 1872
Leptonychotes weddelli (Lesson, 1826)
Weddell's Seal, Foca de Weddel

Description

The body form is stout compared with that of the more streamlined seals. Adults are 210–329 cm long

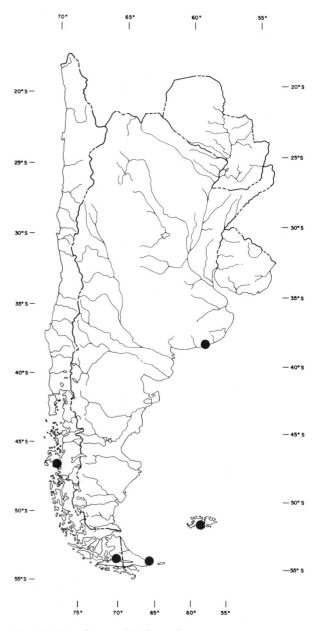

Map 7.5. Distribution of *Hydrurga leptonyx*.

and weigh 318–550 kg, with females slightly larger than males. These seals often lie on their sides, exposing their bellies. This large seal is generally dark brown to black dorsally and lighter on the venter, with the whole body covered with semicircular spots or blotches, dark as well as light. During the summer the color fades to a rusty grayish brown (Erickson and Hofman, in Brown et al. 1974; de Master 1979; King 1983; Sielfeld 1983; Stirling 1971).

Distribution

Leptonychotes weddelli is found on the coasts of Antarctica and adjacent islands. In southern South America animals have been collected or observed on the Uruguayan coast, the southernmost Chilean coast, and on the Malvinas and Juan Fernández Islands (Honacki, Kinman, and Koeppl 1982; de Master 1979; Schneider 1946; Torres, Yáñez, and Cattan 1979; Ximénez, Langguth, and Praderi 1972) (map. 7.6).

Life History

At pupping time females tend to congregate, giving birth on the sea ice from September to November. Litter size is one; birthweight is about 29 kg; lactation lasts five to six weeks; pups change to adult pelage at four to six weeks; and age of first reproduction is three or four years for males and two to four years for females. The underwater vocal repertoire is rich, and local dialects have been recorded (Erickson and Hofman, in Brown et al. 1974; de Master 1979; Scheffer 1958; Sielfeld 1983; Stirling 1971; Thomas and Kueechle 1982).

Ecology

Weddell's seal lives in areas of fast ice and is nonmigratory. It spends most of its time in the water but emerges at intervals to lie on the ice or beaches. Males defend aquatic territories centering on breathing holes. The ability of *L. weddelli* to maintain such breathing holes allows it to use areas of fast ice unavailable to other seals. The aggregating tendency during pupping season is probably due to the limited number of secure areas with recurring tidal cracks. Weddell's seal eats fish and benthic organisms such as crustaceans and cephalopods. It is a superb diver and may stay submerged for up to an hour and reach a depth of 600 m (Erickson and Hofman, in Brown et al. 1974; King 1983; Scheffer 1958; Sielfeld 1983; Stirling 1971).

Genus *Lobodon* Gray, 1844
Lobodon carcinophagus (Hombron and Jacquinot, 1842)
Crab-eating Seal, Foca Cangrejera

Description

Lobodon carcinophagus is similar to *Hydrurga* in shape but does not have as well delineated a neck. Adult males reach 2.57 m and 224 kg, and adult females reach 2.62 m and 227 kg. When disturbed on land this seal characteristically rises on its foreflippers and peers actively about. The dentition is distinctive and allows this species to strain krill from seawater (fig. 7.2). This lithe, long-snouted seal ranges in color from black to silvery gray depending on the individual, age, time of molt, and dampness of pelage. The color is uniform over the entire body (Erickson and Hofman, in Brown et al. 1974; King 1983; Laws 1979b; Scheffer 1958; Sielfeld 1983).

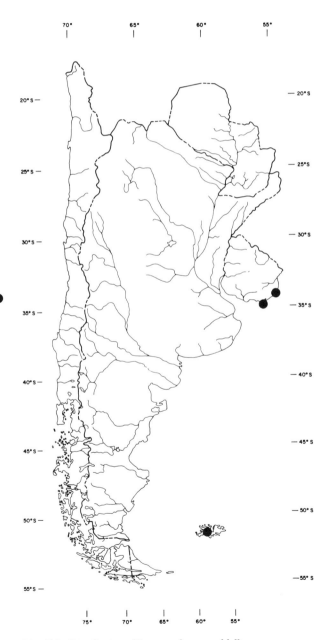

Map 7.6. Distribution of *Leptonychotes weddelli*.

Distribution

Lobodon carcinophagus is found in the Antarctic seas and in southern South America. Individuals have been collected at scattered locations as far north as southern Brazil (Honacki, Kinman, and Koeppl 1982; Scolaro 1976) (map 7.7).

Life History

Pupping apparently occurs on heavy pack ice during the southern spring. The litter size is one; there is an apparent two-month delay in implantation; age of first reproduction is three to six years; and the life span is at least twenty-nine years. Pups are born measuring 1.2 m in length and weighing 20 kg and

have light gray pelage. Lactation apparently lasts one month, and adult size is reached at about four months (Erickson and Hofman, in Brown et al. 1974; Laws 1979b; Sielfeld 1983).

Ecology

The crab-eating seal occurs throughout the Antarctic pack ice, wherever it can haul out. Where it occurs, it is usually the most abundant seal. It is found most commonly at the edges of pack ice and in heavily broken ice, though it moves seasonally as the ice conditions change. It can be found singly or in groups, depending on the season of the year. Even though called the "crab-eating" seal, it consumes mostly krill. It usually feeds at night when the krill are close to the surface, sucking water and krill into the mouth, where the krill are sieved out as the water is expelled through the complicated cheek teeth (fig. 7.2) (Erickson and Hofman, in Brown et al. 1974; King 1983; Laws 1979b; Scheffer 1958; Sielfeld 1983).

Genus *Mirounga* Flemming, 1822
Mirounga leonina (Linnaeus, 1758)
Elephant Seal, Elefante Marino

Description

Mirounga leonina cannot be confused with any other seal in southern South America because of its tremendous size. The males are up to 6 m long and weigh up to 4,000 kg, while the much smaller females measure up to 3.5 m and weigh up to 900 kg. There is considerable variation is size throughout the range of this species. The eyes are very large and the snout is short and equipped, in males, with a well-developed bulbous proboscis (plate 10). Adults are dark gray dorsally and lighter ventrally, weathering to shades of brown (Laws 1979a; Sielfeld 1983).

Distribution

Mirounga leonina is found on the islands of the subantarctic and adjacent continents. In southern South America the range of this species has been decreased because of hunting, but it is still found along the Argentine coast north to Uruguay and on the Chilean coast to Parry Ford. The northernmost breeding locality is in Chubut province, Argentina. The type locality for this species was the Juan Fernández Islands, on which it is now extinct (Daciuk 1973; Sielfeld 1978, 1983; Torres, Yáñez, and Cattan 1979) (map 7.8).

Life History

The breeding season for the southern elephant seal varies with latitude: in Argentina it starts near

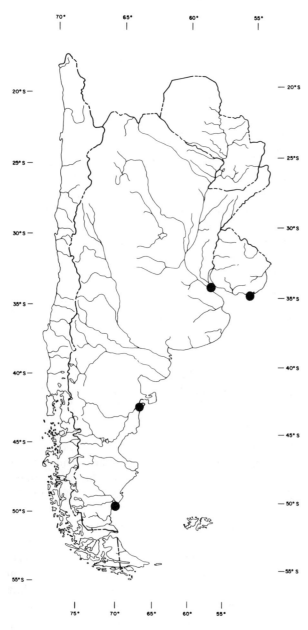

Map 7.7. Distribution of *Lobodon carcinophagus*.

the end of August and lasts until the beginning of November. A single black pup is born weighing between 40 and 46 kg and measuring about 1.27 m. The gestation period is fifty weeks, including a twelve-week delay in implantation. Lactation lasts only twenty-three days, and females may lose 318 kg during this time. Females come into estrus eighteen days after giving birth. The age of sexual maturity varies from locality to locality. In males it ranges from four to six years and in females from two to seven years. Males are polygynous, defending harems of twenty to thirty cows (Erickson and Hofman, in Brown et al. 1974; Laws 1979a; Sielfeld 1983; Torres, Yáñez, and Cattan 1981).

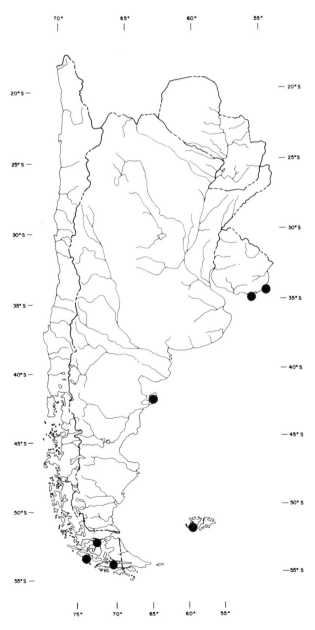

Map 7.8. Distribution of *Mirounga leonina*.

Ecology

Mirounga leonina is thought to be migratory, moving in winter toward the pelagic feeding grounds at the edges of the pack ice. During the breeding season males set up territories on peaceful sand or rock beaches and are followed by the females about three weeks later. Pups begin feeding on amphipods and then switch to the adult diet of cephalopods and fish.

The southern elephant seal was nearly exterminated for its oil in the nineteenth century, but since then it has been increasing. At present the stock at Península Valdés is over twelve thousand animals during the breeding season (Castello 1984; Erickson and Hofman, in Brown et al. 1974; Laws 1979a; Scheffer 1958; Sielfeld 1983).

Genus *Ommatophoca* Gray, 1844
Ommatophoca rossi Gray, 1844
Ross Seal, Foca de Ross

Description

Males and females measure about 2.25 m in total length and weigh up to 230 kg. This seal has proportionally large front flippers with an enlarged first digit. The head is short and broad with a short muzzle, and the nasal openings are directed upward. The teeth are very small, and the eyes are very large. The dorsum is dark with some stripes, and the venter is silver gray. When approached on the ice this seal inflates its throat and produces a unique birdlike call (Erickson and Hofman, in Brown et al. 1974; Sielfeld 1983).

Distribution

Ommatophoca rossi is found on the Antarctic pack ice, where it seems to prefer smaller, smooth-surfaced floes (Honacki, Kinman, and Koeppl 1982; King 1983).

Life History

In Antarctica molting takes place in January or February. The mating peak is after the beginning of February, and the birth peak is in the first half of November. Gestation lasts about nine months and lactation from four to six weeks. Sexual maturity is reached at three years (Skinner 1984).

Ecology

This is one of the rarest of the Antarctic seals. It appears to be mainly solitary. The diet comprises mainly cephalopods (Erickson and Hofman, in Brown et al. 1974; King 1983).

References

Aguayo L., A. 1979. Juan Fernández fur seal. In *Mammals in the seas*, 2:30–38. FAO Fisheries

Series 5. Rome: Food and Agriculture Organization of the United Nations.

Berta, A., C. E. Ray, and A. R. Wyss. 1989. Skeleton of the oldest known pinniped *Enaliarctos mealsi*. *Science* 244:60–62.

Bonner, W. N. 1979. Antarctic (Kenuelen) fur seal. In *Mammals in the seas*, 2:49–51. FAO Fisheries Series 5. Rome: Food and Agriculture Organization of the United Nations.

Brandenburg, F. G. 1938. Notes on the Patagonian sealion. *J. Mammal.* 19:44–47.

Brown, S. G., R. L. Brownell, Jr., A. W. Erickson, R. J. Hofman, G. A. Llano, and N. A. Mackintosh, eds. 1974. *Antarctic mammals*. Antarctic Map Folio Series 18. New York: American Geographical Society.

Castello, H. P. 1984. Registros del elefante marino, *Mirounga leonina* (Carnivora, Phocidae) en las costas del Atlántico S.O., fuera del area de cría. *Rev. Mus. Argent. Cienc. Nat. "Bernardino Rivadavia," Zool.* 13(1–60): 235–43.

Castello, H. P., and M. Rumboll. 1978. Extension of range of the leopard seal, *Hydrurga leptonyx*, for the Argentine coast. *Mammalia* 42:135–37.

Cattan, P., J. Yáñez, D. Torres, M. Gajardo, and J. C. Cardenas. 1982. Censo, marcaje y estructura poblacional del lobo fino antártico *Arctocephalus gazella* (Peters, 1875) en las islas Shetland del Sur, Chile (Pinnipedia: Otariidae). *INACH, Ser. Cient.* 29:31–38.

Daciuk, J. 1973. Notas faunísticas y bioecológicas de Península Valdés y Patagonia. 10. Estudio cuantitativo y observaciones del comportamiento de la población del elefante marino del sur *Mirounga leonina* (Linne) en sus apostaderos de la provincia de Chubut (república Argentina). *Physis*, sec. C, 32(85): 403–22.

———. 1974. Notas faunísticas y bioecológicas de Península Valdés y Patagonia. 12. Mamíferos colectados y observados en la Península Valdés y zona litoral de los Golfos San José y Nuevo (provincia de Chubut, república Argentina). *Physis*, sec. C, 33(86): 23–39.

de Master, D. P. 1979. Weddell seal. In *Mammals in the seas*, 2:130–34. FAO Fisheries Series 5. Rome: Food and Agriculture Organization of the United Nations.

Hofman, R. J. 1979. Leopard seal. In *Mammals in the seas*, 2:125–29. FAO Fisheries Series 5. Rome: Food and Agriculture Organization of the United Nations.

Honacki, J. H., K. E. Kinman, and J. W. Koeppl, eds. 1982. *Mammal species of the world*. Lawrence, Kans.: Allen Press and Association of Systematics Collections.

Housse, R. P. R. 1952. Mamíferos de Chile orden de los pinípedos. *Acad. Chil. Cienc. Nat.* 17, *Anal., Rev. Univ.* (Santiago) 37:163–75.

Hubbs, C. L., and K. S. Norris. 1971. Original teeming abundance, supposed extinction, and survival of the Juan Fernández fur seal. *Antarct. Res. Ser.* 18:35–52.

Iriarte, J. A., and F. M. Jaksić. 1986. The fur trade in Chile: An overview of seventy-five years of export data (1910–1984). *Biol. Conserv.* 38: 243–53.

King, J. E. 1956. The monk seals, genus *Monachus*. *Bull. Brit. Mus. Nat. Hist.* (*Zool.*) 3:203–56.

———. 1964. *Seals of the world*. London: British Museum of Natural History.

———. 1983. *Seals of the world*. 2d ed. London: British Museum of Natural History.

Laws, R. M. 1979a. Southern elephant seal. In *Mammals in the seas*, 2:106–9. FOA Fisheries Series 5. Rome: Food and Agriculture Organization of the United Nations.

———. 1979b. Crabeater seal. In *Mammals in the seas*, 2:115–19. FOA Fisheries Series 5. Rome: Food and Agriculture Organization of the United Nations.

Majluf, P., and F. Trillmich. 1981. Distribution and abundance of sea lions (*Otaria byronia*) and fur seals (*Arctocephalus australis*) in Peru. *Z. Säugetierk.* 46:384–93.

Miller, S. D., J. Rottman, K. J. Raedeke, and R. D. Taber. 1983. Endangered mammals of Chile: Status and conservation. *Biol. Conserv.* 25:335–52.

Oftedal, O. T., D. J. Boness, and R. A. Tedman. 1987. The behavior, physiology, and anatomy of lactation in the Pinnipedia. *Current Mammal.* 1:175–246.

Osgood, W. H. 1943. The mammals of Chile. *Field Mus. Nat. Hist., Zool. Se.* 30:1–268.

Sarich, V. M. 1969. Pinniped phylogeny. *Syst. Zool.* 18:416–22.

Scheffer, V. B. 1958. *Seals, sea lions and walruses: A review of the Pinnipedia*. Stanford: Stanford University Press.

Schneider, C. O. 1946. Catálogo de los mamíferos de la provincia de Concepción. *Bol. Soc. Biol. Concepción* 21:67–83.

Scolaro, J. A. 1976. Nota sobre una foca cangrejera (*Lobodon carcinophagus*) capturada en Puerto Madryn, Chubut (Phocidae, Pinnipedia). *Neotrópica* 22(68): 117–19.

Sielfeld, W. 1978. Algunas consideraciones sobre focidos (*Pinnipedia*) asociados a las costas de Chile. *Ans. Inst. Pat., Punta Arenas* (Chile) 9:153–56.

———. 1983. *Mamíferos marinos de Chile*. Santiago: Ediciones de la Universidad de Chile.

Skinner, J. D. 1984. Research on the Ross seal, *Ommatophoca rossi*, in the King Haakon VII Sea, Antarctica. *S. Afr. J. Sci.* 80:30–31.

Stirling, I. 1971. *Leptonychotes weddelli. Mammal. Species* 6:1–5.

Texera, W. A. 1974. Nuevos antecedentes sobre mamíferos de Magallanes. 2. Hallazgo de *Arctocephalus gazella* (Mammalia: Otariidae) en Isla Hoste, de la región de Magallanes, Anillado en Isla Bird, Georgia del Sur. *Anal. Inst. Patagonia, Punta Arenas* (Chile) 5(1–2): 193–94.

Thomas, J., and V. Kueechle. 1982. Quantitative analysis of Weddell seal (*Leptonychotes weddelli*) underwater vocalizations at McMurdo Sound, Antarctica. *J. Acoust. Soc. Amer.* 72:1730–38.

Torres, D., J. Yáñez, and P. Cattan. 1979. Mamíferos marinos de Chile: Antecedentes y situación actual. *Biol. Pesquera* (Chile) 11:49–81.

Trillmich, F., and P. Majluf. 1981. First observation on colony structure, behavior, and vocal repertoire of the South American fur seal (*Arctocephalus australis* Zimmermann, 1783) in Peru. *Z. Säugetierk.* 46:310–22.

Vaz-Ferreira, R. 1979a. South American sea lion. In *Mammals in the seas*, 2:9–12. FOA Fisheries Series 5. Rome: Food and Agriculture Organization of the United Nations.

———. 1979b. South American fur seal. In *Mammals in the seas*, 2:34–36. FOA Fisheries Series 5. Rome: Food and Agriculture Organization of the United Nations.

Ximénez, A., A. Langguth, and R. Praderi. 1972. Lista sistemática de los mamíferos del Uruguay. *Anal. Muse. Nac. Hist. Nat. Montevideo* 7(5): 1–45.

8   Order Cetacea
(Whales, Dolphins, and Their Allies)

Diagnosis

Except for the Sirenia, the Cetacea exhibit the most complete adaptation for aquatic existence shown within the class Mammalia (see plates 10 and 11). The body is spindle shaped, hair is absent except for a few bristles around the lips, and the skin is smooth. The forelimbs are modified into paddles; the digits cannot be detected externally. Hind limbs are absent, and the tail is flattened dorsoventrally and extended laterally to produce two pointed flukes. The nasal passages open through either a single or a double aperture on top of the head. The teeth are absent or extremely modified, exhibiting no cusp pattern in the adult. All the teeth are very similar in shape within the jaw of any species (homodont); it is impossible to distinguish molars, premolars, and canines. Some standard references on the order's biology and taxonomy are Slijper (1962), Norris (1966), Gaskin (1982), Minasian, Balcomb, and Foster (1984), and Hershkovitz (1966).

Distribution

The order is found worldwide in oceans and seas; some small species ascend rivers or are adapted for a permanent freshwater existence.

History and Classification

Cetaceans are classically divided into three suborders, the Archaeoceti, the Odontoceti, and the Mysticeti. Whether the last two lineages descended from a common ancestor is still open to debate. It is enough to say that the Archaeoceti first appear in the fossil record in the middle Eocene and may have given rise to the extant families. The oldest odontocete fossils are from the late Eocene, and mysticete fossils first appear in the mid-Oligocene. Odontocetes have teeth and are thus often referred to as "toothed" whales. These forms usually feed

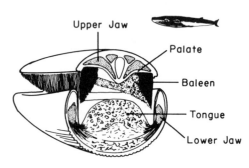

Figure 8.1. Schematic diagram of the baleen plates and their position in the mouth of *Balaenoptera* (modified from Slijper 1962).

on fish and cephalopods (see species accounts). The mysticetes do not have teeth as adults, but there are modified epidermal derivatives hanging from the roof of the mouth in longitudinal plates called baleen (see fig. 8.1), which serve as a straining device. Mysticetes are specialized for feeding on various small crustaceans (krill) that they obtain by swimming through swarms of the small, shrimplike creatures with their mouths partly open, periodically expelling excess water and swallowing the retained crustaceans after scraping them off the baleen plates with the tongue. Baleen whales also eat fish (see species accounts).

Natural History

The great whales and dolphins are entirely aquatic and never come to land unless accidentally stranded or washed ashore during storms. Usually only a single young is born, after an extended gestation period, delivered at sea in a highly advanced state. Odontocetes are specialized for feeding on fish or pelagic squid. Mysticetes, as noted before, are specialized for feeding on planktonic crustaceans. Many of the great mysticete whales make enormous migrations to and away from the poles following plank-

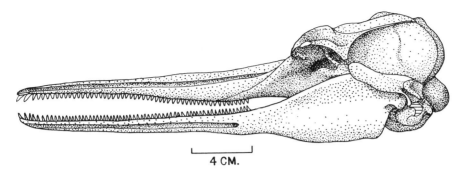

Figure 8.2. Skull of *Pontoporia blainvillei*.

tonic abundance. During their short growing season, the polar regions have high productivity and thus are favorite feeding grounds for mysticete whales during their summer movements. Although some whales are specialized for tropical waters, most cetacean taxa show their greatest species richness in the temperate and Arctic latitudes. Gaskin (1982) is a useful reference on the behavior and ecology of cetaceans. Eisenberg (1986) has reviewed the behavioral convergence demonstrable when terrestrial mammals are compared with cetaceans. Many of the smaller, toothed whales have been studied in aquariums. Some years ago it was recognized that small delphinids could locate objects in the water by emitting high-frequency "sonar pulses." The whole subject of underwater echolocation caught the imagination of scientists, and the pioneer results are included in Kellogg (1961).

SUBORDER ODONTOCETI

Diagnosis
Baleen is absent; some simple teeth are present that are cone shaped or peglike in form (but see the Ziphiidae). A single nasal opening at the top of the head is typical.

FAMILY PLATANISTIDAE
Freshwater Dolphins

Diagnosis
There are four Recent genera, all adapted to freshwater, riverine systems. The size is modest, rarely exceeding 3 m and 180 kg. The beak is long, and the forehead bulges (see fig. 8.2 and plate 10).

Distribution
The family is found in the Ganges, Indus, and Brahmaputra rivers of India, the Yangtze Kiang of China, and the Orinoco, Amazon, and Plata of South America.

Genus *Pontoporia* Gray, 1846
Pontoporia blainvillei (Gervais and d'Orbigny, 1844)
Franciscana Dolphin, Delfín del Plata

Measurements

	Mean	Min.	Max.	N	Loc.	Source[a]
TL	1.33 m	1.22 m	1.47 m	21 m	U	1
	1.53 m	1.37 m	1.71 m	28 f		
Wta	28.20 kg	19.90 kg	32.20 kg	4 m		
	40.30 kg	29.90 kg	52.10 kg	5 f		

[a](1) Kasuya and Brownell 1979.

Description
This small dolphin is easily recognized by its elongated body and long, thin beak (see fig. 8.2). It has a disproportionately small head and 50 or more teeth. The distinct triangular dorsal fin is on the midback and extends as a ridge to the flukes. The paddle-shaped flippers are rounded at the tips. It is pale brown dorsally shading to lighter brown ventrally, but some individuals are nearly white (Minasian, Balcomb, and Foster 1984).

Distribution
Pontoporia blainvillei is found from the Valdés Peninsula, Chubut province, Argentina, to Espírito Santo state in Brazil and well into the Río de la Plata in shallow waters (Borobia and Geise 1984; Ximénez, Langguth, and Praderi 1972) (map 8.1).

Life History
The birth season is from December to January. Gestation is 10.5 months, and lactation lasts about 9 months. Females become sexually mature at between 1.34 and 1.37 m and males at above 1.4 m—at about two to three years (Brownell 1975b; Kasuya and Brownell 1979).

Ecology
This dolphin is found in salt water near the coast and up into the Río de la Plata. It is found in the river from October to April and then seems to migrate north to the Brazilian coast. It apparently

feeds at or near the bottom, and examination of nine stomachs revealed mostly fish and squid (Fitch and Brownell 1971; Minasian, Balcomb, and Foster 1984; Ximénez, Langguth, and Praderi 1972).

FAMILY DELPHINIDAE
Dolphins and Porpoises

Diagnosis
Composed of approximately eighteen genera and sixty-two species, this family includes most of the smaller, toothed whales. The beak is variable in its expression, being prominent in *Tursiops* and virtually absent in *Globicephala*, *Grampus*, and *Faresa*

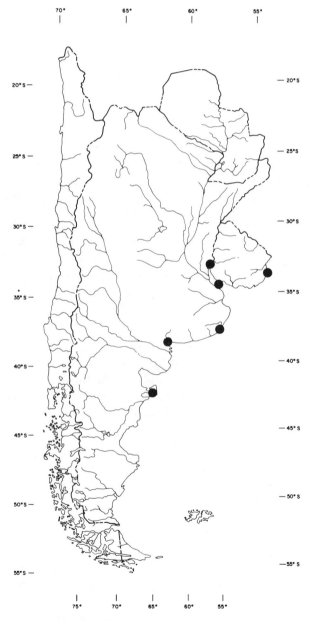

Map 8.1. Distribution of *Pontoporia blainvillei*.

(plate 11, figs. 8.4 and 8.8). The dorsal fin is prominent in all species except the Asiatic genus *Neomeris* and the Southern Hemisphere genus *Lissodelphis*.

Distribution
Dolphins and porpoises are widely distributed in the oceans of the world. Some species, such as *Sotalia*, ascend rivers.

Natural History
These small cetaceans feed on fish and pelagic squid. They are the best studied of the Cetacea, since they can be kept in captivity and are easily trained. Research with captives allowed the experimental demonstration of their echolocating ability (Kellogg 1961).

Highly gregarious, these cetaceans are almost always found in schools. In addition to the echolocation pulses produced from the larynx, they use a wide variety of sounds for underwater communication (Caldwell and Caldwell 1977).

Genus *Cephalorhynchus* Gray, 1846

Description
These dolphins are of modest size (less than 2 m), without a conspicuous beak. There are 25 to 32 teeth on each side of the jaw. The color patterns include varying amounts of white and black on the dorsum (see species accounts and plate 10).

Distribution
This Southern Hemisphere genus has been noted off the shores of southwest Africa, New Zealand, and southern South America. An excellent volume on dolphins for the genus *Cephalorhynchus* has recently come out, edited by Brownell and Donovan (1988). It contains superb chapters on *C. commersonii* and *C. eutropia* that update much of the information presented below. Unfortunately, data from these chapters could not be incorporated into this volume.

Cephalorhynchus commersonii (Lacépède, 1804)
Commerson's Dolphin, Delfín Blanco

Description
This dolphin is characterized by its small size and robust form. The rounded tip of the dorsal fin separates *C. commersonii* from other delphinids. The average greatest length for males from a South Atlantic population is 1.37 m (*n* = 15) and for females 1.46 m (*n* = 10). It is distinctively colored black and white, with two broad areas of black: one on the back starting just anterior to the dorsal fin and extending to the end of the tail, and the second including the pectoral fins and most of the head. There is

also a smaller ventral spot of black. The pigmentation varies individually with age; there is sexual dimorphism in the form and size of the genital black stripe. Calves are both completely brown and become grayish as they age (Brownell, in Brown et al. 1974; Goodall, Galeazzi, and Cameron 1984; Lichter and Hooper 1984; Mermoz 1980b; Perrin and Reilly 1984).

Distribution

Cephalorhynchus commersonii is found in the coastal waters of the western South Atlantic along the coast of Argentina from the Valdés Peninsula south to Tierra del Fuego, off southernmost Chile, and around island groups such as the Malvinas Islands and South Georgia (Aguayo 1975; Brownell, in Brown et al. 1974; Sielfeld 1980) (map 8.2).

Life History

Neonates weigh about 9 kg and are born after a gestation of about one year. This species can live at least seventeen years (Goodall, Galeazzi, and Cameron 1984; Hill 1985; Minasian, Balcomb, and Foster 1984).

Ecology

Cephalorhynchus commersonii is both pelagic and littoral and is seen in groups ranging from three individuals to over two hundred. They prey on schooling fish and seem to follow sardine shoals.

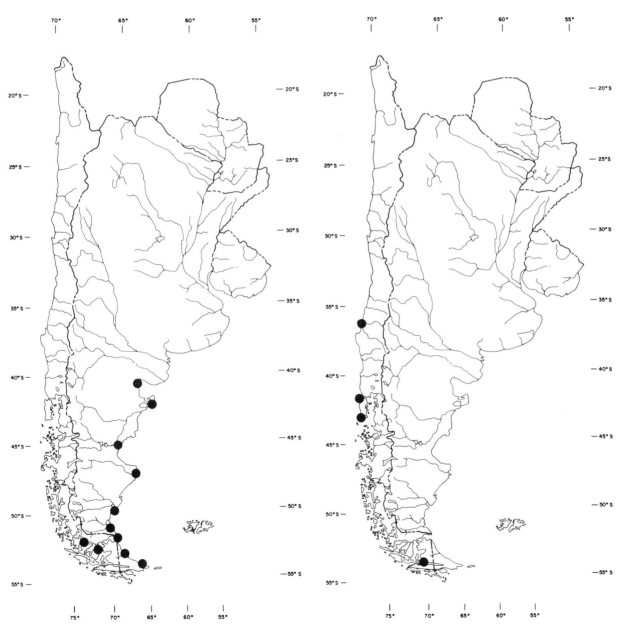

Map 8.2. Distribution of *Cephalorhynchus commersonii*.

Map 8.3. Distribution of *Cephalorhynchus eutropia*.

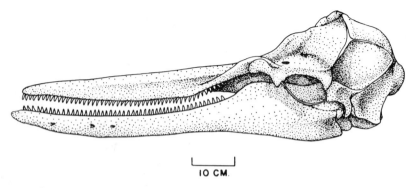

10 CM

Figure 8.3. Skull of *Delphinus delphis*.

Commerson's dolphin appears to breath less frequently than many other small delphinids and does not commonly approach and ride the bow wave of boats (Brownell, in Brown et al. 1974; Brownell and Praderi, n.d.; Gewalt 1979; Mermoz 1980b; Sielfeld 1983).

Cephalorhynchus eutropia Gray, 1846
Chilean Dolphin

Description
This species has 29–30 pairs of teeth per side in the mandible and the maxilla. *C. eutropia* is longer than *C. commersonii* and can reach 1.67 m. Morphologically it is very similar to *C. commersonii*, but it has a more pronounced snout and very different coloration. It is all black above and on the sides and white below, with one white spot behind the pectoral fins and occasionally one on the throat (Goodall et al. 1984; Sielfeld 1983).

Distribution
Cephalorhynchus eutropia is found only in the coastal waters of Chile, from 37° south to about 55°, and perhaps into southernmost Argentina (Aguayo 1975; Goodall et al. 1984) (map 8.3).

Ecology
This littoral species is observed in groups ranging from two to over four hundred individuals (Goodall et al. 1984).

Genus *Delphinus* Linnaeus, 1758
• *Delphinus delphis* Linnaeus, 1758
Common Dolphin, Delfín Común

Description
Delphinus delphis is easily recognized by its distinctive short beak, its low, smoothly sloping head, and its color (figs. 8.3 and 8.4). The beak is 12–15 cm, longer and narrower than that of *Tursiops* and is sharply marked off from the low forehead by a V-shaped groove. There are 40–50 pairs of teeth in the upper and lower jaws. Males from an eastern Pacific population measured 2.08 m total length ($n = 40$), and females measured 1.96 m ($n = 48$). The color is distinctive: the body is black or dark brown dorsally and white on the venter. On each side there are undulating bands or stripes of gray, yellow, or white that give the impression of overlapping one another in the area below the back fin. Black is also found from the end of the mandible to the pectoral fin and from the base of the maxilla to the eye (Aguayo 1975; Baker 1983; Lichter and Hooper 1984; Norman and Fraser 1964; Perrin and Reilly 1984).

Distribution
This dolphin is found in temperate and tropical waters throughout the world. In southern South America it is found off Chile between about 30° and 40° S, off Uruguay, and off Argentina south to about Chubut province (Lichter and Hooper 1984; Tamayo and Frassinetti 1980) (map 8.4).

Life History
The gestation period for the common dolphin is ten to eleven months; the interbirth interval is 1.3 to 2.6 years; lactation lasts five to six months; and the age of first reproduction is about three years (Perrin and Reilly 1984).

Ecology
Delphinus delphis is frequently found in deep water from 60 to 160 km off the coast, where schools

Figure 8.4. Head of *Delphinus delphis*.

of thousands of dolphins have been seen. It eats fish and squid, and most populations appear to follow the movements of their food fish (Baker 1983; Lichter and Hooper 1984; Minasian, Balcomb, and Foster 1984; Sielfeld 1983).

Genus *Globicephala* Lesson, 1828
Globicephala melaena (Trail, 1809)
Blackfish, Long-finned Pilot Whale, Calderón

Description

This species is easily distinguished from other cetaceans by (1) its all-black color, except for a white patch running above the eye, a gray saddle behind the dorsal fin, and a large whitish ventral patch; (2) its bulbous head, which may overhang the tip of the jaw, and its lack of a protruding snout; (3) and its falcate dorsal fin with a wide base, set anterior to the midbody (plate 10). The mouth slants up toward the eye; there are 8–11 teeth in each row, all set near the front of the jaws. Males from an Atlantic population averaged 5.45 m ($n = 12$), while females averaged 3.81 m ($n = 85$). Males can exceed 6 m, and the smaller females can reach 4.6 m. The flippers are long, up to 27% of the body, and sickle shaped (Baker 1983; Brownell, in Brown et al. 1974; Perrin and Reilly 1984; Piñero and Castello 1975).

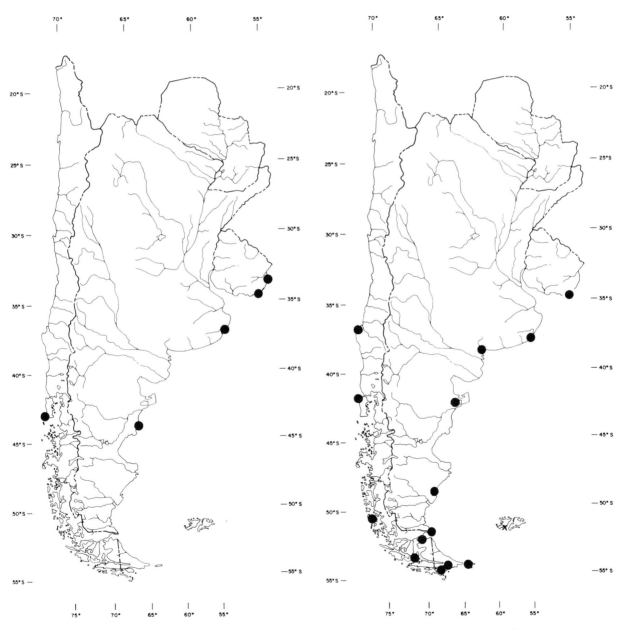

Map 8.4. Distribution of *Delphinus delphis*.

Map 8.5. Distribution of *Globicephala melaena*.

Distribution

Globicephala melaena is found in cold temperate waters of the North Atlantic and southern oceans. In southern South America it is found in cold currents north of the Antarctic convergence, and stranded individuals have been found in Chile, Uruguay, and Argentina (Goodall 1978; Honacki, Kinman, and Koeppl 1982; Lichter and Hooper 1984; Ringuelet and Arámburu 1957; Sielfeld 1980) (map 8.5).

Life History

The species has a gestation period between fifteen and sixteen months; calves first eat solid food at six to nine months, though lactation continues for about twenty months; the interbirth interval is about 3.3 years. Males become sexually mature at twelve years and females at six years. In Chile young are born year round, with a spring peak (Perrin and Reilly 1984; Sielfeld 1983).

Ecology

Blackfish are usually observed in schools of up to three hundred individuals, frequently with other species. Smaller groups are also commonly seen; of eleven schools counted, the average size was 17.6 animals. *G. melaena* is noted for its tendency to become stranded in large groups. One stranded group of seventeen animals consisted of two mature males, two smaller males, three immature males, nine females, and one unsexed animal (Brownell, in Brown et al. 1974; Clarke, Aguayo, and Basulto del Campo 1978; Crespo, Pagnoni, and Pedraza 1984; Piñero and Castello 1975).

Genus *Grampus* Gray, 1828
• *Grampus griseus* (G. Cuvier, 1812)
Risso's Dolphin

Description

This beakless dolphin has a distinct melon: the forehead bulges forward and slopes steeply to the mouth. The mouth is slightly underslung, and there are 3–7 pairs of teeth up to 1 cm in diameter near the tip of the jaw. There are usually no teeth in the upper jaw. Males from a northern Pacific population averaged 2.88 m in total length ($n = 9$), and females averaged 2.76 m ($n = 13$). The body is usually a uniform gray above, with a broad anchor-shaped white patch on the throat, a lighter venter, and sometimes a white head. The fins and tail are darker than the rest of the body. Old dolphins are often heavily scarred (Baker 1983; Norman and Fraser 1964; Perrin and Reilly 1984; Sielfeld 1983).

Distribution

Risso's dolphin is widespread in temperate and tropical waters of the world. Stranded individuals have been found on the Chilean and Argentine coasts (Honacki, Kinman, and Koeppl 1982; Sielfeld 1983) (map 8.6).

Life History

Calves become sexually mature when they reach about 3 m (Minasian, Balcomb, and Foster 1984).

Ecology

This species is usually found in water deeper than 180 m. It is social and schools with *Lissodelphis*,

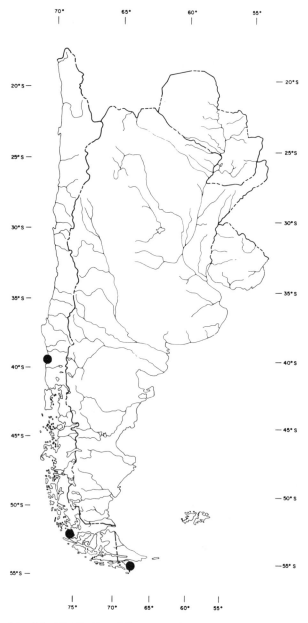

Map 8.6. Distribution of *Grampus griseus*.

Globicephala, and *Lagenorhynchus*. Cephalopods appear to be the major dietary item (Sielfeld 1983).

Genus *Lagenorhynchus* Gray, 1846
White-sided Dolphin

Description

These medium-sized dolphins rarely exceed 3 m in head and body length. The beak is very short but still visible. The venter is white and the dorsum is dark brown. The white of the venter extends to the flanks, but the coloration is variable (see species accounts and plate 10).

Distribution

The genus is oceanic and worldwide in the temperate seas of the Northern and Southern Hemispheres.

Lagenorhynchus australis (Peale, 1848)
Peale's Dolphin, Delfín Griseoblanco

Description

This dolphin has a relatively short beak that grades smoothly into the head, and a distinct depression separates it from the forehead. The dorsal fin is prominent, recurved with a bluntly rounded tip. One female measured 2.16 m. Its slate black color covers the head and body dorsally as well as the snout and eyes. The pectoral fin is black, connected by a thin stripe to the black head region. The tail stock and tail are black, and there is a white patch on the flank. Three main pigment areas distinguish *L. australis* from the other two species of *Lagenorhynchus:* a black patch that covers the lower lip, chin, and throat and extends posteriorly to below the eye; a simple whitish gray flank patch, without the dorsal and ventral flank blazes found in *L. obscurus;* and the white abdominal field pigmentation, which extends above the flipper by way of the axilla (Brownell, in Brown et al. 1974; Lichter and Hooper 1984; Norman and Fraser 1964; Perrin and Reilly 1984; Sielfeld 1983).

Distribution

Lagenorhynchus australis is found in the cold temperate waters of Chile north to Valparaíso, Argentina, and the Malvinas Islands (Honacki, Kinman, and Koeppl 1982; Sielfeld 1983) (map 8.7).

Ecology

This species is most commonly found in coastal waters, apparently preferring fjords and deep bays. It occurs in groups ranging from five to thirty (Goodall 1978; Lichter and Hooper 1984; Sielfeld 1983).

Lagenorhynchus cruciger (Quoy and Gaimard, 1824)
Hourglass Dolphin, Delfín Cruzado

Description

This is the smallest of the species of *Lagenorhynchus*, averaging 1.69 m ($n = 4$) and weighing 88.2 kg ($n = 1$). Its body is not robust, and it has a short beak and a tall, distinctively hooked dorsal fin. It is best distinguished from other species of *Lagenorhynchus* by its lateral color pattern: an hourglass. Two lateral white areas connected with a fine white line separates the black of the dorsum from that of

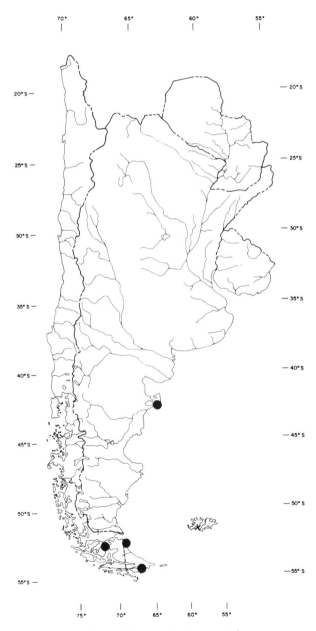

Map 8.7. Distribution of *Lagenorhynchus australis*.

the side below the dorsal fin. The pectoral fin is black, and the anterior part of the venter and a genital patch are white (Baker 1983; R. L. Brownell, unpubl. data; Brownell, in Brown et al. 1974; Lichter and Hooper 1984).

Distribution

Lagenorhynchus cruciger is found in the Antarctic and in cold temperate circumpolar waters of the Atlantic and Pacific. Individuals have been collected off the coasts of southern Chile and Argentina (Aguayo 1975; Baker 1983; Honacki, Kinman, and Koeppl 1982; Lichter and Hooper 1984; Tamayo and Frassinetti 1980) (map 8.8).

Ecology

This is the only dolphin commonly sighted south of the Antarctic convergence. Most groups are small (average 6.6; $n = 5$), but groups up to forty have been seen (R. L. Brownell, unpubl. data; Leatherwood et al. 1982).

Lagenorhynchus obscurus (Gray, 1828)
Dusky Dolphin, Delfín Oscuro

Description

The dusky dolphin has a smooth head with virtually no beak. Its dorsal fin is not markedly hooked,

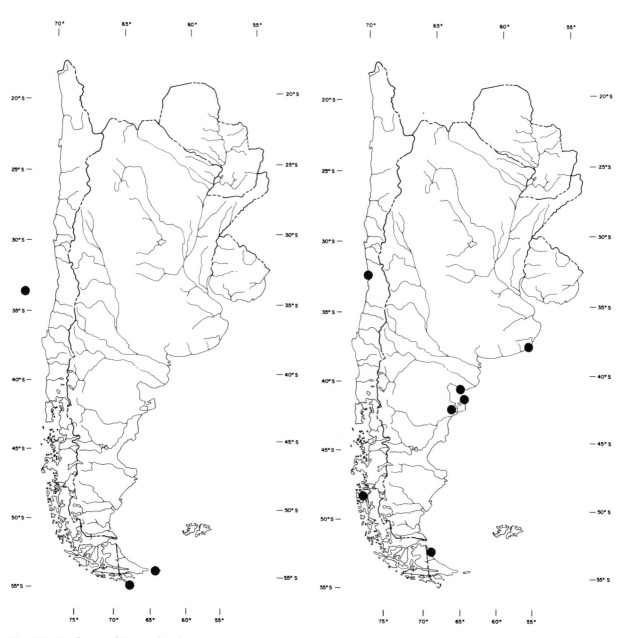

Map 8.8 Distribution of *Lagenorhynchus cruciger*.

Map 8.9. Distribution of *Lagenorhynchus obscurus*.

and the tip is rather blunt. There are 29–35 pairs of teeth in both jaws. Males grow to 2.1 m and females to 1.9 m. This species exhibits a considerable variation in body color, but basically it is bluish gray on the dorsum and tail with a darker band running diagonally across the flanks from below the dorsal fin toward the vent and along the tail stock. The venter is white with the white extending up onto the flanks, and the tip of the snout and the lower jaw are dark. *L. obscurus* can be differentiated from *L. australis* because both dorsal and ventral flank blazes are present; the cephalic end of the dorsal flank blaze extends anterior to the lateral sides of the blowhole; the black lower lip patch is thin; and the dorsal fin has a distinctly dark anterior and light posterior coloration. *L. obscurus* performs spectacular acrobatics (Baker 1983; Brownell, in Brown et al. 1974; Norman and Fraser 1964; Perrin and Reilly 1984).

Distribution

Lagenorhynchus obscurus is found in the coastal temperate waters of the Southern Hemisphere. In southern South America it has been found on the Chilean and Argentinian coasts (Honacki, Kinman, and Koeppl 1982; Lichter and Hooper 1984; Sielfeld 1983) (map 8.9).

Life History

Births take place principally in the spring. Calves are born every two or three years (Minasian, Balcomb, and Foster 1984; Sielfeld 1983).

Ecology

The dusky dolphin, a fish eater, is found mainly in coastal waters in groups that average 10.6 (range 4–20; $n = 5$) but can range up to 100 animals. There is some evidence for long-term association of individuals, and possibly for stable subgroups making up a larger group as subgroups coalesce and diverge. Individuals in a population off the Valdés Peninsula, Argentina, never seemed to go farther than 20 km offshore. At least a portion of this population undertakes seasonal migration, perhaps associated with movement of anchovies (Leatherwood et al. 1982; Sielfeld 1983; Wursig and Bastida 1986).

Genus *Lissodelphis* Gloger, 1841
Lissodelphis peronii (Lacépède, 1804)
Southern Right Whale Dolphin, Delfín Liso

Description

The southern right whale dolphin is called a "right whale" because it lacks a dorsal fin. This, together with its slender, sleek form and dramatic coloration, make *L. peronii* unmistakable. It has a very slight beak and 45–48 teeth in each row in both upper and lower jaws. Males reach 2.1 m and females 2.3 m. The upper part of the head, back, and flukes is black, and the rest of the body, including the flippers, is white. This species travels very fast at the surface and makes low-angled jumps (Baker 1983; Norman and Fraser 1964; Perrin and Reilly 1984).

Distribution

Lissodelphis peronii is found in temperate waters of the Southern Hemisphere. In southern South America it has been found off the coasts of Chile and Argentina (Goodall 1978; Honacki, Kinman, and Koeppl 1982; Sielfeld 1983) (map 8.10).

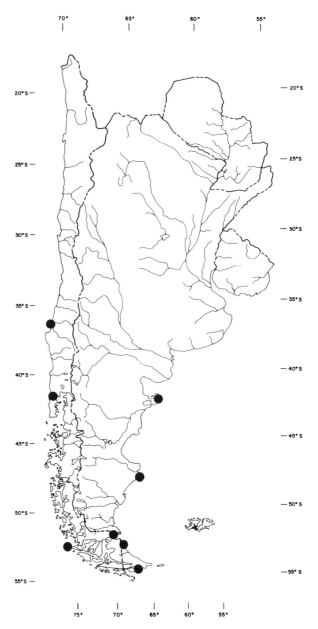

Map 8.10. Distribution of *Lissodelphis peronii*.

Ecology

The southern right whale dolphin forms schools of up to one thousand individuals (Brownell, in Brown et al. 1974).

Genus *Orcinus* Fitzinger, 1860

• *Orcinus orca* (Linnaeus, 1758)
Killer Whale, Orca

Description

This, the largest of the dolphins, is unmistakable because of its large, stout body, short, blunt head, and dramatic black-and-white coloration (plate 11).

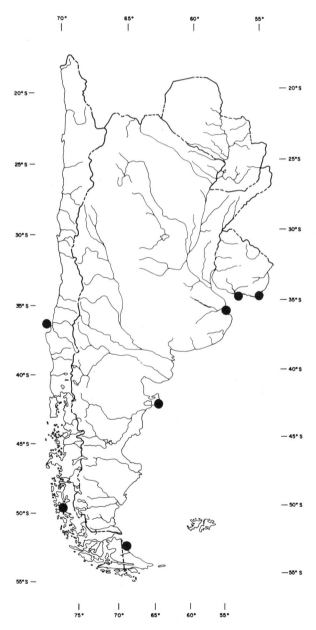

Map 8.11. Distribution of *Orcinus orca.*

There are 11–13 pairs of large conical teeth in each jaw. Males can reach 9.1 m and females 7.6 m. An average female from Antarctic waters measures 6.45 m ($n = 91$). Males have a very tall, pointed dorsal fin that can measure 1.8 m, and females have a dorsal fin about half the height that is slightly hooked. The flippers are large and paddle shaped. The back is black and the belly is white. Just behind and above the eye is a blaze of white, and in some individuals there is a white saddle behind the dorsal fin. The chin is white, and the white of the belly frequently extends onto the flanks halfway between the tail and the flipper. There is great variation in color that can be used to identify individuals (Aguayo 1975; Baker 1983; Brownell, in Brown et al. 1974; Norman and Fraser 1964; Perrin and Reilly 1984).

Distribution

Orcinus is distributed worldwide in all oceans but is most abundant in the cold waters of both hemispheres. In southern South America it has been collected on the coasts of Argentina, Uruguay, and Chile (Honacki, Kinman, and Koeppl 1982; Lichter and Hooper 1984; Schneider 1946; Ximénez, Langguth, and Praderi 1972) (map 8.11).

Life History

Female orcas from Norway reproduce for the first time at about eight years and males at about fifteen years. The interbirth interval is probably three years, the gestation period is about twelve months, and lactation lasts at least twelve months (Christensen 1984; Perrin and Reilly 1984; Sielfeld 1983).

Ecology

This species is found in both pelagic and littoral waters, and in some areas it forms groups averaging 10.3 animals (range 1–30; $n = 9$). Off the Valdés Peninsula, Argentina, the average number of animals per group was only 3.2. Most single animals are males. Groups are often resident in certain areas, and there is evidence that the small social groups of orcas are matriarchies and that females within such groups are related by descent. Schools in adjacent waters may have distinctive dialects, and underwater communication is important in regulating group movements and cohesion.

Orcas have been recorded as eating fish, squid, pinnipeds, birds, other cetaceans, leatherback turtles, and stingrays. Whales hunt cooperatively, and off the Valdés Peninsula, Argentina, they frequently kill seals by chasing them toward a beach. Adult animals have been observed tossing prey to young animals (Baker 1983; Castello 1977; Ford and Fisher 1983; Leatherwood et al. 1982; Lichter and Hooper 1984; López and López 1985; Sielfeld 1983).

Figure 8.5. *Pseudorca crassidens.*

Genus *Pseudorca* Reinhardt, 1862
• *Pseudorca crassidens* (Owen, 1846)
False Killer Whale, Falsa Orca

Description

This species is more slender than *Orcinus orca*, with a rounded snout that projects a short way beyond the tip of the lower jaw. The mouth is underslung and particularly long and contains 8–10 strong teeth in each row of the upper and lower jaws. Males from all regions reach 5.96 m total length, and the smaller females reach 5.06 m. The dorsal fin is small, with the tip pointing toward the tail, and the flippers are fairly long, pointed, and slightly S-shaped (fig. 8.5). Most individuals are entirely black, though occasional animals have a patch of gray along the anterior dorsal surface (Baker 1983; Norman and Fraser 1964; Perrin and Reilly 1984).

Distribution

Pseudorca is found worldwide in tropical to temperate waters. In southern South America specimens have been collected from the coasts of Argentina, Uruguay, and Chile (Honacki, Kinman, and Koeppl 1982; Lichter and Hooper 1984; Schneider 1946; Ximénez, Langguth, and Praderi 1972) (map 8.12).

Ecology

This pelagic, fish-eating species can be found in schools of several hundred (Baker 1983; Sielfeld 1983).

Genus *Stenella* Gray, 1866
Spinner Dolphin

Description

This dolphin genus is of medium body size, attaining 3.5 m. The beak is prominent. The color pattern is variable, but in common with other delphinids the darker dorsum contrasts with a pale venter. A speckled pattern on the flanks charac-

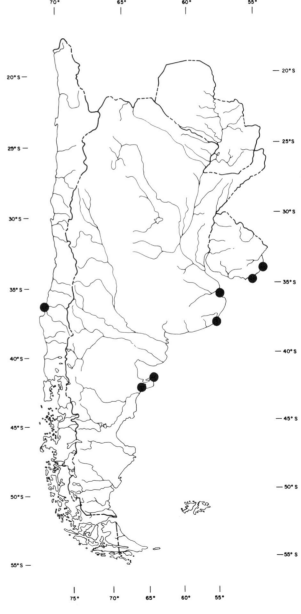

Map 8.12. Distribution of *Pseudorca crassidens.*

terizes some species, while flank stripes are typical of others (see species accounts).

Distribution

The genus is oceanic and worldwide but typically occurs in warmer waters.

• *Stenella attenuata* (Gray, 1846)
Spotted Porpoise, Delfín Pardo

Description

The spotted porpoise has a well-demarcated snout with a moderately long beak. Its dorsal fin, placed halfway along the body, is tall and narrow, and the tip has a sharply concave posterior edge. Males from

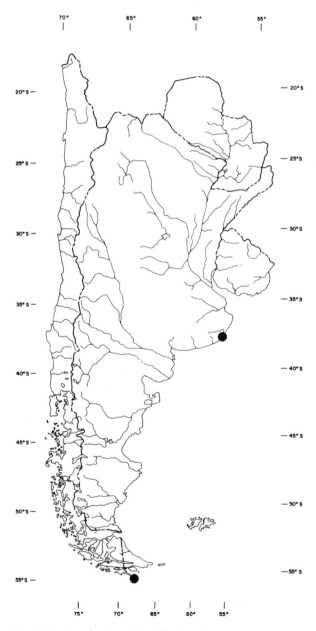

Map 8.13. Distribution of *Stenella attenuata*.

the eastern Pacific average 2.01 m ($n = 730$) total length and females 1.87 m ($n = 1,123$). There is great variation in adult coloration, individually and geographically, but basically the dorsum is a dark gray and the venter is whitish with an overlay of small light spots on the dark areas and dark spots on the light areas. Calves are unspotted at birth (Baker 1983; Perrin 1975; Perrin and Reilly 1984).

Distribution

Stenella attenuata is found in tropical waters of the Indian, Pacific, and Atlantic Oceans. In southern south America there are only two records of this species, from Cape Horn and Buenos Aires province, but it is undoubtedly found off the northern coasts of Chile and Argentina (Honacki, Kinman, and Koeppl 1982; Lichter and Hooper 1984; Sielfeld 1983; Tamayo and Frassinetti 1980) (map 8.13).

Life History

The species has a gestation period of about 11 months, a lactation period averaging 2.4 years; and an interbirth interval of 6.4 years. The age of first reproduction is six to nine years, and the maximum recorded age is forty-nine years (Kasuya 1976; Sielfeld 1983).

Ecology

The spotted porpoise normally occurs in warmer waters, where it is frequently caught in nets set for yellowfin tuna. It lives in groups of twenty to thirty and eats fish and squid (Baker 1983; Lichter and Hooper 1984; Sielfeld 1983).

• *Stenella coeruleoalba* (Meyen, 1833)
Striped Dolphin, Delfín Azul

Description

Stenella coeruleoalba is similar in shape to the common dolphin but has a different color pattern and is larger, with males from all regions reaching 2.56 m ($n = 2,272$) total length and females 2.4 to 2.5 m ($n = 2,239$). In each row of both jaws there are 39–46 small teeth. It has a distinct, relatively short beak. This species is highly variable in color but basically is brownish black dorsally, becoming lighter toward the tail. The beak, lower edge of the mouth, flippers, and tail are all dark, and there are dark stripes extending from the eyes to the flippers and from the eyes to the anus. There may also be a short, light blaze rising from the flanks, behind the eyes and curving up toward the dorsal fin (Baker 1983; Lichter and Hooper 1984; Perrin and Reilly 1984).

Distribution

The striped dolphin is found in temperate and tropical waters of the world. In southern South America specimens have been recorded from the coasts of Argentina and Uruguay, and it probably occurs off Chile as well (Honacki, Kinman, and Koeppl 1982; Lichter and Hooper 1984; Sielfeld 1983) (map 8.14).

Life History

The gestation period of this species is twelve months, lactation lasts fifteen to eighteen months, and the interbirth interval is probably about three years. The age of first reproduction is about nine years and the maximum recorded age is fifty years (Kasuya 1976; Perrin and Reilly 1984).

Ecology

This species is found in warmer waters in groups of five to five hundred, often segregated by age and

sex. It frequently rides the wakes of boats; it is a fish eater (Lichter and Hooper 1984; Sielfeld 1983).

• *Stenella longirostris* (Gray, 1828)
Spinner Dolphin

Description

The spinner dolphin has a very long beak and a tall dorsal fin with a very shallow concave posterior edge. Combining specimens from all regions, males can reach 2.35 m ($n = 2,340$) total length and females 2.04 m ($n = 2,621$). Males from the eastern Pacific average 1.76 m ($n = 594$) and females 1.7 m ($n = 560$). There is a great deal of geographic variation in color, but basically it has a dark gray upper

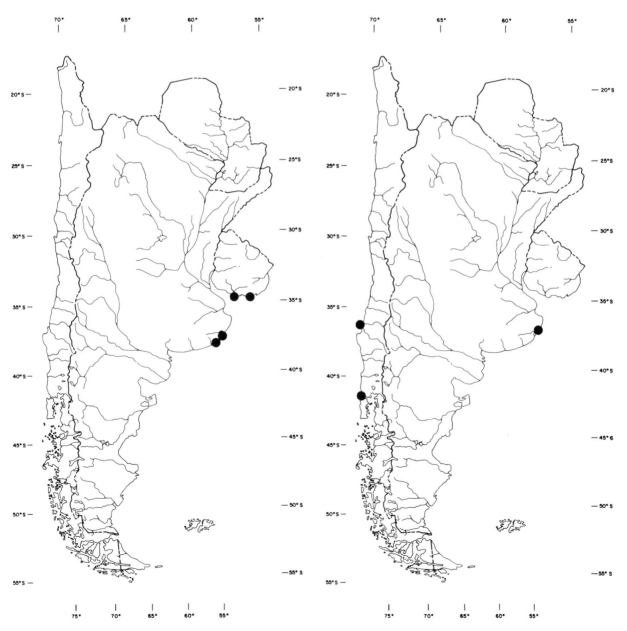

Map 8.14. Distribution of *Stenella coeruleoalba*.

Map 8.15. Distribution of *Stenella longirostris*.

beak, head, and back to about the dorsal fin, and is light gray or tan on the flanks. The lower jaw and belly are white, and there is a dark stripe between the eye and the flipper. The eye is surrounded by a black patch. This species obtains its common name from its habit of jumping well clear of the water and spinning on its longitudinal axis (Baker 1983; Sielfeld 1983).

Distribution

Stenella longirostris is found in the tropical to warm temperate waters of the world. In southern South America it has been recorded from the northern coasts of Chile and Argentina (Honacki, Kinman, and Koeppl 1982; Ringuelet and Arámburu 1957; Tamayo and Frassinetti 1980) (map 8.15).

Life History

The gestation period of this species is 9.5 to 10.7 months, lactation lasts 11 to 34 months, and the interbirth interval is 2.9 to 3.3 years (Perrin and Reilly 1984).

Ecology

Spinner dolphins are commonly caught in nets set for yellowfin tuna. They are found in warmer waters, are strongly social, and eat both fish and squid (Baker 1983; Sielfeld 1983).

Genus *Tursiops* Gervais, 1855
• *Tursiops truncatus* (Montagu, 1821)
Bottle-nosed Dolphin, Tonina Común

Description

This species has a short, well-demarcated beak and is one of the largest of the dolphin species. Males from Florida (USA) average 2.58 m ($n = 13$) in total length and reach 2.9 m ($n = 127$), while females average 2.39 m ($n = 45$) and reach 2.76 m ($n = 143$). The mouth is curved up in a permanent "smile," and the beak contains 21–29 pairs of teeth in each jaw. The dorsal fin is high and well hooked, and the flippers are moderate in size. The dorsum is light gray to blackish, grading to white on the venter. The head and snout are dark, but the edges of the upper lip and the whole lower jar are white (Baker 1983; Norman and Fraser 1964; Perrin and Reilly 1984; Sielfeld 1983).

Distribution

Tursiops is found worldwide in tropical and temperate waters. In southern South America it has been found on the coasts of Argentina, Uruguay, and Chile (Honacki, Kinman, and Koeppl 1982; Lichter and Hooper 1984; Tamayo and Frassinetti 1980) (map 8.16).

Life History

The gestation period of this species is twelve to thirteen months, lactation lasts an average of eighteen to twenty months, and the interbirth interval is eighteen to twenty months with up to eight young produced per lifetime. Longevity is twenty-five to thirty years. Off Argentina calves are born in the spring (Lichter and Hooper 1984; Perrin and Reilly 1984; Sielfeld 1983).

Ecology

The fish-eating *Tursiops* is found in warmer waters closer to the shore, in groups up to thirty. This

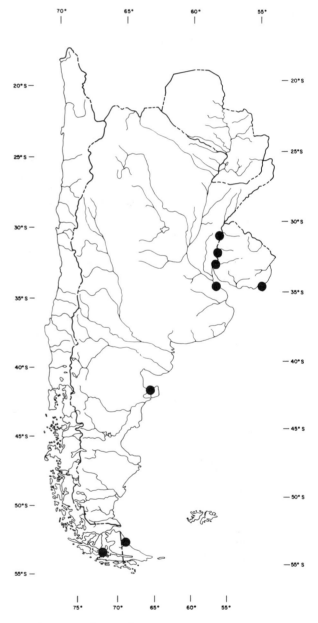

Map 8.16. Distribution of *Tursiops truncatus*.

species forms loosely bonded schools of adults that divide into subgroups. Subgroup composition may be quite stable over two to three years, and units may be composed of mothers and their maturing young. These dolphins are highly gregarious and have a complicated communication system including vocalizations, touching, and visual display. The use of high-frequency pulses of sounds in echolocation underwater was clearly demonstrated in captive individuals (Baker 1983; Caldwell and Caldwell 1966, 1977; Sielfeld 1983).

FAMILY PHOCOENIDAE

Diagnosis
These small dolphins are usually less than 1.5 m long. There is no beak, and the face has a blunt appearance. Some workers include *Phocoenoides* and *Neophocaena* in this family. In *Phocoena* the teeth are compressed laterally and have weakly developed crowns; tooth number ranges from 15/15 to 30/30. *Phocoena* has a gray dorsum and white venter.

Distribution
The family is worldwide in its distribution and generally occurs inshore.

Genus *Phocoena* G. Cuvier, 1817
Phocoena dioptrica Lahille, 1912
Spectacled Porpoise, Marsopa Bicolor

Description
This porpoise has a bluntly pointed snout with no beak and teeth with spade-shaped crowns. Its robust body, reaching about 2 m (average 1.81; range 1.36–2.04; $n = 5$) has small, round-tipped flippers and a large, low, triangular dorsal fin. The dorsal surface and the upper lateral surface to just above the eye are black, but the dorsal keel of the tail stock is white. The dorsal surface of the flukes is black, and the ventral surface is white with a gray border. The eyes are surrounded by a wide black eye patch, and the pectoral fins are white (Baker 1983; Brownell 1975a; Praderi 1971a).

Distribution
Phocoena dioptrica is found in waters off Uruguay, Argentina, and southernmost Chile as well as around offshore islands like the Malvinas and South Georgia (Honacki, Kinman, and Koeppl 1982; Praderi 1971a; Sielfeld 1983) (map 8.17).

Ecology
This is the only species of *Phocoena* found around offshore islands (Brownell 1975a).

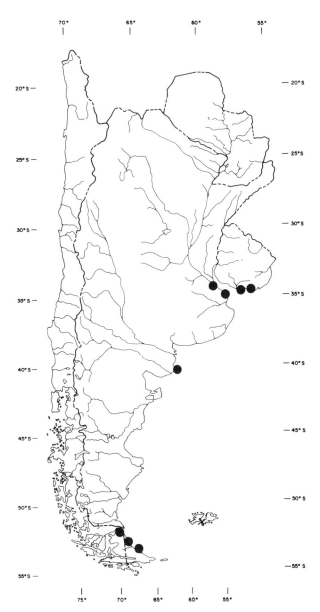

Map 8.17. Distribution of *Phocoena dioptrica*.

Phocoena spinipinnis Burmeister, 1865
Burmeister's Porpoise, Marsopa Espinosa

Description
The tooth number on each side ranges from 13–18 in the upper jaw and 15–20 in the lower. Animals are relatively stout with a small head and short snout and range between 1.5 and 2 m in length. *Phocoena spinipinnis* can be distinguished externally from all other species of *Phocoena* by the slightly convex anterior and posterior borders and the raised blunt tip of the dorsal fin. Horny denticles are present along the anterior border of the dorsal fin. Animals are uniformly dark gray to brown, becoming darker on

the sides and lighter on the venter. The flippers and flukes are the same dark color as the dorsal surface (Brownell and Praderi, n.d.; Lichter and Hooper 1984; Praderi 1971a; Sielfeld 1983).

Distribution

The species is found in the temperate waters of South America from Uruguay around Tierra del Fuego north to Peru (Goodall 1978; Honacki, Kinman, and Koeppl 1982, Praderi 1971a) (map 8.18).

Life History

Calving takes place during the austral fall in the waters off Uruguay (Brownell and Praderi, n.d.).

Ecology

Phocoena spinipinnis in found most commonly in coastal waters and estuaries, usually in groups of six

to eight. It rarely jumps and can be identified during breathing by its typical dorsal fin. Fish and squid were found in the stomach of one individual. It is sold for human consumption, being caught in gill nets set for fish (Brownell and Praderi, n.d.; Sielfeld 1983).

FAMILY ZIPHIIDAE
Beaked Whales

Diagnosis

Most species show a vastly reduced tooth number. The more primitive *Tasmacetus* has a tooth number of 19/27, but in most species it is reduced to 0/2 or 0/1 and that often occurs in males only. A rather long beak and a roundish head characterize

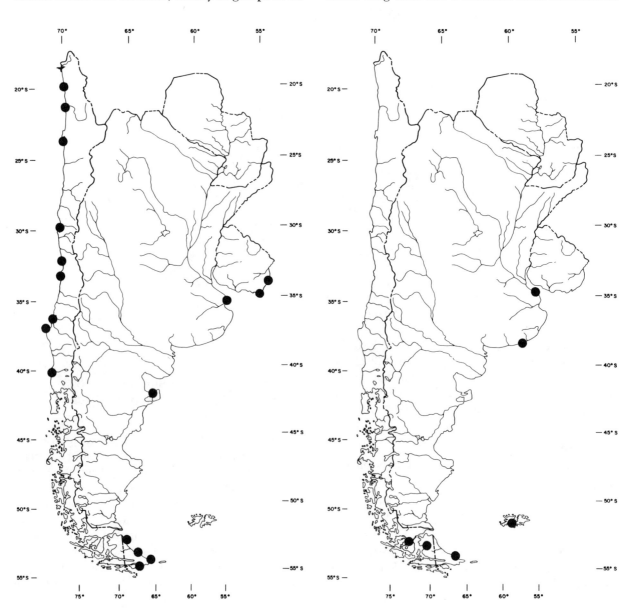

Map 8.18. Distribution of *Phocoena spinipinnis*.

Map 8.19. Distribution of *Berardius arnuxii*.

this family. The beak is not sharply demarcated from the head, and the contours are smooth in *Ziphius* and *Mesoplodon*. The dorsal fin is set back about three-quarters of the total length.

Distribution
The family occurs worldwide and is oceanic.

Natural History
This family is poorly known. Many members are believed to feed on pelagic squid. The standard reference is Moore (1968).

Genus *Berardius* Duvernoy, 1851
Berardius arnuxii Duvernoy, 1851
Arnoux's Beaked Whale, Zifio Marsopa, Berardio do Arnoux

Description
Berardius arnuxii grows to 9.7 m, with females apparently slightly larger than males; it is a robust animal with a prominent beak and a bulging forehead. The lower jaw extends several inches beyond the upper jaw and has two pairs of teeth near the tip, the front pair being larger. Unlike other beaked whales, in females the teeth do emerge above the gums. A pair of gular grooves form an inverted V. The flippers are moderately broad and rounded distally, and the dorsal fin is small and set well back of the center of the body. Dorsally this species is black or dark gray, lightening on the venter. Males frequently are heavily scarred (Baker 1983; Brownell, in Brown et al. 1974; McCann 1975).

Distribution
The species is restricted to temperate and polar waters of the Southern Hemisphere. Individuals have been collected on the coasts of Argentina and Chile (Lichter and Hooper 1984; Mead 1984; Sielfeld 1983) (map 8.19).

Ecology
Berardius arnuxii is presumed to migrate seasonally away from the ice to breed. The scarring on males is most likely due to male-male fighting. This species is probably a squid feeder (McCann 1975, Minasian, Balcomb, and Foster 1984).

Genus *Hyperoodon* Lacépède, 1804
Hyperoodon planifrons Flower, 1882
Southern Bottle-nosed Whale, Zifio Nariz Botella

Description
Males average 7.6 m long and females 6.7 m; an average weight is 3,600 kg. *Hyperoodon planifrons* has a distinct short beak protruding from an inflated, sometimes overhanging forehead. The body is robust, the tail flukes are wide, and the dorsal fin is large for a beaked whale. The snout is shorter than that of *Berardius arnuxii*, and the lower jaw does not protrude as far. In addition, in *H. planifrons* only two teeth are found at the tip of the lower jaw, and teeth do not erupt in females. Adults are gray to black, with pronounced V-shaped throat grooves. The venter and sometimes the head are paler, and there is a distinct shield-shaped light patch under each eye (Baker 1983; Brownell, in Brown et al. 1974; Gianuca and Castello 1976; Lichter and Hooper 1984).

Distribution
Hyperoodon planifrons is found in waters of the Southern Hemisphere, circumpolar in temperate and cold waters. In southern South America individuals have been found on the shores of Argentina,

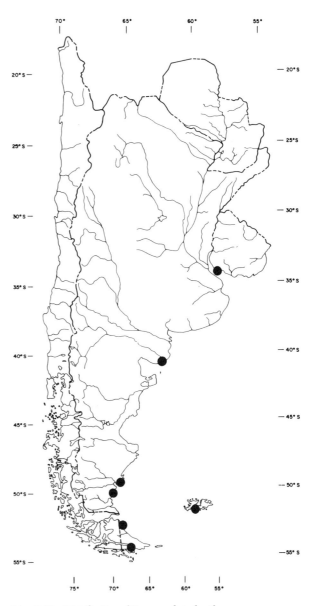

Map 8.20. Distribution of *Hyperoodon planifrons*.

Uruguay, and Chile (Goodall 1978; Honacki, Kinman, and Koeppl 1982; Lichter and Hooper 1984) (map 8.20).

Ecology
This rare beaked whale probably feeds on squid (Baker 1983).

Genus *Mesoplodon* Gervais, 1850
Beaked Whale

Description
The species of the genus rarely exceed 7 m in length. The beak is well developed, and the head melds smoothly into the body, in contrast to *Hyperoodon* (figs. 8.6 and 8.7). The color patterns are highly variable, and the placement of the teeth in males varies (see species accounts).

Distribution
The genus is found worldwide and is oceanic.

Genus *Mesoplodon* Gervais, 1850
Mesoplodon grayi Von Haast, 1876
Gray's Beaked Whale, Zifio Negro

Description
Males have a small vertical triangular tooth on each side of the lower jaw, set well back from the lip, and a row of 17–22 very small teeth in the upper jaw that are not attached to the gums. This beaked whale has a long slender body reaching 5.6 m in males and 5.3 m in females. The head is small, with a long, pointed beak and no clearly discernible bulge on the forehead. A hooked dorsal fin is set well toward the

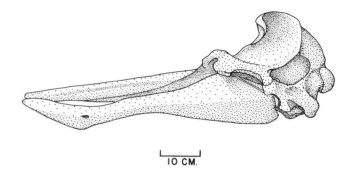

10 CM.

Figure 8.7. Skull of *Mesoplodon* sp.

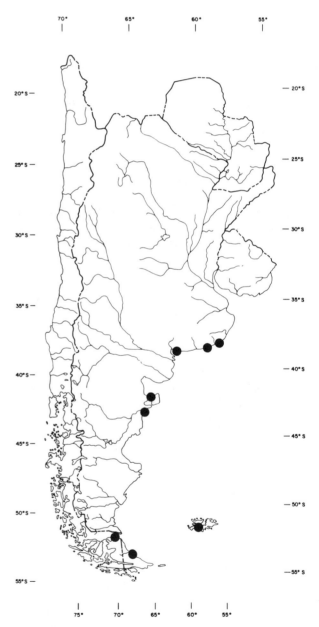

Map 8.21. Distribution of *Mesoplodon grayi*.

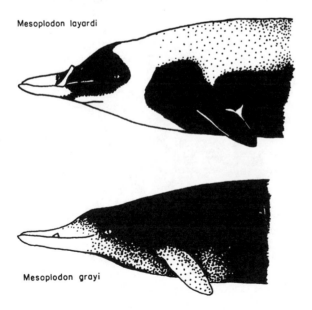

Mesoplodon layardi

Mesoplodon grayi

Figure 8.6. Head of two species of *Mesoplodon* (modified from Baker 1983).

tail, and there is a sharp ridge on the back between the dorsal fin and the tail. *M. grayi* is dark bluish gray dorsally and paler ventrally. The beak and the front part of the forehead and lower jaws are white, and individuals may be heavily scarred (Baker 1983; Mead 1984).

Distribution

Mesoplodon grayi is found in temperate southern oceans and cold temperate waters. In southern South America it has been found on the coasts of Argentina and Chile (Honacki, Kinman, and Koeppl 1982; Lichter and Hooper 1984; Sielfeld 1983) (map 8.21).

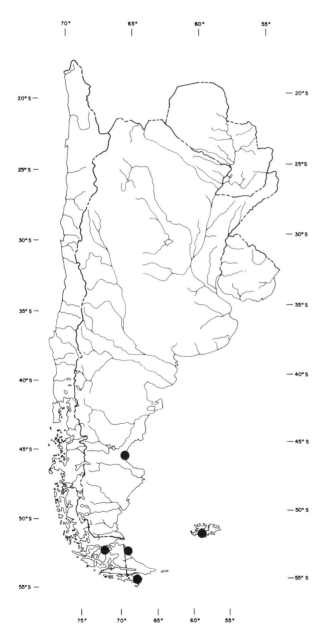

Map 8.22. Distribution of *Mesoplodon hectori*.

Ecology

This species feeds mainly on squid, though it also eats fish (Sielfeld 1983).

Mesoplodon hectori (Gray, 1871)
Hector's Beaked Whale

Description

A single pair of teeth is placed 1 or 2 cm back from the tip of the lower jaw. In adult males the teeth are shaped roughly like isoceles triangles and stand just over 3 cm high. This species is the smallest of the beaked whales, males reaching 4.3 m and females 4.5 m. It has a relatively small head, long and deep midbody, and short tail. The beak is short, and the forehead slopes fairly steeply over a slightly bulging forehead. The dorsal fin is small and shaped like a low, rounded triangle. Adult males are dark gray dorsally and paler ventrally, with a white beak and lower jaw (Baker 1983; Mead 1981, 1984).

Distribution

Mesoplodon hectori is found in both Northern and Southern Hemispheres; in southern South America individuals have been found on the coasts of southernmost Argentina and Chile and on the Malvinas Islands (Goodall 1978; Honacki, Kinman, and Koeppl 1982, Sielfeld 1983) (map 8.22).

Mesoplodon layardii (Gray, 1865)
Strap-toothed Mesoplodon, Zifio de Layard

Description

Males are easily recognized because of two strap-shaped teeth about 30 cm long, one on each side of the lower jaw. These teeth are hardly developed in females, but in some males they may grow so as to prevent the mouth from opening completely. This is the largest of the *Mesoplodon* (except perhaps for *M. pacificus*); males reach 5.9 m and females 6.2 m. It has a small head with a pronounced snout. The dorsal fin is deeply concave on the posterior side, and the pectoral fins are reduced. *M. layardii* is dark blue black on the back and flanks, with a white beak and throat and a white genital patch. In adults there is a grayish "cape" over the back between the head and the dorsal fin (Baker 1983; Lichter and Hooper 1984; Mead 1984).

Distribution

Mesoplodon layardii is found in temperate waters of the Southern Hemisphere and has been found on the coasts of Uruguay, Chile, and Argentina (Goodall 1978; Honacki, Kinman, and Koeppl 1982, Sielfeld 1983; Ximénez, Langguth, and Praderi 1972) (map 8.23).

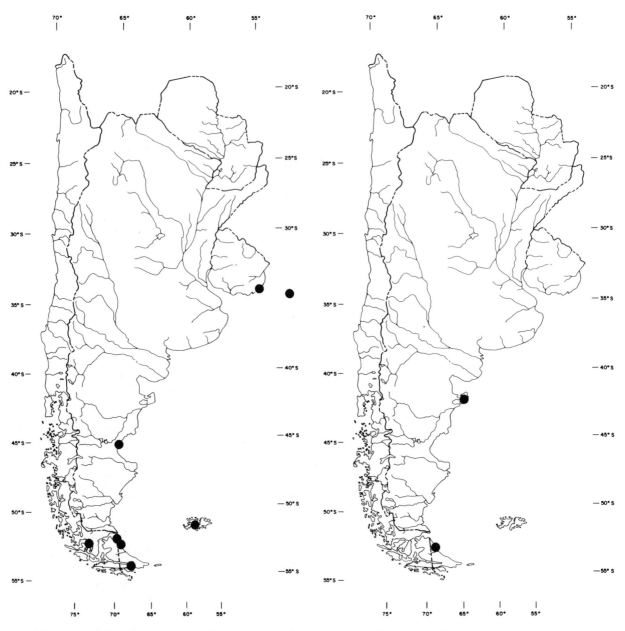

Map 8.23. Distribution of *Mesoplodon layardii*.

Map 8.24. Distribution of *Tasmacetus shepherdi*.

Ecology

Mesoplodon layardii is found in small family groups (Minasian, Balcomb, and Foster 1984).

Genus *Tasmacetus* Oliver, 1937
Tasmacetus shepherdi Oliver, 1937
Shepherd's Beaked Whale

Description

Tasmacetus shepherdi is characterized by a large number of teeth: most beaked whales have only a few teeth or none. This species has 17–21 in the upper jaw and 23–26 in the lower, including a pair of enlarged teeth at the tip. This beaked whale reaches 6.6 m in males and 7.0 m in females. The body

is robust, with small fins, a moderately developed melon, and a noticeably sharp beak. The dorsum is dark brownish black, as are the beak, flippers, and tail. The melon, belly, and irregular stripes on the sides are much lighter (Baker 1983; Lichter and Hooper 1984; Mead 1984; Mead and Payne 1975).

Distribution

Tasmacetus shepherdi is found in waters of the Southern Hemisphere and has been stranded on the coasts of Chile and Argentina (Honacki, Kinman, and Koeppl 1982) (map 8.24).

Ecology

The stomach of one animal contained bottom fish (Minasian, Balcomb, and Foster 1984).

Genus *Ziphius* G. Cuvier, 1823
• *Ziphius cavirostris* G. Cuvier, 1823
Goose-beaked Whale, Zifio Común,
Ballena de Cuvier

Description
This species has a short beak and no pronounced
forehead. The mouth has an upward curve; there is
one pair of cylindrical, sharp teeth at the tip of the
lower gums, and up to 34 small vestigial teeth are
sometimes present in the gums of the lower jaw (fig.
8.8). In females all teeth are hidden in the gums.
Males measure to 5.5 m and females to 5.8 m. The
color varies markedly from purplish black to fawn
above, lighter on the sides, and white below, except
toward the tail, where it is brown. Some individuals

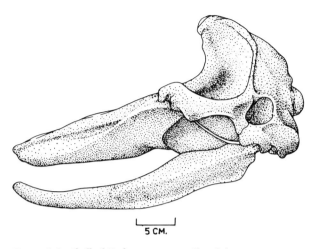

Figure 8.8. Skull of *Ziphius cavirostris* (female).

may be heavily marked with white oval blotches and
scars (Baker 1983; Mead 1984; Norman and Fraser
1964; Sielfeld 1983).

Distribution
Ziphius cavirostris occurs worldwide. In southern South America individuals have been found on
the coasts of Argentina, Uruguay, and Chile (Goodall 1978; Honacki, Kinman, and Koeppl 1982; Praderi 1971b, 1981; Sielfeld 1983; Ximénez, Balcomb,
and Foster 1972) (map. 8.25).

Ecology
The species in basically pelagic and is usually
found in groups of four to six, ranging to fifteen. It
eats deepwater fish and cephalopods (Lichter and
Hooper 1984; Sielfeld 1983).

FAMILY PHYSETERIDAE
Sperm Whales, Cachalotes

Diagnosis
The extant species of this family are included in
two genera, *Physeter* and *Kogia*. They are so different that we will reserve description for the generic
level. Neither has a beak, and the great size of *Physeter* immediately distinguishes it (plate 11). The un-
dershot jaw of *Kogia* separates it from the beakless
delphinids.

Genus *Kogia* Gray, 1846
Dwarf Sperm Whale

Description
Length does not exceed 2.7 m, and the skull is
about 400 mm long. The snout is blunt as in the
larger sperm whale, but the head is not nearly as
large proportionately. There is no beak, and the
lower jaw is set well back from the front of the head.
The blowhole opens far back on the forehead. The

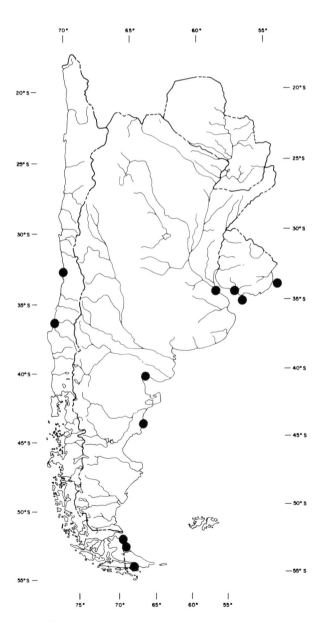

Map 8.25. Distribution of *Ziphius cavirostris*.

Figure 8.9. Two species of *Kogia*.

small dorsal fin is set about two-thirds of the way back from the head (fig. 8.9). The dorsum is dark gray to black, and the venter is paler (Handley 1966; Nagorsen 1985).

Distribution

The genus is found in all oceans but does not extend to the polar regions.

Life History and Ecology

Dwarf sperm whales are believed to feed on pelagic cuttlefish and squid. The gestation period is approximately eleven months, and a single calf is born (Nagorsen 1985).

Kogia breviceps (Blainville, 1836)
Pygmy Sperm Whale, Cachalote Pigmeo

Description

Kogia breviceps has a very short, underslung jaw with 12–16 pairs of sharp, backward-curved teeth in the lower jaw. It is longer and heavier than *K. simus*, reaching 2.7 to 3.4 m and weighing 318 to 408 kg (fig. 8.9). The body is robust, and the snout is heavy and squared off. The blowhole is on the top left of the head, and there is a small, low dorsal fin set behind the center of the body. The body is uniformly gray, with darker flanks and a lighter venter (Baker 1983; Handley 1966; Lichter and Hooper 1984; Sielfeld 1983).

Distribution

This pygmy sperm whale is found worldwide in tropical and temperate waters. In southern South America animals have been found on the coasts of Uruguay and Argentina (Castello, Erize, and Lichter 1984; Honacki, Kinman, and Koeppl 1982; Ximénez, Langguth, and Praderi 1972) (map 8.26).

Life History

Calves are thought to be born in late spring after an eleven-month gestation period. Males reach sex-

ual maturity at 2.7 to 3 m and females at 2.6 to 2.7 m (Minasian, Balcomb, and Foster 1984).

Ecology

This timid whale is found in groups apparently ranging from three to six. It feeds on squid, shrimp, crabs, and fish (Baker 1983; Ferreira and Praderi 1973; Sielfeld 1983).

• *Kogia simus* (Owen, 1866)
Dwarf Sperm Whale

Description

The dwarf sperm whale can be differentiated from its congener *K. breviceps* by its smaller size; adults range from 2.1 to 2.7 m and 136 to 272 kg (fig. 8.9). The mouth is typically small and set well below the squared-off head. There are 7–12 teeth, fewer than in *K. breviceps*. Another distinguishing character is that its dorsal fin is high and near the center of the back. The pectoral fins are short, and the dorsum is black, grading to gray on the venter (Baker 1983; Handley 1966; Sielfeld 1983).

Distribution

Kogia simus is found in tropical and warm temperate waters of the world. Individuals have been found on the coast of Chile (Crovetto and Toro 1983; Honacki, Kinman, and Koeppl 1982; Sielfeld 1983) (map 8.27).

Life History

The gestation period is about eleven months, and females may give birth in successive years (Minasian, Balcomb, and Foster 1984).

Ecology

There may be age- and sex-related segregation. Animals are often seen individually or in small groups (Minasian, Balcomb, and Foster, 1984).

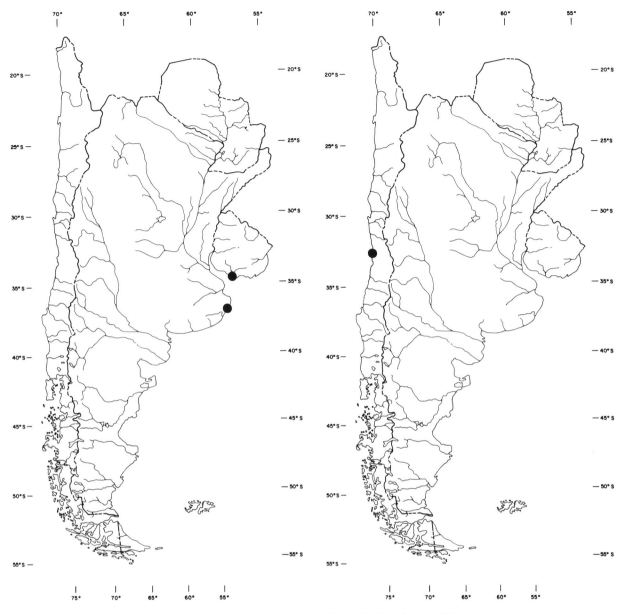

Map 8.26. Distribution of *Kogia breviceps*.

Map 8.27. Distribution of *Kogia simus*.

Genus *Physeter* Linnaeus, 1758

Description

The lower jaw is extremely slender and contains the only functional teeth, up to 60 in total number; the teeth of the upper jaw do not erupt through the gums. In this strongly dimorphic species, males may be nearly twice as large as females; adult males range from 15 to 18 m in length, and females range from 10 to 11 m. The foreflippers are about 2 m long. The most striking feature of these whales is the enormous head, nearly one-third of the total length. The head is blunt (plate 11), and the blowhole is near the tip of the rounded forehead. The rounded dorsal fin is set midway along the back. The color is gray to black, but males appear to become paler with age.

Distribution

This genus is distributed worldwide. Strongly migratory in its movements, it may be seasonally abundant in tropical waters.

• *Physeter macrocephalus* Linnaeus, 1758
Sperm Whale, Cachalote

Description

The sperm whale is unmistakable because of its shape and size. The most massive part of the body is the head, which constitutes up to one-third of the total length and is equipped with a long, narrow lower jaw set with 20–28 large teeth. Adults reach 18 m and 35 metric tons; males are larger than females. The pectoral fins are poorly developed, and the dorsal fin is not pronounced. Sperm whales are dark gray, darker on the head and lighter on the sides, paling to a silvery gray or white on the venter (Kawakami 1980; Lichter and Hooper 1984; MacKintosh and Brown, in Brown et al. 1974; Norman and Fraser 1964).

Distribution

Physeter is found in all oceans, moving seasonally. In southern South America individuals have been found off the coasts of Argentina, Uruguay, and Chile (Honacki, Kinman, and Koeppl 1982; Lichter and Hooper 1984; Schneider 1946) (map 8.28).

Life History

The gestation period in sperm whales is fifteen to sixteen months, and the interbirth interval is five to

six years. Calves are born 4 m long and take solid food within the first year. Females lie at the surface when calves are nursing. There is some indication of communal suckling, and animals thirteen years old have been found with milk in their stomachs. Males reproduce when they are 9 to 12.5 m long. In the Southern Hemisphere the peak breeding season is between September and December. Females calve in southerly latitudes between 28° and 36° S, and they and their calves do not migrate as far north as the males do (Best, Canham, and Macleod 1984; Caldwell, Caldwell, and Rice 1966; Clarke, Aguayo, and Basulto del Campo 1978).

Ecology

Sperm whales are animals of deep, blue waters, and they migrate seasonally. Males move farther north than females with calves, which tend to stay in warmer waters. Pods are formed by females, calves, and immature animals, while males are usually solitary or found in bachelor pods. The association between a breeding bull and a group of females is apparently very short. It seems males fight using their teeth, which do not penetrate the gums until sexual maturity and so are probably not primarily food-gathering organs. Females appear to communally defend calves against predators such as killer whales and oceanic sharks. The average group size for thirty-four groups was six.

Sperm whales are excellent divers, and bulls can stay down more than an hour. Squid forms the mainstay of the diet of *Physeter*: in one study seven species of squid formed 80% of the diet, and up to three tons may be eaten each day. These squid are frequently large deepwater species; one squid removed from a whale stomach was 10.5 m long and weighed 184 kg. After squid, fish is the most important food (Best, Canham, and Macleod 1984; Caldwell, Caldwell, and Rice 1966; Clarke, Aguayo, and Basulto de Campo 1978, 1980; Kawakami 1980).

SUBORDER MYSTICETI

As indicated in the description of the order Cetacea, the mysticetes include the baleen whales. There are three families, the Eschrichtidae, the Balaenidae, and the Balaenopteridae. Only the last two families occur within the range covered by this volume.

FAMILY BALAENOPTERIDAE

Rorquals, Ballenas con Aleta

Diagnosis

The teeth appear embryologically but are functionally replaced in the adult by epidermally derived plates called baleen, which hang from the roof of the

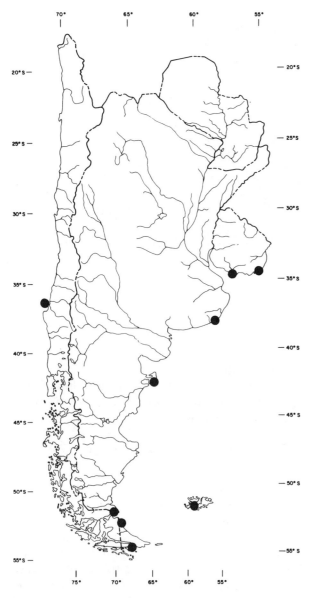

Map 8.28. Distribution of *Physeter macrocephalus*.

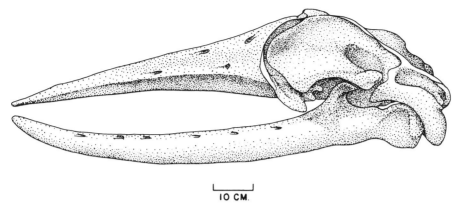

Figure 8.10. Skull of *Balaenoptera acutorostrata*.

mouth (fig. 8.1). These large whales range from 9 to 30 m in total length. There is size dimorphism, and females generally exceed males by about 1 m. The body is elongated and streamlined, and in the region of the throat and chest there are numerous longitudinal furrows that allow the buccal cavity to distend during feeding (plate 11). The skull is vastly modified to accommodate the baleen plates, and its structure is unmistakable (fig. 8.10).

Distribution
The family is found worldwide and is oceanic.

Genus *Balaenoptera* Lacépède, 1804
Fin Whales and Relatives

Description
The body is long and streamlined, with the dorsal fin rather small and set far back. The female is larger than the male. The species of this genus are distinguishable by their size and form a graded series. The color pattern is dark brown to black above and white below.

Distribution
These whales are oceanic and found worldwide. Gaskin (1982) is a useful summary of their biology.

• *Balaenoptera acutorostrata* Lacépède, 1804
Minke Whale, Rorcual Menor

Description
This is the smallest of the rorquals, averaging about 8.5 m in length (range 7.5 to 10.7) and weighing 6,000 to 7,000 kg. It is less gracile than other *Balaenoptera* species; its head is less tapered and is pointed and crowned with a keel. The dorsal fin is on the last third of the body, and the pectoral fin is about one-eighth of the total length. Minke whales have fifty to sixty throat furrows that, as in *B. physalus*, do not reach the umbilicus. They are blue gray,

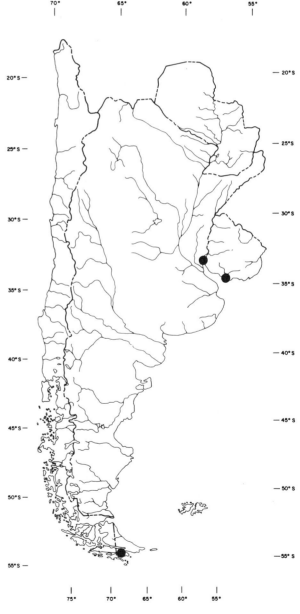

Map 8.29. Distribution of *Balaenoptera acutorostrata*.

darker above and lighter below, and some have a white patch on the outer surface of the flipper. The baleen is yellowish white (Lichter and Hooper 1984; Norman and Fraser 1964; Sielfeld 1983; Williamson 1975).

Distribution

Balaenoptera acutorostrata is found worldwide in subtropical to Arctic waters. In southern South America individuals of this species have been found on the coasts of Argentina and Uruguay (Honacki, Kinman, and Koeppl 1982; Praderi 1981; Sielfeld 1983; Ximénez, Langguth, and Praderi 1972) (map 8.29).

Life History

In the Southern Hemisphere the peak of conceptions takes place from August to September and the peak of births from May to June. The gestation period is about ten months, and the interbirth interval is twelve to fourteen months. Neonates range from 2.69 to 2.84 m and suckle for about four months, being weaned when they are about 4.5 m. First reproduction takes place when males are 7.2 m and females are 8 m in length (Lockyer 1984; Stewart and Leatherwood 1985).

Ecology

In the Southern Hemisphere Minke whales migrate annually from near the equator to the Antarctic, where young are born and suckled. Immature whales precede adults in the northward migration. This species is characteristically divided into groups dominated by one sex or age class. In Antarctic oceans, where animals are frequently seen near the ice edge, the average group size was only 1.9 (*n* = 50). There are other reports of groups of over two hundred.

In northern waters this species feeds on euphausiids, copepods, krill, and fish. Their feeding behavior is similar to that of other species of rorquals (Leatherwood et al. 1982; Sielfeld 1983; Stewart and Leatherwood 1985; Williamson 1975).

• *Balaenoptera borealis* Lesson, 1828
Sei Whale, Rorcual Mediano

Description

This whale reaches 20 m in length and 12,000 kg. Females are slightly larger than males. It has thirty-two to sixty-two ventral grooves that terminate well before the navel. The dorsal fin, which has a deeply concave hind margin, is relatively large, and the pectoral fins are relatively small (about one-eleventh of the total length). The dorsum is blackish blue, and the venter is white. When breaching the sei whale shows its head and a large part of its dorsum, including the dorsal fin. Its spout is like an inverted cone,

about 3 m high (Gambell 1985a; Lichter and Hooper 1984; Lockyer 1984; Norman and Fraser 1964; Sielfeld 1983).

Distribution

Balaenoptera borealis is found worldwide in subtropical to subarctic waters. It is not distributed as far toward polar waters as other rorquals. In southern South America stranded individuals have been found on the beaches of Argentina, Uruguay, and Chile (Gambell 1985a; Honacki, Kinman, and Koeppl 1982; Lichter and Hooper 1984; Oporto 1984) (map 8.30).

Life History

In Antarctica the peak month for conceptions is July, and the peak month for births is June. The ges-

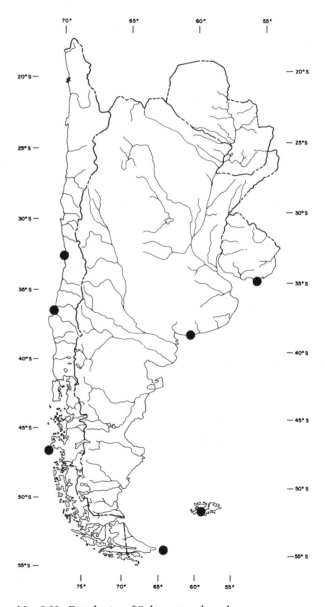

Map 8.30. Distribution of *Balaenoptera borealis*.

tation period is 11 to 11.5 months, and the interbirth interval is probably three years. Calves are born measuring 4.5 m, suckle for about six months, and are weaned at about 8 m. The first reproduction takes place when males are 13.6 m long and females 14 m (Gambell 1985a; Lockyer 1984).

Ecology

Sei whales winter in temperate waters, returning to polar waters in summer to feed. Mating and conception takes place in polar waters. This species is usually observed alone or in pairs, though aggregations of up to fifty animals can be found. Sei whales are primarily krill feeders (Gambell 1985a; Minasian, Balcomb, and Foster 1984; Sielfeld 1983).

• *Balaenoptera edeni* Anderson, 1878
Bryde's Whale

Description

Bryde's whales reach 15.5 m in length and weigh 12,000 kg. Females are slightly larger than males of the same age. These are the second smallest balaenopterids. They have a large, flat rostrum with a keel and a delicate body. On the throat there are thirty to sixty grooves. The pectoral fins measure one-tenth to one-eleventh of the total length. This species can be distinguished from Sei whales by the three prominent ridges usually present on the head. The dorsum is dark gray blue, and the throat is lighter (Cummings 1985; Lichter and Hooper 1984; Lockyer 1984; Minasian, Balcomb, and Foster 1984; Sielfeld 1983).

Distribution

Balaenoptera edeni is found worldwide in temperate to tropical waters. It has been found on the coasts of Brazil and Chile, where it is particularly common between 35° and 40° south (Honacki, Kinman, and Koeppl 1982; Lichter and Hooper 1984; Pastene and Acevedo 1984) (map 8.31).

Life History

In an offshore South African population the peak month of conception is March, and births are most common between February and March. The gestation period is twelve months, young are born about 4 m long, and lactation lasts about six months. Females attain sexual maturity at about 12 m or ten years of age (Cummings 1985; Lockyer 1984).

Ecology

This is a whale of warmer waters. Off South Africa there are two populations: an inshore one that is present year round, and an offshore one that migrates. This may also be the case off South America. Long migrations are probably not typical of this species, though there are apparently some seasonal

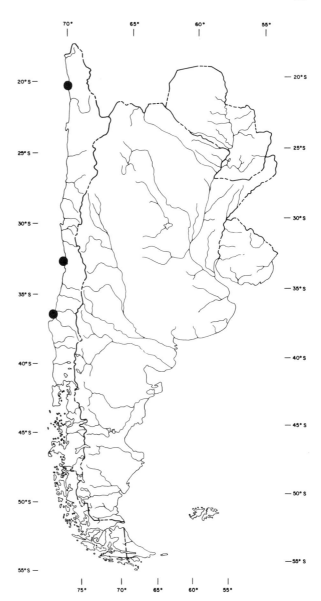

Map 8.31. Distribution of *Balaenoptera edeni*.

shifts. Bryde's whale has never been of major importance to whalers. It feeds on pelagic fish, crustaceans, and cephalopods (Summings 1985; Lockyer 1984; Sielfeld 1983).

• *Balaenoptera musculus* (Linnaeus, 1758)
Blue Whale, Rorcual Azul

Description

The blue whale is the largest living mammal: males average 26 m in length ($n = 11$) and females average 28 m ($n = 8$); individuals weigh 80,000 to 130,000 kg. The dorsal fin is particularly low and small and is set far back on the body. The long, tapering flippers measure about one-seventh of the total body length. The head is flattened on the anterior portion; the mandible is longer than the maxilla, and there are eighty to one hundred gular furrows. Dorsally this species is dark blue sprinkled with small gray stripes; the tips of the fins and the belly are lighter blue. There is individual and age-related variation in color. The spout of the blue whale is tall (8–15 m) and narrow (Lichter and Hooper 1984; MacKintosh and Brown, in Brown et al. 1974; Nishiwaki 1950; Norman and Fraser 1964; Omura, Ichihara, and Kasuya 1970; Sielfeld 1983).

Distribution

Balaenoptera musculus is found worldwide in subtropical to subarctic waters. In southern South America individuals have been found on the coasts of Uruguay, Argentina, Chile, and the Malvinas Islands (Honacki, Kinman, and Koeppl 1982; Lichter and Hooper 1984; Praderi 1981; Sielfeld 1983; Ximénez, Langguth, and Praderi 1972) (map 8.32).

Life History

In Antarctica most conceptions take place between June and July, and the birth peak is in May. There are probably two breeding seasons, a main one during the winter and a lesser one in the summer. The gestation period is eleven months, and calves are born 7 m long. Lactation lasts about seven months, and calves are weaned at 12.8 m. First reproduction takes place when males are 22.6 m in length and females 24 m (Lockyer 1984).

Ecology

Blue whales are animals of the deep water, where they feed on pelagic crustaceans. They make great migratory movements to productive feeding grounds. They are more solitary than other species in the genus and are often found in groups of two or three. Because of extensive hunting their numbers have been greatly reduced (MacKintosh and Brown, in Brown et al. 1974; Norman and Fraser 1964; Sielfeld 1983).

• *Balaenoptera physalus* (Linnaeus, 1758)
Fin Whale, Rorcual Común

Description

Balaenoptera physalus is smaller than *B. musculus:* males average 23 m in length ($n = 4$) and females average 24 m ($n = 6$). Adults weigh 35,000 to 45,000 kg. In addition, the species can be distinguished by the wedge-shaped head, the distinct ridge on the back toward the tail, and the higher dorsal fin. The pectoral fins are one-ninth to one-tenth of the total length. There are 68 to 114 ventral throat grooves. The head color is asymmetrical: the lower mandible is white on the right and dark on the left. The dorsum is dark to light gray, darker than in *B. musculus*, and the venter is pure white. The spout is 3 to 4.5 m high (Lichter and Hooper 1984; Nishiwaki 1950; Norman and Fraser 1964; Sielfeld 1983).

Distribution

The species is found worldwide in tropical and Arctic waters. In southern South America individuals have been found on the coasts of Argentina, Uruguay, and Chile (Honacki, Kinman, and Koeppl 1982; Lichter and Hooper 1984; Sielfeld 1983) (map 8.33).

Life History

In Antarctica the peak period of conception is June to July, and the peak month for births is May. The gestation period is 11.25 months, and calves are

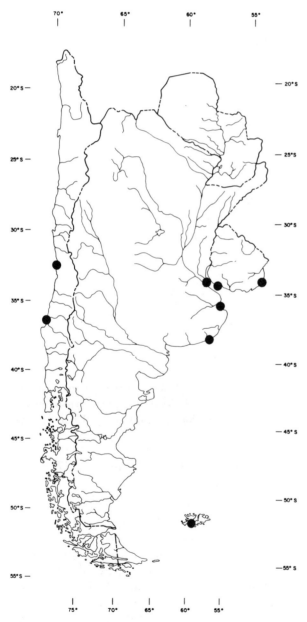

Map 8.32. Distribution of *Balaenoptera musculus*.

born 6.4 m long. After a lactation period of seven months, calves are weaned at 11.5 m. Females reach sexual maturity at a length of 19.9 m and males at 19.2 m. Fin whales can live a hundred years (Gambell 1985b; Lockyer 1984).

Ecology

The fin whale migrates from temperate waters, where it mates and calves, to polar waters. Larger and older animals move farther south. Males tend to precede females south, and pregnant females migrate before the other sex classes. This species is sometimes found singly or in pairs, but it commonly forms groups of three to twenty that may in turn coalesce into concentrations of over a hundred ani-mals. It feeds mainly on planktonic crustaceans but also eats some fish and cephalopods.

Fin whales were prime targets of whalers after the development of the explosive grenade harpoon. Before intensive whaling began, estimates of fin whale abundance in the Southern Hemisphere reached almost 500,000. Now the population is slightly over 100,000 (Gambell 1985b; Norman and Fraser 1964; Sielfeld 1983).

Genus *Megaptera* Gray, 1846
• *Megaptera novaeangliae* (Borowski, 1781)
Humpback Whale, Ballena Yubarta, Jorobada

Description

The humpback whale is easily identified by its extraordinarily long flippers with irregular wavy anterior edges, measuring nearly one-third of the total length. Males average 14.6 m in length and females 15.2 m, and weight ranges from 30,000 to 40,000 kg. This species has a robust body with a large head; in the center of the head is a row of irregular protuberances, extending onto the chin. It has fifteen to twenty furrows on the throat that extend to the genital opening. The humpback is dark gray dorsally with a white throat; its pectoral fins are white beneath and mottled black and white above. This species frequently jumps well out of the water as well as floating in the water with one of the long pectoral fins pointing toward the sky. Its spout is 3 m high (Lichter and Hooper 1984; Norman and Fraser 1964).

Distribution

Megaptera is found worldwide in tropical and Arctic waters. In southern South America individuals have been stranded on the coasts of Argentina and Chile (Honacki, Kinman, and Koeppl 1982; Lichter and Hooper 1984; Sielfeld 1980) (map 8.34).

Life History

In Antarctica the peak of conceptions occurs in July and August and the peak of births is in July and August. Whales generally calve along coastal areas or near islands. The gestation period is 11.5 months, and calves are born 4.3 m long. Lactation lasts 10.5 to 11 months, and young are weaned at 8.8 m. First reproduction takes place when males are 11.5 m and females 12 m long (Lockyer 1984; Sielfeld 1983).

The underwater vocalization of males has been the subject of great interest. Adult males apparently select traditional locations for calling, and on a given year they develop "songs" that may be copied by subadult males. Dialects may change from year to year, but clearly females are attracted to the singing grounds (Payne and McVay 1971; Payne, Tyack, and Payne 1983).

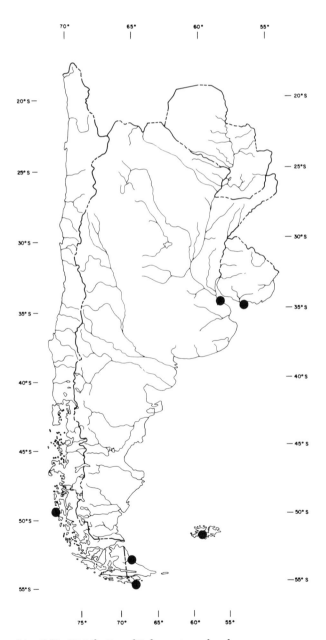

Map 8.33. Distribution of *Balaenoptera physalus*.

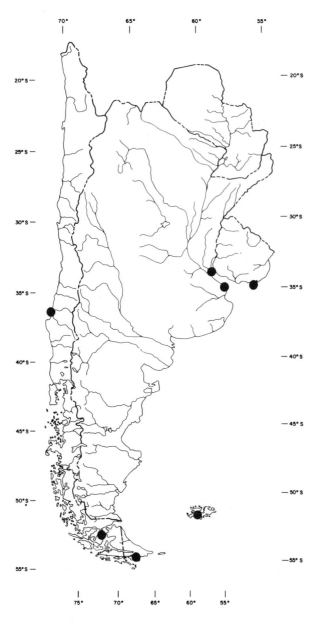

Map 8.34. Distribution of *Megaptera novaeangliae*.

Ecology

Humpback whales move seasonally, and in the Southern Hemisphere they spend the austral summer in Antarctic waters. They are filter feeders and are usually found in groups of only a few individuals. The average size for seven groups was 2.1 (Leatherwood et al. 1982; Lichter and Hooper 1984; Sielfeld 1983; Tamayo and Frassinetti 1980).

FAMILY BALAENIDAE
Baleen Whales

Diagnosis

These whales are mysticetes (see subordinal account). The cervical vertebrae are fused, and the rostrum is narrow but highly arched. The baleen plates are extremely long and narrow, and the lower jaw is massive. There are no throat grooves, distinctly contrasting with the Balaenopteridae. There are two extant genera, *Balaena* and *Caperea,* which differ markedly in size (see plate 11 and species accounts).

Distribution

Species of this family are oceanic and tend to be most abundant toward the poles, from 45° north latitude and 30° south latitude.

Genus *Balaena* Linnaeus, 1758
Balaena glacialis Muller, 1776
Right Whale, Ballena Franca Austral

Description

The right whale is easily recognized by the proliferation of protuberances (corneal callosities) on top of its head and on its chin. Its body is robust, constricting sharply near the tail, and averages 15 m in total length (plate 11). An average weight is 54,000 kg. Compared with other mysticetes, right whales are very large in girth relative to their length. The head is large, almost one-quarter of the total length. These whales are black, frequently mottled with brown and marked with scars and lines. The calves are lighter colored. The spray is V-shaped and up to 5 m high. Right whales frequently jump well out of the water (Cummings 1985; Lichter and Hooper 1984; Sielfeld 1983).

Distribution

Right whales are found worldwide in temperate and subarctic waters. In southern South America individuals have been found on the shores of Argentina, Chile, and Uruguay. Off the Valdés Peninsula, Chubut province, Argentina, right whales congregate seasonally to breed (Honacki, Kinman, and Koeppl 1982; Lichter and Hooper 1984; Daciuk 1974) (map 8.35).

Life History

In the Southern Hemisphere the peak in conceptions occurs from August to October and the peak in births from May to August. The gestation period is ten months, lactation lasts six to seven months, and the interbirth interval is about three years. Calves are born between 5.5 and 6.1 m long. Sexual maturity is reached at 15 m in males and at 15.5 m in females. Courtship occurs in sheltered bays (Cummings 1985; Lockyer 1984).

Ecology

Right whales move from subpolar waters in winter to lower latitudes, staying in continental waters or near island masses. This species appears off the Valdés Peninsula, Argentina, from the end of May to December to calve. There concentrations of up to two hundred animals court, mate, and calve. During this period they appear not to feed. Several males have been seen courting a single female. Off Chile near the Arauco Gulf is apparently another calving area. In areas of reproduction whales are often seen in groups of three to seven. This species is a filter feeder, concentrating on krill (Bastida and Bastida 1984a,b; Cummings 1985; Harris and García 1984; Mermoz 1980a; Payne 1983; Schusterman, Thomas, and Wood 1986; Sielfeld 1983).

Genus *Caperea* Gray, 1864
Caperea marginata (Gray, 1846)
Pygmy Right Whale, Ballena Pigmea

Description

This small right whale has a slender, lithe body, with both sexes reaching 5–6 m. There apparently is sexual dimorphism, with females longer than males. Individuals weigh about 4,500 kg (plate 11). The mandible is bowed like the maxilla; there is a short, thin pectoral fin and a small dorsal fin with a concave posterior edge. The tail flukes are full and are separated by a well-formed median notch. The line of the

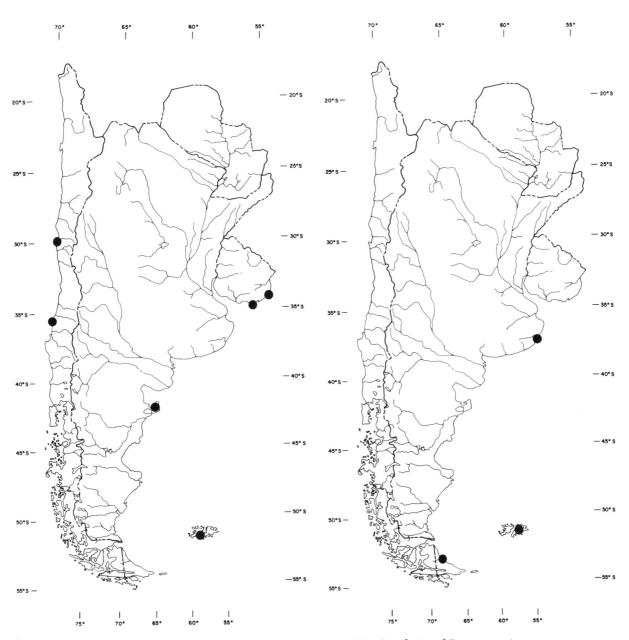

Map 8.35. Distribution of *Balaena glacialis*.

Map 8.36. Distribution of *Caperea marginata*.

mouth extends past the eye. They are blue-gray to black dorsally and much lighter ventrally. There is a dark band from the pectoral fin to the eye and a light band across the dorsum, behind and between the pectoral fins. The spout of this species is small and poorly visible (Baker 1983; Lichter and Hooper 1984; Ross, Best, and Donnelly 1975; Sielfeld 1983).

Distribution
Caperea is found in temperate waters of the Southern Hemisphere. In southern South America individuals have been found on the Malvinas Islands and the Argentine coast (Honacki, Kinman, and Koeppl 1982; Lichter and Hooper 1984; Sielfeld 1983) (map 8.36).

Life History
In Antarctica there appears to be an extended birth season and a gestation period of about twelve months. Calves are born between 1.5 and 2.2 m long and nurse for about six months, being weaned at a length of 3.2 to 3.6 m. Juveniles apparently disperse to coastal waters in the spring (Lockyer 1984; Tamayo and Frassinetti 1980).

Ecology
Pygmy right whales are frequently found in sheltered shallow bays. Because of its smaller size and less noticeable blow, this species is relatively inconspicuous. It takes only short dives and apparently never goes deeper than several meters below the surface. Copepods are the only food to have been found in stomachs of *Caperea* (Ross, Best, and Donnelley 1975; Sielfeld 1983).

References

Aguayo, L., A. 1975. Progress report on small cetacean research in Chile. *J. Fish. Res. Board Can.* 32:1123–43.

Baker, A. N. 1983. *Whales and dolphins of New Zealand and Australia.* Wellington: Victoria University Press.

Bastida, R., and V. L. de Bastida. 1984a. Informe preliminar sobre los estudios de ballena franca austral (*Eubalaena australis*) en la zona de la Península Valdés (Chubut, Argentina). *Rev. Mus. Argent. Cienc. Nat. "Bernardino Rivadavia," Zool.* 13(1–60): 197–210.

———. 1984b. Avistajes de cetáceos realizados por buques balleneros en aguas argentinas. *Rev. Mus. Argent. Cienc. Nat. "Bernardino Rivadavia," Zool.* 13(1–60): 211–24.

Best, P. B., P. A. S. Canham, and N. Macleod. 1984. Patterns of reproduction in sperm whales, *Physeter macrocephalus.* In *Reproduction in whales, dolphins and porpoises,* ed. W. F. Perrin, R. L. Brownell, Jr., and D. P. DeMaster, 51–79. Reports of the International Whaling Commission, Special Issue 6. Cambridge: International Whaling Commission.

Borobia, M., and L. Geise. 1984. Registro de *Pontoporia blainvillei* (Cetacea, Platanistidae) no Espírito Santo, Brasil. In *Primera reunion de trabajo de expertos en mamíferos acuáticos de América del Sud* (resúmenes) 8. Buenos Aires.

Brown, S. G., R. L. Brownell, Jr., A. W. Erickson, R. J. Hofman, G. A. Llano, and N. A. Mackintosh, eds. 1974. *Antarctic mammals.* Antarctic Map Folio Series 18. New York: American Geographical Society.

Brownell, R. L., Jr. 1975a. *Phocoena dioptrica. Mammal. Species* 66:1–3.

———. 1975b. Progress report on the biology of the Franciscana dolphin, *Pontoporia blainvillei,* in Uruguayan waters. *J. Fisheries Res. Board Can.* 32(7): 1073–78.

Brownell, R. L., Jr., and G. P. Donovan, eds. 1988. *Biology of the genus* Cephalorhynchus. Reports of the International Whaling Commission, Special Issue 9. Cambridge: International Whaling Commission.

Brownell, R. L., Jr., and R. Praderi. n.d. Taxonomy and distribution of Commerson's dolphin *Cephalorhynchus commersonii.* Unpublished manuscript.

Caldwell, D. K., and M. C. Caldwell. 1977. Cetaceans. In *How animals communicate,* ed. T. A. Sebeok, 794–801. Bloomington: Indiana University Press.

Caldwell, M. C., and D. K. Caldwell. 1966. Epimeletic (care giving) behavior in Cetacea. In *Whales, dolphins and porpoises,* ed. K. S. Norris, 755–89. Berkeley: University of California Press.

Caldwell, D. K., M. C. Caldwell, and D. W. Rice. 1966. Behavior of the sperm whale, *Physeter catodon* L. In *Whales, dolphins, and porpoises,* ed. K. S. Norris, 677–717. Berkeley: University of California Press.

Castello, H. P. 1977. Food of a killer whale: Eagle sting-ray, *Myliobatis,* found in the stomach of a stranded *Orcinus orca. Sci. Repts. Whales Res. Inst.* 29:107–11.

Castello, H. P., F. Erize, and A. Lichter. 1984. Primer registro del cachalote pigmeo, *Kogia breviceps* (Blainville, 1838) para las costas de la rep. Argentina. In *Primera reunión de trabajo de expertos en mamíferos acuáticos de América del Sud,* 16. Buenos Aires.

Christensen, I. 1984. Growth and reproduction of killer whales, *Orcinus orca,* in Norwegian coastal waters. In *Reproduction in whales, dolphins and porpoises*, ed. W. F. Perrin, R. L. Brownell, Jr., and D. P. DeMaster, 253–58. Reports of the International Whaling Commission, Special Issue 6. Cambridge: International Whaling Commission.

Clarke, M. R., N. MacLeod, H. P. Castello, and M. C. Pinedo. 1980. Cephalopod remains from the stomach of a sperm whale stranded at Rio Grande do Sul in Brazil. *Marine Biol.* 59: 235–39.

Clarke, R., A. Aguayo, and S. Basulto del Campo. 1978. Whale observation and whale marking off the coast of Chile in 1964. *Sci. Repts. Whales Res. Inst.* 30:117–77.

Crespo, E. A., G. Pagnoni, and S. N. Pedraza. 1984. Estructura social de un grupo de ballenas piloto (*Globicephala melaena edwardi*) varadas en el litoral Patagónico. In *Primera reunión de trabajo de expertos en mamíferos acuáticos de América del Sud*, 18. Buenos Aires.

Crovetto, A., and H. Toro. 1983. Présence de *Kogia simus* (Cetacea, Physeteridae) dans les eaux chiliennes. *Mammalia* 47(4): 591–93.

Cummings, W. C. 1985. Right whales *Eubalaena glacialis* (Muller, 1776) and *Eubalaena australis* (Desmoulins, 1922). In *Handbook of marine mammals*, 3:275–303. New York: Academic Press.

Daciuk, J. 1974. Notas faunísticas y bioecológicas de Península Valdés y Patagonia. 12. Mamíferos colectados y observados en la Península Valdés y zona litoral de los Golfos San José y Nuevo (provincia de Chubut, república Argentina). *Physis*, sec. C, 33(86): 23–39.

Eisenberg, J. F. 1986. Dolphin behavior and cognition: Evolutionary and ecological aspects. In *Dolphin cognition and behavior: A comparative approach*, ed. R. Buhr, R. Schusterman, J. Thomas, and F. Wood, 261–70. Hillsdale, N.J.: Lawrence Erlbaum.

Ferreira, R. V., and R. Praderi. 1973. Un nuevo ejemplar de *Kogia breviceps* (Blainville), (Cetacea, Physeteridae) del Atlántico sudoccidental: Caracteres y notas. *Trav. V Congr. Latinoamer. Zool.* 1:261–77.

Fitch, J. E., and R. L. Brownell, Jr. 1971. Food habits of the Franciscana *Pontoporia blainvillei* (Cetacea: Platanistidae) from South America. *Bull. Marine Sci.* 21:626–36.

Ford, J. K. B., and D. Fisher. 1983. Group-specific dialects of killer whales (*Orcinus orca*) in British Columbia. In *Communication and behavior of whales*, ed. R. Payne, 129–62. Boulder, Colo.: Westview Press.

Gambell, R. 1985a. Sei whale *Balaenoptera borealis* Lesson, 1928. In *Handbook of marine mammals*, pp. 155–69. New York: Academic Press.

———. 1985b. Fin whale *Balaenoptera physalus* (Linnaeus, 1758). In *Handbook of marine mammals*, pp. 171–92. New York: Academic Press.

Gaskin, D. E. 1982. *The ecology of whales and dolphins*. London: Heinemann.

Gewalt, W. 1979. The Commerson's dolphin (*Cephalorhynchus commersonii*): Capture and first experiences. *Aquatic Mammals* 7(2): 37–40.

Gianuca, N. M., and H. P. Castello. 1976. First record of the southern bottlenose whale, *Hyperoodon planifrons* from Brazil. *Sci. Repts. Whales Res. Inst.* 28:119–26.

Goodall, R. N. P. 1978. Report on the small cetaceans stranded on the coasts of Tierra del Fuego. *Scie. Repts. Whales Res. Inst.* 30:197–230.

Goodall, R. N. P., A. R. Galeazzi, and I. S. Cameron. 1984. La tonina overa, *Cephalorhynchus commersonii*, en Tierra del Fuego. In *Primera reunión de trabajo de expertos en mamíferos acuáticos de América del Sud*, 22. Buenos Aires.

Goodall, R. N. P., K. S. Norris, I. S. Cameron, and J. A. Oporto. 1984. El delfín chileno, *Cephalorhynchus eutropia* (Gray 1846). In *Primera reunión de trabajo de expertos en Mamíferos Acuáticos de América del Sud*, 23. Buenos Aires.

Handley, C. O., Jr. 1966. A synopsis of the genus *Kogia* (Pygmy sperm whales). In *Whales, dolphins, and porpoises*, ed. K. S. Norris, 62–69. Berkeley: University of California Press.

Harris, G., and C. García. 1984. Censos de ballenas francas (*Eubalaena australis*) en el Golfo San José, Chubut, Argentina: Años 1982–1983. In *Primera reunión de trabajo de expertos en mamíferos acuáticos de América del Sud*, 24. Buenos Aires.

Hershkovitz, P. 1966. Catalog of living whales. *Bull. U.S. Nat. Mus.* 246:1–259.

Hill, J. 1985. Commerson's dolphin born at Sea World. *AAZPA Newsletter* 26(5): 23.

Honacki, J. II., K. E. Kinman, and J. W. Koeppl, eds. 1982. *Mammal species of the world*. Lawrence, Kans.: Allen Press and Association of Systematics Collections.

Kasuya, T. 1976. Reconsideration of life history parameters of the spotted and striped dolphins based on cemental layers. *Sci. Repts. Whales Res. Inst.* 28:73–106.

Kasuya, T., and R. L. Brownell, Jr. 1979. Age determination, reproduction, and growth of the Franciscana dolphin, *Pontoporia blainvillei*. *Sci. Repts. Whales Res. Inst.* 31:45–67.

Kawakami, T. 1980. A review of sperm whale food. *Sci. Repts. Whales Res. Inst.* 32:199–218.

Kellogg, W. N. 1961. *Porpoises and sonar.* Chicago: University of Chicago Press.

Leatherwood, S., F. S. Todd, J. A. Thomas, and F. T. Awbrey. 1982. Incidental records of cetaceans in southern seas, January and February 1981. *Repts. Int. Whaling Comm.* 32:515–20.

Lichter, A., and A. Hooper. 1984. *Guía para el reconocimiento de cetáceos del Mar Argentino.* Buenos Aires: Fundación Vida Silvestre Argentina.

Lockyer, C. 1984. Review of baleen whale (Mysticeti) reproduction and implications for management. In *Reproduction in whales, dolphins and porpoises,* ed. W. F. Perrin, R. L. Brownell, Jr., and D. P. DeMaster, 27–51. Reports of the International Whaling Commission, Special Issue 6. Cambridge: International Whaling Commission.

López, J. C., and D. López. 1985. Killer whales (*Orcinus orca*) of Patagonia, and their behavior of intentional stranding while hunting near shore. *J. Mammal.* 66:181–83.

McCann, C. 1975. A study of the genus *Berardius* Duvernoy. *Sci. Repts. Whales Res. Inst.* 27:111–37.

Mead, J. G. 1981. First records of *Mesoplodon hectori* (Ziphiidae) from the Northern Hemisphere and a description of the adult male. *J. Mammal.* 62:430–32.

———. 1984. Survey of reproductive data for the beaked whales (Ziphiidae). In *Reproduction in whales, dolphins and porpoises,* ed. W. F. Perrin, R. L. Brownell, Jr., and D. P. DeMaster, 91–96. Reports of the International Whaling Commission, Special Issue 6. Cambridge: International Whaling Commission.

Mead, J. G., and R. S. Payne. 1975. A specimen of the tasman beaked whale, *Tasmacetus shepherdi,* from Argentina. *J. Mammal.* 56:213–18.

Mermoz, J. F. 1980a. Preliminary report on the southern right whale in the southwestern Atlantic. *Repts. Int. Whaling Comm.* 30:183–86.

———. 1980b. A brief report on the behaviour of Commerson's dolphin, *Cephalorhynchus commersonii,* in Patagonian shores. *Sci. Repts. Whales Res. Inst.* 32:149–53.

Minasian, S. M., K. Balcomb III, and L. Foster. 1984. *The world's whales.* Washington, D.C.: Smithsonian Institution Press.

Moore, J. C. 1968. Relationships among the living genera of beaked whales with classifications, diagnoses and keys. *Fieldiana: Zool.* 53(4): 209–98.

Nagorsen, D. 1985. *Kogia simus. Mammal. Species* 239:1–6.

Nishiwaki, M. 1950. On the body weight of whales. *Sci. Repts. Whales Res. Inst.* 4:184–209.

Norman, J. R., and F. C. Fraser. 1964. *Giant fishes, whales and dolphins.* New York: Putnam.

Norris, K. S., ed. 1966. *Whales, dolphins, and porpoises.* Berkeley: University of California Press.

Omura, H., T. Ichihara, and T. Kasuya. 1970. Osteology of pygmy blue whale with additional information on external and other characteristics. *Sci. Repts. Whales Res. Inst.* 22:1–27.

Oporto, J. A. 1984. Observaciones de cetáceos en los canales del sur de Chile. In *Primera reunión de trabajo de expertos en mamíferos acuáticos de América del Sud,* 31. Buenos Aires.

Pastene, L., and M. Acevedo. 1984. Nota preliminar sobre grandes ballenas en las costas Chilenas, con énfasis en la ballena de Bryde. In *Primera reunión de trabajo de expertos en mamíferos acuáticos de América del Sud,* 32. Buenos Aires.

Payne, K., P. Tyack, and R. Payne. 1983. Progressive changes in the songs of humpback whales (*Megaptera novaeangliae*): A detailed analysis of two seasons in Hawaii. In *Communication and behavior of whales,* ed. R. Payne, 9–57. Boulder, Colo.: Westview Press.

Payne, R. S., ed. 1983. *Communication and behavior of whales.* Boulder, Colo.: Westview Press.

Payne, R. S., and S. McVay. 1971. Songs of humpbacked whales. *Science* 173:585–97.

Perrin, W. F. 1975. Distribution and differentiation of populations of dolphins of the genus *Stenella* in the eastern tropical Pacific. *J. Fish. Res. Board Can.* 32:1058–67.

Perrin, W. F., and S. B. Reilly. 1984. Reproductive parameters of dolphins and small whales of the family Delphinidae. In *Reproduction in whales, dolphins and porpoises,* ed. W. F. Perrin, R. L. Brownell, Jr., and D. P. DeMaster, 97–133. Reports of the International Whaling Commission, Special Issue 6. Cambridge: International Whaling Commission.

Piñero, M. E., and H. P. Castello. 1975. Sobre "ballenas piloto," *Globicephala melaena edwardi* (Cetacea, Delphinidae) varadas en Isla Trinidad, provincia de Buenos Aires. *Rev. Mus. Argent. Cienc. Nat. "Bernardino Rivadavia," Zool.* 12(2):13–24.

Praderi, R. 1971a. Contribución al conocimiento del género *Phocoena* (Cetacea, Phocenidae). *Rev. Mus. Argent. Cienc. Nat. "Bernardino Rivadavia," Zool.* 7(2): 251–66.

———. 1971b. Sobre la presencia de *Ziphius cavirostris* G. Cuvier (Cetacea, Hyperoodontidae) en las costas Uruguayas del Río de la Plata. *Bol. Soc. Zool. Uruguay* 1:52–54.

———. 1981. Varamientos ocasionales de cetáceos

en costas del Río de la Plata. *Res. Com. Jorn. C. Nat.* (Montevideo) 2:13–14.

Ringuelet, R. A., and R. H. Arámburu. 1957. *Enumeración sistemática de los vertebrados de la provincia de Buenos Aires.* La Plata: Ministerio de Asuntos Agrarios.

Ross, G. J. B., P. B. Best, and B. G. Donnelly. 1975. New records of the pygmy right whale (*Caperea marginata*) from South Africa, with comments on distribution, migration, appearance, and behavior. *J. Fish. Res. Board Can.* 32(7): 1005–17.

Schneider, C. O. 1946. Catálogo de los mamíferos de la provincia de Concepción. *Bol. Soc. Biol. Concepción* 21:67–83.

Schusterman, R. J., J. A. Thomas, and F. G. Wood, eds. 1986. *Dolphin cognition and behavior: A comparative approach.* Hillsdale, N.J.: Lawrence Erlbaum.

Sielfeld, W. 1980. Mamíferos marinos en colecciones y museos de Chile. *Anal. Inst. Patagonia, Punta Arenas* (Chile) 11:273–80.

———. 1983. *Mamíferos marinos de Chile.* Santiago: Ediciones de la Universidad de Chile.

Slijper, E. J. 1962. *Whales.* London: Hutchinson.

Stewart, B. S., and S. Leatherwood. 1985. Mink whale *Balaeonoptera acutorostrata* Lacépède, 1804. In *Handbook of marine mammals*, 3:91–136. New York: Academic Press.

Tamayo, M., and D. Frassinetti. 1980. Catálogo de los mamíferos fósiles y vivientes de Chile. *Mus. Nac. Hist. Nat., Chile* 37:323–99.

Williamson, G. R. 1975. Minke whales off Brazil. *Sci. Repts. Whales Res. Inst.* 27:37–59.

Wursig, B., and R. Bastida. 1986. Long-range movement and individual associations of two dusky dolphins (*Lagenorhynchus obscurus*) off Argentina. *J. Mammal.* 67(4): 773–74.

Ximénez, A., A. Langguth, and R. Praderi. 1972. Lista sistemática de los mamíferos del Uruguay. *Anal. Mus. Nac. Hist. Nat. Montevideo* 7(5): 1–45.

9 Order Perissodactyla (Odd-toed Ungulates)

Diagnosis

These ungulates characteristically have an enlarged middle digit on both the forefeet and hind feet. The major weight-bearing axis is on the middle digit, in contrast to the Artiodactyla, where the major toes are reduced to two and the axis of weight passes between them. Among the Perissodactyla toe reduction reaches its most extreme form in the horse family (Equidae), where there is only one functional toe. The digestive system does not show extreme modification of the stomach, but the cecum, a blind pouch at the union of the small and large intestines, is enlarged and serves as a fermentation chamber. Tooth reduction is not pronounced, and the incisors are retained. Horns in extant forms, if they are developed at all, occur as epidermal derivatives and are situated in the midline of the nasal bones (e.g., Rhinocerotidae).

Distribution

The Recent distribution includes South America, Africa, Europe, and Asia. Recently extinct in North America (ca. 8,000 B.P.), the order was reintroduced by humans in the form of the domestic horse and ass (Equidae).

History and Classification

There are three extant families, the Rhinocerotidae, the Equidae, and the Tapiridae. The last, the tapirs, is the only extant family in South America, although the horse has been reintroduced by Europeans. This order has had a long evolutionary history, first appearing in the Eocene. It radiated rapidly in Asia and North America to include twelve families by the end of the Eocene, but their diversity diminished until only four families existed in the late Miocene.

FAMILY TAPIRIDAE
Genus *Tapirus* Brunnich, 1772
Tapirs, Dantas, Antas

Diagnosis

The dental formula is I 3/3, C 1/1, P 4/3–4, M 3/3 (fig. 9.1). Head and body length averages 2 m; the short tail is less than 10 cm. The nasal bones of the skull are short, and the animal has a distinct proboscis formed from the nostrils and the upper lip, which overhangs the lower lip. The forefeet bear four toes, although a vestige of the thumb can be detected on dissection. The hind foot bears three toes (plate 12) (see also Eisenberg, Groves, and Mackinnon 1987).

Distribution

Species of this family were once widely distributed in North America and Asia. Tapirs crossed into South America during the Pliocene; they persist to-

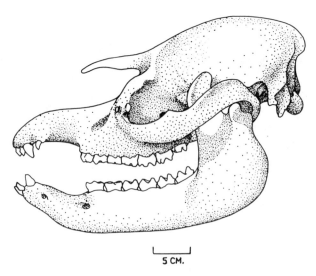

5 CM.

Figure 9.1. Skull of *Tapirus*.

day in South and Central America and in Southeast Asia.

• *Tapirus terrestris* (Linnaeus, 1758)
Tapir, Anta, Danta

Measurements

	Mean	Min.	Max.	N	Loc.	Source[a]
TL	2,142.1	1,803	2,570	14	Br, cap, P, V	1
HB	2,142.4	1,727	2,576	9		
T	82.6	60	100	5		
HF	144.0	140	152	3		
E	126.5	120	146	4		
Wta	177.2 kg	112 kg	220 kg	9		

[a](1) Allen 1916; Mallinson 1969; Mondolfi 1971; Wilson and Wilson 1973a,b; PCorps.

Description

The tapir is the largest land mammal in southern South America and is unmistakable because of its size and shape. It is an ungainly-looking ungulate with a robust body set on proportionally short, slender legs. Tapirs stand an average of 1,048 mm at the shoulder (range 955–1,128; $n = 6$). The tail is short. The large head is equipped with the tapir's most distinctive feature, an elongated, flexible nose that resembles a short trunk (plate 12). *T. terrestris* can be differentiated from other tapirs by its more pronounced mane, though it is still short, and more pronounced sagittal crest. Newborn tapirs have horizontal stripes and dots of yellowish white; this coloration persists past seven months, disappearing last from the legs. Adults are dull brownish (Hershkovitz 1954; Wilson and Wilson 1973a,b; pers. obs.).

Distribution

Tapirus terrestris is found from Venezuela south through Paraguay to northern Argentina. Its range has contracted in Argentina because of hunting and habitat destruction—formerly it ranged south into Tucumán province, but it is now found only across the northernmost fringe of the country (Hershkovitz 1954; Ojeda and Mares 1982; UConn) (map 9.1).

Life History

The gestation period of this tapir is about 383 days; only a single young is born, and the average birthweight for four young was 2.2 kg. Tapirs in captivity can reproduce for the first time at two years, though three years is the average, and can have young every fourteen months (range 14–28; $n = 4$). Both in captivity and in the wild there appears to be no seasonality to reproduction. In captivity one individual lived thirty-two years (Crespo 1982; Hershkovitz 1954; Mallinson 1969; Wilson and Wilson 1973a,b).

Ecology

Tapirs have usually been thought of as animals of the moist tropical forest, but in southern South America they are found in xeric parts of the Paraguayan and Argentine Chaco. They are also found in the tropical forest of northeastern Argentina and at up to 2,000 m in Jujuy province. In the Brazilian state of Mato Grosso they are most abundant in the gallery forests and in low-lying deciduous and secondary forests and have been calculated at a crude biomass of 96.4 kg/km².

Tapirs are browsers and frugivores. When water

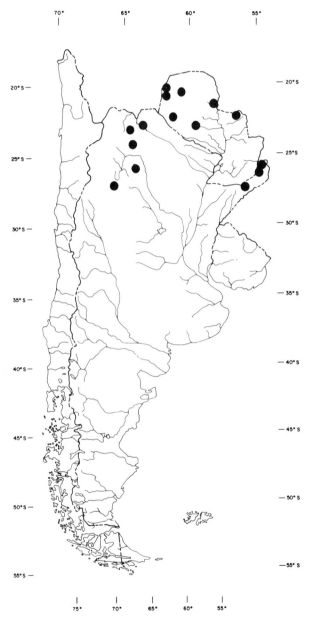

Map 9.1. Distribution of *Tapirus terrestris*.

is available they frequently defecate in it, and their large globular feces are obvious.

In captivity tapirs were found to produce four vocalizations: a shrill fluctuating squeal during pain and fear responses; a sliding squeal during exploratory behavior, which may also serve as a contact call; a clicking noise; and the characteristic snorting noise that is frequently heard in the wild when an animal has detected a human and is about to flee. They are solitary (Hunsaker and Hahn 1965; Miller 1930; Olrog 1979; Richter 1966; Schaller 1983; pers. obs.).

References

Allen, J. A. 1916. Mammals collected on the Roosevelt Brazilian expedition, with field notes by Leo E. Miller. *Bull. Amer. Mus. Nat. Hist.* 34: 559–610.

Crespo, J. A. 1982. Ecología de la comunidad de mamíferos del Parque Nacional Iguazú, Misiones. *Rev. Mus. Argent. Cienc. Nat. "Bernardino Rivadavia," Ecol.* 3(2): 45–162.

Eisenberg, J. F., C. P. Groves, and K. Mackinnon. 1987. Tapire. In *Grzimeks Encyclopädie*, vol. 4, *Säugetiere*, 598–609. Munich: Kindler.

Hershkovitz, P. 1954. Mammals of northern Colombia, preliminary report no. 7: Tapirs (genus *Tapirus*), with a systematic review of American species. *Proc. U.S. Nat. Mus.* 103:465–96.

Hunsaker, D., II, and T. C. Hahn. 1965. Vocalizations of the South American tapir *Tapirus terrestris. Anim. Behav.* 13:69–78.

Mallinson, J. J. C. 1969. Reproduction and development of Brazilian Tapir *Tapirus terrestris. Dodo* 6:47–51.

Miller, F. W. 1930. Notes on some mammals of southern Mato Grosso, Brazil. *J. Mammal.* 11: 10–22.

Mondolfi, E. 1971. La danta o tapir. *Defensa Nat.* 1(4): 13–19.

Ojeda, R. A., and M. A. Mares. 1982. Conservation of South American mammals: Argentina as a paradigm. In *Mammalian biology in South America*, ed. M. A. Mares and H. H. Genoways, 505–22. Pymatuning Symposia in Ecology 6. Special Publication Series. Pittsburgh: Pymatuning Laboratory of Ecology, University of Pittsburgh.

Olrog, C. C. 1979. Los mamíferos de la selva húmeda, Cerro Calilegua, Jujuy. *Acta Zool. Lilloana* 33:9–14.

Richter, W. von. 1966. Untersuchung über angeborene Verhaltensweise den Shabrackentapirs und Flachland Tapirs. *Zool. Bieträge* (Nuremberg) 1(12): 67–159.

Schaller, G. B. 1983. Mammals and their biomass on a Brazilian ranch. *Arq. Zool. São Paulo* 31(1): 1–36.

Wilson, R., and S. Wilson. 1973a. *Tapirs in captivity, 1973.* Claremont, Calif.: Tapir Research Institute.

———. 1973b. *Tapirs in the United States.* Claremont, Calif.: Tapir Research Institute.

10 Order Artiodactyla (Even-toed Ungulates)

Diagnosis

These mammals are especially adapted for feeding on fallen fruit and nuts or on grasses and leaves. In New World forms the cheek teeth may be either high crowned or low crowned. In the latter case, the species either is omnivorous (e.g., Tayassuidae) or is adapted for browsing rather than grazing and browsing (e.g., Cervidae, Bovidae). The premolars are usually simple in structure compared with the molars. The grazing and browsing forms exhibit a loss of the upper incisors. Enlarged canines or tusks are present in the Suiformes, Tragulidae, and some species of Cervidae. The main axis of weight passes between the third and fourth digits of the foot; the first, second, and fifth digits are reduced or lost in the more specialized forms. A corresponding reduction in metacarpals and metatarsals to some extent parallels the reduction in toe number. In males of many species of this order, bony excrescences develop on the forehead (Cervidae and Bovidae). Generally there are two types of hornlike structures, either antlers that are cast annually or true horns, which are an extended growth of the frontal bone covered with an epidermally derived cap.

Distribution

The Recent natural distribution is worldwide except for the Australian area, the Oceanic islands, and Antarctica. Domesticated and wild members of this order have been introduced into Australia and New Zealand.

History and Classification

The earliest artiodactyls appear in the early Eocene in both North America and Europe. This group rapidly differentiated into three specialized forms that are reflected in the subordinal classification of the Artiodactyla: the Suiformes, today represented by the hippopotamuses, peccaries, and swine; the Tylopoda, represented by the family Camelidae; and the suborder Ruminantia, which today includes the giraffes (Giraffidae), pronghorn antelopes (Antilocapridae), mouse deer (Tragulidae), musk deer (Moschidae), true deer (Cervidae), and bovines (Bovidae).

The artiodactyls had their origin in the northern continents, and since North America and Asia have been in contact off and on, faunal interchange was possible. Faunal interchange between the Northern Hemisphere and Southern Hemisphere increased as Africa became connected to Eurasia. The artiodactyls did not appear in South America until the late Pliocene, when the Panamanian land bridge became complete, linking North and South America. All three suborders are now represented in South America; they have arrived and differentiated only since the Pliocene. The three families in South America at present are the Tayassuidae, or peccaries; the Camelidae, including the vicuna and the llama; and the Cervidae, including some eleven species currently grouped into six genera. Members of the family Bovidae did not reach South America until Europeans transported them in the sixteenth century. The Cervidae underwent a rapid adaptive radiation in the Pleistocene of South America to fill some niches that are occupied by bovines on the more contiguous continental land masses (Eisenberg 1987; Eisenberg and McKay 1974).

FAMILY TAYASSUIDAE

Peccaries, Pecarís, Báquiros

Diagnosis

These piglike Artiodactyla have a dental formula of I 2/3, C 1/1, P 3/3, M 3/3. The canines are modified into tusks but are small; the upper canines are directed downward and rub against the lower ones, thus maintaining a sharp cutting edge (fig. 10.1). The molars are low crowned. Head and body length ranges from 700 to 1,000 mm, the tail averages 20 mm, and the weight averages about 24 kg. There are four visible toes on the forefeet and three on the hind feet, but only two toes on each foot bear the weight (see ordinal account). Although the stomach is modified into three chambers (Langer 1979), peccaries are not ruminants.

Distribution

Species of this family are distributed from Texas, in the United States, to northern Argentina. There are currently three species, all occurring within the area covered by this volume. Two species are grouped into the genus *Tayassu*, including *T. tajacu* and *T. pecari*. The second genus, *Catagonus*, is restricted to Paraguay and adjacent parts of Argentina. Although described as a fossil, it was rediscovered as a living form by Wetzel in 1974 (Wetzel et al. 1975; Wetzel 1977a,b).

Genus *Catagonus* Ameghino, 1904
Catagonus wagneri (Rusconi, 1930)
Chacoan Peccary, Pecarí Quimilero, Taguá

Measurements

	Mean	Min.	Max.	N	Loc.	Source[a]
TL	1,081.4	957.0	1,161.0	24 m	P	1
	1,100.5	1,030.0	1,170.0	24 f		
T	59.7	24.0	102.0	25 m		
	64.2	45.0	100.0	24 f		
HF	230.3	206.0	250.0	26 m		
	223.3	222.0	257.0	24 f		
E	112.6	100.0	122.0	29 m		
	114.2	100.0	120.0	28 f		
Wta	34.8 kg	29.5 kg	40.0 kg	18 m		
	34.6 kg	30.5 kg	38.5 kg	10 f		

[a](1) Mayer and Brandt 1982.

Description

Males of *Catagonus* stand 567.8 mm at the shoulder (range 520–690; n = 22). *Catagonus* differs from the other peccaries in the following characteristics: it is larger, has longer dorsal fur, has no dewclaws, and has a longer, more concave rostrum. Long bristles give the animal a shaggy appearance. The head is larger, and the ears, legs, and tail are longer (plate 13). The Chacoan peccary is colored very much like *Tayassu tajacu*, with a grizzled, brownish gray coat, a dark middorsal stripe, and a faint collar of white hairs that crosses up over the shoulders. The hair on the ears and legs is longer and paler than in *Tayassu* (Mayer and Brandt 1982; Wetzel 1977b). Chromosome number: $2n = 20$ (Benirschke and Kumamoto 1989; Wetzel et al. 1975b).

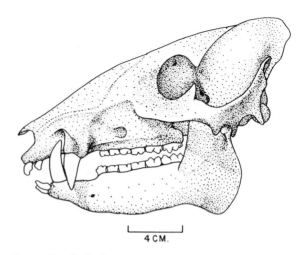

Figure 10.1 Skull of *Tayassu tajacu*.

Distribution

Catagonus wagneri is found in the Gran Chaco of Paraguay, Argentina, and Bolivia. In Paraguay it is found throughout the Chaco, and in Argentina it has been collected in Chaco, Santiago del Estero, and Salta provinces (M. Ludlow, unpubl. data; Olrog, Ojeda, and Bárquez 1976) (map 10.1).

Life History

The average number of embryos per female is 2.7 (range 2–4; mode 2), and the average litter size is 2.5 (range 1–4; mode 2–3). Mating occurs from April to May, and farrowing takes place from early September to early December. Females have four pairs of mammae. Young achieve adult coloration at between three and four months of age (M. Ludlow, unpubl. data; Mayer and Brandt 1982).

Ecology

Catagonus is found in the xeric Chaco of southern South America in areas of high temperature and low rainfall. It is most prevalent in low to moderate stature thorn forests of the central Paraguayan Chaco, where it can reach densities of 9.24/km². Herds are stable social units, composed of males and females, and average 3.8 members (range 1–10; n = 148). During winter *Catagonus* is active diurnally, between the hours of 0800 and 1100. Animals are inactive at night except during the hottest and coldest weather, when they are active for one to two hours in the middle of the night.

Like other peccaries, *Catagonus* does a great deal of scent marking using the large dorsal gland, from which scent can be rubbed or squirted. Scent marking is more prevalent in areas of high herd use. Members of the herd also frequently scent mark each other. Defecation is usually done at scat stations.

In the central Paraguayan Chaco, home ranges of

three groups of *Catagonus* had a stable core area of about 6 km², but with excursions they covered an area of 15 km². These ranges are exclusive and probably are territories.

During the winter cacti form the major dietary item, though bromeliads are also eaten. Soil is frequently eaten from the ground, at "salt licks," or from leaf-cutter ant (*Atta*) mounds.

Catagonus is heavily hunted throughout its range, for meat and to a limited extent for its hide. Curiosity, use of salt licks, and a tendency to remain near fallen herd members all contribute to the ease with which this species can be killed. The distribution of

Catagonus appears to be patchy and discontinuous, perhaps owing to the intense hunting that is taking place as the Chaco is developed (Ludlow, unpubl. data; Mayer and Brandt 1982; Mayer and Wetzel 1986; Sowls 1984; A. Taber, unpubl. data; pers. obs.).

Comment

The living Chacoan peccary was unknown to Western science until 1974, when it was discovered in the Paraguayan Chaco by Wetzel and his colleagues.

Genus *Tayassu* G. Fisher, 1814
• *Tayassu pecari* (Link, 1795)
White-lipped Peccary, Pecarí Labiado

Measurements

	Mean	Min.	Max.	N	Loc.	Source[a]
TL	1,126.4	961.0	1,390.0	15 m	P	1
	1,059.9	905.0	1,250.0	14 f		
T	37.9	10.0	55.0	14 m		
	43.0	20.0	65.0	14 f		
HF	215.4	165.0	233.0	15 m		
	217.4	181.0	250.0	14 f		
E	78.5	73.0	85.0	11 m		
	74.8	64.0	80.0	5 f		
Wta	29.3 kg	25.0 kg	34.0 kg	6	Br, P	2
	33.5 kg	20.5 kg	49.5 kg	13	P	3

[a](1) Mayer and Brandt 1982; (2) Allen 1916; Husson 1978; Mayer and Brandt 1982; Schaller 1983; PCorps; (3) K. Hill, unpubl. data (only individuals > 15 kg included).

Description

Adult white-lipped peccaries stand about 530 mm at the shoulder ($n = 1$). They are easily distinguished from the other peccaries by their coloration. Adult animals are dark brown or black, with a sharply contrasting blaze of white on the lower jaw that extends up onto the cheeks. Some animals have additional white near the tip of the muzzle and under the eye as well as in the pelvic region. The ears have white hairs on the insides, and the legs are grizzled black and tan. Young are born reddish and become mixed black and tan as they mature, but adult coloration is not achieved until the second year. Because of the strikingly different coloration, immature animals have been mistaken for another species (Mayer and Brandt 1982; Sowls 1984).

Chromosome number: $2n = 26$ (Sowls 1984).

Distribution

Tayassu pecari is found from southern Mexico to northern Argentina. In southern South America it is found in both eastern and western Paraguay and across the northern provinces of Argentina (Honacki, Kinman, and Koeppl 1982; Olrog and Lucero 1981; UM) (map 10.2).

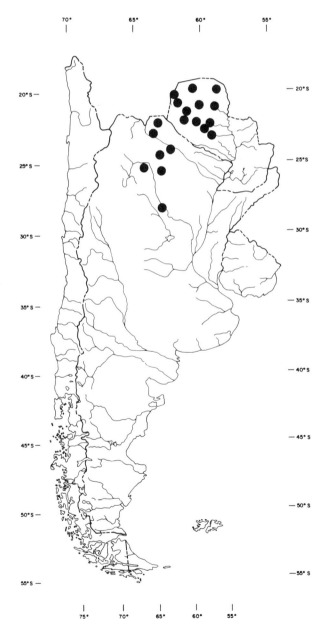

Map 10.1. Distribution of *Catagonus wagneri*.

Life History

White-lipped peccaries usually produce twins after a gestation period of about 158 days. In captivity there appear to be two annual breeding periods, and females reproduce for the first time at about eighteen months. In the Brazilian Pantanal a female was seen with young in September; in Peru mating was observed between July and August, and in the Paraguayan Chaco two females with two fetuses each were collected in July (Miller 1930; Roots 1966; Sowls 1984; UConn).

Ecology

White-lipped peccaries, though usually thought of as animals of the moist tropical forest, are also found in the seasonally xeric parts of southern South America. In the Brazilian Pantanal this species was found at a crude biomass of 44.0 kg/km², and in the Paraguayan Chaco they are found at a density of 1.06/km². The only field study of this species, done in the Peruvian rain forest, concluded that if indeed white-lipped peccary herds had home ranges at all they were probably between 60 and 200 km².

Tayassu pecari forms the largest herds of any South American terrestrial mammal. There are few good counts of herd size: in Peru the minimum number of individuals in five herds averaged 106 (range 90–138); in Mato Grosso, Brazil, one herd of at least 200 was counted; in Misiones province, Argentina, one group was estimated at 250; in the Paraguayan Chaco groups averaged much smaller, with 25–60 animals per herd. Work in Peru has demonstrated that these large groups of peccaries move over large areas in what may be a seasonal pattern, utilizing large patches of mast fruiting palms. Anecdotal evidence, particularly from indigenous hunters, corroborates this pattern of movement, and further evidence comes from southeastern Brazil, where it was reported that a herd used an area extensively and then did not return for six months. Kiltie and Terborgh (1983) have suggested that the large group size may be due to use of patchy food resources by white-lipped peccaries and to the ease with which a herd of peccaries can be found by predators, so that a large group can be a mutual defense system.

The herds are composed of both males and females. *T. pecari* has a more elaborate repertoire of vocalizations and a closer affiliation to other herd members than does *T. tajacu*; there is a great deal of vocalizing and reciprocal scent-gland rubbing within a herd. In fact herds are often found by their noise and strong musky smell. An alerted herd is often detected by the loud clacking the canines produce when rubbed together.

In the Paraguayan Chaco white-lipped peccaries eat *Prosopis* pods and other plant material; in Brazil stomachs have been found to contain figs and palm nuts (*n* = 3); and in Peru the stomachs of thirty-four animals contained reproductive plant parts, especially palm seeds (61% volume, 97% occurrence), vegetative plant parts (39% volume, 97% occurrence, and animal parts, including snail opercula, tissue from vertebrates, and parts of invertebrate adults and larvae (trace % volume, 82% occurrence). *T. pecari* is crespecular and nocturnal in Paraguay and is known to be a prey of pumas. It is one of the most common mammals taken by indigenous hunters throughout its range (Crespo 1982; Davis 1947; Kiltie 1981a,b; Kiltie and Terborgh 1983; Mayer and Brandt 1982; Mayer and Wetzel 1987; Miller 1930; Redford and Robinson 1987; Schaller 1983; Sowls 1984; Yost and Kelley 1983).

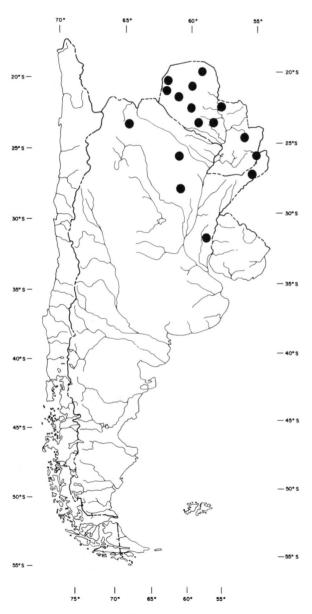

Map 10.2. Distribution of *Tayassu pecari*.

• *Tayassu tajacu* (Linnaeus, 1758)
Collared Peccary, Pecarí de Collar

Measurements

	Mean	Min.	Max.	N	Loc.	Source[a]
TL	931.7	840	1,040.0	45 m	P	1
	940.3	788	1,060.0	36 f		
T	38.6	10	106.0	42 m		
	38.3	10	60.0	36 f		
HF	196.4	163	253.0	45 m		
	196.6	170	235.0	36 f		
E	82.4	27	98.0	27 m		
	77.9	27	93.0	23 f		
Wta	22.3 kg	21 kg	23.0 kg	3 m		
	19.4 kg	17 kg	23.0 kg	9	Br, P	2
	19.7 kg	16 kg	23.5 kg	8	P	3

[a](a) Mayer and Brandt 1982; (2) Mayer and Brandt 1982;
Schaller 1983; PCorps; (3) K. Hill, unpubl. data (only animals >
15 kg included).

Description

The canine teeth grow continuously for the first
four to five years. Males stand about 4,500 mm at
the shoulder; there is no obvious external sexual dimorphism, though in the United States males are
slightly heavier than females. There are four front
and three hind digits. The coloration of the collared
peccary is very similar to that of *Catagonus*, but it is
smaller and has shorter hair and darker legs (plate
13). The young achieve adult coloration by two to
three months.
Chromosome number: $2n = 30$ (Mayer and Brandt
1982; Sowls 1984).

Distribution

Tayassu tajacu is found from the southern United
States south into northern Argentina. In southern
South America it is found in eastern and western
Paraguay and at present in the northern Argentinian
provinces south to at least Tucumán (Lucero 1983;
Sowls 1984; BA, UM) (map 10.3).

Life History

In the United States the collared peccary breeds
year round, and females exhibit a postpartum estrus. Females conceive at the end of their first year;
the gestation period is 145 days, and litter size varies
from one to four with a mode of two. Young are born
averaging 665 g ($n = 10$) and nurse for six to eight
weeks. In Paraguay the average litter size is similar
(range 1–4; average 1.98). In northeastern Argentina collared peccaries reproduce from August to
October and produce litters of one or two, though in
Amazonian Peru there is no evidence of seasonal
breeding. In Brazil one litter was found in a *Priodontes* burrow. In both wild and captive herds,
young will nurse from females other than their own
mothers (Byers and Beckoff 1981; Crespo 1982;
Lochmiller, Hellgren, and Grant 1984; Mayer and
Brandt 1982; Miller 1930; Sowls 1984).

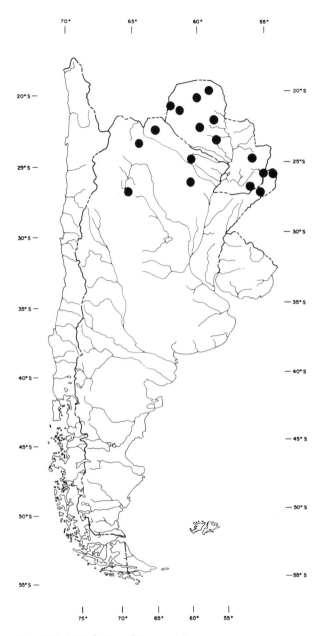

Map 10.3. Distribution of *Tayassu tajacu*.

Ecology

Collared peccaries occur in a wide variety of
habitats from moist tropical forest to the xeric Paraguayan Chaco. In the Brazilian Pantanal they occurred at a crude biomass of 14.1 individuals per
square kilometer, and in Peru there was an estimate
of 1.2 groups per square kilometer. They can be active at any time during the day or night depending
on the weather, season, and availability of food. *T.
tajacu* will hide in armadillo burrows or fallen tree
trunks, and animals are frequently driven into these
shelters by dogs belonging to hunters, who then dig
or chop them out.

Mixed-sex bands move along trails, some of which
may be traditional. There does not appear to be any

one leader of the band, and the members usually disperse while feeding. Group size varies geographically, with an average in Arizona, United States, of 7.7, in Texas, United States, of 13.6, in the llanos of Venezuela of 6.5 ($n = 65$), in the Brazilian Pantanal of 3.3, and in Peru of 5.6 ($n = 18$). The basic social unit is a herd of one or more males and several females with young. The herd may divide into smaller groups or may, in some circumstances, temporarily combine with other herds. This latter grouping, termed an aggregation, seems to occur most commonly when there is a great abundance of food, such as mast fruiting of mesquite or palm trees. Aggregations of more than fifty animals have been seen.

Collared peccary home ranges seem to be fairly small, and in Arizona they are defended against neighboring herds. In the Venezuelan llanos, dry-season home ranges varied from 38 to 45 ha and wet-season home ranges from 100 to 126 ha. The core area of home ranges was associated with a permanent defecation station. In Costa Rica, dry-season home ranges were larger than those during the wet season. Wallows and mineral licks are frequently used. As with other peccaries, there is a great deal of mutual rubbing within a herd, usually oriented toward the dorsal scent gland. Herd cohesiveness is maintained through auditory and olfactory signals.

Tayassu tajacu eats fruits, underground tubers, green grass, shoots of plants, fruit and stems of cacti, and growing points of bromelaids and agave. They will dig shallow holes in search of underground plant parts and buried seeds. In Peru a study of seventeen stomachs revealed reproductive plant parts (71% volume, 100% occurrence), vegetative plant parts (29% volume, 94% occurrence), and animal parts (trace % volume, 82% occurrence) (Byers and Beckoff 1981; Castellanos 1983, 1986; Crespo 1982; Díaz 1978; Eddy 1961; Green and Grant 1984; Kiltie 1981a; Kiltie and Terborgh 1983; Mayer and Brandt 1982; McCoy and Vaughan 1986; Miller 1930; Robinson and Eisenberg 1985; Schaller 1983; Schweinsburg 1971; Smythe 1970; Sowls 1984).

FAMILY CAMELIDAE

Diagnosis

These artiodactyls have a selenodont dentition, and the upper incisors and canines are retained. The dental formula for the genus *Lama* is I 1/3, C 1/1, P 3/1–2, M 3/3 (fig. 10.2). They are ruminants with a three-chambered stomach. The toes bear nails rather than hooves, and in the extant species there are no hornlike structures on the head.

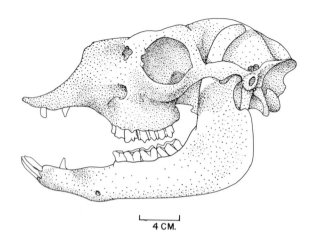

Figure 10.2. Skull of *Lama guanicoe*.

Distribution

The family originated in North America but has survived to the present in the Old World (Africa and Asia) and in South America. The two extant wild species in South America occur in Peru, Bolivia, Chile, and Argentina.

Genus *Lama* G. Cuvier, 1800
Lama guanicoe (Muller, 1776)
Guanaco

Measurements

	Mean	S.D.	N	Loc.	Source[a]
TL	1,898.0	64	8 m	A	1
	1,915.0	78	11 f		
HF	508.0	15	24 m		
	494.0	15	17 f		
E	132.0	8	9 m		
	134.0	11	11 f		
Wta	118.7 kg	91	8 m		
	121.3 kg	143	11 f		

[a](1) Raedeke 1979.

Description

Guanacos are large ungulates with slender necks, long legs with broad, modified hooves, and thick, long pelage. They stand between 1,100 and 1,200 mm at the shoulder. The head is camellike, with long, pointed ears and deeply cleft, highly mobile lips. Males have much longer canines than females. The general color of guanacos is reddish brown, with a grayish head and white underparts. Guanacos are distinguishable from vicunas by being larger, by having callosities on the inner sides of the forelimbs, and by lacking the characteristic white or yellowish bib of the vicuna (plate 13) (Franklin 1982; Miller, Rottman, and Taber 1973; Osgood 1943; Raedeke 1979).

Distribution

The guanaco was originally found from northern Peru, and perhaps even southern Colombia, south to the southern tip of Chile and throughout Argentina. Now it is found from southern Peru south along the Andean zone of Chile and Argentina and thence to Tierra del Fuego and Navarino Island. It also occurs in far western Paraguay. It is also found from the mountains to the ocean in southern Argentina south of about 40° (Gilmore 1950; Honacki, Kinman, and Koeppl 1982; Olrog and Lucero 1981; Tamayo and Frassinetti 1980; Tonni and Politis 1980; UConn) (map 10.4).

Life History

In southern Argentina there is a restricted breeding season with the peak in February. Females are apparently induced ovulators. The birth season is in the spring months of December to mid-February, after a gestation period of about eleven months. The single young is born weighing 8–15 kg, and lactation lasts eleven to fifteen months. With good nutrition females can breed in their second year, though full adult body size is not achieved until about three years (Franklin 1983; Raedeke 1979).

Ecology

The historical range of guanacos was much larger than the current range. Hunting, habitat destruction, and apparently climatic change all contributed to driving the herds off the lower pampas and into foothills and mountains. The major exception is at the southern tip of the continent, where guanacos are still reasonably common. Guanacos inhabit both warm and cold grasslands and shrublands from sea level to over 4,000 m. In some areas they inhabit forests during the winter. The vicuna, confined to the altiplano above 3,500 m, is much less catholic in its habitat requirements than the guanaco, which can occupy drier habitat because it can go for long periods without drinking. Guanacos are also more versatile foragers, since they browse and graze whereas vicunas only graze. In southern Argentina 61.5% of the food guanacos ate was grass and 15.4% was browse.

Guanacos are found in three principal types of social groups: family bands, male troops, and solitary males. A family band consists of a single breeding male with several females and their young. In southern Argentina the average band consisted of 5.5 females, and the largest had 18 females and a total of 25 individuals. Band males limit the size of their groups by driving off females who try to join and driving out young males at six to twelve months of age. Each family group occupies a territory de-

fended by the male. Family bands occupy the best available habitat and use communal defecation piles that can be quite large and serve to demarcate their territory.

Herd males compose only about 18% of the total number of males in a southern Argentine population of guanacos. The rest either are solitary or form troops of males. The solitary males are mostly sexually mature. The male troops are not stable in size or composition and occupy peripheral habitat. Males spend their first three to four years in all-male groups, where they develop fighting ability using

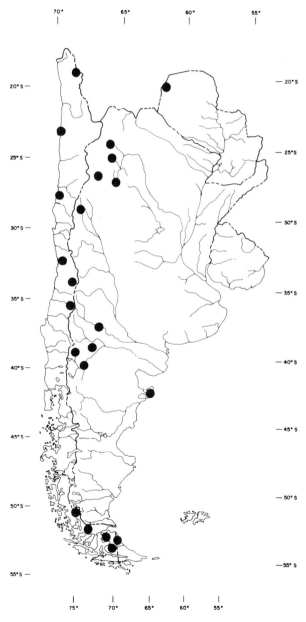

Map 10.4. Distribution of *Lama guanicoe*.

"chest rams" and play fighting. Guanaco fighting is similar to that of equids in that fights proceed in an unpredictable manner, with opponents trying to unbalance one another and bite each other's legs.

Guanacos exhibit a very flexible social structure that varies from habitat to habitat. In some areas fixed territories are defended throughout the year and in others the animals migrate seasonally, spending the winter in more sheltered areas. In San Juan province, Argentina, territories are defended from December to April; outside this time animals combine into larger groups that move into protected areas during the winter.

Young guanacos are particularly vulnerable to predation by pumas. In southern Argentina, where the puma has been reduced in numbers, starvation is a major source of mortality. Before European contact guanacos were very important to several of the indigenous peoples, who based their lives on pursuit of the large herds. Guanacos have remained an important source of income to settlers: between 1976 and 1984, skins exported from Argentina totaled 401,193 (Cajal 1983b, 1986; Franklin 1982, 1983; Miller, Rottman, and Taber 1973; Raedeke 1979; Tonni and Politis 1980; Torres 1985; Wilson 1984; Wilson and Franklin 1985).

Genus *Vicugna* Lesson, 1842
Vicugna vicugna (Molina, 1782)
Vicuna, Vicuña

Description

The lower incisors are very long and slender and have open roots. This is the only ungulate reputed to have continuously growing incisors. Adults weight from 45 to 55 kg. This camelid is similar to the guanaco but is smaller, with finer pelage and a conspicuous white or yellowish "apron." The body is a uniform rich cinnamon with or without a long white chest bib and with a white venter (plate 13) (Franklin 1980, 1983; Osgood 1943).

Distribution

The vicuna is found from southern Peru to northern Chile and northwestern Argentina. In Chile it is found only in Tarapacá and Antofagasta provinces, and in Argentina it is found in the high Andes of the five northwestern states (Boswell 1972; Cajal 1983b, d; Honacki, Kinman, and Koeppl 1982; Millter, Rottman, and Taber 1973; Pefaur et al. 1968) (map 10.5).

Life History

Vicunas are seasonal breeders, with births taking place in the season of greatest plant productivity. A single young is born after a gestation period of 330–50 days, and females breed within two weeks of giving birth. Lactation lasts six to eight months. The age of first reproduction for females is twelve to fourteen months, though most do not breed until two years of age (Franklin 1980, 1983).

Ecology

Vicunas are found only on the altiplano above about 3,500 m. They inhabit areas of open vegetation and on the best range can reach densities of twenty-one per square kilometer. Unlike guanacos, vicunas never migrate and exhibit year-round territorial defense, though there are reports of animals' moving off their territories to graze on other pastures. Sleeping territories, in sheltered areas, are separate from feeding territories.

Vicunas are polygnous and are found in family groups (one male, several females with offspring), as

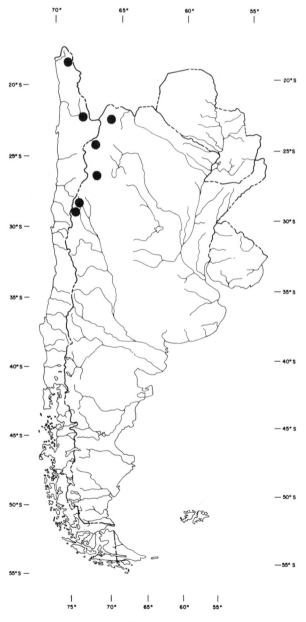

Map 10.5. Distribution of *Vicugna vicugna*.

solitary males, or in all-male groups. The average family group is composed of a single breeding male, three females, and two young. The largest group recorded by Franklin consisted of nineteen animals. The breeding group occupies a feeding territory that usually includes water and is defended by the male against other individuals. Some family groups occupy peripheral habitat with no water and may attempt to intrude on established territories. Family group size is strongly related to forage production on the territory. Young males and females are expelled from family groups before they are one year old. There is a very constant group size and composition, and females show little independent behavior.

Solitary males are sexually mature and are attempting to establish territories. Male groups are usually composed of five to ten animals but may have more than a hundred; 50% to 80% of the males in a population are in male groups. Males generally become resident at three to four years and are able to defend a territory for at least six years.

Vicunas are grazers, feeding mainly on grass, and their continuously growing incisors seem to be an adaptation for close cropping of small forbs and perennial grasses. In one study of a Peruvian population the diet consisted of two-thirds monocots and one-third dicots. Vicunas of both sexes, all ages, and all social groups defecate only on traditional dung piles.

Many unsuccessful attempts have been made to domesticate the vicuna because of the potential value of its wool. The Incas reserved this wool for royalty and high officials and managed the wild vicunas by engaging in huge roundups that sometimes covered an area 30 km in diameter. Old females and males were killed for meat and hides, while young adults were shorn and released. Since the Incas, the range of the vicuna has decreased because of hunting for the meat and the very valuable wool (Bosch and Svendsen 1987; Cajal et al. 1981; Cajal and Sánchez 1983; Franklin 1980, 1983; Gilmore 1950; Koford 1957; Ménard 1982, 1984; Miller, Rottman, and Taber 1973; Miller et al. 1983).

FAMILY CERVIDAE
Deer, Venados

Diagnosis
The dental formula is I 0/3, C 0/1, P 3/3, M 3/3 for deer found in South America. The upper incisors are lacking (fig. 10.3). Deer have four-chambered stomachs and are specialized for browsing. Deer are true ruminants, and after feeding they lie up in a sheltered area, eructate, and remasticate the rumen contents. The first, second, and fifth digits are greatly

reduced. Males grow antlers that are usually cast and renewed annually (but see species accounts; figs. 10.4 and 10.5). Adults show a uniform dorsal pelage color, but newborns are usually spotted (plates 12 and 13).

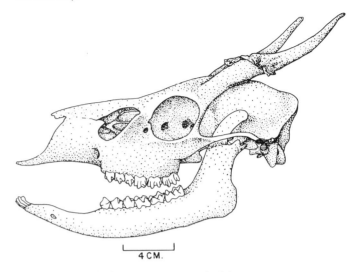

Figure 10.3. Skull of *Mazama americana* (male).

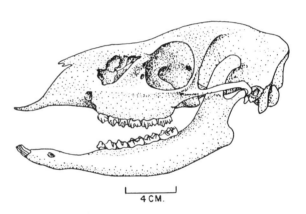

Figure 10.4. Skull of *Mazama americana* (female).

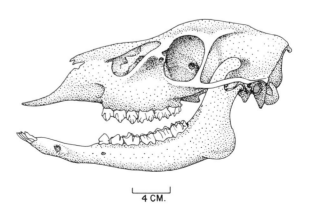

Figure 10.5. Skull of *Blastocerus dichotomus* (female).

Distribution

The Recent distribution of the deer family includes North and South America, Europe, and Asia; they are present in Africa only north of the Sahara and are absent from Australia and New Zealand except where introduced by man.

Natural History

Males of the species constituting the family Cervidae typically bear antlers that are usually shed annually and grown anew. The only exceptions are the musk deer (*Moschus*), often placed in their own family, and the Chinese water deer (*Hydropotes*) of Asia, which are antlerless. Most species are adapted for browsing or for mixed browsing and grazing. Deer originated in Europe and Asia and spread to North America in the Pliocene. Their spread was rapid, and upon entering South America in the late Pliocene they underwent an adaptive radiation.

Cervids evolved in the high latitudes of the Northern Hemisphere, and thus reproduction was highly synchronized to ensure that fawns would be born at a time of year that would favor their survival. In the North, antler growth of the males, breeding, and the timing of birth tends to be highly seasonal. When deer adapt to the equatorial latitudes births tend to be less sharply seasonal, as does antler casting in males (Brock 1965). Most studies of deer living near the equator have reported the following: males still cast their antlers annually, though in any given area the population may show weak synchrony (Brokx 1972; Branan and Marchinton 1987; Blouch 1987). Conception and birth of fawns, though not tightly synchronized, clearly are timed so that the birthdate favors survival. No sharp birth peak may be shown, because the factor determining survival is plant productivity, which in turn is tied to rainfall. In Suriname there are two peaks of rainfall. In both areas births, though scattered, still reflect a compromise that probably favors fawn survival (Brokx 1972; Blouch 1987; Branan and Marchinton 1987). In temperate South America, or at high altitudes, reproductive synchrony may be marked, paralleling the tendency in the Northern Hemisphere (Merkt 1987; Polvilitis 1979).

The small deer (*Pudu*, *Mazama chunyi*, and *Mazama rufina*) all live in dense underbrush. Their small stature and reduced antler size could be an adaptation for efficient locomotion in dense vegetative cover. Herbivorous mammals from different taxa worldwide have converged in body form and antipredator behavior to occupy this ecological niche (Bourlière 1973; Dubost 1968; Eisenberg and McKay 1974).

Genus *Blastocerus* Wagner, 1844
Blastocerus dichotomus (Illiger, 1815)
Marsh Deer, Ciervo de los Pantanos, Guazú Pucú

Measurements

	Mean	Min.	Max.	N	Loc.	Source[a]
TL	1,803.7	1,190	2,100	6	A, B, P, Peru	1
HB	1,637.7	1,069	1,910			
T	137.0	120	165	5		
E	196.3	165	215	4		
Wta	108.6 kg	89 kg	125 kg	5	Br	2

[a](1) Hofmann, Ponce del Prado, and Otte 1975–76; Miranda-Ribeiro 1919; FM; (2) Schaller 1983.

Description

The marsh deer is the largest of the South American deer and is unmistakable because of its size, its large ears and feet, and its heavy, multitined antlers (plate 12). These deer stand 1,000 to 1,200 mm at the shoulder and are bright rufous chestnut in summer and darker brown in the winter. The lower legs are darker, and there is a black band on the muzzle. The ears are conspicuously large and have white hairs on the inside. The antlers are very thick compared with their length (about 600 mm long) and normally have five points per side (Almeida 1976; Hofmann, Ponce del Prado, and Otte 1975–76; Whitehead 1972).

Distribution

Blastocerus is found from Amazonian Peru to central Brazil and south through Paraguay to northern Argentina. Its distribution within southern South America is limited to areas of suitable habitat in the Río Paraguay and Paraná basins. This species is apparently extinct in Uruguay (González Romero, Moreno Ortiz, and Calabrese 1978; Hofmann, Ponce del Prado, and Otte 1975–76; Honacki, Kinman, and Koeppl 1982; G. Schaller and A. Tarak, unpubl. data; Ximénez, Langguth, and Praderi 1972) (map 10.6).

Life History

In captivity does exhibit a postpartum estrus and stags may keep their antlers up to twenty-one months. In the Pantanal of Brazil the mating season is apparently very long: bucks with fully developed antlers were collected through most of the year, and the birth season is apparently from at least May to September. Fawns associate with their mothers until almost one year and may continue the association as yearlings (Almeida 1976; Frädrich 1987; Miller 1930; Schaller and Vasconcelos 1978).

Ecology

Blastocerus, true to its common name, is found only in areas with standing water and dense cover, though it may move away from these areas to feed. In the Brazilian Pantanal it prefers habitats of reeds or bunchgrass in or near water not over 60 cm deep. As the water levels change seasonally the deer move, inhabiting their preferred habitat. In the wet season they are dispersed widely, but in the dry season they are concentrated near the remaining water. In Formosa province, Argentina, marsh deer are found in dense stands of *Cyperus*, 2 m or more tall, during the dry season and in the surrounding grassland during the annual floods.

In different parts of its range, *Blastocerus* is reported to be diurnal, crepuscular, and nocturnal, probably changing with season and extent of hunt-

ing. This species is often solitary, though in some places small family groups are reported. Stomachs from five animals contained leaves, particularly those of water lilies, grass, and some browse. In another study the diet was found to consist 50% of grasses and 31% of legumes.

Marsh deer are very susceptible to cattle diseases, and their rarity is undoubtedly due to this as well as to hunting pressure and habitat destruction (González Romero, Moreno Ortiz, and Calabrese 1978; Hofmann, Ponce del Prado, and Otte 1975–76; Schaller 1983; G. Schaller and A. Tarak, unpubl. data; Schaller and Vasconcelos 1978; Tomas 1986 a,b,c; Voss et al. 1981).

Genus *Hippocamelus* Leukart, 1816
Huemul

Description
These medium-sized deer have a head and body length of 1.4 to 1.6 m. The males bear a bifurcate set of antlers that are cast annually. The dorsum is medium brown, contrasting with the white venter and white limbs (see plate 12 and fig. 10.6).

Distribution
The two species of this high-altitude genus are disjunctly distributed in the Andes of Peru, Bolivia, Chile, and Argentina.

Hippocamelus antisensis (d'Orbigny, 1834)
Huemul Andino, Taruca

Measurements

	Mean	Min.	Max.	N	Loc.	Source[a]
TL	1,385.7	1,280	1,460	7	A, Peru	1
HB	1,270.5	1,170	1,330	6		
T	116.2	100	130			
HF	335.0	320	365			
E	158.3	140	178			

[a](1) Crespo 1974; Pearson 1951.

Description
Hippocamelus antisensis is slightly smaller and paler than the southern species *H. bisulcus*. It stands about 730 mm at the shoulder and is generally sandy gray. The antlers branch once to form two points per side and are 200–300 mm long (fig. 10.6). Most individuals have a dark Y-shaped facial pattern. The throat and neck are whitish, as are the fronts of the forelegs. The tail has a dark patch at the base and is white on the underside (Pearson 1951; Roe and Rees 1976; Whitehead 1972).

Distribution
The species is found in the Andes from Ecuador to northwestern Argentina south to at least La Rioja

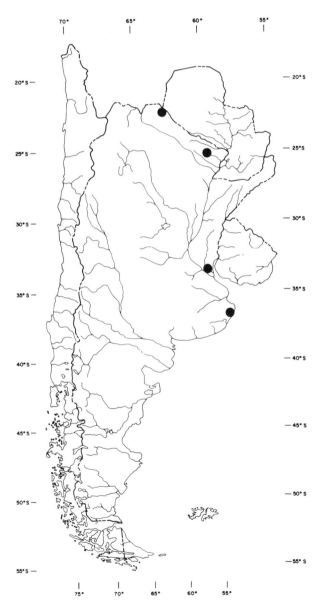

Map 10.6. Distribution of *Blastocerus dichotomus.*

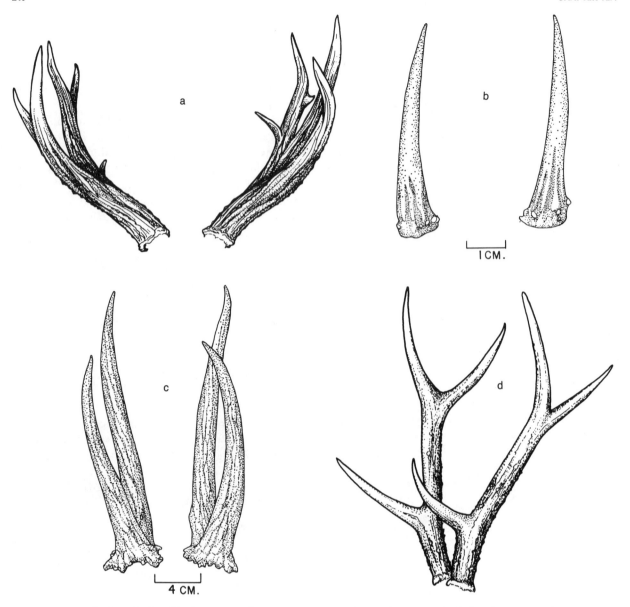

Figure 10.6. Antler form displayed by four genera of South American cervids: (*a*) *Blastocerus*; (*b*) *Pudu*; (*c*) *Hippocamelus*; (*d*) *Ozotoceros*.

province, and in Chile south probably to at least Tignamar (Cajal 1983a,d; Honacki, Kinman, and Koeppl 1982; Pine, Miller, and Schamberger 1979) (map 10.7).

Life History

In southern Peru young are born from February to April, the end of the rainy season, after an apparent gestation period of 240 days. Mating is most commonly observed in June, during the dry season. Males shed their antlers in September and October, during the onset of the rains. Both fawning and the antler cycle are highly synchronized, reflecting the highly seasonal nature of the habitat (Merkt 1987; Pearson 1951).

Ecology

Hippocamelus antisensis is a deer of the open country of the high Andes, where it is found between 2,500 and 5,000 m. In some parts of its range it is frequently found above timberline during the summer, moving into the woods during winter. In southern Peru this species occurs in high altitude open grassland and scrub.

Animals are rarely seen singly; most are in groups averaging 6.4 in southern Peru, with a maximum of 40. Adult males spend most of their time in mixed-sex groups except during fawning. Groups composed of only females or only males are also occasionally seen. In northwestern Argentina this species is reported to congregate in larger groups during the winter, splitting into smaller groups of one male and an average of two females in September. In this area *H. antisensis* is known to be preyed upon by pumas (Cajal 1983a; Crespo 1974; Merkt 1987; Roe and Rees 1976; Tamayo and Frassinetti 1980).

Hippocamelus bisulcus (Molina, 1882)
Huemul Patagónico

Measurements

	Mean	Min.	Max.	N	Loc.	Source[a]
TL	1,620.2	1,549	1,727	5	A, C	1
	1,630.0	1,400	1,750	6 m	C	2
	1,510.0	1,400	1,570	3 f		
HB	1,497.5	1,440	1,555	5	A, C	1
T	127.5	125	130			
	120.0	90	160	4 m	C	2
	130.0	110	160	3 f		
HF	450.0	430	470	2	A, C	1
	430.0	400	470	3 m	C	2
	380.0	360	400	3 f		
E	160.0	140	170	3 m		
	170.0	160	180	3 f		
Wta	70.0 kg			1	C	1

[a](1) Allen 1903; Osgood 1943; Texera 1974; (2) Polvilitis 1979.

Description

Hippocamelus bisulcus is a stout deer that averages 800–900 mm at the shoulder, with males larger and heavier. Males are further distinguished from females by a dark band that runs along the top of the muzzle, often extending and forking above the eyes. The bifurcated antlers are about 300 mm long. The pelage is dense, coarse, brittle, and relatively long. Between spring and autumn the coat of this species is dark rusty brown to dark coffee colored, becoming

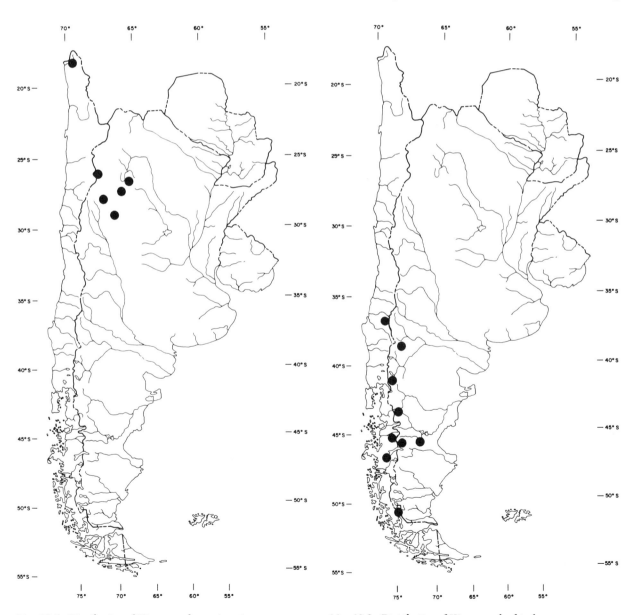

Map 10.7. Distribution of *Hippocamelus antisensis*.

Map 10.8. Distribution of *Hippocamelus bisulcus*.

lighter in the winter with a grayish yellow tone. The inguinal and anal areas, as well as the interior of the large ears and areas around the eyes, are whitish (Drouilly 1983; Polvilitis 1979).

Chromosome number: $2n = 70$ (Spotorno, Brum, and Di Tomaso 1987).

Distribution

The species is found in the Andes of southern Chile and southern Argentina. It formerly was found between 33° and 54° south latitude but now occurs in less than half of this area. In Chile it is now found in Aisén and Magallanes provinces with isolated populations in Ñuble and continental Chiloé provinces, and in Argentina it is found in the southwestern province of Neuquén (Drouilly 1983; Tamayo and Frassinetti 1980; BA) (map 10.8).

Life History

The gestation period of this species is six to seven months, with births in November and December. Usually only one apparently unspotted fawn is born. Males first show antlers at about eighteen months of age (Drouilly 1983; Polvilitis 1979; Rau 1980).

Ecology

The Patagonian huemul is a deer of the mountains and is found mostly at or just below the tree line. Its preferred habitat is steep, rocky, north-facing slopes with irregular topography, dense shrub cover, and clearings in the forest created by wind and fire. It moves downslope in winter and is active primarily by day but occasionally at night as well. In areas where it occurs, *H. bisulcus* is found at densities of 0.02 to 1.2/km².

In one area this species was found to use brushy areas 65% of the time and the forest 20% of the time. The deer feed primarily on herbaceous plants and shrubs and have an average minimum home range of 340 ha. Group size at one site in Chile ranged from 1 to 3 with an average of 1.6, while in another area the average group size was 4.8.

The southern huemul has almost been exterminated, and there are an estimated 1,300 animals left—more than a 99% decline from levels before the Europeans arrived. The reasons for this precipitous decline include hunting, predation by dogs, susceptibility to cattle diseases, and competition with cattle and introduced deer (Drouilly 1983; Miller et al. 1983; Polvilitis 1979, 1982, 1983; Prichard 1902; Rau 1980; Texera 1974).

Genus *Mazama* Rafinesque, 1817
Brocket, Corzuela

Description

These small deer have a shoulder height greater than 370 mm but less than 710 mm. There are four species; two are dwarfs (*M. rufina* and *M. chunyi*) and may possibly be confused with pudus. *M. chunyi* is restricted to southern Peru and thus is not within the range covered by this book (Hershkovitz 1959). It averages about 380 mm in shoulder height but does not co-occur with any known species of pudu. *M. rufina* averages about 450 mm at the shoulder and is reddish brown. *M. americana* is reddish brown and the largest species, some 710 mm at the shoulder. It can co-occur with the second of the two larger species, *M. gouazoubira*. Over most of its range, *M. gouazoubira* is smaller than *M. americana;* approximately 610 mm would be an average in the northern part of South America (Hershkovitz 1982). Its coat tends to be gray to gray brown. Males of all species bear a single spike antler that is usually cast annually. As far as can be determined from captive studies, fawns of all species of *Mazama* have spots at birth (Frädrich 1974, 1975).

Distribution

The genus *Mazama* is distributed from southern Mexico to central Argentina. It is adapted to a variety of habitats including montane forests, lowland rain forests, and tree savannas.

Life History and Ecology

These small deer are adapted to habitats with suitable cover. In savanna situations they can hide effectively in tall grass, but they are absent from shortgrass areas unless they have access to gallery forests. Forest-edge habitats seem to be excellent in providing shrubs for browsing and shelter. These browsers and frugivores occupy a wide variety of elevations and forest types. They do not form large aggregations and are typically seen singly or as courting pairs. After a gestation period of seven months the newborn spotted fawn is concealed in a sheltered place, and the female returns at intervals to nurse it. The fawn does not begin to follow its mother until it is several weeks old (Thomas 1975).

• *Mazama americana* (Erxleben, 1777)
Red Brocket, Corzuela Roja

Measurements

	Mean	Min.	Max.	N	Loc.	Source[a]
TL	1,208.9	1,040.0	1,340.0	11	A	1
HB	1,082.1	910.0	1,205.0			
T	126.8	95.0	145.0			
HF	674.4	580.0	800.0	5		
E	99.0	86.0	105.0	11		
Wta	28.9	18.5	33.9	10	P	2

[a](1) BA; (2) K. Hill, unpubl. data.

Description

Mazama americana is easily distinguished from the other species of *Mazama* by its larger size and brilliant reddish brown coat (plate 13). The venter is white or cream. In other parts of their range red brockets can be much heavier than the weights reported from eastern Paraguay. In Suriname the maximum recorded weight was 65 kg (Branan and Marchinton 1987).

Distribution

The species is found from southern Mexico south through eastern Paraguay to northern Argentina (Honacki, Kinman, and Koeppl 1982; Lucero 1983; Olrog 1979; UM) (map 10.9).

Life History

In captivity red brockets have a gestation period of about 225 days and produce a single young weigh-

ing between 510 and 567 g. The young are born spotted and lose their spots after two to three months. In zoos in the Northern Hemisphere males do not show a seasonal rut and irregularly shed and regrow antlers, in some cases retaining a pair of hard antlers for over a year. Males can breed in soft or hard antlers.

Does in the wild exhibit a postpartum estrus. In northeastern Argentina *M. americana* reproduces from August to October. In Suriname two of eighteen does had twin embryos, and fawning took place over at least seven months, while in Guyana fawning was reported throughout the year. The antler cycle was one year, and the quantity of sperm present was not related to antler condition. The youngest reproducing doe was thirteen months old and the youngest stag twelve months or less (Branan and Marchinton 1987; Brock 1965; Crespo 1982; Gardner 1971; MacNamara and Eldridge 1987; Thomas 1975).

Ecology

The red brocket is apparently an animal of thick forest and seems to prefer moist areas. It is a good swimmer and readily takes to the water to escape. In Suriname this species was found at a density of approximately one per square kilometer and fed on over sixty plant species. Fungi formed an important component of the diet during the wet season. Fruit was preferred when available, with leaves predominating at the end of the wet season when fruit was scarce (Branan and Marchinton 1987; Branan, Werkhoven, and Marchinton 1985; Crespo 1982; Miller 1930; Schaller 1983).

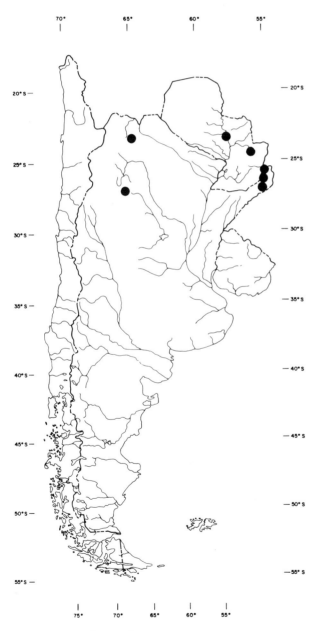

Map 10.9. Distribution of *Mazama americana*.

• *Mazama gouazoubira* (G. Fischer, 1814)
Gray Brocket, Corzuela Parda

Measurements

	Mean	Min.	Max.	N	Loc.	Source[a]
TL	1,034.0	910	1,190.0	32	A, P, U	1
HB	923.6	819	1,105.0			
T	109.8	80	152.0	33		
HF	267.7	175	319.0	29		
E	107.5	90	120.0	21		
Wta	16.3 kg	13 kg	20.5 kg	20		

[a](1) Barlow 1965; Crespo 1974; PCorps, UConn.

Description

Mazama gouazoubira is easily distinguishable from the other species of *Mazama* by its gray to gray brown color. It is smaller and slighter than *M. americana* and larger than *M. rufina* (plate 13). Its single-spike antlers are 700–100 mm long (Almeida 1976; pers. obs.).

Distribution

The gray brocket is found from Panama south to northern Argentina. In southern South America it is found in the Chaco of Paraguay, and in Argentina it occurs across the northern quarter of the country south to the province of Catamarca (Barlow 1965; Honacki, Kinman, and Koeppl 1982; BA, UConn) (map 10.10).

Life History

In captivity the gestation period is 271 days, there is an immediate postpartum estrus, and a stag can keep its antlers for 2.5 years. In the Brazilian Pantanal stags with hard antlers have been collected from May to December and those in velvet from January to June, indicating a long or irregular rut. In

the Chaco of Paraguay males with hard antlers were similarly seen throughout the year. Spotted fawns were also seen all year, reflecting year-round reproduction. Does were found to be pregnant and lactating, indicating postpartum estrus in this species. The litter size was always one. In Guyana a similar pattern of aseasonal reproduction was observed (Almeida 1976; Brock 1965; Frädrich 1987; Miller 1930; J. R. Stallings, unpubl. data; UConn).

Ecology

Unlike the red brocket, *M. gouazoubira* is not confined to thick forest. It is common in more open

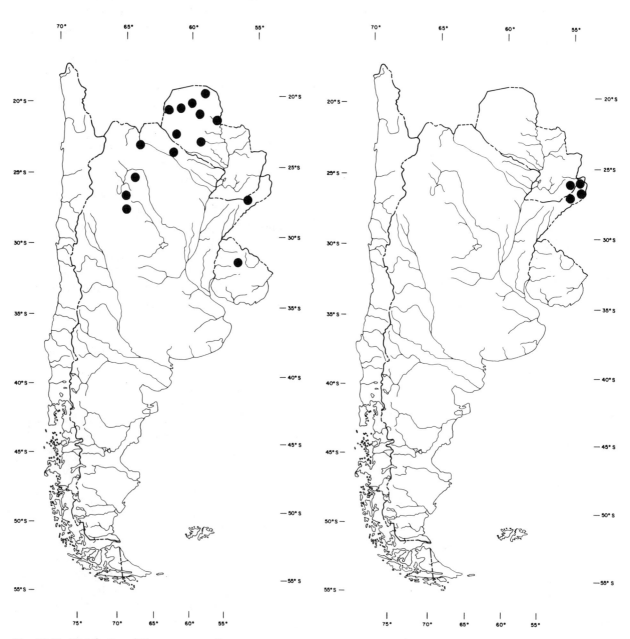

Map 10.10. Distribution of *Mazama gouazoubira*.

Map 10.11. Distribution of *Mazama rufina*.

areas such as the thorn scrub of the Paraguayan Chaco and is found in very dry areas. The gray brocket is a grazer, browser, and frugivore. In the Chaco of Paraguay an examination of twenty-three rumens showed that this species tracks fruit availability as diet varies seasonally, particularly eating fruit of *Zizyphus* and *Caesalpinia*. It feeds heavily on hard, dry fruit in the dry season and soft, fleshy fruit during the wet season and also takes leaves, fruits, buds, flowers, twigs, and roots (Schaller 1983; Stallings 1984).

• *Mazama rufina* (Bourcier and Pucheran, 1852)
Dwarf Red Brocket, Corzuela Enana

Measurements

	Mean	Min.	Max.	N	Loc.	Source[a]
TL	853.4	720	1,000	10	A, P	1
HB	775.8	660	905			
T	77.6	60	97			
E	82.8	75	90	9		
Wta	8.2 kg			1		

[a]Crespo 1950, 1982; BA; UM.

Description

Mazama rufina is easily distinguished from the other southern South American species of *Mazama* by its small size. It is similar in color to *M. americana*, reddish brown, but much smaller (plate 13). It can weigh up to 15 kg (Crespo 1982; pers. obs.).

Distribution

The species is reported from two populations, one in Venezuela, Ecuador, Peru, and Bolivia and the other, apparently disjunct, in southeastern Brazil, southeastern Paraguay, and northeastern Argentina (Crespo 1950; Honacki, Kinman, and Koeppl 1982; UM) (map 10.11).

Life History

In northeastern Argentina a single spotted fawn is born between September and February. A male with antlers in velvet was collected in July in eastern Paraguay (Crespo 1982; UM).

Ecology

In eastern Paraguay *M. rufina* has been collected in moist forest with thick bamboo understory (UM).

Comment

The *M. rufina* found in southern South America may be a different species from that found in the northern part of South America. If this is so, it could be called *M. nana* (Cabrera 1960; Miranda-Ribeiro 1919).

Genus *Ozotoceros* Ameghino, 1891
Ozotoceros bezoarticus (Linnaeus, 1758)
Pampas Deer, Ciervo de la Pampa, Venado

Measurements

	Mean	Min.	Max.	N	Loc.	Source[a]
TL	1,327.7	1,145	1,409	6	A	1
HB	1,229.3	1,193	1,284	4		
T	133.8	115	155			
E	149.0			1		

[a](1) Cabrera 1943.

Description

The pampas deer resembles the white-tailed deer (*Odocoileus*): it is longer legged than deer of the genus *Mazama* (standing 700–750 mm at the shoulder) and smaller and more slender than the marsh deer, *Blastocerus* (weighing up to 35 kg) (plate 12). The branched antlers are moderate in size and usually bear three points per side, a brow tine and a simple upper bifurcation (fig. 10.6). However, up to eight tines per side have been reported. The coat is smooth and short haired, yellowish gray, with white underparts and white on the insides of the ears and sides of the muzzle and a white eye-ring (Almeida 1976; Jackson and Langguth 1987; pers. obs.).
Chromosome number: $2n = 68$; FN = 74 (Spotorno, Brum, and Di Tomaso 1987).

Distribution

Ozotoceros is found from central Brazil south into Bolivia, Paraguay, Uruguay, and northeastern and north-central Argentina. The range of this species has contracted greatly owing to habitat destruction and hunting (Honacki, Kinman, and Koeppl 1982; Jackson 1978; Jackson, Landa, and Langguth 1980; Ojeda and Mares 1982) (map 10.12).

Life History

The gestation period of *Ozotoceros* is a little longer than seven months, after which a single fawn weighing 2,100 g is born. In central Brazil births occur from July to December, with a peak in October to November. In Argentina the reproductive activity is from December to February, with most births occurring between September and November. Fawns are spotted until about one month of age. Stags are in hard antlers from November to July. In the Brazilian Pantanal all stags observed had antlers in velvet between June and August. The general pattern of antler growth is for one-year-old males to have spikes, two-year-olds to have a total of four tines, and three-year-old and older males to have a total of six tines (Frädrich 1981; Jackson and Langguth 1987; Redford 1987; Tomas 1988a).

Ecology

The pampas deer inhabits the open vegetation formations of central and southern South America. It was once found in most of the natural grasslands of eastern South America below the equator, but it has been exterminated from the vast majority of its range. In San Luis province, Argentina, this species is a selective grazer, depending on fresh green material from a few plant species. Its diet overlaps considerably with those of cattle, horses, European hare, and viscachas. As of 1984 there were only a few known remaining populations of the southern subspecies *O. b. celer*, with about four hundred deer in Argentina and about a thousand in Uruguay.

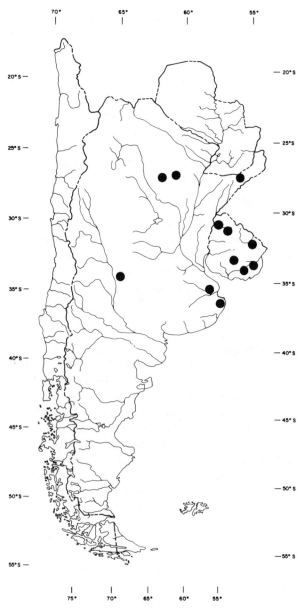

Map 10.12. Distribution of *Ozotoceros bezoarticus*.

In central Brazil the mean group size was 1.36, though on better pasture and during the rainy season groups were considerably larger. In Argentina groups averaged 5–6 animals and included both males and females, but group size and composition were both fluid. In the Brazilian Pantanal densities ranged from 1.6/km² in cerrado to 4.6/km² in open grassland. Historical documents indicate that even when numerous the pampas deer lived in small mixed-sex groups rarely exceeding five individuals, except when congregating on feeding grounds. These deer produce a very pungent odor from the interdigital gland on the hindfeet that is strongest in adult males.

During the late nineteenth century there was intensive commercial hunting of the pampas deer, and this, combined with conversion of grassland to cattle pasture and agricultural fields, was responsible for the remarkable decrease in population size. In 1880, there were 61,401 pampas deer skins exported from Buenos Aires (Jackson 1978, 1985, 1986, 1987; Jackson and Giulietti 1988; Jackson and Langguth 1987; Langguth and Jackson 1980; Miller 1930; Redford 1987; Tomas 1988b).

Genus *Pudu* Gray, 1852
Pudu puda (Molina, 1782)
Pudu, Pudú

Measurements

	Mean	Min.	Max.	N	Loc.	Source[a]
TL	817.0	743.0	867.0	5	C	1
HB	775.0	700.0	825.0			
	771.0	696.0	825.0	7		2
T	42.0	35.0	50.0	5		1
	37.0	31.0	43.0	7		2
HF	190.0	145.0	205.0	5		1
	163.0	165.0	205.0	6		2
E	86.0	85.0	88.0	4		1
	72.0			4		2
Wta	9.1 kg	7.5 kg	10.9 kg	9		1
	10.4 kg	8.3 kg	13.4 kg	?		2

[a](1) Courtin, Pacheco, and Eldridge 1980; Osgood 1943; Santiago, Spotorno and Fernández-Donoso 1975; Wolffsohn 1923; (2) Hershkovitz 1982.

Description

This is the smallest of the southern South American deer, unmistakable with its very short legs and small, low-slung body. It is a rich reddish brown agouti with darker head and legs. The spike antlers are usually less than 100 mm long. The fawn of this species of *Pudu* is spotted at birth, unlike the fawn of *P. mephistophiles*. *P. puda* is also larger (Hershkovitz 1982; pers. obs.).

Chromosome number: $2n = 70$; FN = 74 (Spotorno and Fernández-Donoso 1975).

Distribution

Pudu puda is found in southern Chile from the province of Maule south nearly to the Strait of Magellan. In Argentina it is found from southwestern Neuquén province south along the foothills to southwestern Santa Cruz province (Hershkovitz 1982) (map 10.13).

Life History

In Chile the gestation period in captivity is 207 to 223 days, after which a single young is born weighing about 370 g. The fawn loses its spots after two months, and young males grow their first antlers at nine months. Sexual maturity for does can be as early as six months ($n = 2$) and for stags 1.5 years ($n = 1$). Births were recorded from November to February, and the rut lasted from April to mid-May. Antlers were carried about seven months and cast in mid-July. Captive animals in Argentina gave birth between November and February. The young are reported to hide away from the mother for the first ten days of life (Bruzone 1984; Feer 1984; MacNamara and Eldridge 1987; Vanoli 1967).

Ecology

Pudus occur in a wide variety of habitats throughout Chile but are characteristically found in thick forests from sea level to 1,000 m, often with a thick understory of bamboo. These thickets are used to escape predators, the most important of which is the puma. When the bamboo flowers and dies pudus are forced to emigrate to areas of suitable cover.

As for other forest ungulates, scent is an important mode of communication. Scent marking, more common in males than in females, is accomplished by secretions from the preorbital and frontal glands, as well as by site-specific defecation and urination.

Pudus are active both day and night, and in Chile they occupy home ranges varying from 16 to 26 ha ($n = 4$). They are versatile in their food habits, eating bark, twigs, buds, blossoms, fruit, and berries, but they feed predominantly on herbaceous vegetation. Animals will actually "climb" by trampling stalks of bamboo, thereby foraging for leaves while completely off the ground. In captivity pudus are highly aggressive toward like-sexed conspecifics, and in the wild they are almost never found in groups larger than three—usually they occur singly or in pairs.

Habitat destruction, competition with livestock, and hunting by humans and dogs have all contributed to the decline in numbers (Courtin, Pacheco, and Eldridge 1980; Eldridge, MacNamara, and Pacheco 1987; Feer 1984; Hershkovitz 1982; MacNamara and Eldridge 1987; Miller et al. 1983; Wetterberg 1972).

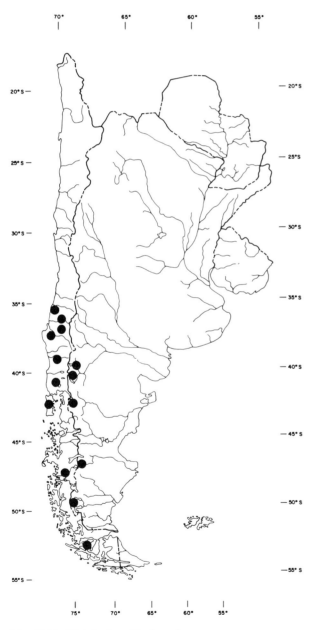

Map 10.13. Distribution of *Pudu puda*.

References

Allen, J. A. 1903. Descriptions of new rodents from southern Patagonia, with a note on the genus *Euneomys coues*, and an addendum to article IV, on Siberian mammals. *Bull. Amer. Mus. Nat. Hist.* 19:185–96.

———. 1916. Mammals collected on the Roosevelt Brazilian expedition, with field notes by Leo E. Miller. *Bull. Amer. Mus. Nat. Hist.* 34:559–610.

Almeida, A. de. 1976. *Jaguar hunting in the Mato Grosso.* England: Stanwill Press.

Barlow, J. C. 1965. Land mammals from Uruguay: Ecology and zoogeography. Ph.D. diss., University of Kansas.

Benirschke, K., and A. T. Kumamoto. 1989. Further studies on the chromosomes of three species of peccary. In *Advances in Neotropical mammalogy*, ed. K. H. Redford and J. F. Eisenberg, 309–16. Gainesville, Fla.: Sandhill Crane Press.

Blouch, R. A. 1987. Reproductive seasonality of the white-tailed deer on the Colombian llanos. In *Biology and management of the Cervidae*, ed. C. M. Wemmer, 339–43. Washington, D.C.: Smithsonian Institution Press.

Bosch, P. C., and G. E. Svendsen. 1987. Behavior of male and female vicuna (*Vicugna vicugna* Molina 1782) as it relates to reproductive effort. *J. Mammal.* 68(2): 425–29.

Boswall, J. 1972. Vicuna in Argentina. *Oryx* 11(6): 449–53.

Bourlière, F. 1973. The comparative ecology of rain forest mammals in Africa and tropical America. In *Tropical forest ecosystems in Africa and South America: A comparative review*, ed. B. J. Meggers, E. S. Ayensu, and W. D. Duckworth, 279–92. Washington, D.C.: Smithsonian Institution Press.

Branan, W. V., and R. L. Marchinton. 1987. Reproductive ecology of white-tailed and red brocket deer in Suriname. In *Biology and management of the Cervidae*, ed. C. Wemmer, 344–51. Washington, D.C.: Smithsonian Institution Press.

Branan, W. V., M. C. M. Werkhoven, and R. L. Marchinton. 1985. Food habits of brocket and white-tailed deer in Suriname. *J. Wildl. Manage.* 49(4): 972–76.

Brock, S. E. 1965. The deer of British Guiana. *J. British Guiana Mus. Zoo Roy. Agric. Comm. Soc.* 40: 18–25.

Brokx, P. A. J. 1972. A study of the biology of Venezuelan white-tailed deer (*Odocoileus virginianus gymnotis* Wiegmann, 1833), with a hypothesis on the origin of South American cervids. Ph.D. diss., University of Waterloo, Ontario.

Bruzone, J. H. 1984. Estación de recría *Pudu pudu*, Isla Victoria, Parque Nacional Nahuel Huapi: Historia. *Rev. Mus. Argent. Cienc. Nat. "Bernardino Rivadavia," Zool.* 13(1–60): 281–92.

Byers, J. A., and M. Beckoff. 1981. Social, spacing, and cooperative behavior of the collared peccary, *Tayassu tajacu. J. Mammal.* 62: 767–85.

Cabrera, A. 1943. Sobre la sistemática del venado y su variación individual y geográfica. *Rev. Mus. La Plata*, n.s., 3(18): 5–41.

———. 1960. Catálogo de los mamíferos de América del Sur. 2 (Sirenia-Perissodactyla-Artiodactyla-Lagomorpha-Rodentia-Cetacea). *Rev. Mus. Argent. Cienc. Nat. "Bernardino Rivadavia," Zool.* 4(2): 309–732.

Cajal, J. L. 1983a. Über den Bestand des Nord-Andenhirsches (*Hippocamelus antisensis*) in der argentinischen Provinz La Rioja. *Bongo* (Berlin) 7: 83–90.

———. 1983b. La vicuña en Argentina: Pautas para su Manejo. *Interciencia* 8(1): 19–22.

———. 1983c. *La situación del taruca en la provincia de La Rioja, república Argentina.* Buenos Aires: Subsecretaría de Ciencia y Tecnología, Programa Nacional de Recursos Naturales Renovables.

———. 1983d. *Estructura social y area de acción del guanaco* (Lama guanicoe) *en la Reserva de San Guillermo (provincia de San Juan).* Buenos Aires: Subsecretaría de Ciencia y Tecnología, Programa Nacional de Recursos Naturales Renovables.

———. 1986. El recurso fauna en la Argentina: Antecedentes y cuadro de situación actual. Conferencia presentada en las Primeras Jornadas Ambientales Sanjuaninas 16 de agosto de 1984, Fundación Ambientalista Sanjuanina. Ministerio de Educación y Justicia, Secretaría de Ciencia y Técnica, Programa Nacional de Recursos Naturales Renovables.

Cajal, J. L., N. E. López, A. Reca, and J. C. Pujalte. 1981. *La estrategia de la conservación de los camélidos en Argentina con especial referencia a la vicuña.* Buenos Aires: Subsecretaría de Ciencia y Technología.

Cajal, J. L., and E. Sánchez. 1983. Marcha de los censos de vicuña (*Vicugna vicugna*), guanaco (*Lama guanicoe*), y ñandú cordillerano (*Pterocnemia pennata*), en la Reserva de San Guillermo provincia de San Juan (2/1977–2/1979). Buenos Aires: Subsecretaría de Ciencia y Tecnología, Programa Nacional de Recursos Naturales Renovables.

Castellanos, H. G. 1983. Aspectos de la organización social del báquiro de collar, *Tayassu tajacu* L., en el estado Guárico-Venezuela. *Acta. Biol. Venez.* 11(4): 127–43.

———. 1986. Home range size and habitat selection of the collared peccary in the state of Guárico, Venezuela. In *Proceedings of the peccary workshop*, ed. R. A. Ockenfels, G. I. Day, and V. C. Supplee, 50. Phoenix: Arizona Game and Fish Department.

Courtin, S. L., N. V. Pacheco, and W. D. Eldridge. 1980. Observaciones de alimentación, movimientos y preferencias de hábitat del puma, en el Islote Rupanco. *Medio Ambiente* (Valdivia, Chile) 4(2): 50–55.

Crespo, J. A. 1950. Nota sobre mamíferos de Misiones nuevos para Argentina. *Comun. Inst. Nac.*

*Invest. Cienc. Nat., Mus. Argent. Cienc. Nat.
"Bernardino Rivadavia," Zool.* 1(14): 1–14.

———. 1974. Comentarios sobre nuevas localidades para mamíferos de Argentina y de Bolivia. *Rev. Mus. Argent. Cienc. Nat. "Bernardino Rivadavia," Zool.* 11(1): 1–31.

———. 1982. Ecología de la comunidad de mamíferos del Parque Nacional Iguazú, Misiones. *Rev. Mus. Argent. Cienc. Nat. "Bernardino Rivadavia," Ecol.* 3(2): 45–162.

Davis, D. E. 1947. Notes on the life histories of some Brazilian mammals. *Bol. Mus. Nac., Zool.* 76: 1–8.

Díaz, G. A. C. 1978. Social behavior of the collared peccary (*Tayassu tajacu*) in captivity. *CEIBA* 22(2): 73–126.

Drouilly, P. 1983. *Recopilación de antecedentes biológicos y ecológicos del huemul chileno y consideraciones sobre su manejo.* Boletín Técnico 5. Santiago: Corporación Nacional Forestal.

Dubost, G. 1968. Les niches écologiques des forêts tropicales sud-américaines et africaines, sources de convergences remarquables entre ronqeurs et artiodactyles. *Terre et Vie* 1: 3–28.

Eddy, T. A. 1961. Foods and feeding patterns of the collared peccary in southern Arizona. *J. Wildl. Manage.* 25(3): 248–57.

Eisenberg, J. F. 1987. Evolutionary history of the Cervidae with special reference to the South American radiation. In *Biology and management of the Cervidae,* ed. C. M. Wemmer, 60–64. Washington, D.C.: Smithsonian Institution Press.

Eisenberg, J. F., and G. M. McKay. 1974. Comparison of ungulate adaptations in the New World and Old World tropical forests with special reference to Ceylon and the rainforests of Central America. In *The behaviour of ungulates and its relation to management,* 2 vols., ed. V. Geist and F. Walther, 2: 585–602. Publications n.s. 24. Morges, Switzerland: IUCN.

Eldridge, W. D., M. M. MacNamara, and N. V. Pacheco. 1987. Activity patterns and habitat utilization of pudus (*Pudu puda*) in south-central Chile. In *Biology and management of the Cervidae,* ed. C. M. Wemmer, 352–70. Washington, D.C.: Smithsonian Institution Press.

Feer, F. 1984. Observations ethologiques sur *Pudu pudu* (Molina, 1782) en captivité. *Zool. Gart.,* n.s., 54(1/2): 1–27.

Frädrich, H. 1974. Notizen über seltener gehaltene Cerviden. Part 1. *Zool. Gart.,* n.s., 44(4): 189–200.

———. 1975. Notizen über seltener gehaltene Cerviden. Part 2. *Zool. Gart.,* n.s., 45(1): 67–77.

———. 1981. Beobachtungen am Pampashirsch, *Blastoceros bezoarticus* (L., 1758). *Zool. Gart.,* n.s., 51: 7–32.

———. 1987. The husbandry of tropical and temperate cervids in the West Berlin Zoo. In *Biology and management of the Cervidae,* ed. C. M. Wemmer, 422–48. Washington, D.C.: Smithsonian Institution Press.

Franklin, W. L. 1980. Territorial marking behavior by the South American vicuna. In *Chemical signals: Vertebrates and aquatic invertebrates,* ed. D. Muller-Schwarze and R. M. Silverstein, 53–66. New York: Plenum Press.

———. 1982. Biology, ecology, and relationship to man of the South American camelids. In *Mammalian biology in South America,* ed. M. A. Mares and H. H. Genoways, 457–489. Pymatuning Symposia in Ecology 6. Special Publication Series. Pittsburgh: Pymatuning Laboratory of Ecology, University of Pittsburgh.

———. 1983. Contrasting socioecologies of South America's wild camelids the vicuna and the guanaco. In *Advances in the study of mammalian behavior,* ed. J. F. Eisenberg and D. G. Kleiman, 573–629. Special Publication 7. Shippensburg, Pa.: American Society of Mammalogists.

Gardner, A. L. 1971. Postpartum estrus in a red brocket deer, *Mazama americana,* from Peru. *J. Mammal.* 52: 623–24.

Gilmore, R. 1950. Fauna and ethnozoology of South America. In *Handbook of South American Indians,* vol. 6, ed. J. Steward, 264–345. Washington, D.C.: Government Printing Office.

González Romero, N., H. Moreno Ortiz, and P. Calabrese. 1978. El ciervo de los pantanos o guazú pucú (*Blastocerus dichotomus*) en el Paraguay. *Informes Cient.* 1(1).

Green, G. E., and W. E. Grant. 1984. Variability of observed group sizes within collared peccary herds. *J. Wildl. Manage.* 48(1): 244–48.

Hershkovitz, P. 1959. A new species of South American brocket, genus *Mazama* (Cervidae). *Proc. Biol. Soc. Washington* 72: 45–54.

———. 1982. Neotropical deer (Cervidae). Part 1. Pudus, genus *Pudu* Gray. *Fieldiana: Zool.,* n.s., 11: 1–86.

Hofmann, R. K., C. F. Ponce del Prado, and K. C. Otte. 1975–76. Registro de dos nuevas especies de mamíferos para el Perú, *Odocoileus dichotomus* (Illiger–1811) y *Chrysocyon brachyurus* (Illiger–1811) con notas sobre su habitat. *Rev. Forest. Perú* 6: 61–81.

Honacki, J. H., K. E. Kinman, and J. W. Koeppl, eds. 1982. *Mammal species of the world.* Lawrence, Kans.: Allen Press and Association of Systematics Collections.

Husson, A. M. 1978. *The mammals of Suriname.* Leiden: E. J. Brill.

Jackson, J. E. 1978. Part 1: The IUCN threatened deer programme, 1. Species close to extinction in the wild: The Argentinian pampas deer or venado (*Ozotoceros bezoarticus celer*). In *Threatened deer.* Morges, Switzerland: IUCN.

———. 1985. Behavioural observations on the Argentinian pampas deer (*Ozotoceros bezoarticus celer* Cabrera, 1943). *Z. Säugetierk.* 50:107–16.

———. 1986. Antler cycle in pampas deer (*Ozotoceros bezoarticus*) from San Luis, Argentina. *J. Mammal.* 67:175–76.

———. 1987. *Ozotoceros bezoarticus. Mammal. Species* 295:1–5.

Jackson, J. E., and J. D. Giulietti. 1988. The food habits of pampas deer *Ozotoceros bezoarticus celer* in relation to its conservation in a relict natural grassland in Argentina. *Biol. Conserv.* 45:1–10.

Jackson, J. E., P. Landa, and A. Langguth. 1980. Pampas deer in Uruguay. *Oryx* 15:267–72.

Jackson J. E., and A. Langguth. 1987. Ecology and status of the pampas deer in the Argentinian pampas and Uruguay. In *Biology and management of the Cervidae,* ed. C. M. Wemmer, 402–9. Washington, D.C.: Smithsonian Institution Press.

Kiltie, R. A. 1981a. Stomach contents of rain forest peccaries (*Tayassu tajacu* and *T. pecari*). *Biotropica* 13:234–36.

———. 1981b. The function of interlocking canines in rain forest peccaries (Tayassuidae). *J. Mammal.* 62:459–69.

Kiltie, R. A., and J. Terborgh. 1983. Observations on the behavior of rain forest peccaries in Peru: Why do white-lipped peccaries form herds? *Z. Tierpsychol.* 62:241–55.

Koford, C. B. 1957. The vicuna and the puna. *Ecol. Monogr.* 27:153–219.

Langer, P. 1979. Adaptational significance of the forestomach of the collared peccary, *Dicotyles tajacu* (L., 1758) (Mammalia: Artiodactyla). *Mammalia* 43(2): 235–45.

Langguth, A., and J. Jackson. 1980. Cutaneous scent glands in pampas deer *Blastoceros bezoarticus* (L. 1758). *Z. Säugetierk.* 45:82–90.

Lochmiller, R. L., E. C. Hellgren, and W. E. Grant. 1984. Selected aspects of collared peccary (*Dicotyles tajacu*) reproductive biology in a captive Texas herd. *Zoo Biol.* 3:145–49.

Lucero, M. M. 1983. Lista y distribución de aves y mamíferos de la provincia de Tucumán. *Miscelánea* 75:5–53.

McCoy, M. B., and C. Vaughan. 1986. Movement, activity, and diet of collared peccaries in Costa Rican dry forest. In *Proceedings of the peccary workshop,* ed. R. A. Ockenfels, G. I. Day, and V. C. Supplee, 52–53. Phoenix: Arizona Game and Fish Department.

McCoy, M. B., C. Vaughan, and V. Villalobos. 1983. An interesting feeding habit for the collared peccary (*Tayassu tajacu* Bangs) in Costa Rica. *Brenesia* 21:456–57.

MacNamara, M., and W. D. Eldridge. 1987. Behavior and reproduction in captive pudu (*Pudu puda*) and red brocket (*Mazama americana*): A descriptive and comparative analysis. In *Biology and management of the Cervidae,* ed. C. M. Wemmer, 371–87. Washington, D.C.: Smithsonian Institution Press.

Mayer, J. J., and P. N. Brandt. 1982. Identity, distribution, and natural history of the peccaries, Tayassuidae *Mammalian biology in South America,* ed. M. A. Mares and H. H. Genoways, 433–56. Pymatuning Symposia in Ecology 6. Special Publication Series. Pittsburgh: Pymatuning Laboratory of Ecology, University of Pittsburgh.

Mayer, J. J., and R. M. Wetzel. 1986. *Catagonus wagneri. Mammal. Species* 259:1–5.

———. 1987. *Tayassu pecari. Mammal. Species* 293:1–7.

Ménard, N. 1982. Quelques aspects de la socioécologie de la vigogne *Lama vicugna. Rev. Ecol. (Terre et Vie)* 36:15–35.

———. 1984. Le régime alimentaire des vigognes (*Lama vicugna*) pendant une période de sécheresse. *Mammalia* 48(4): 529–39.

Merkt, J. R. 1987. Reproductive seasonality and grouping patterns of the north Andean deer or taruca (*Hippocamelus antisensis*) in southern Peru. In *Biology and management of the Cervidae,* ed. C. M. Wemmer, 388–401. Washington, D.C.: Smithsonian Institution Press.

Miller, F. W. 1930. Notes on some mammals of southern Mato Grosso, Brazil. *J. Mammal.* 11:10–22.

Miller, S. D., J. Rottman, K. J. Raedeke, and R. D. Taber. 1983. Endangered mammals of Chile: Status and conservation. *Biol. Conserv.* 25:335–52.

Miller, S. D., J. Rottman, and R. D. Taber. 1973. Dwindling and endangered ungulates of Chile: Vicugna, llama, hippocamelus, and pudu. *Trans. N. Amer. Wildl. Conf.* 38:55–68.

Miranda-Ribeiro, A. de. 1919. Os veados do Brasil, segundo as coleções Rondon e de vários museus nacionais e estrangeiros. *Rev. Mus. Paulista* 11:1–99.

Ojeda, R. A., and M. A. Mares. 1982. Conservation of South American mammals: Argentina as a paradigm. In *Mammalian biology in South America,* ed. M. A. Mares and H. H. Genoways, 505–22. Pymatuning Symposia in Ecology 6. Special Publication Series. Pittsburgh: Pymatuning Laboratory of Ecology, University of Pittsburgh.

Olrog, C. C. 1979. Los mamíferos de la selva húmeda, Cerro Calilegua, Jujuy. *Acta Zool. Lilloana* 33:9–14.

Olrog, C. C., and M. M. Lucero. 1981. *Guía de los mamíferos argentinos.* Tucumán, Argentina: Ministerio de Cultura y Educación Fundación Miguel Lillo.

Olrog, C. C., R. A. Ojeda, and R. M. Bárquez. 1976. *Catagonus wagneri* (Rusconi) en el noroeste argentino (Mammalia, Tayassuidae). *Neotrópica* 22:53–56.

Osgood, W. H. 1943. The mammals of Chile. *Field Mus. Nat. Hist., Zool. Ser.* 30:1–268.

Pearson, O. P. 1951. Mammals in the high-lands of southern Peru. *Bull. Mus. Comp. Zool.* 106(3): 117–74.

Pefaur, J., W. Hermosilla, F. Di Castri, R. González, and F. Salinas. 1968. Estudio preliminar de mamíferos silvestres Chilenos: Su distribución, valor económico e importancia zoonótica. *Rev. Soc. Med. Vet.* (Chile) 18(1–4): 3–15.

Pine, R. H., S. D. Miller, and M. L. Schamberger. 1979. Contributions to the mammalogy of Chile. *Mammalia* 43:339–76.

Polvilitis, A. J. 1979. The Chilean huemul project (1975–1976): Huemul ecology and conservation. Ph.D. diss., Colorado State University, Fort Collins.

———. 1982. The huemul in Chile: National symbol in jeopardy? *Oryx* 17:34–40.

———. 1983. Social organization and mating strategy of the huemul (*Hippocamelus bisulcus*). *J. Mammal.* 64:156–58.

Prichard, H. 1902. Field-notes upon some of the larger mammals of Patagonia, made between September 1900 and June 1901. *Proc. Zool. Soc. London* 1902:272–77.

Raedeke, K. J. 1979. Population dynamics and socioecology of the guanaco (*Lama guanicoe*) of Magallanes, Chile. Ph.D. diss., University of Washington.

Rau, J. 1980. Movimiento, hábitat y velocidad del huemul del sur (*Hippocamelus bisulcus*) (Artiodactyla, Cervidae). *Not. Mens. Mus. Nac. Hist. Nat.* (Santiago) 24(281/82): 7–9.

———. 1982. Situación de la bibliografía e información relativa a mamíferos chilenos. *Mus. Nac. Hist. Nat. Publ. Ocas.* 38:29–51.

Redford, K. H. 1987. The pampas deer (*Ozotoceros bezoarticus* in central Brasil. In *Biology and management of the Cervidae*, ed. C. M. Wemmer,

410–16. Washington, D.C.: Smithsonian Institution Press.

Redford, K. H., and J. G. Robinson. 1987. The game of choice: Patterns of Indian and colonist hunting in the Neotropics. *Amer. Anthropol.* 89:650–67.

Robinson, J. G., and J. F. Eisenberg. 1985. Group size and foraging habits of the collared peccary *Tayassu tajacu. J. Mammal.* 66:153–55.

Roe, N. A., and W. E. Rees. 1976. Preliminary observations of the Taruca (*Hippocamelus antisensis:* Cervidae) in southern Peru. *J. Mammal.* 57:722–30.

Roots, C. G. 1966. Notes on the breeding of white-lipped peccaries *Tayassu albirostris* at Dudley Zoo. *Int. Zoo Yearb.* 6:198–99.

Schaller, G. B. 1983. Mammals and their biomass on a Brazilian ranch. *Arq. Zool. São Paulo* 31(1): 1–36.

Schaller, G. B., and J. M. C. Vasconcelos. 1978. A marsh deer census in Brazil. *Oryx* 14(4): 341–51.

Schweinsburg, R. E. 1971. Home range, movements, and herd integrity of the collared peccary. *J. Wildl. Manage.* 35:455–60.

Smythe, N. D. E. 1970. Ecology and behavior of the agouti (*Dasyprocta punctata*) and related species on Barro Colorado Island, Panama. Ph.D. diss., University of Maryland.

Sowls, L. K. 1984. *The peccaries.* Tucson: University of Arizona Press.

Spotorno O., A. E., N. Brum, and M. Di Tomaso. 1987. Comparative cytogenetics of South American deer. In *Studies in Neotropical mammalogy: Essays in honor of Philip Hershkovitz*, ed. B. D. Patterson and R. M. Timm, 473–83. *Fieldiana: Zool.*, n.s., 39:1–506.

Spotorno O., A. E., and R. Fernández-Donoso. 1975. The chromosomes of the chilean dwarf-deer "pudu" *Pudu pudu* (Molina). *Mammal. Chrom. Newsl.* 16(1): 17.

Stallings, J. R. 1984. Notes on feeding habits of *Mazama gouazoubira* in the Chaco Boreal of Paraguay. Biotropica 16(2): 155–57.

Tamayo, M., and D. Frassinetti. 1980. Catálogo de los mamíferos fósiles y vivientes de Chile. *Mus. Nac. Hist. Nat., Chile* 37:323–99.

Texera, W. A. 1974. Algunos aspectos de la biología del huemul (*Hippocamelus bisulcus*) (Mammalia: Artiodactyla, Cervidae) en cautividad. *Anal. Inst. Patagonia, Punta Arenas* (Chile) 5(1–2): 155–88.

Thomas, W. D. 1975. Observations on captive brockets *Mazama americana* and *M. gouazoubira. Int. Zoo Yearb.* 15:77–78.

Tomas, W. M. 1986a. Hábitos alimentares do cervo-do-pantanal *Blastoceros dichotomus* (Illiger, 1811)

(Cervidae) no pantanal matogrossense, Poconé-Mt. XIII Congresso Brasileiro de Zoologia.

———. 1986b. Considerações sobre os ambientes frequentados pelo cervo-do-pantanal *Blastoceros dichotomus* (Illiger, 1811) (Cervidae) no pantanal matogrossense, Poconé-Mt. XIII Congresso Brasileiro de Zoologia.

———. 1986c. Padrão diário de atividade e estrutura de grupos do cervo-do-pantanal *Blastoceros dichotomus* (Illiger, 1811) (Cervidae) no pantanal matogrossense, Poconé-Mt. XIII Congresso Brasileiro de Zoologia.

———. 1988a. Observações preliminares sobre densidades e estrutura de grupos de veado campeiro no pantanal da Nhecolândia, Corumbá, MS. XV Congresso Brasileiro de Zoologia.

———. 1988b. Nota sobre a troca das galhadas pelo cervo-do-pantanal (*Blastoceros dichotomus*) e pelo veado campeiro (*Ozotoceros bezoarticus*). XV Congresso Brasileiro de Zoologia.

Tonni, E. P., and G. C. Politis. 1980. La distribución del guanaco (Mammalia, Camelidae) en la provincia de Buenos Aires durante el Pleistoceno tardío y Holoceno: Los factores climáticos como causas de su retracción. *Ameghiniana* 17(1): 53–66.

Torres, H. 1985. Distribution and conservation of the guanaco (*Lama guanicoe*). International Union for Conservation of Nature and Natural Resources, Species Survival Commission, South American Camelid Specialist Group, Special Report 2:1–37. Gland, Switzerland: IUCN.

Vanoli, T. 1967. Beobachtungen an Pudus, *Mazama pudu* (Molina, 1782). *Säugetierk. Mitt.* 15: 155–63.

Voss, W. A., F. R. dos Santos Breyer, G. C. Mattes, and H. G. Konrad. 1981. Constatação e observação de uma população residual de *Blastoceros dichotomus* (Illiger, 1811) (Mammalia, Cervidae). *Iheringia* (*Ser. Zool.*) *Porto Alegre* (59): 25–36.

Wetterberg, G. B. 1972. Pudu in a Chilean National Park. *Oryx* 11(5): 347–51.

Wetzel, R. M. 1977a. The extinction of peccaries and a new case of survival. *Ann. New York Acad. Sci.* 288:538–44.

———. 1977b. The Chacoan peccary *Catagonus wagneri* (Rusconi). *Bull. Carnegie Mus. Nat. Hist.* 3:1–36.

Wetzel, R. M., R. E. Dubos, R. L. Martin, and P. Myers. 1975. *Catagonus*, an "extinct" peccary, alive in Paraguay. *Science* 189:379–81.

Whitehead, G. K. 1972. *Deer of the world.* New York: Viking Press.

Wilson, P. 1984. Puma predation on guanacos in Torres del Paine National Park, Chile. *Mammalia* 48(4): 515–22.

Wilson, P., and W. L. Franklin. 1985. Male group dynamics and intermale aggression of guanacos in southern Chile. *Z. Tierpsychol.* 69:305–28.

Wolffsohn, J. A. 1923. Medidas máximas y mínimas de algunos mamíferos chilenos colectados entre los años 1896 y 1917. *Rev. Chil. Hist. Nat.* 27: 159–67.

Ximénez, A., A. Langguth, and R. Praderi. 1972. Lista sistemática de los mamíferos del Uruguay. *Anal. Mus. Nac. Hist. Nat. Montevideo* 7(5): 1–45.

Yost, J. A., and P. M. Kelley. 1983. Shotguns, blowguns, and spears: The analysis of technological efficiency. In *Adaptive responses of native Amazonians*, ed. R. B. Hames and W. T. Vickers, 189–224. New York: Academic Press.

11   Order Rodentia
        (Rodents, Roedores)

Diagnosis

The dental formula is distinctive: a single pair of upper and lower incisors; no canines; premolars not exceeding two pair per side; molars not exceeding three per side of the upper and lower jaws. There is a distinct gap or diastema between the incisors and the cheek teeth (fig. 11.1). The incisors are ever growing and have enamel on the anterior surface. Premolar and molar patterns of cusps are diverse and indicate both phylogenetic relationship and adaptation for particular feeding strategies.

The articulation of the jaw with the skull is loose, permitting both lateral and back-to-front motion. The lower jaw can be moved forward into a gnawing phase, occluding the incisors, and retracted to permit occlusion of the molars and premolars during chewing.

Distribution

Rodents are distributed over the entire world except Antarctica. They were originally absent from the Oceanic islands and New Zealand but have been introduced by humans. In South America the rodents derive from two separate colonizations. On the one hand the hystricognath or caviomorph rodents are well represented from the early Oligocene to the present. The sigmodontine rodents (Muridae), Sciuridae, and Geomyoidea entered South America with a later invasion, probably in the Pliocene.

History and Classification

One of the oldest fossil rodent groups is known from the late Paleocene of North America. These forms have been assigned to the family Paramyidae, and modern squirrels derive from them. From a modest beginning rodents speciated widely during the Eocene in North America and Eurasia. Hystricognath

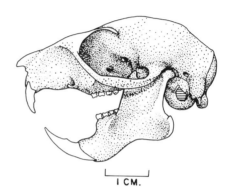

Figure 11.1 Skull of *Sciurus aestuans*.

rodents first appear in the Oligocene of South America, and this group shows certain affinities with African hystricognath rodents.

Attempts at classifying rodents into natural groupings at the level of subclasses have been fraught with difficulty. In the older literature such terms as Sciuromorpha, Myomorpha, and Hystricomorpha are frequently encountered (Simpson 1945), referring to taxonomic groupings based on the way the masseter muscles are attached anteriorly on the rostrum. The hystricomorph rodents have an enlarged infraorbital foramen with a slip from the masseter passing through it to attach anteriorly to the rostrum (fig. 11.8). The lower jaw is often sharply flared posteriorly, referred to as the hystricognathous condition. The sciuromorph rodents show no enlarged infraorbital foramen, and the masseter attaches anteriorly on the rostrum and beneath the zygomatic arch (fig. 11.1). The typical myomorph rodent has a small slip of masseter muscle passing through a slightly enlarged infraorbital foramen (fig. 11.5), but most of the anterior attachment of the masseter is similar to that described for the Sciuromorpha.

The tripartite classification below the ordinal level has not proved entirely satisfactory because

Table 11.1   Key to the Suborders of Rodents and
              Families of the Sciuromorpha in Southern
              South America

1 Infraorbital canal not modified for transmitting any
   part of medial masseter muscle (Sciuromorpha) . . . Sciuridae
1′ Infraorbital canal enlarged for transmitting
   medial masseter muscle 2
2 Infraorbital foramen greatly enlarged . . . . . . Hystricognatha
                                                      (Caviomorpha)
2′ Infraorbital foramen moderately enlarged
   or minute . . . . Myomorpha

Map 11.1. Distribution of *Sciurus aestuans*.

some rodent taxa exhibit intermediate characters or different mixtures of the major features of jaw form and masseter muscle attachment. For this reason there is a tendency to refer to the superfamily as the highest taxonomic category while recognizing that groups of superfamilies may be conveniently subsumed under yet higher subordinal groupings, but no consistent nomenclature has yet been agreed upon (Simpson 1959; Wood 1955, 1974). A key to the suborders is included in table 11.1.

The New World hystricognath rodents include the superfamilies Erethizontoidea, Cavioidea, Chinchilloidea, and Octodontoidea (Woods 1982). The South American sciuromorph rodents include the superfamilies Sciuriodea and, marginally, Geomyoidea. The native South American myomorph genera are grouped under the superfamily Muroidea and include a single subfamily, the Sigmodontinae (Reig 1980; Carleton and Musser 1984).

SUPERFAMILY SCIUROIDEA
FAMILY SCIURIDAE
Squirrels and Marmots, Ardillas

Diagnosis
The dental formula is I 1/1, C 0/0, P 1–2/1, M 3/3 (fig. 11.1). The cheek teeth usually show prominent cusps and ridges arranged in a triangular pattern in the upper molars (see fig. 11.18). The infraorbital foramen is minute, and no muscle passes through it. There are five digits on the hind feet and four on the forefeet, each bearing a long claw. The tail is always haired and may look bushy (plate 16).

Distribution
Species of this family are distributed on all continents except Antarctica and Australia.

Genus *Sciurus* Linnaeus, 1758
• *Sciurus aestuans* Linnaeus, 1776
Brazilian Squirrel, Ardilla Gris

Measurements

	Mean	Min.	Max.	N	Loc.	Source[a]
TL	363.3	360	367	3	A	1
	367.2	350	380	11	Br	2
HB	188.7	187	192	3	A	1
	202.4	170	350	11	Br	2
T	174.7	168	180	3	A	1
	180.3	165	192	11	Br	2
HF	41.5	35	48	2	A	1
	51.2	47	55	11	Br	2
E	22.0	20	24	2	A	1
	21.7	20	24	11	Br	2
Wta	800.0					1

[a] (1) Crespo 1982b; BA; (2) FM.

Description

This squirrel has a medium brown to agouti dark brown dorsum (individually variable) and is lighter on the venter, ranging from gray to orange. The chin often has a white patch. The long tail is well furred and has a prominent terminal tuft. The ends of the tail hairs are often orange, creating an impression of indistinct orange bands (see plate 16).

Distribution

Sciurus aestuans ranges from Venezuela to northern Argentina, where it is found in Misiones province. It may also be found on the Paraguayan side of the Río Paraná (Honacki, Kinman, and Koeppl 1982; BA) (map 11.1).

Ecology

In addition to feeding on fruits and nuts, this species is thought to prey on eggs and young of birds (Husson 1978).

• *Sciurus spadiceus* Olfers, 1818
Ardilla Roja

Description

Measurements of two males were TL 503–22; T 266–67; HF 67–71; E 33–34; Wt 570–660 g (MVZ). This squirrel is easily separated from *S. aestuans* by its larger size and reddish brown dorsum, although some individuals may have white venters. It is rufous on the sides, hips, forearms, and sides of the face and cream colored to reddish on the venter. The tail is brown at the base, becoming red distally. There is much variation in color (pers. obs.).

Distribution

Sciurus spadiceus is distributed from southern Venezuela to southern Bolivia and southeastern Peru and possibly to the Argentine provinces of Jujuy and Salta (Patton 1984).

Sciurus ignitus (Gray, 1867)

Description

One male specimen collected in Peru measured TL 369; T 178; HF 52; E 23; Wta 240 g (MVZ). The dorsum is agouti brown and the tail is essentially the same but more reddish. The differential degree of reddish on the hair tips yields an almost banded pattern. The venter is reddish tan.

Distribution

Sciurus ignitus is found in the Andes of Bolivia and Peru, thence south to northwestern Argentina (Yepes 1944; Ojeda and Mares 1989) (map 11.2).

Ecology

This squirrel can be found up to 2,600 m and has been observed in the alder forests of Jujuy province (Ojeda and Mares 1989; Olrog 1979).

Comment

There seems to be some confusion regarding the specific identity of the larger squirrels in lowland northwestern Argentina. For a discussion of the identity of *Sciurus igniventris* and *S. spadiceus*, see Patton (1984). These two large squirrels co-occur in much of Amazonas and are superficially similar.

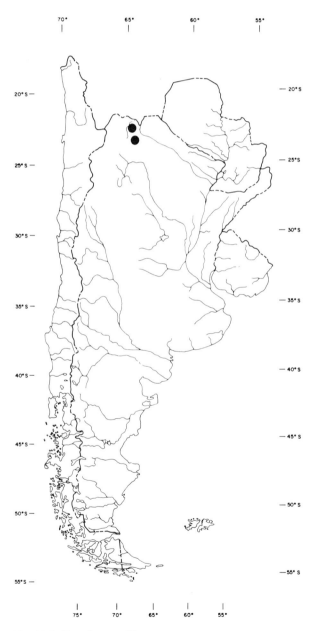

Map 11.2. Distribution of *Sciurus ignitus*.

SUPERFAMILY MUROIDEA
FAMILY MURIDAE
SUBFAMILY SIGMODONTINAE
(HESPEROMYINAE)
New World Rats and Mice

Diagnosis

The dental formula I 1/1, C 0/0, P 0/0, M 3/3 distinguishes this group from the Sciuridae and Geomyoidea (see fig. 11.2). *Daptomys oyapocki* is exceptional with two upper and lower molars per side. The cheek teeth are variable in cusp pattern and may be laminate, prismatic, or cuspidate. When cusps are present they are arranged in two longitudinal rows in both upper and lower molars (see fig. 11.8). The infraorbital foramen is rather small; generally a strip of masseter muscle passes through to attach to the rostrum. The bulk of the masseter attaches on the zygomatic arch and the rostrum. There is no postorbital process on the frontal bone.

We follow Carleton and Musser (1984) in their classification and terminology.

Distribution

Sigmodontine rodents are found from northern Canada throughout North America to Patagonia in South America. This is a New World subfamily.

History and Classification

The sigmodontine rodents are listed here as a subfamily of the family Muridae (Carleton and Musser 1984; Reig 1980). They are frequently referred to as the New World cricetines. For this account we consider the gerbils to be within their own subfamily, the Gerbillinae, and the voles to be in their own subfamily, the Arvicolinae. The New World Sigmodontinae have as their nearest relatives in the Old World the hamsters (Cricetinae) of Europe and Asia and the pouched rats (Cricetomyinae) of Africa. Cricetine rodents first appear in the Oligocene of Europe, and by the mid-Oligocene they are represented in North America. Rodents that can be referred to the subfamily Sigmodontinae are identifiable in the Miocene of North America.

The classification of these rodents above the generic level, into tribes, has been difficult. It seems fair to say that the original stock was adapted to forested environments, but with the increasing drying during the Miocene and the creation of extensive grasslands, the animals began to adapt to more open grassland and xeric habitats (Hershkovitz 1962). How often this happened during the early divergence of the ancestral stock is unknown.

If we consider only the terrestrial mammals of South America, then rodents make up 42% of the species described to date. Approximately half of these rodent species may be assigned to the sub-

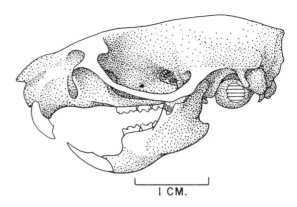

Figure 11.2. Skull of *Oryzomys buccinatus*.

family Hesperomyinae of the Cricetidae or the subfamily Sigmodontinae of the Muridae. Reig (1986) has proposed that the South American cricetids be included in the subfamily Sigmodontinae and be subdivided into seven tribes: Oryzomyini, Ichthyomyini, Akodontini, Scapteromyini, Wiedomyini, Phyllotini, and Sigmodontini. This grouping does not include the Central American sigmodontines that enter parts of northwestern South America from Panama. In addition, four South American genera cannot be placed with certainty in any of the seven tribes. These genera are *Zygodontomys*, *Rhagomys*, *Punomys*, and *Abrawayaomys* (Reig 1986).

Reig (1986) summarizes his previous work and that of others concerning the origin of the tribes of sigmodontine rodents, concluding that some sigmodontines crossed into South America as early as the Miocene. The four largest tribes in South America are the Oryzomyini, Akodontini, Phyllotini, and Sigmodontini. Reig believes these taxa originated in South America and radiated into many genera before the Pleistocene. Some genera of the Sigmodontini and the Oryzomyini subsequently reinvaded Central America and North America via the Panamanian land bridge.

Earlier workers considered the neotomine-peromyscine group of the Hesperomyinae or Sigmodontinae to be North American in origin and a clearly defined taxonomic unit. The remaining sigmodontines were thought to have evolved primarily in the Isthmus and the southern United States (Hooper and Musser 1964). Recently the synonymy of the fossil *Bensonomys* of North America with the extant South American *Calomys* suggests that the major stocks of the South American sigmodontines (cricetines) may have differentiated in North America before a Pliocene transfer to South America (Baskin 1978).

In an effort to better understand the relationships among the sigmodontine rodents, numerous characters have been evaluated. Early workers such as

Cabrera (1960), Moojen (1952a), Hershkovitz (1962, 1966), and Pearson (1958) undertook important descriptions and attempts at classification. Teeth have proved useful for delineating some groups but have provided equivocal evidence of relationship for others. Studies have compared blood proteins, and extensive work has been done on chromosomal evolution, especially by Reig, Olivo, and Kiblisky (1971), Pearson and Patton (1976), and Gardner and Patton (1976).

Hooper and Musser (1964) studied the morphology of the baculum and phallus and concluded that the neotomine-peromyscine group was distinguishable from the predominantly South American radiation by the structure of the penis. Implicit in their hypothesis was the assumption that the sequence of evolution went from a complex baculum to a simplified one. Recent workers have challenged this concept of the primitive morphological characters and conclude that the primitive character was probably a simple baculum (Carleton 1980).

The most recent attempt to analyze the problem of relationships was made by Carleton (1980), who first ascertained the primitive and derived conditions for each morphological character. He then performed a multivariate analysis for over seventy characters for the neotomine-peromyscine complex, with some comparisons of South American forms.

He concluded that the neotomine-peromyscine group is not necessarily a natural unit, that the polarity of characters assumed by other workers is open to question, and that the assumption about the derived state of the simple baculum was probably erroneous. Furthermore, Carleton proposed that the neotomine-peromyscine group is divisible into at least four units or supergeneric groupings. These groupings were not proposed as formal taxonomic units but were useful in an informal sense. They include the peromyscine rodents that are for the most part North American in distribution and include, among others, the deer mice (*Peromyscus*), harvest mice (*Reithrodontomys*), grasshopper mice (*Onychomys*), and golden mice (*Ochrotomys*). The neotomine group includes the wood rats (*Neotoma*) and their relatives, again a primarily North American group. The baiomyine group includes the pygmy mice (*Baiomys*) and brown mice (*Scotinomys*); and finally, the tylomyine group includes the climbing rats (*Tylomys* and *Ototylomys*) of Central America (fig. 11.3).

Following Carleton in his groupings of genera, apparently the closest living relatives of the tylomyines are what may be referred to as the nyctomyines, or vesper mice. The nyctomyines are Central American forms adapted to tropical forests. They are excellent climbers with long tails and form a

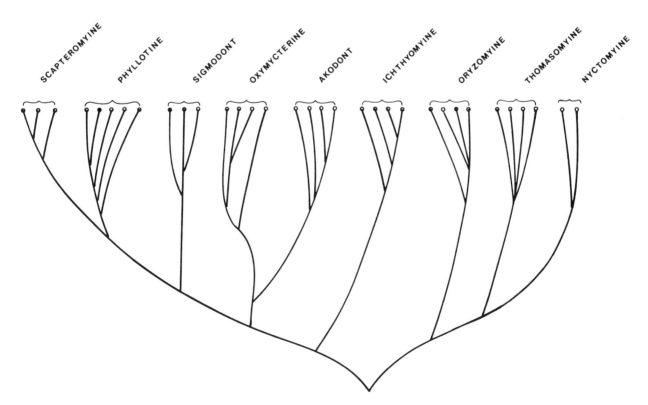

Figure 11.3. Phylogenetic tree indicating the relationships of the South American sigmodontine rodents (modified from Carleton 1980).

phenetic link between the North American neotomine-peromyscines and the South American thomasomyines. These latter mice also are predominantly long-tailed forms occupying forested habitats and are excellent climbers. In northern South America the genus *Thomasomys* is often confined to Temperate-Zone, high-altitude forests in the Andes, whereas the closely related *Rhipidomys* inhabits lower-altitude moist rain forests.

The closest living relatives of the thomasomyines are the oryzomyines, and in fact Reig (1980) combined them into the same tribe. These rice rats include six genera. All are moderately long tailed; some are adapted for semiaquatic life and others are predominantly terrestrial and scansorial.

There is considerable divergence in the remaining South American sigmodontines. The akodonts show a range of adaptations and include five genera. They are predominantly, terrestrial and show tendencies toward exploiting more open habitats, though some members of this group are still confined to somewhat moist, forested conditions. The key to their adaptive radiation is terrestriality. The tribe most nearly related to the akodonts is the oxymycterines. The oxymycterines have carried terrestriality to further extremes, but many of them have become specialized for feeding on insects as well as seeds. This trend culminates in the burrowing, insect-feeding forms adapted for niches similar to those of the North American shrews and moles. This syndrome includes the genus *Blarinomys* of southeastern Brazil.

The tribe Zygodontomyini contains the distinctive genus *Zygodontomys* and exhibits a superficial resemblance to the akodonts. Somewhat related to the oxymyceterine-akodont groupings are the so-called fishing rats, or ichthyomyines. The genera composing this set exhibit various adaptations to an aquatic life, and the most specialized members are adapted for feeding on crustaceans and fish. The Ichthyomyini are confined to northern South America (see Voss 1988).

The next three suprageneric groupings show a

close relation to one another. The sigmodont genera clearly are adapted for feeding on the green parts of plants, and some of them appear to occupy an almost volelike niche in South America. This assemblage may be artificial. The phyllotines are a diverse group of genera that have colonized many open habitats from high-altitude to dry, low-altitude areas and open grasslands. They show an array of dietary specializations from eating seeds to feeding almost completely on herbaceous vegetation and perhaps demonstrate the greatest diversity of all the tribes of South American cricetid rodents. The last group, the scapteromyines, appear to be the most derived genera and consist of the pampas water rats and their relatives from southeastern Brazil, Argentina, Uruguay, and parts of Paraguay.

This brief review of taxonomy should bring home the point that in South America the adaptive radiation of the sigmodontines has allowed them to occupy many ecological niches. We can identify a set of aquatic fish-eating, insect-eating, and crustacean-eating adaptations and an aquatic niche for larger rodents that are omnivorous or more herbivorous. There is an adaptive zone occupied by scansorial rodents that are able climbers and feed on fruits and seeds. We can identify many terrestrial niches that exhibit a trophic diversity from partially insectivorous forms to extreme insectivores that burrow. There are terrestrial forms that are granivores and, finally, terrestrial forms that are predominantly herbivores. It is clear that the adaptive radiation into different feeding niches in South America has occurred several times, and different lineages can have species in similar feeding niches but geographically separated. This is especially true of those species occupying the grassland niches in South America. There are clear cases of convergent evolution brought about through geographical separation when one compares sigmodonts (sensu stricto) with phyllotines (Eisenberg 1984).

In North America the grassland niches were predominantly filled by another subfamily of murid rodents that entered North America late but did not

Table 11.2   Characters of the "Tribes" of Sigmodontine Rodents Found in Southern South America

"Tribe"	Penis and Baculum Complex in Structure	Tail Visibly Longer Than Head and Body	Hind Feet Elongated	Pinnae Relatively Large	Molars Sigmodont	Molars "Pentalophodont"	Molars "Tetralophodont"	M_3 Half to Three-fourths Length of M_2
Thomasomyini	+	+	+	+	−	+	−	+
Oryzomyini	+	+/−	+/−	+/−	−	+/−	−	+
Akodontini	+	−	−	−	−	−	+	−
Sigmodontini	+	−	−	−	+	−	+	+
Phyllotini	+	+/−	+/−	+	−	−	+	+
Scapteromyini	+	−	−	−	−	+	−	+/−

Note: +/− = variable within taxa; + = character expressed; − = character not typical.

reach South America. These North American rodents belong to the subfamily Arvicolinae (= Microtinae), exemplified by the lemmings and voles.

The most conservative members of the various taxa, or those that have retained conservative traits, are to some extent scansorial and have rather long tails. Deviations from this early body form involved specialization for more terrestrial habits, with reduction in tail length, or specialization for aquatic habitats, with no necessary reduction in tail length but with modifications of the hind feet, including webbing, and lateral flattening of the tail for better propulsion. All these rodents are of rather modest size; head and body lengths range from 44 mm to over 280 mm. The only exceptions are the extinct *Megalomys* of the Lesser Antilles and the extinct *Megaoryzomys* of the Galápagos, whose head and body length exceeded 330 mm. A phenetic matrix for the Sigmodontinae is included in table 11.2 (Eisenberg 1984, 1989).

TRIBE THOMASOMYINI
South American Climbing Rats

Description
In southern South America two genera enter the range covered by this volume, *Thomasomys* and *Rhipidomys*. The molars are complex (pentalophodont) (fig. 11.18), and the M_3 is large (one-half to three-fourths the length of the M_2). These rodents have a complex baculum, and a short, broad palate; the tail exceeds the head and body in length; and they have relatively long hind feet and moderate to long ears. They may be confused with oryzomyine rodents over much of their range but are generally distinguishable by their size and their relatively long tails, usually with somewhat of a tuft (see species accounts).

Distribution
Predominately a South American group, these rats are replaced in the north by Central American climbing rats of the genus *Tylomys*.

Genus *Rhipidomys* Tschudi, 1844
Climbing Rat

Description
This genus is closely related to *Thomasomys* and shares many of the features listed above. *Rhipidomys* has several characters that distinguish it from *Thomasomys*: in *Rhipidomys* the eye is relatively large, the dorsal color is usually sharply set off from the ventral color, and the tail tip has elongated penicillate hairs (plate 15). The interorbital region of the skull of *Rhipidomys* has a distinct ridge.

Distribution
The genus is typically found in the lowland, forested regions of tropical South America.

• *Rhipidomys leucodactylus* (Tshudi, 1845)
Colilargo Peludo

Measurements

	Mean	Min.	Max.	N	Loc.	Source[a]
TL	324.7	305	335.0	3	A	1
	369.8	332	410.0	11	Peru	2
HB	148.7	140	154.0	3	A	1
	160.7	136	190.0	11	Peru	2
T	176.0	165	183.0	3	A	1
	209.1	187	230.0	11	Peru	2
HF	30.1	29	31.4	3	A	1
	36.1	34	38.0	11	Peru	2
E	23.3	21	25.0	3	A	1

[a](1) Thomas 1921b; BA; (2) FM.

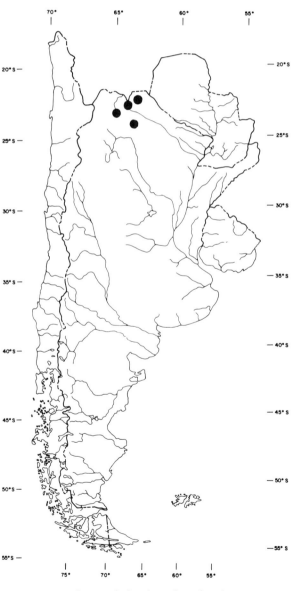

Map 11.3. Distribution of *Rhipidomys leucodactylus*.

Description

This mouse is identifiable by its large size and very long, penicillate tail. It is light reddish brown to yellowish brown mixed with black hairs dorsally, with more yellowish sides and a white to yellowish gray venter. The vibrissae are very long and stout (Gyldenstolpe 1932; pers. obs.).

Distribution

Rhipidomys leucodactylus is widely distributed from southern Venezuela to the very northwestern corner of Argentina in the provinces of Jujuy and Salta (Honacki, Kinman, and Koeppl 1982; Ojeda and Mares 1989; Thomas 1921b) (map 11.3).

Life History

In Argentina this species has been trapped in lower montane forest and transitional forest (Ojeda and Mares 1989).

Genus *Thomasomys* Coues, 1884

Description

The molars are complex (pentalophodont), and the M_3 is longer than the M_2. These medium-sized, ratlike rodents have rather long tails. The baculum is complex in structure, the hind feet are rather long, the pinnae are conspicuous, and the eyes are reduced in size. The dorsal color is variable, ranging

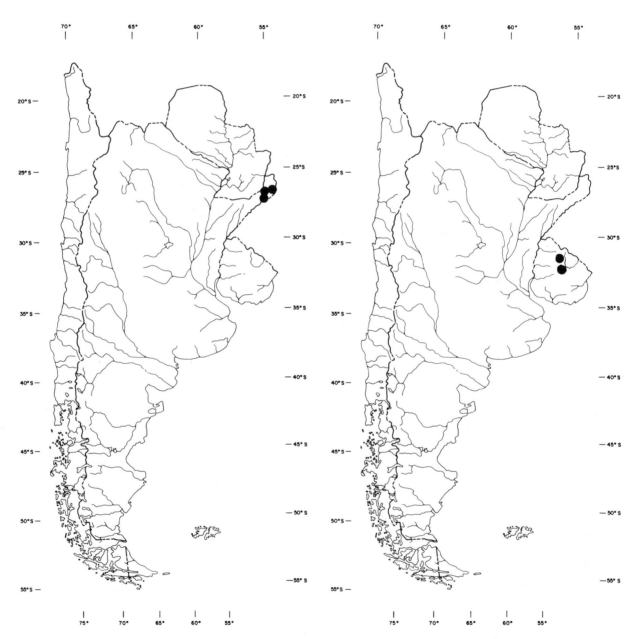

Map 11.4. Distribution of *Thomasomys dorsalis*.

Map 11.5. Distribution of *Thomasomys oenax*.

from gray brown to dark brown. The venter is paler than the dorsum, but there is no sharp demarcation (plate 15).

Distribution

The genus is found throughout most of tropical and subtropical South America, south to northern Argentina. Most species are found in northern South America, where many range to 3,000 m.

Comment

The southern species of *Thomasomys* are more similar in appearance to *Rhipidomys* than to the northern species of *Thomasomys* and are often placed in their own genus, *Delomys*.

Thomasomys dorsalis (Hensel, 1872)

Measurements

	Mean	Min.	Max.	N	Loc.	Source[a]
TL	242.2	219	265	9	A, Br	1
HB	129.7	122	140			
T	112.6	96	127			
HF	29.1	27	32			
E	20.0	20	20	2		

[a](1) BA, FM.

Description

This mouse has an agouti brown dorsum, and many individuals have an indistinct medial line of black from the shoulders to the base of the tail. The sides often have a yellowish wash, and the venter is white in southern forms with the hair bases showing through. The ears are fairly large and scantily haired. The tail, virtually naked, is weakly bicolored, with the scale pattern evident (pers. obs.).

Distribution

Thomasomys dorsalis is found in southern and eastern Brazil and into northeastern Argentina in the province of Misiones (Honacki, Kinman, and Koeppl 1982; BA) (map 11.4).

Thomasomys oenax Thomas, 1928

Measurements

	Mean	Min.	Max.	N	Loc.	Source[a]
TL	300.6	280.0	315.0	5	U	1
HB	127.0	112.0	145.0			
T	173.6	161.0	186.0			
HF	31.8	29.0	35.0			
E	18.7	12.0	22.0	3		
Wta	46.8	34.6	61.3	4		

[a](1) Barlow 1969; Pine 1980; Vaz-Ferreira 1958.

Description

This species resembles a large *Oryzomys* except for the extremely long tail. The dorsum is pale gray to agouti brown with a yellowish cast, and the venter is buffy white. The most striking feature is the color of the nose: the tip of the muzzle and the lightly haired ears are bright ochraceous, and there is another ochraceous patch on the rump. The long tail is haired and uniformly buffy (plate 15) (Barlow 1969; Pine 1980; pers. obs.).

Distribution

Thomasomys oenax is found in southern Brazil and in Uruguay (Honacki, Kinman, and Koeppl 1982; Pine 1980) (map 11.5).

Ecology

In Uruguay this species is apparently limited to patches of dense subtropical woodland. Specimens

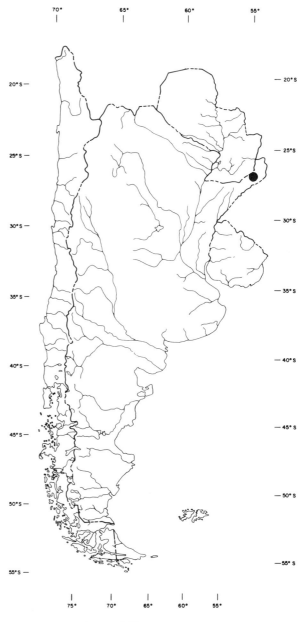

Map 11.6. Distribution of *Thomasomys pictipes*.

have been caught in trees at night (Barlow 1965, 1969; Vaz-Ferreira 1958).

Thomasomys pictipes Osgood, 1933
Ratón Gris

Measurements

	Mean	Min.	Max.	N	Loc.	Source[a]
TL	182.7	175	197	3	A, Br	1
HB	87.3	81	100			
T	95.3	94	97			
HF	21.0	21	21			
E	13.5	12	15			

[a](1) Osgood 1933; FM.

Description

Thomasomys pictipes is the smallest of the *Thomasomys* in southern South America. Its tail is slightly shorter than the head and body, naked, bicolored, and slightly penicillate. It is warm, reddish brown to brownish orange dorsally and warm tan on the sides, with a white venter. The feet are distinctly bicolored in some individuals (Osgood 1933; Pine 1980; pers. obs.).

Distribution

The species is found in southeastern Brazil and in the northeastern Argentine province of Misiones (Honacki, Kinman, and Koeppl 1982; Pine 1980) (map 11.6).

Ecology

In southern Brazil two individuals were found nesting off the ground in bamboo and in bromeliads (Pine 1980).

Comment

Massoia (1980a) records *T. dorsalis* from Argentina, but the specific identity of the form needs to be checked.

TRIBE ORYZOMYINI

Description

The molars are generally not complex (pentalophodont; but see below), and the third molar is large (one-half to three-fourths the length of the M_2) (figs. 11.2 and 11.18). In common with the thomasomyine rodents, the genera of this tribe usually have a tail length longer than the head and body. The ears are modest to rather large, and the tail is seldom penicillate. Some genera (*Neacomys*) have semispinescent pelage.

Distribution

This tribe has members extending from coastal New Jersey in the United States to southern Argentina.

Life History and Ecology

These rodents usually occupy mesic habitats. Some species are highly arboreal, others may exhibit aquatic specializations.

Genus *Oryzomys* Baird, 1858
Rice Rat

Description

The genus is highly variable in size, color, and tail length. The M_3 ranges from one-half to three-fourths the length of the M_2, and the molar structure may be simple (tetralophodont) or complex (pentalophodont). In terrestrial forms the sparsely haired tail is equal to or slightly less than the head and body length. The baculum is complex in structure, and

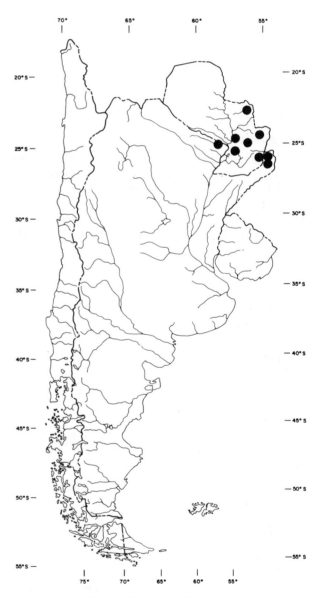

Map 11.7. Distribution of *Oryzomys buccinatus*.

the pinnae are conspicuous. The dorsum is usually some shade of brown with a contrasting ventral color, usually white to cream.

Distribution

This genus ranges from the southeastern United States through Central and South America to northern Chile and Argentina.

Oryzomys buccinatus (Olfers, 1818)
Colilargo Rojizo

Measurements

	Mean	Min.	Max.	N	Loc.	Source[a]
TL	358.7	349	375	8	A	1
HB	163.5	140	185			
T	195.3	186	205			
HF	36.8	35	38			
E	21.2	20	23	9		
Wta	100.3	75	120	8		

[a](1) BA, UM.

Description

This is a large *Oryzomys* with a long tail; it is smaller than *O. ratticeps*, with more dark flecking, less tan, and with a generally dark belly. Dorsally it is tan agouti with some black and has a tendency to darken on the head and rump and along the middorsum. Some individuals have more orangish than others. The dorsal color lightens on the sides and grades into a grayish white venter. The tail is long and slightly bicolored: in general it is gray on the top and whitish below except for the proximal few centimeters, which are often yellowish (plate 15).

Distribution

Oryzomys buccinatus is distributed in eastern Paraguay and northeastern Argentina (Honacki, Kinman, and Koeppl 1982) (map 11.7).

Ecology

In Paraguay this species was usually caught under bromeliads at the ecotone between fields and forests (UM).

Comment

Oryzomys buccinatus is very similar to *O. sub-flavus* of the forests and forest edges of southeastern Brazil (Myers 1982) and may in fact belong to this latter species (P. Myers, pers. comm.). *O ratticeps* was included in *O. buccinatus* by Hershkovitz (1959), but see Honacki, Kinman, and Koeppl (1982).

•*Oryzomys capito* (Olfers, 1818)
Colilargo Acanelado

Measurements

	Mean	Min.	Max.	N	Loc.	Source[a]
TL	261.6	236	283	13	A, P	1
HB	133.5	118	146			
T	129.3	114	156			
HF	30.8	29	33			
E	21.8	20	23			
Wta	66.3	46	87			

[a](1) Massoia 1976b; UM.

Description

This smaller, stouter-bodied *Oryzomys* has a proportionally much shorter, thinner tail. The dorsum is chestnut, grading to paler on the sides and sharply demarcated from a grayish white belly. The face is generally gray with whitish gray at the base of the vibrissae, and it has a white chin. The tail is not bicolored but may sometimes be weakly so at the base (pers. obs.).

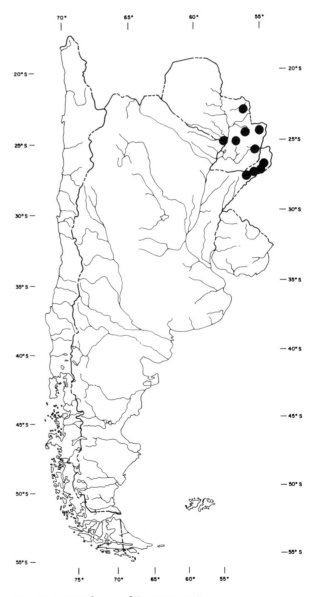

Map 11.8. Distribution of *Oryzomys capito*.

Distribution

Oryzomys capito is found from northern South America to northwestern Argentina and eastern Paraguay (Carleton 1980; Honacki, Kinman, and Koeppl 1982; Massoia 1980a) (map 11.8).

Life History

In the laboratory *O. capito (laticeps)* has a gestation of twenty-five days, an average litter size of 3.5 (range 1–6), and a range of birthweights of 3.7 to 4.0 g. The eyes of the young open after seven to eight days; young mice reproduce at fifty days and live three years. In the lab, breeding is continuous (Worth 1967).

Ecology

In Misiones province, Argentina, mice of this species were caught in tropical forest, and in Paraguay they have been caught on and off the ground in the forest. These mice are omnivorous, feeding on fruits, seeds, adult insects, insect larvae, and fungi. Densities can vary between 0.5 and 5.0 individuals per hectare in Venezuela. In French Guiana they are strongly frugivorous in second-growth forests. They do not burrow but build a leaf nest within a natural crevice (Guillotin 1982; Massoia 1976b; O'Connell 1981; UM).

Oryzomys chacoensis Myers and Carleton, 1981

Measurements

	Mean	Min.	Max.	N	Loc.	Source[a]
TL	223.4	185	280	90	P	1
T	129.0	105	150			
HF	24.8	18	30			
E	16.6	13	19			
Wta	31.0	25	40	9		2

[a](1) Myers and Carleton 1981; (2) UM.

Description

This medium-sized species of the subgenus *Oligoryzomys* is unique in its whitish underside, extending to the base of the chin, relatively long ears heavily furred on the inside, with short or absent dark basal bands, and small but distinctive tufts of orangish hair in front of the ears. The weakly bicolored tail averages 131% of the head and body length, and the feet average 25% of it. The dorsum is clay colored, heavily lined with black, darkest over the shoulders. The head is grayer, and the cheeks are brown or gray, occasionally with some orange (Myers and Carleton 1981).

Chromosome number: $2n = 58$ (Myers and Carleton 1981).

Distribution

Oryzomys chacoensis is found in the Chaco of Paraguay, Bolivia, and Argentina (Honacki, Kinman, and Koeppl 1982) (map 11.9).

Life History

In the Paraguayan Chaco ten females were found to contain an average of 4.6 embryos (range 2–5) (Myers and Carleton 1981).

Ecology

In the northern and western Chaco of Paraguay this species is common in thorn scrub and grassland, reaching its highest densities in grassland. In the lower, wetter Chaco it found in forest and dry grassland but not in wet marshes, where *O. fornesi* occurs (Myers and Carleton 1981).

Comment

Oryzomys chacoensis is closely related to the eastern form *O. nigripes* and to the Andean *O. longicaudatus*.

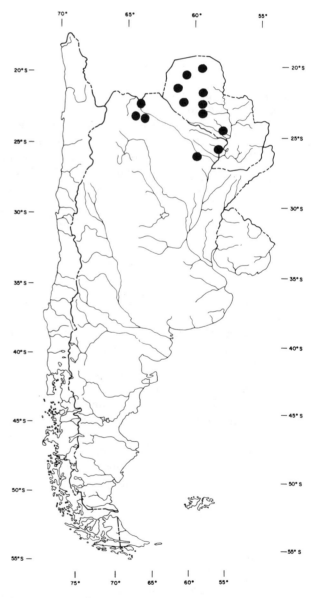

Map 11.9. Distribution of *Oryzomys chacoensis*.

• *Oryzomys concolor* (Wagner, 1845)
Colilargo Bayo

Measurements

	Mean	Min.	Max.	N	Loc.	Source[a]
TL	302.5	275	333	12	B, P	1
HB	138.3	118	149			
T	164.3	140	184			
HF	26.2	20	30			
E	18.8	15	22	9		
Wta	63.0	56	70	2		

[a](1) FM, UM.

Description

Individuals of this species from Paraguay are easily separated from other larger *Oryzomys* because of their color. Dorsally they are grayish tan, shading to lighter gray on the sides, with a narrow wash of orangish demarcating the venter. This orange extends onto the hind legs, the base of the tail, the forelegs, and the base of the ears, and up onto the chin. The venter is dirty white. The tail is unicolored, stout and quite long (pers. obs.).

Distribution

Oryzomys concolor is found from Costa Rica south to Paraguay and northern Argentina, where it has been caught in Chaco province (Honacki, Kinman, and Koeppl 1982; Massoia and Fornes 1965b) (map 11.10).

Life History

In Venezuela litter size ranges from two to four (O'Connell 1981).

Ecology

In Paraguay this species is arboreal and has been captured in thorn forest. In central Brazil 29% of captures were on the ground, and most individuals were caught in or near fern thickets. The mean distance males moved was 45 m, and females moved 26 m (Myers 1982; Myers and Wetzel 1979; Nitikman and Mares 1987).

Oryzomys delticola Thomas, 1917
Colilargo Isleño

Measurements

	Mean	Min.	Max.	N	Loc.	Source[a]
TL	229.3	190.0	264	20	U	1
HB	102.8	85.0	119	36	A	2
T	127.6	102.0	151	20	U	1
	137.1	112.0	155	37	A	2
HF	26.3	25.0	29	20	U	1
	26.1	24.5	28	34	A	2
E	18.1	15.0	20	20	U	1
	18.1	16.0	21	37	A	2
Wta	29.4	15.4	41	20	U	1

[a](1) Barlow 1965; (2) Langguth 1963.

Description

This is a medium-sized *Oryzomys*, closely resembling *O. longicaudatus* but larger. In adults the dorsum is grayish brown to reddish brown and the venter is whitish or grayish, never ochraceous. The sides are sometimes washed with yellow. The tail is long and bicolored (Barlow 1965; Massoia 1973b; Thomas 1917).

Distribution

Oryzomys delticola, as the name suggests, is found in a restricted range centering on the delta of the Río Paraná, and including all of Uruguay and the Argentine provinces of Entre Ríos, Misiones, and Buenos Aires (Barlow 1965) (map 11.11).

Life History

In Uruguay *O. delticola* breeds in late summer and early autumn, and females have two to four embryos (*n* = 5) (Barlow 1969).

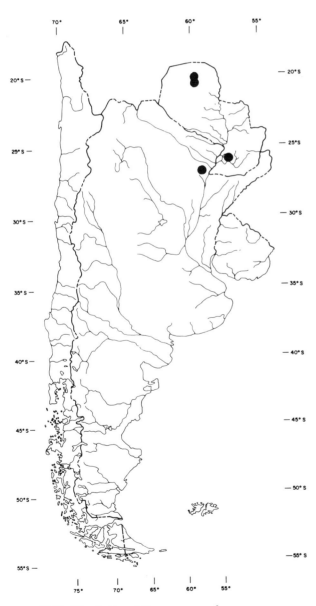

Map 11.10. Distribution of *Oryzomys concolor*.

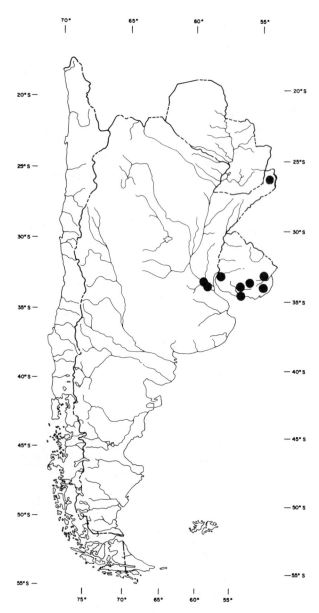

Map 11.11. Distribution of *Oryzomys delticola*.

T	111.0	97.0	130.0	12 m	A	1
	101.6	88.0	117.0	10 f		
	112.4	102.0	127.0	18	U	2
HF	23.9	22.0	25.0	13 m	A	1
	23.2	22.0	24.0	11 f		
	26.5	23.0	28.0	18	U	2
E	14.3	13.0	16.5	13 m	A	1
	14.2	13.0	16.0	10 f		
	14.3	12.0	15.0	18	U	2
Wta	32.0	25.0	39.0	2 m	A	1
	21.4	18.0	26.0	5 f		
	21.0	15.5	30.0	18	U	2
	18.9			56	A	3

[a](1) Fornes and Massoia 1965; (2) Barlow 1965; (3) Crespo et al. 1970.

Description

The tail is never longer than 138 mm. Dorsally this mouse is brown mixed with yellowish and tinged with black. The venter is whitish yellow, sometimes mixed with a little gray. The sides and rump are sometimes heavily washed with orange. Some individuals look mottled yellowish brown (Barlow 1965; Gyldenstolpe 1932; Massoia 1973b).

Distribution

Oryzomys flavescens ranges from southern Brazil south through Uruguay to northern and central Argentina to at least Buenos Aires province (Contreras and Berry 1983; Contreras and Rosi 1980c; Honacki, Kinman, and Koeppl 1982) (map 11.12).

Life History

In Uruguay the reproductive peak is from April to May. The average number of embryos per female is 5.1 (range 3–7; $n = 10$) (Barlow 1969).

Ecology

Oryzomys flavescens appears to be found typically near water. In Uruguay it has been captured in stands of tall grass in marshes, along rivers and streams, and occasionally away from water. In central Argentina it is found near water in brushy arid country. *O. flavescens* is nocturnal. Ten stomachs were found to contain principally green plant material, though five held some invertebrate remains (Barlow 1969; Fornes and Massoia 1965).

Oryzomys fornesi Massoia, 1973

Measurements

	Mean	Min.	Max.	N	Loc.	Source[a]
TL	188.0	165	212	50	P	1
T	105.5	93	118			
HF	23.7	21	26			
E	13.8	12	16			
Wta	17.6	14	27	16		2

[a](1) Myers and Carleton 1981; (2) UM.

Description

This is the smallest *Oryzomys* found in Paraguay and Argentina. The tail is equal to about 128% of the

Ecology

This is the common woodland mouse of Uruguay, found in mesic subtropical forest with a closed canopy and a scanty understory as well as in more xeric, open thorn woods. It is primarily nocturnal, climbs well, and is primarily herbivorous, though it eats some insects (10%; $n = 4$) (Barlow 1969).

Oryzomys flavescens (Waterhouse, 1837)
Colilargo Chico

Measurements

	Mean	Min.	Max.	N	Loc.	Source[a]
TL	192.4	179.0	223.0	12 m	A	1
	176.6	156.0	206.0	10 f		
	202.0	187.0	223.0	18	U	2

head and body length, and the hind feet are relatively long (28.5% of head and body). The pinnae are densely furred inside. The dorsum is umber to russet, usually heavily lined with black. The orangish lateral stripe seen in some individuals of *O. nigripes* is absent or indistinct. The throat is variably white to gray, and the venter is whitish or grayish, occasionally washed with buffy. The cheeks are orangish brown or gray, and the tail is weakly bicolored (Massoia 1973b).

Chromosome number: $2n = 62-66$ (Myers and Carleton 1981).

Distribution

The species is found from Bolivia east to southern Brazil and south to northern Argentina in the prov-ince of Corrientes. It occurs throughout eastern Paraguay and into the Chaco (Contreras 1982a; Honacki, Kinman, and Koeppl 1982; Myers and Carleton 1981) (map 11.13).

Ecology

Oryzomys fornesi is strongly associated with marshes and wet grasslands in Paraguay and Argentina. In the Paraguayan Chaco it has been taken only in marshy habitats, both wet and dry. Some individuals have been taken on floating mats of grass (Dietz 1983; Massoia 1973b; Myers 1982; Myers and Carleton 1981; UM).

Comment

Olds and Anderson (1987) state that *Oryzomys fornesi* is a junior synonym of *O. microtis*.

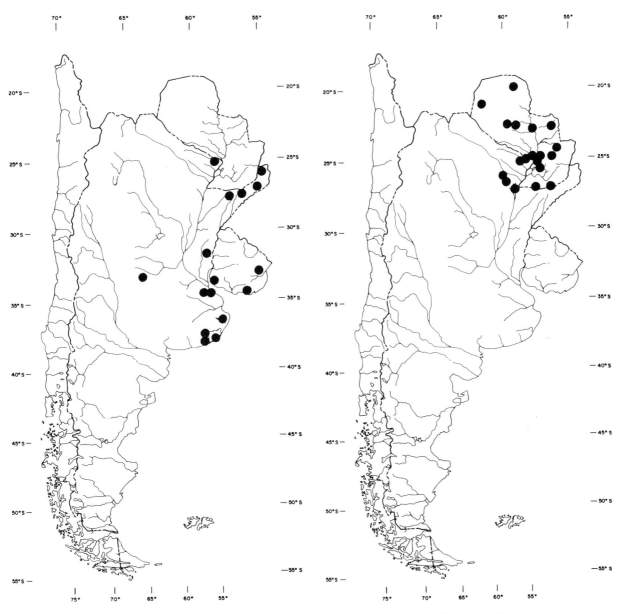

Map 11.12. Distribution of *Oryzomys flavescens*.

Map 11.13. Distribution of *Oryzomys fornesi*.

Oryzomys longicaudatus (Bennett, 1832)
Colilargo Común

Measurements

	Mean	Min.	Max.	N	Loc.	Source[a]
TL	222.5			15 m	C	1
	215.2			21 f		
HB	101.0	85	121	45 m	A	2
	93.2	77	111	45 f		
	98.3	85	119	56	C	3
T	127.3			15 m	C	1
	121.6			21 f		
HF	27.0			15 m		
	26.5			21 f		
E	15.5			15 m		
	15.1			21 f		
Wta	23.8			15 m		
	21.7			21 f		
	37.7	23	60	54 m	A	2
	27.8	15	43	58 f		

[a](1) Pefaur, Yáñez, and Jaksić 1979; (2) Pearson 1983; (3) Jaksić and Yáñez 1979.

Description

Oryzomys longicaudatus is easily distinguished among the Chilean fauna because of its very long tail and hind feet. The tail is strongly bicolored and sparsely haired and decreases in relative length with decreasing latitude. The ears are small and also sparsely haired. The dorsum is buffy with fine lines of blackish to light brown, with the gray at the bases of the hairs occasionally showing through. The venter is gray white. In Argentina the dorsum is often washed with ochraceous, which can be prominent along the sides (Mann 1978; Osgood 1943; pers. obs.).

Distribution

The species is distributed in Chile from approximately 25° south latitude to the islands off Tierra del Fuego. It is found in Argentina from the northwesternmost province of Jujuy south along the Andes to about Santa Cruz province, where it appears to occur across the whole width of the continent (Honacki, Kinman, and Koeppl 1982; Olrog and Lucero 1981; Tamayo and Frassinetti 1980; Thomas 1898a) (map 11.14).

Life History

In Argentina the average number of fetuses per female was 4.9 (range 2–11; *n* = 44). In Malleco province (Chile) females reproduce from November to February and average 4.9 embryos (range 3–9; *n* = 43). In Chile females start reproducing when only a few months old, have litters of three to five, and can breed three times a year (Greer 1966; Mann 1978; Pearson 1983).

Ecology

Oryzomys longicaudatus shows a remarkable flexibility in habitat choice, although it requires free water. In one area in north-central Chile it was caught in cloud forest, in brush forest transition, and in open brushy habitat, though it was by far most common in the cloud forest, where it represented 59% of all the small captures. In the Patagonian forests of Argentina it occurs occasionally in dense forest but seems to prefer brushy areas, edges of clearings, and roadsides with blackberry and wild rose tangles. In Salta province, Argentina, this species was captured in some areas only near permanent water, while in other areas it was most common in dense second growth though occasionally taken in mature subtropical forest away from standing water. In general *O. longicaudatus* seems to prefer moister areas with abundant cover.

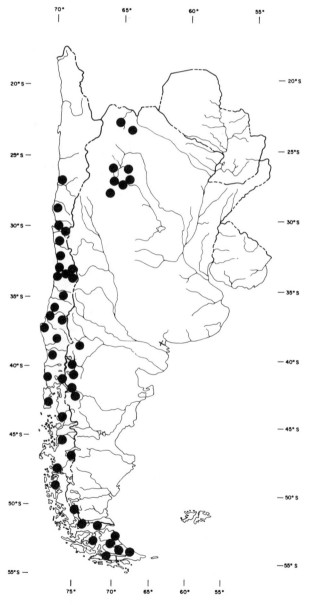

Map 11.14. Distribution of *Oryzomys longicaudatus*.

In Patagonian forests this mouse reaches densities of only 5.4/ha; in central Chile densities of 24–47/ha have been recorded; and in another area in central Chile during a population outbreak in cultivated fields, the incredible densities of 1,710 to 1,802 individuals per hectare were recorded. In southern Argentina the average home range for adult males was 1,007 m², and for adult females it was 681 m².

As suggested by its long tail and long hind feet, *O. longicaudatus* is a good climber and has been captured 3 m off the ground. It builds its nests in bushes and trees and will use abandoned bird nests. It is also a good jumper.

In one area *O. longicaudatus* was strongly granivorous during the dry season (seeds = 72.7% by volume), and in the wet season it specialized on flowers, pollen, and foliage of *Chenopodium* (flowers and pollen = 53.3% by volume; foliage = 29.4%). Arthropods were of very little significance in the diet. This species' fondness for seeds applies in forests as well as in brushy habitats, but in one area insects formed 14.7% of the diet measured year round. When bamboos undergo mass flowering, populations of *O. longicaudatus* have been reported to undergo dramatic increases (Contreras 1972a; Glanz 1977a,b; Greer 1966; Mann 1978; Mares 1973, 1977a; Meserve 1981a,b; Murúa and González 1979, 1981, 1982; Murúa, González, and Jofre 1982; Ojeda 1979; Pearson 1983; Pearson and Pearson 1982; Pefaur, Yáñez, and Jaksić 1979; Pereira 1941; Schamberger and Fulk 1974).

Comment
Some investigators believe that the southernmost members of *Oryzomys longicaudatus* belong to a separate species, *O. magellanicus*.

Oryzomys nigripes (Olfers, 1818)

Measurements

	Mean	Min.	Max.	N	Loc.	Source[a]
TL	224.8	161.0	270.0	80	P	1
	216.1	199.0	240.0	10	A	2
HB	91.6	79.0	107.0			
T	125.5	100.0	149.0	80	P	1
	124.5	110.0	137.0	10	A	2
HF	24.8	21.0	28.0	80	P	1
	23.9	23.0	25.0	10	A	2
E	16.7	14.0	20.0	80	P	1
	15.3	14.0	16.0	10	A	2
Wta	23.9	23.0	25.0			
	21.4	13.5	29.5	7	A	3

[a](1) Myers and Carleton 1981; (2) Massoia and Fornes 1967; (3) Mares, Ojeda, and Kosco 1981.

Description
Oryzomys nigripes is a medium-sized species of the subgenus *Oligoryzomys*. The lightly bicolored tail equals about 126% of the head and body length, and the relatively small tan feet equal about 24.5% of head and body length. The ears are small and sparsely haired. The animal is characterized by a grayish white venter with white-tipped hairs, often with an orangish pectoral band. It has a light brown dorsum. In Paraguay *O. nigripes* can be distinguished from *O. chacoensis*, which has a paler, yellowish dorsal stripe, white venter, and no pectoral band. Also, *O. chacoensis* has hairs on the throat that are usually white to the base; there is a complete or almost complete absence of a dark basal band on the same hairs inside the ear (prominent in *O. nigripes*). Some individuals have a pronounced V of black hairs originating at the nose and passing over the eyes.

In Argentina this species is similar in size to *O. flavescens* but always has a light gray venter and dark ears like *O. delticola*. It differs from *O. longicaudatus* in pelage color and texture (Massoia 1973b; Myers and Carleton 1981; pers. obs.).

Distribution
The species is found from eastern Brazil to Argentina. In Paraguay it occurs east of the Río Paraguay; it occurs throughout Uruguay; and in Argentina it is found south to at least Buenos Aires province (Honacki, Kinman, and Koeppl 1982; Massoia and Fornes 1967b; Myers and Carleton 1981) (map 11.15).

Life History
Animals from a Brazilian population taken into the laboratory had an average litter size of 3 (*n* = 18), an average neonatal weight of 3.3 g, and an estrous cycle of ten days; females reproduced for the first time at fifty-eight days. In southern Brazil animals reproduced all year, with two peaks from September to November and February to April. Gestation lasted twenty-five days; litters were produced every 2.5 to 3 months, and the litter size was three to four. In Paraguay the average number of embryos was 3.6 (range 2–5) (Dalby 1975; Mello 1978a; Myers and Carleton 1981; Veiga-Borgeaud 1982).

Ecology
In Paraguay *O. nigripes* primarily inhabits forests and areas of second growth, though very high densities can occur in newly cleared fields. In southern Brazil this species was found in cultivated fields, forest borders, and areas of second growth; in northeastern Argentina it was caught along trails and roads and in fields of thick grass; in northwestern Argentina it has been trapped in mesic forests, gallery forests, old fields, and areas with *Chusquea* bamboo. In one Argentinian study *O. nigripes* was found at a density of 5/ha. In Minas Gerais, Brazil, this

species was reasonably uncommon in forested areas and was a good climber, making extensive use of low shrubs, where it built nests about 1 m off the ground.

These animals are nocturnal and are caught on the ground as well as in trees. They build ball nests, usually in trees, though they are also reported to burrow under logs. Stomachs were found to contain leaves, grass seeds, fungal hyphae, and insects. In southeastern Brazil insects were found in abundance in the stomachs examined (Contreras and Berry 1983; Crespo 1982a; Dalby 1975; Fonseca 1988; Mares, Ojeda, and Kosco 1981; Myers and Carleton 1981; Stallings 1988a,b; Vaz-Ferreira 1958; Veiga-Borgeaud 1982; Villafañe et al. 1973).

Comment

Oryzomys nigripes includes *O. eliurus.* Dalby (1975) mistakenly called the species he worked with *O. nigripes* instead of the proper *O. flavescens* (P. Myers, pers. comm.).

Oryzomys nitidus (Thomas, 1884)

Measurements

	Mean	Min.	Max.	N	Loc.	Source[a]
TL	273.6	255.0	286.0	14	A	1
T	148.8	140.0	155.0			
HF	33.3	32.7	34.0			
E	24.9	24.1	25.8			
Wta	55.2	43.0	80.0			

[a](1) Mares, Ojeda, and Kosco 1981.

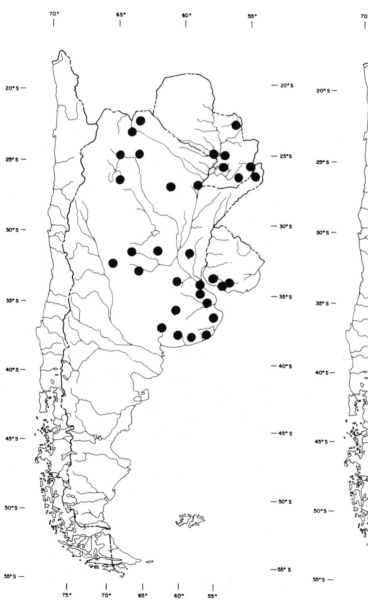

Map 11.15. Distribution of *Oryzomys nigripes*.

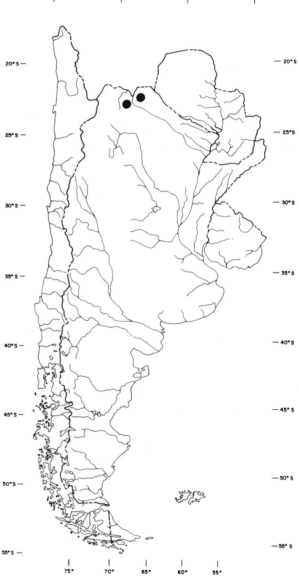

Map 11.16. Distribution of *Oryzomys nitidus*.

Description

This large *Oryzomys* is orangish brown on the dorsum and lighter on the sides, with a sharply contrasting white or gray white venter. The tail is long and bicolored; the ears are large and sparsely haired (pers. obs.).

Distribution

Oryzomys nitidus is distributed from Ecuador south to northwestern Argentina, where it occurs in Salta province (Honacki, Kinman, and Koeppl 1982; Mares, Ojeda, and Kosco 1981) (map 11.16).

Ecology

In Salta province *O. nitidus* occurs only in mesic and transitional forests. It is not abundant in any area but is most common in forests with little undergrowth. Individuals of this species have been taken in dense second growth along streams and roads. They inhabit burrows in the forest floor. Stomachs have been found to contain berries, unidentified plant remains, and insects (Mares, Ojeda, and Kosco 1981; Ojeda and Mares 1989).

Comment

Oryzomys nitidus includes *O. legatus.*

Oryzomys ratticeps (Hensel, 1873)

Measurements

	Mean	Min.	Max.	N	Loc.	Source[a]
TL	393.5	355	424	10	P	1
HB	185.0	164	194			
T	208.5	191	230			
HF	37.9	35	41			
E	24.9	22	26			
Wta	144.0	120	157	4		

[a](1) UM.

Description

This is a very large *Oryzomys*. The ears are large and sparsely haired; the tail is very long and not bicolored. The dorsum is orangish brown and the cream-colored venter is sharply demarcated (see account for *O. buccinatus*). The head is darker gray with a wash of white around the nose of some specimens, and the feet are whitish above (pers. obs.).

Distribution

Oryzomys ratticeps is found in southern Brazil, northeastern Paraguay, and eastern Paraguay, though it has been collected at a few localities in the Paraguayan Chaco (Honacki, Kinman, and Koeppl 1982; Massoia 1980a; Myers 1982; UM)((map 11.17).

Ecology

This is a forest-dwelling species, though it has also been caught at the ecotones between forests and fields (Myers 1982; UM).

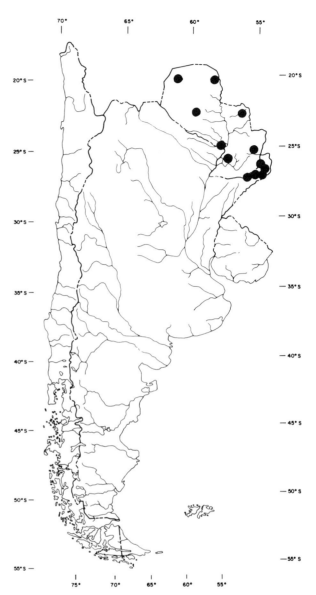

Map 11.17. Distribution of *Oryzomys ratticeps.*

Genus *Nectomys* Peters, 1861
• *Nectomys squamipes* (Brants, 1827)
Neotropical Water Rat, Rata Nadadora

Measurements

	Mean	Min.	Max.	N	Loc.	Source[a]
TL	419.9	390.0	497.0	15	A, P	1
HB	206.9	181.0	252.0			
	192.0	166.0	221.0	7	A	2
T	213.0	198.0	245.0	15	A, P	1
	223.0	197.0	246.0	7	A	2
HF	50.8	48.3	53.5	15	A, P	1
	50.0	47.0	52.0	7	A	2
E	22.0	20.0	24.0			
Wta	255.0	170.0	350.0	15	Br	1
	236.0	153.0	310.0	4	A	2
	281.5	222.0	380.0	10	P	3

[a](1) Hershkovitz 1944; (2) Massoia 1976a; (3) UM.

Description

This large rat is unmistakable because of its long, glossy coat, dark brown dorsum, and grayish venter. The sides are lighter and often show an orangish wash (plate 15). *Nectomys* can be confused with rats of the genus *Holochilus*, but species of *Nectomys* generally have smaller hind feet. Also, in general the coat is softer and more like otter (*Lutra*) fur, and the vibrissae are longer (Hershkovitz 1944; Massoia 1976a; pers. obs.).

Distribution

Nectomys squamipes is distributed from the Guianas to Paraguay and northeastern Argentina. In

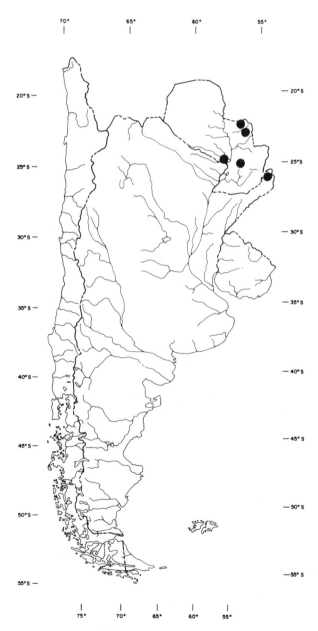

Map 11.18. Distribution of *Nectomys squamipes*.

southern South America it is found in eastern Paraguay and northeastern Argentina (Ernest 1986; Hershkovitz 1944; Honacki, Kinman, and Koeppl 1982; Massoia 1976a; UM) (map 11.18).

Life History

In Misiones province this species reproduces from October to November with a litter size of four or five (Crespo 1982b).

Ecology

In general this semiaquatic species is confined to streams within forests, but it is occasionally found in other moist areas. In central Brazil home ranges were calculated as 2,200 m². This nocturnal species spends considerable time foraging in the water and eats fungi, plant matter, fruits, seeds, invertebrates, and vertebrates. It builds nests under old logs or brush heaps or in tangled roots. When released from a trap, individuals almost always flee to the water (Ernest 1986; Hershkovitz 1944).

Comment

Nectomys squamipes is probably a composite species containing at least five named forms (Reig 1984, 1986).

Genus *Pseudoryzomys* Hershkovitz, 1962
Pseudoryzomys wavrini (Thomas, 1921)
Rata de Estero

Measurements

	Mean	Min.	Max.	N	Loc.	Source[a]
TL	229.0	211	280.0	12	A, B, P	1
T	116.0	94	140.0	13		
HF	28.0	25	33.0	10		
E	16.0	15	18.0	8		
Wta	40.9	30	56.5	9	A, P	2

[a](1) Pine and Wetzel 1975; (2) Massoia 1976b; Pine and Wetzel 1975; UConn.

Description

Externally, this species resembles some species of *Oryzomys* (*Oligoryzomys*). The digits of the hind feet may be connected at the base by rudimentary membranes. Dorsally and laterally it is brown, darker on the back and streaked with black, sometimes with a faint flecking of orange that is more pronounced on the sides and cheeks. Ventrally it is pale plumbeous washed with tawny buff. The cheeks may tend to be washed with white. The tail is gray on top and white below.

Distribution

Pseudoryzomys wavrini is found in Bolivia, the Chaco of Paraguay, and the northern Argentine provinces of Formosa and Chaco (Honacki, Kinman, and Koeppl 1982; Massoia 1976b; Myers 1982) (map 11.19).

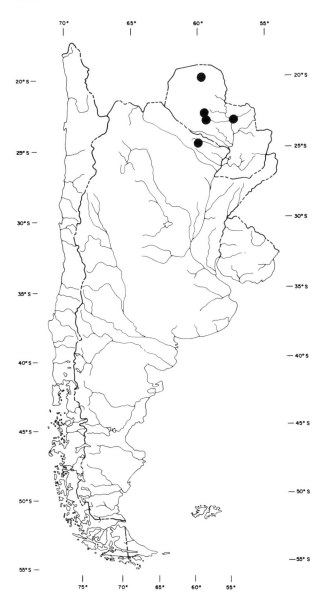

Map 11.19. Distribution of *Pseudoryzomys wavrini*.

Ecology

In the Paraguayan Chaco this species is apparently restricted to mesic savannas and grasslands; it has been taken in a grassy patch surrounded by thorn forest, in grass on a small floodplain, in grass at the edge of thorn scrub, and in tall bunchgrass pasture (Myers 1982; Pine and Wetzel 1975).

TRIBE PHYLLOTINI

The tribe Phyllotini, though not formally diagnosed, was first proposed as a taxon by Vorontzov (1960). Olds and Anderson (1989) have offered a diagnosis for the group, along with the following caveats. There is no one character that is unique for the tribe. The

genera *Reithrodon*, *Neotomys*, and *Euneomys* form a distinct subgroup within this tribe. The tribe exhibits the following character states: the heel of the hind foot tends to be haired, the pinnae are moderately long to rather long; the incisive foramina are long; the zygomatic notch is deeply excised; the molars are tetralophodont; and the M_3 is more than half the length of the M_2 (this character distinguishes the phyllotines from most akodonts). This diagnosis does not include *Pseudoryzomys* (see previous section). The genera *Reithrodon*, *Neotomys*, and *Euneomys* have grooved upper incisors. In *Euneomys* the molars approximate an S shape, a condition exhibited to a lesser extent to *Reithrodon* and *Neotomys*, reminiscent of the Sigmodontini (Olds and Anderson 1989).

Genus *Andalgalomys* Williams and Mares, 1978

Description

Members of this phyllotine rodent genus have large ears and a tail roughly equal to or longer than the head and body, which range from 86 to 115 mm in length while the tail is 97–130 mm. The eyes are not noticeably reduced in relative size. The tail is bicolored, sparsely haired, and reddish brown to gray brown, contrasting with the white venter and feet. The feet are marked, distinguishing the genus from the similar *Eligmodontia*.

Distribution

The genus is found in xeric and grassland habitats of Argentina and Paraguay.

Andalgalomys olrogi Williams and Mares, 1978
Laucha Colilarga Gris

Measurements

	Mean	Min.	Max.	N	Loc.	Source[a]
HB	100.0	86.0	113.0	5	A	1
T	127.0	125.0	130.0			
HF	24.4	23.8	24.9			
E	20.6	19.6	22.0			

[a](1) Williams and Mares 1978.

Description

Andalgalomys olrogi is similar to the sympatric *Eligmodontia typus*, but this moderately small mouse has naked-soled, cushionless feet and greatly inflated auditory bullae. Externally it resembles a small *Graomys griseoflavus*. The tail is relatively long and moderately penicillate, the pinnae are large and broad, and the vibrissae are well developed. It has moderately long, lax yellowish brown fur, a small whitish subauricular spot, and a small

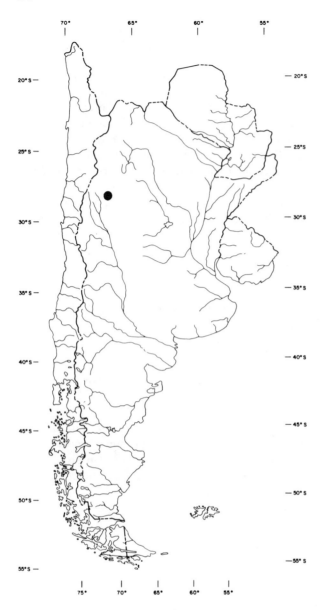

Map 11.20. Distribution of *Andalgalomys olrogi*.

whitish postauricular patch. The hairs on the venter and the tops of the feet are white to their bases (Williams and Mares 1978; pers. obs.).

Distribution

This species is known only from three localities in Catamarca province, Argentina (Williams and Mares 1978) (map 11.20).

Ecology

Andalgalomys olrogi appears to be relatively specialized for the semidesert habitat where it occurs. All known specimens were trapped in or near *Larrea* (creosote bush) stands on fine-textured soils. The only other cricetines caught were *Calomys musculinus* and *Eligmodontia typus* (Williams and Mares 1978).

Andalgalomys pearsoni (Myers, 1977)

Measurements

	Mean	Min.	Max.	N	Loc.	Source[a]
TL	202.3	163	227	14	P	1
HB	93.9	72	109			
	97.0	90	115	5	A	2
T	108.4	91	130	14	P	1
	108.0	92	117	5	A	2
HF	24.9	23	26	14	P	1
	23.6	22	25	5	A	2
E	18.1	16	20	14	P	1
	19.2	18	20	5	A	2
Wta	25.4	16	35	8	P	1

[a](1) PCorps; (2) Williams and Mares 1978.

Description

Andalgalomys pearsoni is larger than the sympatric *Calomys* and smaller than *Graomys griseoflavus*, with proportionally smaller ears and a shorter, nonpenicillate tail that is longer than the head and body. The dorsum is yellowish brown overwashed with black-tipped hairs. The bicolored tail is brownish dorsally and sparsely haired. The hairs on the venter and the feet are white; the face is brownish and not noticeably lighter than the dorsum, but without the black wash. The sides are often washed with orange (Myers 1977; Williams and Mares 1978; pers. obs.).

Distribution

The species is limited to the Chaco of western Paraguay (Williams and Mares 1978) (map 11.21).

Ecology

This mouse inhabits dry grasslands that occur in islands in the Chaco of western Paraguay. The surrounding thorn scrub is inhabited by *Graomys griseoflavus* (Myers 1977).

Genus *Andinomys* Thomas, 1902
Andinomys edax Thomas, 1902
Rata Andina

Measurements

	Mean	Min.	Max.	N	Loc.	Source[a]
TL	261.2	222	300	13	A, C, Peru	1
HB	140.8	106	162			
T	24.2	22	30	12		
HF	27.7	26	30	13		
E	24.2	22	30	12		
Wta	57.5	41	91			

[a](1) Pearson 1951; Pine, Miller, and Schamberger 1979; Spotorno 1976; BA, Santiago.

Description

This stout-bodied rat has very dense fur that is long, soft, and fine. The ears are fairly large, and the tail is bicolored and nonpenicillate. The color varies from uniformly dark buffy to uniformly brown dorsally and gray ventrally (Gyldenstolpe 1932; Pearson 1951; pers. obs.).

Distribution

Andinomys edax is found at high altitudes in Peru, Bolivia, northwestern Argentina, and northern Chile. In Argentina it occurs along the Andes south to La Rioja province, and in Chile it is found in Arica province (map 11.22).

Life History

In southern Peru one trapped female had three embryos (Pearson 1951).

Ecology

This high-altitude animal has been caught at between 1,800 and 5,100 m. Throughout this range it appears to be found mainly in dense vegetation along watercourses, although it has also been trapped in rocky terrain (Fonollat 1984; Olrog 1979; Pearson 1951; Pine, Miller, and Schamberger 1979; Tamayo and Frassinetti 1980; Thomas 1919a).

Genus *Auliscomys* Osgood, 1915

Description

Members of this phyllotine rodent genus possess a complex baculum, a tail shorter than the head and body, relatively short hind feet, a large external ear, a relatively large M_3, and a simplified or tetralophodont molar. Head and body length ranges from 96 to 127 mm and tail length from 53 to 85 mm. The vibrissae are long, the tail tends to be bicolored, and

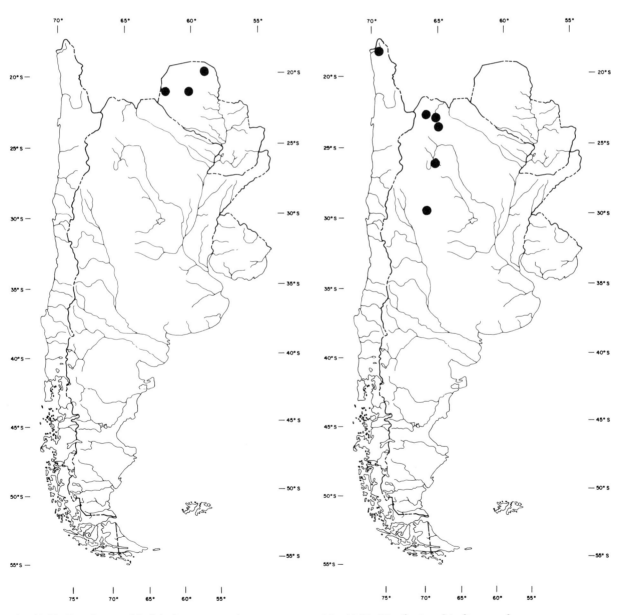

Map 11.21. Distribution of *Andalgalomys pearsoni*.

Map 11.22. Distribution of *Andinomys edax*.

the dorsum is a shade of brown contrasting with the white to gray venter. The juveniles of some species have gray dorsal pelage.

Distribution

The genus is found in Peru, Bolivia, Argentina, and Chile. It prefers high-elevations, occurring up to 6,000 m.

Auliscomys boliviensis (Waterhouse, 1846)

Measurements

	Mean	Min.	Max.	N	Loc.	Source[a]
TL	202.0	182	215	8 m	Peru	1
	193.0	180	213	4 f		
	213.6	185	214	5	C	2
T	85.0	68	95	8 m	Peru	1
	84.0	76	90	4 f		
	86.0	81	94	5	C	2
HF	27.0	26	30	8 m	Peru	1
	26.0	26	27	4 f		
	27.2	26	28	5	C	2
E	26.0	23	27	8 m	Peru	1
	26.0	24	27	4 f		
	26.6	26	27	5	C	2
Wta	54.6	41	70			

[a](1) Pearson 1951; (2) Santiago.

Description

The upper incisors are pale and ungrooved. This stout-bodied mouse has a tail slightly shorter than its head and body, long, lax fur, and very large ears. It is buffy lined with tan dorsally and has a creamy white venter. There is a patch of yellow fur in front of each ear. The soles of the hind feet are blackish, and the tail may be slightly bicolored (Gyldenstolpe 1932; Mann 1978; Osgood 1943; Pearson 1951, 1958; pers. obs.).

Distribution

The species is found in western Bolivia, northern Chile, and southern Peru at high elevations. In Chile it is found in Tarapacá province (Honacki, Kinman, and Koeppl 1982) (map 11.23).

Life History

In Peru females have three to four embryos. While in captivity females will produce three litters a year of three to five young each (Pearson 1951; Pine, Miller, and Schamberger 1979).

Ecology

Auliscomys boliviensis is a high-altitude animal and has been trapped up to almost 6,000 m. It prefers open country and is commonly found on boulder-strewn slopes with sparse vegetation, on rock slides or stone walls, and in *Ctenomys* burrows and viscacha colonies. It is diurnal and is broadly granivorous-frugivorous, with green plant material but not fungus composing a significant part of the diet (Mann 1978; Pearson 1951, 1958; Pine, Miller, and Schamberger 1979; Tamayo and Frassinetti 1980).

Auliscomys micropus (Waterhouse, 1837)

Measurements

	Mean	Min.	Max.	N	Loc.	Source[a]
TL	221.0	212	237	8 m	C	1
	221.0	215	235	10 f		
HB	127.1	110	157	38	—	2
T	92.0	85	100	8 m	C	1
	93.0	85	100	10 f		
	95.8	85	121	38	—	2
HF	29.0	27	30	8 m	C	1
	29.0	28	30	10 f		
	28.9	27	32	38	—	2
E	20.2	18	22	5	C	3
Wta	72.8	50	103	20 m	A	4
	72.6	45	105	25 f		

[a](1) Allen 1903; (2) Pearson 1958; (3) Santiago; (4) Pearson 1983.

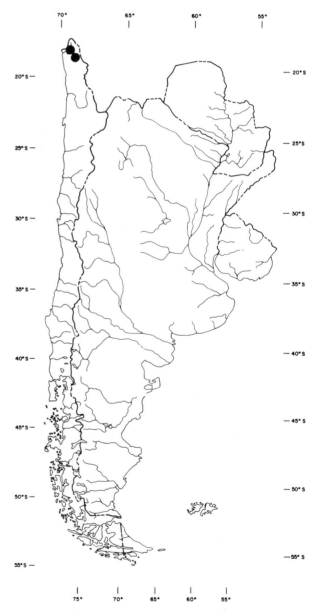

Map 11.23. Distribution of *Auliscomys boliviensis*.

Description

This medium-sized, robust species has an unfurred, bicolored tail shorter than its head and body, relatively small ears, and broad, ungrooved incisors. The dorsum varies from gray to brownish and the venter is lighter, often washed with white (plate 14) (Mann 1978; Osgood 1943; Pearson 1958; pers. obs.).

Distribution

The species is found in southern Chile and Argentina. In Chile it ranges along the Andes from about Talca province south to Magallanes province, and in Argentina north to Tucumán province (Honacki, Kinman, and Koeppl 1982; Mann 1978; Tamayo and Frassinetti 1980; CM) (map 11.24).

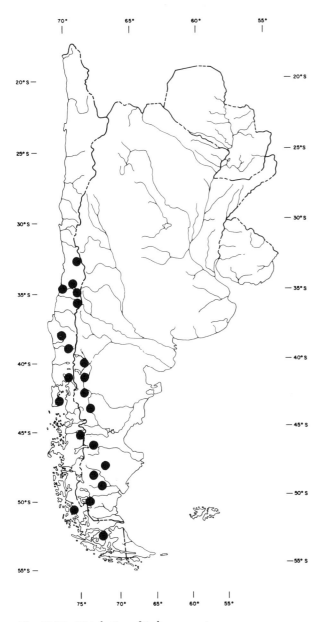

Map 11.24. Distribution of *Auliscomys micropus*.

Life History

In Chile the average number of embryos per female was 4.8 (range 4–5; $n = 5$); in Argentina the average was 4.1 (range 1–7; $n = 27$) (Greer 1966; Pearson 1983; Pine, Miller, and Schamberger 1979).

Ecology

Auliscomys micropus has been caught in a wide variety of habitats from precordilleran steppe to forests. Its presence seems tied to sufficient cover; in forests it occurs only where there is well-developed ground cover, and in grasslands only where there are many bushes. In the Patagonian forests of Argentina it reaches densities of 5.1/ha. *A. micropus* ranges from sea level to 3,000 m, is nocturnal, and burrows but also climbs well. It is primarily herbivorous; stomachs have been found to contain fungus, grass seeds, and fruit (Greer 1966; Mann 1978; Murúa, Gonzalez, and Jofre 1982; Pine, Miller, and Schamberger 1979; Pearson 1983; Pearson and Pearson 1982; Reise and Venegas 1974; Tamayo and Frassinetti 1980).

Auliscomys sublimis Thomas, 1900
Pericote Andino

Measurements

	Mean	Min.	Max.	N	Loc.	Source[a]
TL	163.9	140	178	12	A, Peru	1
HB	111.3	96	125			
T	52.7	43	62			
HF	22.5	21	25			
E	21.4	20	23	8		
Wta	38.5			1		

[a](1) Mares, Ojeda, and Kosco 1981; Pearson 1951; FM.

Description

There is a faint groove on the anterior face of the upper incisors. This is a volelike species of medium size with a short tail. It has long, thick fur and comparatively small ears. The dorsum is dull grayish fawn to tan flecked with gray and may be sharply demarcated from the white venter. Its feet and tail are pale above. Individuals from some populations may have considerable orange on the sides and rump (Gyldenstolpe 1932; Mares, Ojeda, and Kosco 1981; Pearson 1951; pers. obs.).

Distribution

Auliscomys sublimis is found on the altiplano from southern Peru to northeastern Chile and northern Argentina (Honacki, Kinman, and Koeppl 1982; Pearson 1958) (map 11.25).

Ecology

This high-altitude species is found from 4,000 to 6,000 m. At 6,000 m it sets one of the highest altitude records for a mammal in the Western Hemisphere. *A. sublimis* is an animal of open habitats,

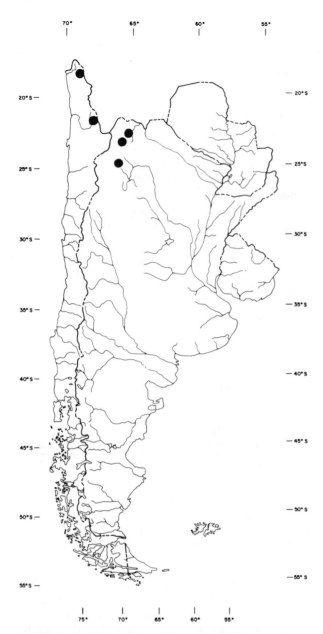

Map 11.25. Distribution of *Auliscomys sublimis*.

preferring areas with bunchgrass and rocks. It shelters under rocks or rock walls or in *Ctenomys* burrows. It is apparently gregarious and is often caught with other individuals of the same species. In southern Peru it disappeared from October to December, perhaps hibernating (Koford 1955; Mares, Ojeda, and Kosco 1981; Pearson 1951, 1958; Sanborn 1950; Tamayo and Frassinetti 1980).

Genus *Calomys* Waterhouse, 1837

Description

There are seven species recorded in this genus, which typically differ in size when sympatric. The molars are tetralophodont, and the M_3 ranges from one-half to three-fourths the length of the M_2 (see fig. 11.4). Head and body length is 62–123 mm, and the tail is 31–92 mm. In these phyllotine rodents, males have a complex baculum, the tail is equal to or less than the head and body length, and the external ears are moderate but not visibly reduced in size. Species of *Calomys* superficially resemble the introduced urban *Mus musculus*, but figure 11.18 clearly points out the differences. The dorsum ranges from light brown to gray brown, and the venter from white to gray.

Distribution

The genus is predominately found in southern South America (Peru, Bolivia, Chile, Argentina, Paraguay, and southeastern Brazil) but is also found in isolated "pockets" in Colombia and Venezuela.

Calomys callosus.(Rengger, 1830)
Laucha Grande

Measurements

	Mean	Min.	Max.	N	Loc.	Source[a]
TL	182.8	175.0	192.0	5	A	1
HB	98.2	65.0	124.0	82	A	2
T	87.2	82.0	92.0	5	A	1
	78.9	60.0	100.0	82	A	2
HF	21.1	18.8	22.5	5	A	1
	20.4	19.0	26.0	81	A	2
E	19.5	18.5	20.5	5	A	1
	16.8	14.0	21.0	81	A	2
Wta	30.9	20.0	40.9	5	A	1

[a](1) Mares, Ojeda, and Kosco 1981; (2) Massoia and Fornes 1965a.

Description

This largest species of *Calomys* is dark gray brown dorsally and gray ventrally. Its sparsely haired tail is bicolored, and its feet are tan (pers. obs.).
Chromosome number: $2n = 36$; FN = 48 (Reig 1986).

Distribution

Calomys callosus is found in eastern and southwestern Brazil, Bolivia, Paraguay, and northern

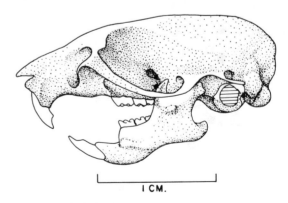

Figure 11.4. Skull of *Calomys laucha*.

Argentina. In southern South America it is found in the Paraguayan Chaco, and in Argentina across the central and northern provinces (Honacki, Kinman, and Koeppl 1982; Lucero 1983; Myers 1982; Thomas 1916a,b) (map 11.26).

Life History

Animals from a central Brazilian population have a gestation period averaging 21.8 days (range 20–23; $n = 43$), a litter size averaging 4.5 (range 2–9; $n = 13$), and an average neonatal weight of 2.3 ($n = 42$). Young open their eyes on the sixth or seventh day, and lactation lasts fifteen to seventeen days. Females undergo a postpartum estrus. Lab animals from Argentina reproduced at thirty-seven days of age for males and thirty for females. In northern Argentina two trapped females had eight and six embryos (Hodara et al. 1984; Mares, Ojeda, and Kosco 1981; Mello 1978b).

Ecology

Calomys callosus is a species of open vegetation formations, especially common in areas disturbed by humans. In Paraguay it was trapped in palm savannas, in bunchgrass meadows, and in dry marshes. In the Brazilian caatinga it is found in the later seral stages of old-field succession and in low thorn scrub, and in central Brazil it is most common in areas where forest has been cleared and grass planted. In northern Argentina cultivated fields, areas of secondary growth, forest edges, orchards, stream edges, and riverbanks are common places to trap this species. Individuals are often captured from burrows under rocks or in tangled roots; in central Brazil nests are generally found aboveground in clumps of grass or in the branches of dead trees (Mares, Ojeda, and Kosco 1981; Massoia and Fornes 1965a; Mello 1978b; Ojeda 1979; Streilein 1982a; UM).

Calomys laucha (Olfers, 1818)
Laucha Chica

Measurements

	Mean	Min.	Max.	N	Loc.	Source[a]
TL	127.0	110	142.0	13	U	1
T	54.9	49	68.0			
HF	16.5	16	18.0			
E	13.1	12	14.0			
Wta	12.6	9	15.5			
	13.3			41	A	2

[a](1) Barlow 1965; (2) Crespo et al. 1970.

Description

This very small rodent looks like a house mouse. The tail is only about 40% of the total length. The dorsum is light brown to tawny mixed with blackish, and the venter is grayish white. Its distinguishing characteristic is a small white patch behind each ear (see plate 14) (Barlow 1965; Mares, Ojeda, and Kosco 1981; Massoia et al. 1968; pers. obs.).
Chromosome number: $2n = 62$; FN = 82 (Reig 1986).

Distribution

Calomys laucha is found in southeastern Brazil, Paraguay, Uruguay, central Argentina, and southern Bolivia. In southern South America it occurs in the south and west of the Paraguayan Chaco, through much of Uruguay, and in Argentina south to about Río Negro province (Barlow 1965; Honacki, Kinman, and Koeppl 1982; Myers 1982; Olrog and Lucero 1981; Vallejo and Gudynas 1981) (map 11.27).

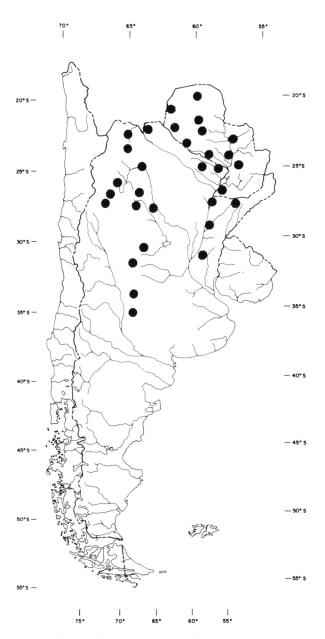

Map 11.26. Distribution of *Calomys callosus*.

Ecology

This little cricetine is common in many of the savanna and grassland areas of southern South America. It is one of the few Argentine rodents tested so far that can live for extended periods with no free water. It has been caught in small shrubs near coastal areas and rivers, in old fields, on rocky hillsides, and near dwellings. It has been reported at a maximum density of 87/ha. Nests have been found under boards and rocks, in crevices in the ground, and even in trees. Typically it uses subterranean galleries or depressions beneath objects to build its grass nests. *C. laucha* is primarily herbivorous, climbs well, and periodically experiences dramatic population increases (Barlow 1965, 1969; Kravetz and Villafañe 1981; Lucero 1983; Mares 1973; Mares, Ojeda, and Kosco 1981; Myers 1982; Vallejo and Gudynas 1981).

Calomys lepidus (Thomas, 1884)
Laucha Andina

Measurements

	Mean	Min.	Max.	N	Loc.	Source[a]
TL		113	140	28	Peru	1
T		35	55			
HF		17	19			

[a](1) Sanborn 1950.

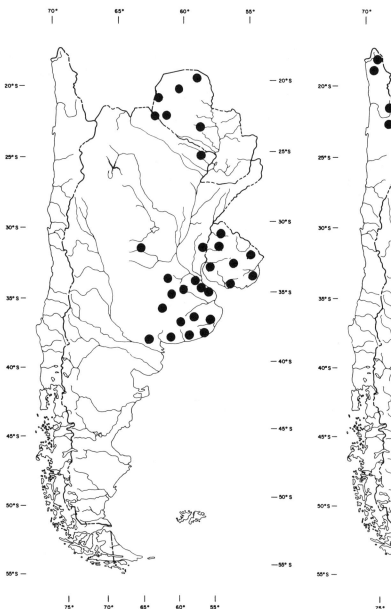

Map 11.27. Distribution of *Calomys laucha*.

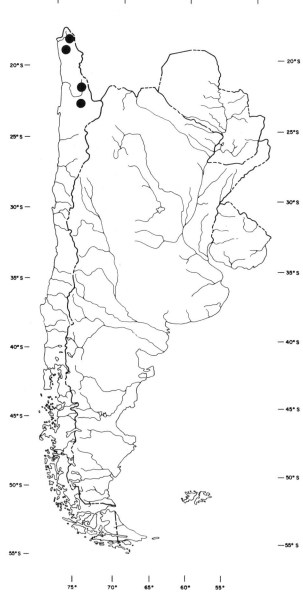

Map 11.28. Distribution of *Calomys lepidus*.

Description

Calomys lepidus is a small mouse with dense fur, grizzled tan and brown on the dorsum, which is darker along the spine, grading to lighter brown on the sides, and with a white or gray venter. The cheeks are white or gray. Its short tail, entirely tan, is less than 30% of the total length, distinguishing this species from the longer-tailed *Eligmodontia* (Mann 1978; pers. obs.).

Chromosome number: $2n = 36$; FN $= 68$ (Reig 1986).

Distribution

This high-altitude mouse is distributed along the Andes from western Bolivia south to the northernmost Chilean province of Antofagasta and the northwestern Argentine province of Jujuy (Honacki, Kinman, and Koeppl 1982; Tamayo and Frassinetti 1980) (map 11.28).

Ecology

This species lives in high-altitude grasslands from 3,300 to 5,000 m (Dorst 1971; Koford 1955; Tamayo and Frassinetti (1980).

Calomys musculinus (Thomas, 1913)
Laucha Bimaculada

Measurements

	Mean	Min.	Max.	N	Loc.	Source[a]
HB	78.5	65.0	96.0	57	A	1
T	73.4	61.0	91.0	55		
HF	17.8	15.5	20.5	57		
E	13.4	11.0	16.0			
Wta	13.5	8.2	21.2			
	17.5			633	A	2

[a](1) Contreras and Rosi 1980b; (2) Crespo et al. 1970.

Description

This small mouse is slightly larger than a house mouse and longer than the conspecific *C. laucha*. The dorsum is agouti brown, washed with orange on the sides and cheeks, with contrasting grayish white on the venter. Its bicolored tail is about 50% of total length, and there is a white stripe behind each ear (Massoia, et al. 1968; pers. obs.).

Distribution

Calomys musculinus is found in most of central and northwestern Argentina (Honacki, Kinman, and Koeppl 1982; Olrog and Lucero 1981; Contreras and Justo 1974) (map 11.29).

Life History

In the laboratory gestation ranges from 21 to 24.5 days; average litter size is 5.4 (range 1–15); and for females the average age of first reproduction is reported to range from 37 to 72.5 days. In the Ar-

gentinian pampas females reproduce mainly from November to April, and the average number of embryos per female is seven (range 2–11; $n = 37$). In a field study in Córdoba maximum life expectancy was six to eight months, and overwintering juveniles reproduced in the spring (Crespo et al. 1970; Hodara et al. 1984; Villafañe 1981b; Villafañe and Bonaventura 1987).

Ecology

Calomys musculinus appears to prefer open vegetation formations and is the dominant species in some parts of the Argentine pampas. In agricultural areas densities may be elevated. In Mendoza prov-

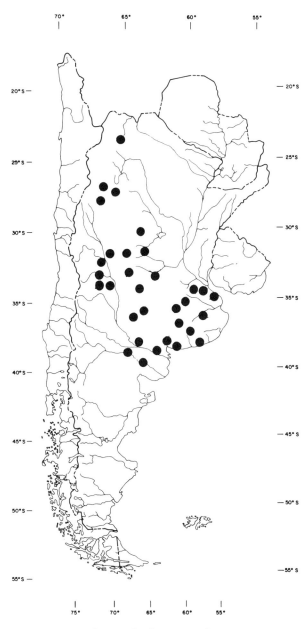

Map 11.29. Distribution of *Calomys musculinus*.

ince these mice reached a density of 37.6/ha with an average home range of 366 m². In Córdoba they are commonly found in sorghum and alfalfa fields and are a reservoir for Argentine hemorrhagic fever.

This species can tolerate a water-free diet in the laboratory but loses weight. Despite its physiological capability for a xeric existence, in the wild it seems to avoid xeric microhabitats (Contreras and Rosi 1980b; Crespo et al. 1970; Kravetz and Villafañe 1981; Mares 1977d; Villafañe 1981b; Villafañe and Bonaventura 1987).

Comment

Reig (1986) has separated a segment of *Calomys musculinus* and placed it in a separate species, *C. murillus*.

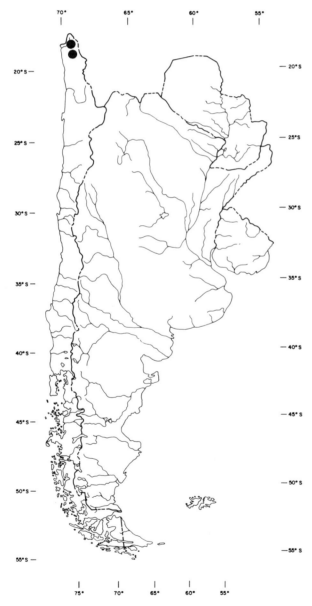

Map 11.30. Distribution of *Chinchillula sahamae*.

Genus *Chinchillula* Thomas, 1898
Chinchillula sahamae Thomas, 1898
Laucha Chinchilla

Measurements

	Mean	Min.	Max.	N	Loc.	Source[a]
TL	263.4	230.0	300	11	Peru	1
HB	160.9	140.0	182			
T	103.8	90.0	118			
HF	34.7	33.5	36			
E	35.8	34.0	38			
Wta	139.5	124.0	184	4		2

[a](1) FM; (2) MVZ.

Description

This beautiful Andean rat has very long, soft, silky fur. The tail is of medium length, well haired and distinctly penicillate. The ears, eyes, and feet are large. It is strikingly and unmistakably marked with a white venter and a dorsum of ashy brown, washed with blackish and intermixed with very long hairs that are black tipped on the back and white tipped on the front. The black markings are particularly evident along the dorsum and on the hips. The hips and rump are also white, with a conspicuous band of black and a blaze of white (plate 14) (Gyldenstolpe 1932; Pearson 1951; pers. obs.).

Distribution

Chinchillula sahamae is distributed at high elevations from southern Peru to northern Chile and reportedly into northwestern Argentina (Tamayo and Frassinetti 1980; Ojeda and Mares 1989) (map 11.30).

Ecology

This is an animal of the altiplano, ranging from 4,000 to 4,600 m. It is usually caught among boulders and along stone walls, often in association with the viscacha, *Lagidium*. *Chinchillula* is nocturnal and herbivorous; stomachs usually contain leaves, seeds, and a few insects. It has been trapped for its beautiful skin (Dorst 1971, 1972; Mann 1978; Pearson 1951; Tamayo and Frassinetti 1980).

Genus *Eligmodontia* F. Cuvier, 1837
Eligmodontia typus F. Cuvier, 1837
Laucha Colilarga Bayo

Measurements

	Mean	Min.	Max.	N	Loc.	Source[a]
TL	168.4	151.0	195.0	20	A, C, Peru	1
HB	80.9	68.0	95.0			
T	87.5	66.0	115.0			
HF	22.3	20.0	24.0			
E	16.4	13.0	19.0	19		
Wta	21.4	13.5	33.0	15		
	20.6	13.8	41.2	45	A	2

[a](1) Daciuk 1974; Greer 1966; Pearson 1951; (2) Mares 1977b.

Description

This small, pale, soft-pelaged mouse has long hind legs and spade-shaped hind feet with the distinctive character of hairy cushions on the soles (plate 14). The venter is wholly or partly white. Animals from highland populations are much larger than those from the lowlands and have shorter tails and longer, laxer pelage. Females are larger than males. There is considerable geographic variation in coloration, tail length, and karyotype (Gyldenstolpe 1932; Mann 1978; Mares 1973, 1977b; Osgood 1943; Pearson 1951; Pearson, Martin, and Bellati 1987).

Distribution

The species is found in western Bolivia, southern Peru, northern Chile, and Argentina. In northern Chile it occurs on the altiplano south to the province of Antofagasta, and in the southern part of the country it is found in Malleco, Aisén, and Ultima Esperanza provinces. In Argentina this species is distributed along the Andes from at least Salta province south to La Pampa and Buenos Aires provinces and then across the whole country south to the Strait of Magellan (Contreras and Justo 1974; Honacki, Kinman, and Koeppl 1982; Mares, Ojeda, and Kosco 1981; Rau, Yáñez, and Jaksić 1978; Tamayo and Frassinetti 1980; BA) (map 11.31).

Life History

In Argentina, near the town of Bariloche, reproduction begins in October and lasts until the end of April. Males and females breed in the same season when they are born. The age of first reproduction is six to eight weeks, the number of embryos averages 5.9 ($n = 20$), and longevity is less than one year. In another study gestation was determined to last eighteen days (Pearson, Martin, and Bellati 1987; Mares 1988).

Ecology

Eligmodontia typus is an animal of sandy soils and cool temperatures, occupying habitats that on other continents are occupied by kangaroo rats and gerbils. In the xeric areas where it is found it is usually the most abundant rodent species. It is found in areas of open sand and low bushes on the Chilean altiplano; in sandy areas and grassy pastures along the coast of Chubut province, Argentina; in rocky areas with grass and flat areas in Malleco province, Chile; and in rolling sand hammocks with halophytic vegetation or creosote bush flats in the higher altitudes of Salta province, Argentina.

In many respects these mice are comparable to the heteromyid rodents of North America; they appear to be approaching bipedal locomotion, they are typically found in sandy, open vegetation areas; they probably forage in and around bushes and bound

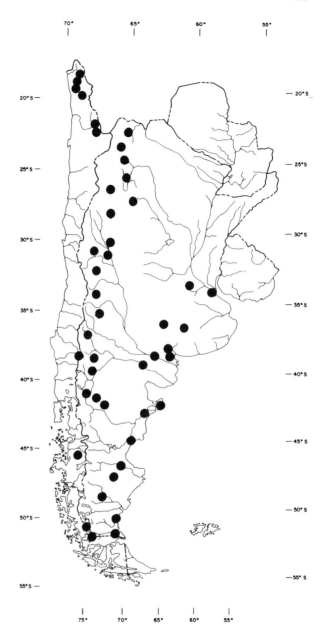

Map 11.31. Distribution of *Eligmodontia typus*.

rapidly across open areas; they can exist for periods with no free water; they store food; and they will go torpid in traps on cold mornings. This species is nocturnal; it will dig its own burrows but also uses burrows dug by other animals, such as *Ctenomys*.

In one study in Argentina, *E. typus* reached a density of 3.5/ha in a grazed area, though in seasons when it was scarce the density could fall to 0.4/ha; home-range diameter averaged 31 m. In this area it is nocturnal, nests and retreats underground, and is primarily granivorous, feeding on *Berberis* seeds and *Prosopis* (Daciuk 1974; Greer 1966; Mann 1978; Mares 1973, 1977b; Mares, Ojeda, and Kosco 1981; Pearson 1951; Pearson, Martin, and Bellati 1987).

Comment

Eligmodontia has been included in *Phyllotis*, and *E. typus* includes *E. hypogaeus*, *E. puerulus*, and *E. elegans*. *E. typus* undoubtedly contains two or more distinct species (Ojeda and Mares 1989; O. P. Pearson, pers. comm.; Ortells et al. 1989).

Genus *Euneomys* Coues, 1874

Description

These phyllotine rodents share many characters with *Phyllotis:* short tail, complex baculum, and relatively large M_3. In contrast, however, *Euneomys* has rather short ears and altogether seems to be more fossorial than other phyllotines. The upperparts are a reddish brown, contrasting with gray or buff underparts. The tops of the forefeet and hind feet are white.

Distribution

The genus is recorded from Chile and Argentina. Pearson (1987) presents data suggesting that in Argentina this genus is a microhabitat specialist that inhabits crevices but forages on windswept scree. It was a dominant species at the close of the Pleistocene, as evinced by owl pellets dating to 10^4 B.P. There is considerable discussion about species-level taxonomy, and some authors (Yáñez et al. 1987) suggest there is only one species, *E. chinchilloides*.

Euneomys chinchilloides (Waterhouse, 1839)
Ratón Peludo Castaño

Measurements

	Mean	Min.	Max.	N	Loc.	Source[a]
TL	213.7	175.0	258.0	11	A, C	1
HB	134.7	115.0	156.0			
T	79.0	59.0	103.0			
HF	29.9	25.4	33.0			
E	23.6	22.0	31.0	9		
Wta	84.7	57.4	122.5	6		

[a](1) Allen 1903; Greer 1966; Pine, Angle, and Bridge 1978; Yáñez et al. 1987.

Description

The incisors are deeply grooved anteriorly. This species is volelike in general form, with long, soft, thick fur. The tail is of moderate length, bicolored and well haired; the blackish ears are small and well haired; and the vibrissae are long and stout. The dorsum is dark gray brown varied with blackish and fulvous, the sides and venter are creamy or white or gray washed with yellow, and the feet are white (Allen 1903; Gyldenstolpe 1932; Mann 1978; Osgood 1943; pers. obs.).

Distribution

Euneomys chinchilloides is found in central and southern Chile and in adjacent Argentina (Greer 1966; Honacki, Kinman, and Koeppl 1982; Tamayo and Frassinetti 1980; Yáñez et al. 1987) (map 11.32).

Life History

One female with four embryos was caught (MVZ).

Ecology

In Chile this species is often found in sandy areas with rocks, grass, and shrubs. In Malleco province it has been trapped in *Nothofagus antarctica* and grassland habitat at approximately 1,600 m. In one Argentine study site *E. chinchilloides* was found only on relatively bare, windswept hilltops with an appreciable expanse of unvegetated fine scree. It was completely absent from forest, brush, or even sparsely vegetated Patagonian steppe and was found far above

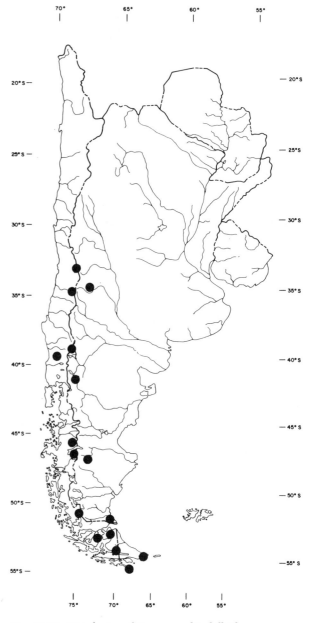

Map 11.32. Distribution of *Euneomys chinchilloides*.

the tree line. In this area it appears that the species ranged over larger areas thousands of years ago as the retreat of glaciers created a great deal of suitable habitat. As vegetation covered these rocky habitats, *Euneomys* retreated.

The species is herbivorous and can live in colonies, easily recognizable by accumulations of the distinctive feces. These mice are agile rock climbers and good diggers (Greer 1966; Mann 1978; Pearson 1987; Tamayo and Frassinetti 1980).

Comment

Yáñez et al. (1987) suggest that there is only one valid species of *Euneomys*, namely, *E. chinchilloides*. The following two "species," *E. mordax* and *E. noei*, may be only subspecifically distinct. According to O. P. Pearson (pers. comm.), *E. fossor* is a composite.

Euneomys mordax Thomas, 1912
Ratón Peludo Oscuro

Description
Measurements from one specimen in Argentina were TL 225; T 78; HF 28; E 31.

Distribution
Euneomys mordax is known only from the west-central province of Mendoza in Argentina and from adjacent Chile (Honacki, Kinman, and Koeppl 1982; Roig 1965; but see Yáñez et al. 1987) (map 11.33).

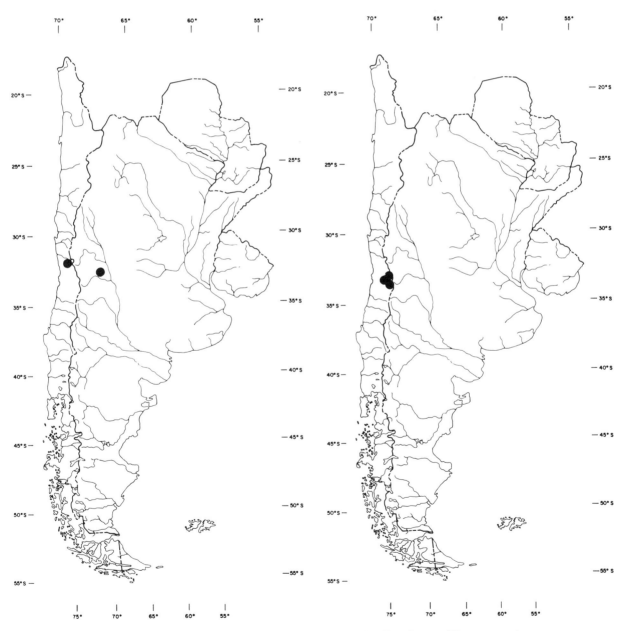

Map 11.33. Distribution of *Euneomys mordax*.

Map 11.34. Distribution of *Euneomys noei*.

Euneomys noei Mann, 1944

Measurements

	Mean	Min.	Max.	N	Loc.	Source[a]
TL	221.0	200	237	4	C	1
		224	268	6	C	2
T	85.0	80	88	4	C	1
		61	74	6	C	2
HF	30.0	27	32	4	C	1
		24	29	6	C	2
E	25.8	24	27	4	C	1
		13	24	6	C	2
Wta	82.0	78	86	2	C	1

[a](1) Pine, Miller, and Schamberger 1979; Yáñez et al. 1987;
(2) Yáñez et al. 1987.

Description

These mice are buffy gray dorsally with a bicolored tail. The venter, throat, forefeet, and hind feet are white. Females have eight mammae.

Distribution

Euneomys noei is known only from the vicinity of Santiago City, Chile (Mann 1978; Tamayo and Frassinetti 1980; Yáñez et al. 1987) (map 11.34).

Ecology

Mice of this species have been caught in rocky areas with sandy soils at altitudes from 2,400 to 3,000 m. They are herbivorous and are probably good rock climbers and diggers that fashion their own burrows in the sandy soil (Mann 1944, 1978; Pine, Miller, and Schamberger 1979; Tamayo and Frassinetti 1980).

Genus *Galenomys* Thomas, 1916
Galenomys garleppi (Thomas, 1898)

Measurements

	Mean	Min.	Max.	N	Loc.	Source[a]
TL	157.7	154.0	161.0	3	B	1
T	40.5	38.0	44.0			
HF	24.3	24.0	25.0			
E	22.0	22.0	22.0	2		
Wta	59.3	58.9	59.6			

[a](1) Gyldenstolpe 1932; AMNH.

Description

The ungrooved incisors are very slender and have anterior faces of pale yellow. This attractive mouse is unmistakable because of its stout body, very short white tail, large ears, and close, thick, yellow-tan fur. The ears are smaller than in *Phyllotis*, and the soles of the hind feet are hairy for the posterior two-thirds. Females have eight mammae. The dorsum is grayish buff finely lined with black, becoming clear buff on the rump and sides; the venter is white and sharply demarcated from the sides (Gyldenstolpe 1932; Thomas 1898b).

Distribution

Galenomys garleppi is found at high altitudes from southern Peru to northeastern Chile (Tamayo and Frassinetti 1980) (map 11.35).

Ecology

This species is found in arid and semiarid areas from 3,800 to 4,500 (Tamayo and Frassinetti 1980).

Genus *Graomys* Thomas, 1916

Description

These phyllotine rodents share many general characters with *Phyllotis*, but in *Graomys* the tail is

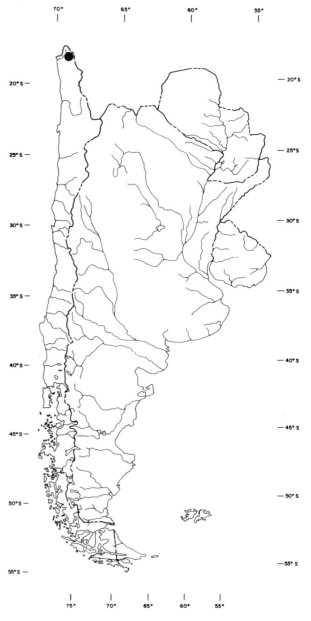

Map 11.35. Distribution of *Galenomys garleppi*.

longer than the head and body. They are ratlike in size, with total lengths ranging from 235 to 327 mm. The tail is frequently bicolored and covered with fine, short hairs, and the tip is usually penicillate. The dorsum is usually some shade of brown, and the venter is paler.

Distribution

The genus is recorded from Bolivia, Paraguay, and Argentina.

Graomys domorum (Thomas, 1902)
Pericote Pálido

Measurements

	Mean	Min.	Max.	N	Loc.	Source[a]
TL	302.1	280.0	327	7	A, B	1
HB	142.4	125.0	166			
T	159.7	149.0	172			
HF	29.9	28.6	31			
E	26.5	22.6	29			
Wta	73.2	61.0	95	5		

[a](1) CM, UM.

Description

This large, stout-bodied *Graomys* has a long tail, penicillate and strongly bicolored. The face has long vibrissae and no distinct markings. It is brown dorsally, lightening to tan on the sides, and whitish gray ventrally. The feet are white.

Distribution

Graomys domorum is found in the Andes of Bolivia and northwestern Argentina. Its southernmost distribution occurs in Salta province, Argentina (Honacki, Kinman, and Koeppl 1982; Ojeda and Mares 1989) (map 11.36).

Ecology

In Salta province, Argentina, this species is limited to transitional forest, occurring on very low mountains in the central part, but it also may occur in drier portions of the subtropical forests of the province. It was trapped in thick grass along road cuts and from areas of second growth (Mares, Ojeda, and Kosco 1981).

Graomys edithae Thomas, 1919

Description

This small species of *Graomys* has a white venter (Gyldenstolpe 1932). One specimen from Argentina measured TL 235; T 127; HF 25; E 20.

Distribution

The species appears to be confined to La Rioja, and Catamarca provinces, Argentina (Gyldenstolpe 1932; Honacki, Kinman, and Koeppl 1982; Thomas 1919a) (map 11.37).

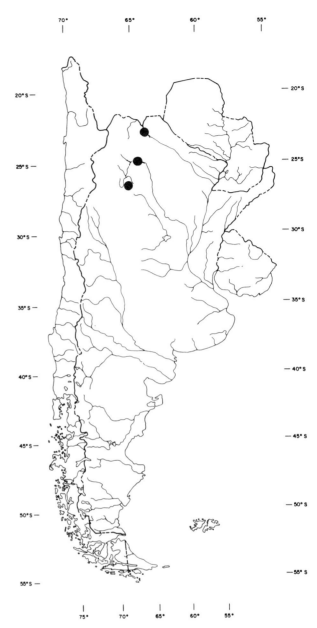

Map 11.36. Distribution of *Graomys domorum*.

Graomys griseoflavus (Waterhouse, 1837)
Pericote Común

Measurements

	Mean	Min.	Max.	N	Loc.	Source[a]
TL	262.1	230.0	325	21	A, P	1
	278.0	261.0	302	40	A	2
T	141.5	123.0	180	19	A, P	1
	144.2	132.0	160	40	A	2
HF	28.7	24.0	35	21	A, P	1
	29.5	25.5	32	40	A	2
E	23.0	21.0	25	21	A, P	1
	24.1	21.5	26	40	A	2
Wta	59.4	42.0	80	20	A, P	1
	62.4	45.6	100	37	A	2

[a](1) Daciuk 1974; PCorps; (2) Rosi 1983.

Description

This large, long-tailed rat resembles a small North American Woodrat (*Neotoma*). It has a strongly bicolored, penicillate long tail, and large ears. The long, soft fur is brown or grayish on the dorsum, lighter on the sides, and white on the venter (Gyldenstolpe 1932; Mares 1973; pers. obs.).

Distribution

The species is found in Argentina, Bolivia, Paraguay, and perhaps southwestern Brazil. In southern South America it is found in the south and west of the Paraguayan Chaco and in Argentina, at least in the provinces of Tucumán, Chaco, Salta, La Pampa,

Chubut, Buenos Aires, and Mendoza (Allen 1901; Contreras 1972b, 1982a,c; Contreras and Justo 1974; Daciuk 1974; Lucero 1983; Honacki, Kinman, and Koeppl 1982; Mares, Ojeda, and Kosco 1981; Myers 1982; Rosi 1983) (map 11.38).

Ecology

Graomys griseoflavus is found in diverse locations that are typically xeric but include cultivated fields, semiarid sandy areas with rocks, hillsides with boulders, rivers with associated gallery forest, and orchards. Its grass nests have been found in large cacti, thorn brushes, or *Microcavia* burrows.

These mice are reasonably good at conserving

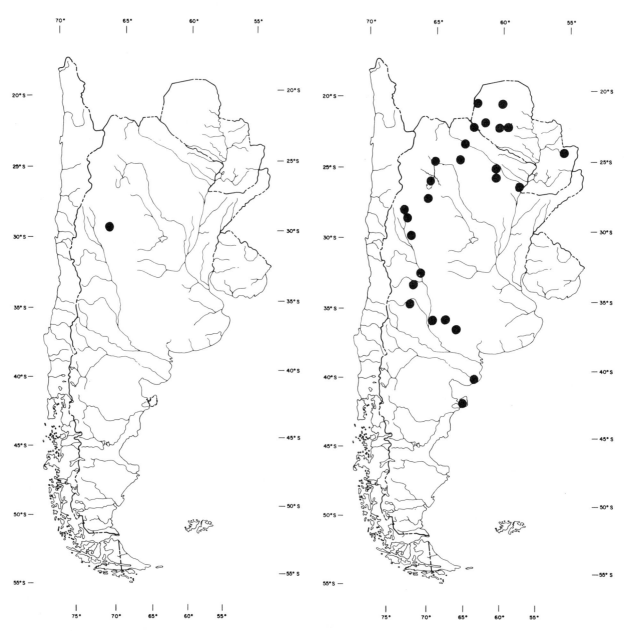

Map 11.37. Distribution of *Graomys edithae*.

Map 11.38. Distribution of *Graomys griseoflavus*.

water and appear to be almost strictly herbivorous. They are good climbers and have been seen foraging up to 15 m high in trees. Groups of three to six may forage together. Individuals are strong and aggressive and bite readily (Mares 1973, 1977c; Mares, Ojeda, and Kosco 1981; Rosi 1983).

Genus *Irenomys* Thomas, 1919
Irenomys tarsalis (Philippi, 1900)
Colilargo Oreja Negra

Measurements

	Mean	Min.	Max.	N	Loc.	Source[a]
TL	293.8	260.0	326	12	C	1
HB	125.3	116.0	138			
	107.2	103.0	116	4 m	A	2
	115.1	95.0	135	8 f		
T	168.5	133.0	188	12	C	1
HF	30.1	27.0	32			
E	22.0	20.0	23	5		
Wta	45.4	36.6	57	5		
	44.4	31.0	65	11 m	A	2
	41.9	30.0	58	13 f		

[a](1) FM; (2) Pearson 1983.

Description

This medium-sized, slender mouse has grooved upper incisors. Its very long tail makes it unmistakable. It has fairly thick fur, grayish brown washed with yellowish or rufous dorsally and dirty yellow or white ventrally. The ears are moderately large and are blackish, in contrast to the dorsal color. The tail is weakly bicolored and distinctly penicillate (plate 14). *Irenomys tarsalis* can be distinguished from the similarly shaped *O. longicaudatus* by its grooved incisors and its larger size (Gyldenstolpe 1932; Mann 1978; Osgood 1943; Pearson 1983; Thomas 1919c; pers. obs.).

Distribution

The species is distributed in central and southern Chile, including Chiloé Island, and adjacent regions of Argentina (Honacki, Kinman, and Koeppl 1982; Olrog and Lucero 1981; Tamayo and Frassinetti 1980) (map 11.39).

Life History

In Argentina, where breeding takes place in the spring, two females that were caught had three and four embryos (Pearson 1983).

Ecology

Irenomys tarsalis lives in the thick, wet forests of the southern Andes. It is found only in or near forests with a well-developed understory of *Chusquea* bamboo, where it reaches densities of 5.1/ha, though occasionally it can be captured at the edges of forest. Its habitat requirements and distribution appear to

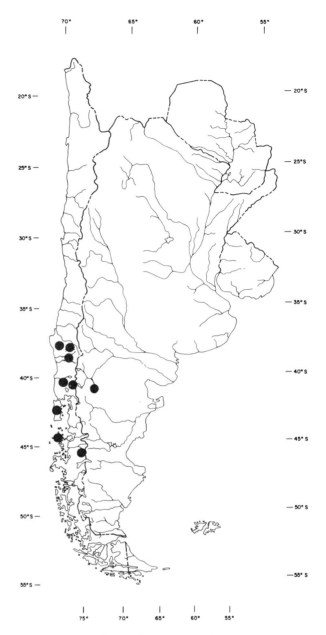

Map 11.39. Distribution of *Irenomys tarsalis*.

be very similar to those of the marsupial *Dromiciops australis*.

This mouse has the long tail and long hind feet of a climber and appears to climb readily and with skill. It is frequently trapped arboreally. Stomachs of this nocturnal mouse were found to contain leaves and fungal hyphae as well as grass seeds. It apparently eats the seeds of the *Chusquea* bamboo and may experience population increases when the bamboo produces seeds. In one study in Chile, seeds and fruits composed 45.5% of the overall diet, with foliage and epidermal and vascular plant tissue composing 40.8% ($n = 10$ stomachs) (Greer 1966; Mann 1978; Meserve, Lang, and Patterson 1988; Murúa,

González, and Jofre 1982; Osgood 1943; Pearson 1983; Pearson and Pearson 1982; Pine, Miller, and Schamberger 1979; Reise and Venegas 1974; Tamayo and Frassinetti 1980).

Genus *Neotomys* Thomas, 1894
Neotomys ebriosus Thomas, 1894
Ratón Ebrio

Measurements

	Mean	Min.	Max.	N	Loc.	Source[a]
TL	198.6	185.0	213	8	A, C	1
HB	122.8	105.0	134			
T	75.9	62.0	83			
HF	23.0	21.0	25			
E	18.5	17.5	19			
Wta	64.5	63.0	69	4		

[a](1) BA, Santiago.

Description
This mouse is unmistakable because of its grizzled gray brown dorsum and the bright cinnamon orange tip of its muzzle. The venter is grayish white to gray, the rump is washed with reddish, and the short tail is well haired and bicolored (Bárquez 1983; Gyldenstolpe 1932; Sanborn 1947b; pers. obs.).

Distribution
The species is found at high altitudes from Peru to northern Chile and northwestern Argentina. In Chile is has been trapped in Tarapacá (Arica) province, and in Argentina it has been found in the provinces of San Juan, Jujuy, Catamarca, and Tucumán (Bárquez 1983; Pine, Miller, and Schamberger 1979; Tamayo and Frassinetti 1980) (map 11.40).

Life History
Embryo counts from Peruvian specimens ranged from one to two (MVZ).

Ecology
Neotomys ebriosus is found at altitudes of 3,300 to 5,000 m in dense grasslands, along streams with dense cover, and in marshes (Bárquez 1983; Pearson 1951; Pine, Miller, and Schamberger 1979; Tamayo and Frassinetti 1980).

Genus *Phyllotis* Waterhouse, 1837

Description
For a full diagnosis of the tribe see p. 273. Species of *Phyllotis* have a simplified molar dentition (tetratophodont), and the M_3 tends to be one-half to three-fourths of the M_2 (fig. 11.5). Head and body length ranges from 70 to 153 mm, and the tail is from 45 to 150 mm. These mice have very large external ears (see plate 14), a tail roughly equal to or shorter than the head and body, and relatively short hind feet (*Eligmodontia* is an exception within the Phyllotini).

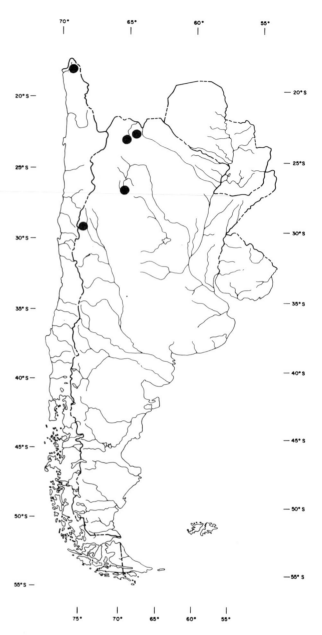

Map 11.40. Distribution of *Neotomys ebriosus*.

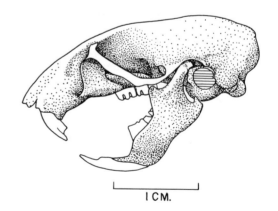

Figure 11.5. Skull of *Phyllotis darwini*.

The dorsal pelage varies in color from gray brown to reddish brown; the venter may be dirty white to buff.

Distribution

The genus occurs from Peru to Argentina, including Bolivia, Chile, and Ecuador, usually at moderate to high elevations (sea level to 5,000 m).

Much ongoing work on the genetics of several *Phyllotis* species, such as *P. darwini*, suggests that there may be several distinct forms that need to be separated out (A. Spotorno, pers. comm.).

Life History and Ecology

These mice, mainly nocturnal, are extremely omnivorous. Seeds, arthropods, lichens, and forbs have been recorded in their diet (see species accounts). In spite of this variety, most species are strongly herbivorous, in contrast to the importance of insects in the diet of the often sympatric species of *Akodon*.

Phyllotis caprinus Pearson, 1958
Pericote Anaranjado

Measurements

	Mean	Min.	Max.	N	Loc.	Source[a]
HB	117.6	102	140	29	Peru	1
T	135.1	116	151	27		
HF	26.1	21	28	29		
E	24.5	23	26	17		

[a](1) Pearson 1958.

Description

This large, relatively short-furred *Phyllotis* is sympatric only with *P. darwini*.

Distribution

Phyllotis caprinus is found from southern Bolivia to northern Argentina on the eastern slope of the Andes (Pearson 1958) (map 11.41).

Ecology

This species is found in brush and thorn scrub above 2,400 m (Pearson 1958).

Phyllotis darwini (Waterhouse, 1837)
Pericote Panza Gris

Measurements

	Mean	Min.	Max.	N	Loc.	Source[a]
TL	219.2	186.0	251.0	4 f	north A	1
HB	128.2	122.0	142.0	11	south A	2
	112.2	99.0	120.0	28	A, B	
T	106.7	91.0	125.0	4 f	north A	1
	120.2	112.0	126.0	8	south A	2
	123.7	112.0	146.0	26	A, B	
HF	25.3	24.1	26.5	4 f	north A	1
	27.8	25.0	30.0	11	south A	2
	24.8	23.0	26.0	25	A, B	
E	26.8	24.6	27.5	4 f	north A	1
	27.2	25.0	29.0	10	south A	2
	24.8	22.0	27.0	25	A, B	
Wta	57.5			39	north A	3

[a](1) Mares, Ojeda, and Kosco 1981; (2) Pearson 1958; (3) Mares 1973.

Description

In this robust mouse the head and body are almost always longer than 90 mm, and the tail is almost always longer than the head and body. The ears are large, and the vibrissae are luxuriant. The species has a very large geographic range; throughout this range the fur tends to be paler in arid regions and darker and richer in color in humid regions. The fur is thick and fluffy in high or cold regions and flatter and less fluffy in specimens from the coast of central

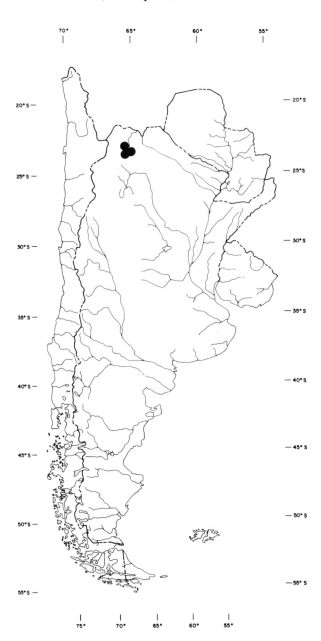

Map 11.41. Distribution of *Phyllotis caprinus*.

Chile and in relatively warm areas. The fur is predominantly in tones of coffee, often with considerable gray, though juveniles are grayer. The venter is whitish, grayish, or buffy, with or without a buffy pectoral streak (plate 14) (Mann 1978; Osgood 1943; Pearson 1958; pers. obs.).

Distribution

Phyllotis darwini is distributed from central Peru, south discontinuously through south-central Chile, and in Argentina along the eastern side of the Andes south to Mendoza, and then east to Buenos Aires province; from there it extends across the width of Argentina and south to Tierra del Fuego, where

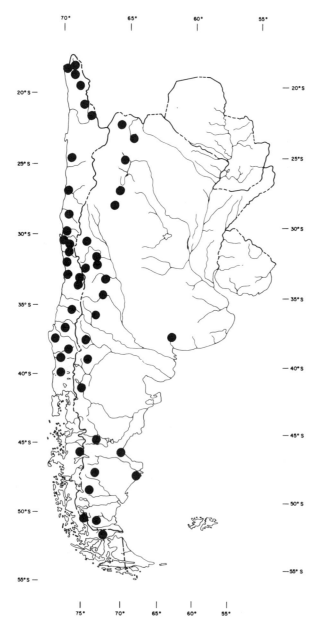

Map 11.42. Distribution of *Phyllotis darwini*.

it may enter Chile again (Honacki, Kinman, and Koeppl 1982; Mann 1978; Tamayo and Frassinetti 1980) (map 11.42).

Life History

In Peru breeding coincides largely with the wet season, but there is no clear peak in reproduction. The average number of embryos per female is 3.7. In Chile the number of embryos per female varies from four to eight. In northern Chilean semiarid thorn scrub *P. darwini* can produce more than one litter per season. It can undergo *ratales* or irruptions of mice after heavier than normal rainfall. During outbreak years, rapid increases in numbers are apparently achieved by rapid maturation of young and by multiple short-interval pregnancies (Mann 1978; Meserve and Le Boulengé 1987; Pearson 1975).

Ecology

Over its entire range this species is found in a very large variety of habitats, from near sea level to 4,500 m and from marsh edges to thorn brush to Araucaria forests. In Salta province, Argentina, *P. darwini* is found on steep rocky slopes where cacti are common and shows a preference for burrowing under rocks. It is also found in riparian areas in mesic forests as well as gallery forest in lowland deserts, but always when rocks are present. In Coquimbo province, Chile, it is found in riverine scrub and brush forest transitions, but most commonly in open brushy habitats. In these areas densities can range from 41 to 207 individuals per hectare, whereas in Mendoza province, Argentina, it occurs at a much lower density of 5.4 per hectare. *P. darwini* is not found in extremely arid areas or in very moist forests.

These mice are good climbers, and their nests have been found above ground in cacti and shrubs, though also among rocks. They are nocturnal and very flexible in their feeding habits. In Chile they are herbivorous in spring and largely granivorous in summer, although the extent of granivory varies with vegetation type. In Peru the diet was found to consist of 13% grass, 37% forbs, 11% seeds, and 39% insects ($n = 42$). Virtually no insects in the diet have been recorded from Chile. This species is a relatively good conserver of water, though it cannot live without free water (Contreras and Rosi 1980a; Fulk 1975; Glanz 1977a; Greer 1966; Mann 1978; Mares 1973, 1977c; Mares, Ojeda, and Kosco 1981; Meserve 1978, 1981b; Pearson 1951, 1958, 1975; Pizzimenti and de Salle 1980; Reise and Venegas 1974; Schamberger and Fulk 1974; Schneider 1946).

Comment

O. P. Pearson (pers. comm.) comments that the present taxonomic situation, as promoted by Spotorno, is that *Phyllotis darwini* sensu stricto is a species of the central coast of Chile, that it meets *P. xanthorhinus* near Santiago without intergrading, and that they do not interbreed in the lab. This means that many, if not most or all, of the subspecies that were formerly considered subspecies of *P. darwini* would now be listed as subspecies of *P. xanthorhinus* (Spotorno and Walker 1983; Walker, Spotorno, and Arrau 1984).

Phyllotis magister Thomas, 1912

Measurements

	Mean	Min.	Max.	N	Loc.	Source[a]
HB	132.0	112	152	15	C, Peru	1
T	131.0	117	158	13		
HF	28.5	27	32	15		
E	26.3	15	27	13		
Wta	68.5	50	87	2	C	2

[a](1) Pearson 1958; (2) Spotorno 1976.

Description

This species is distinguished from *P. darwini*, which it closely resembles, by its larger size and larger feet. The dorsum is light brown to grizzled gray slightly suffused with buff, and the venter is whitish gray. The tail is bicolored, and the ears are sparsely haired (Gyldenstolpe 1932; pes. obs.).

Distribution

Phyllotis magister is found in the Andes of southern Peru and northern Chile from sea level (Honacki, Kinman, and Koeppl 1982; Pearson 1975; Tamayo and Frassinetti 1980) (map 11.43).

Ecology

Brushy areas from 2,000 to 4,200 m are the preferred habitat of this species. It has been trapped among rocks, in stone walls, and along the banks of small streams. Densities of 2.06/ha have been recorded in mountain scrub in Peru. Animals from Peru had 7% grass, 60% forbs, 11% seeds, and 22% insects in their stomachs ($n = 5$). (Pearson 1958; Pearson and Ralph 1978; Pine, Miller, and Schamberger 1979; Pizzimenti and de Salle 1980; Tamayo and Frassinetti 1980).

Phyllotis osgoodi Mann, 1945

Measurements

	Mean	Min.	Max.	N	Loc.	Source[a]
TL	207.3	180.0	245.0	12	C	1
HB	109.5	84.0	135.0	13		
T	98.7	80.0	110.0	12		
HF	25.8	24.0	29.0	13		
E	23.9	19.6	28.0			
Wta	45.1	27.0	57.5	11		

[a](1) Mann 1945; Spotorno 1976; Santiago.

Description

This species is very similar to *P. darwini*. Dorsally it varies from buffy coffee to buffy brown with black hairs, and ventrally it is dirty gray (Mann 1945, 1978; Spotorno 1976).
Chromosome number: $2n = 40$ (Spotorno 1976).

Distribution

Phyllotis osgoodi is found only in the altiplano of Arica province, Chile (Honacki, Kinman, and Koeppl 1982; Spotorno 1976) (map 11.44).

Ecology

This species has been captured at 4,400 m (Tamayo and Frassinetti 1980).

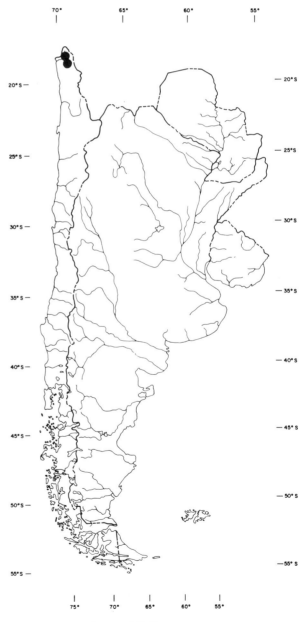

Map 11.43. Distribution of *Phyllotis magister*.

Phyllotis osilae J. A. Allen, 1901
Pericote Osilae

Measurements

	Mean	Min.	Max.	N	Loc.	Source[a]
HB	115.6	90	140	21	Peru	1
T	120.5	105	140			
HF	25.6	22	27	19		
E	21.7	18	24			
Wta	57.0			22	A	2

[a](1) Pearson 1958; (2) Mares 1977c.

Description

This long-tailed mouse is very similar to *P. darwini* in some areas, separable only by skull charac-ters. The dorsum is buff lined with black, the cheeks and sides are pale buffy orange, and the chest has a median longitudinal streak of buff on a white or gray white venter. The fur is thick and rich (Gyldenstolpe 1932; Pearson 1958; pers. obs.).

Distribution

Phyllotis osilae is found from southeastern Peru to northern Argentina, where it occurs in Jujuy, Salta, Tucumán, and Catamarca provinces (Honacki, Kinman, and Koeppl 1982; Mares 1977c; Olrog and Lucero 1981; Pearson 1958; FM) (map 11.45).

Ecology

This species frequents the high Andean bunch-grass areas from 3,000 to 4,500 m, though it can occur at lower elevations in a few localities. In Tucumán province, Argentina, where it is the only *Phyllotis* present, it can reach very high numbers. In *icha* habitat in Peru this species was found at a density of 1.47/ha. *P. osilae* is strictly nocturnal, confined to more mesic habitats because it is a poor conserver of water and eats mostly plant material. Eleven stomachs from Peru contained about 31% grass, 22% forbs, 26% seeds, and 20% insects (Dorst 1971, 1972; Lucero 1983; Mares 1977c; Pearson 1951, 1958; Pearson and Ralph 1978; Pizzimenti and de Salle 1980).

Genus *Reithrodon* Waterhouse, 1837
Reithrodon physodes (Olfers, 1818)
Rata Conejo

Measurements

	Mean	Min.	Max.	N	Loc.	Source[a]
TL	244.9	225.0	270.0	14	U	1
	238.9	221.0	269.0	7	A, C, U	2
HB	142.4	127.0	169.0			
T	106.7	92.0	120.0	14	U	1
	96.4	91.0	102.0	7	A, C, U	2
HF	32.2	29.0	34.5			
E	24.9	23.0	29.0	14	U	1
	22.6	15.0	28.0	9	A, C, U	2
Wta	78.6	58.5	105.0	14	U	1
	81.7	55.0	105.0	15	C	3

[a](1) Barlow 1965; (2) Dalby and Mares 1974; González 1973; Thomas 1927; Wolffsohn 1923; (3) Atalah, Sielfeld, and Venegas 1980.

Description

This species includes *R. auritus*. The incisors are distinctly grooved. This relatively short-tailed, stout-bodied mouse (plate 14) has a large head, large eyes, and large, rounded, well-haired ears. The fur is long and soft, buffy gray to brownish with a yellowish cast on the dorsum, becoming more blackish on the crown and more richly yellowish on the sides. The venter is yellowish to orangish white (Allen 1903; Barlow 1965; Mann 1978; Osgood 1943; pers. obs.).

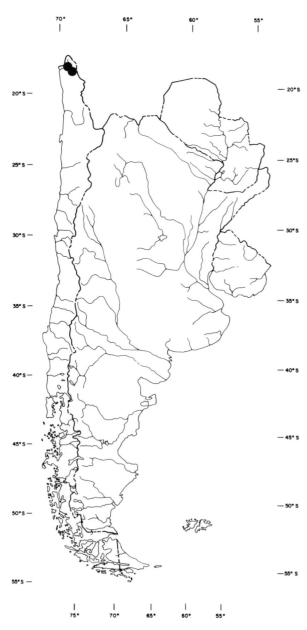

Map 11.44. Distribution of *Phyllotis osgoodi*.

Distribution

The species is found from Uruguay south through Argentina to Tierra del Fuego, where it is also found in lowland Chile, except for the drier Chaco regions (Honacki, Kinman, and Koeppl 1982; Olrog and Lucero 1981; Tamayo and Frassinetti 1980) (map 11.46).

Life History

In Uruguay a female with three embryos was taken. In southern Patagonia, Argentina, females become pregnant in the same season when they are born, probably by two months of age. Females averaged 4.5 fetuses (range 1–8; $n = 17$) (González 1973; Pearson 1988).

Ecology

Reithrodon physodes is an animal of open grassy and brushy areas up to 3,000 m and appears to be restricted to such habitats. It is an ecological equivalent of some of the small, open-country rabbits of the Holoarctic. In Uruguay it is usually found in overgrazed pasture, among rocky outcrops, or on well-drained slopes with scanty vegetation. In northern Patagonia, Argentina, these mice occurred at densities of 10.6, 9.3, and 9.7 animals per hectare. The diet of this species is composed mostly of grass, though it also eats other plants with tuberous rhizomes and roots. Animals need large quantities of food (an individual will eat its own weight in green grass in one day).

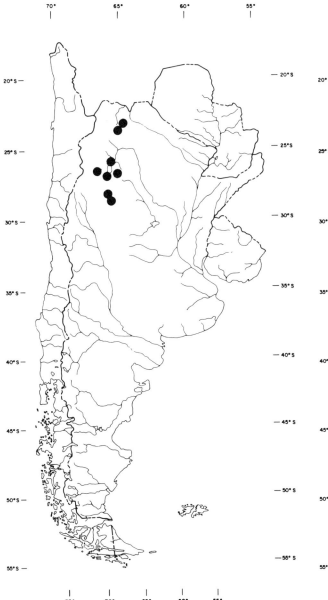

Map 11.45. Distribution of *Phyllotis osilae*.

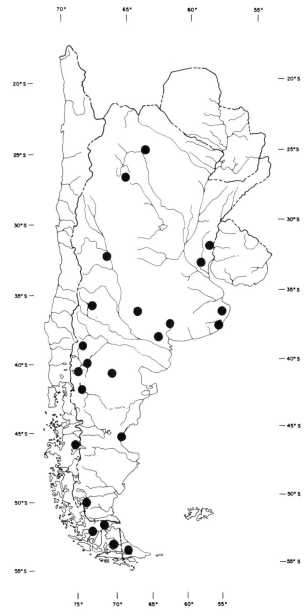

Map 11.46. Distribution of *Reithrodon physodes*.

The mice are commonly caught in active and in-active *Ctenomys* burrows but will also dig their own burrows up to 2 m long. In addition, nests have been found under rocks or fallen tree trunks. Many in-dividuals use the same runways or burrows in the same night; these may be any combination of adults of either sex or juveniles. They sometimes feed diur-nally but are usually nocturnal, either taking food into the burrows or feeding in specific shelters away from the main burrow. *R. physodes* can remain ac-tive under a cover of snow (Barlow 1965, 1969; Dalby and Mares 1974; Lucero 1983; Mann 1978; Pearson 1988; Pine, Miller, and Schamberger 1979).

TRIBE AKODONTINI

These small to medium-sized rodents are generally adapted for terrestrial life; some are semifossorial. Thus the tail is usually shorter than the head and body, the ears and eyes are of modest size, and the claws are long (plate 14). The digestive system is relatively unspecialized, reflecting omnivory or in-sectivory. The molars are not high crowned, and their cusps are crested or secondarily planed (see fig. 11.6). The molar mesoloph and mesolophid are usually small or vestigial. The karyotype does not exceed 54 (2n) (see table 11.2; Reig 1987).

Many rodents tend to be flexible in their diets, and Landry (1970) has made an excellent case for omnivory as an underlying factor in their adaptive radiation. The high level of arthropod predation, as evinced by the stomach contents of many akodonts, suggests that the tribe is preadapted to invade insec-tivore niches. If the members of the genera *Oxy-mycterus*, *Blarinomys*, and *Notiomys* (among others) derive from the ancestor of the extant genus *Akodon*, we may well view this evolutionary development as the ultimate occupancy of the semifossorial, insecti-vore niche in South America by the rodent invasion after the formation of the Pliocene land bridge (see also the section "The 'Mole Mice' of South Amer-ica," pp. 324–27).

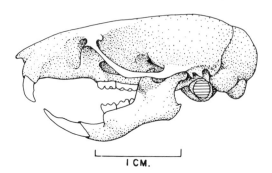

Figure 11.6. Skull of *Akodon* sp.

Table 11.3   Akodontine Rodents according to Reig (1987)

Cricetidae
  Sigmodontinae
    Akodontini
      Bolomys (includes *Cabreramys*; *Thalpomys* is a synonym)
      Akodon
          subgenera
          Akodon (includes *Thaptomys*)
          Abrothrix
          Chroeomys
          Deltamys
          Hypsimys
        Microxus
        Blarinomys
      **Oxymycterus*
        Lenoxus
        Juscelinomys
        Podoxymys
        Chelemys
        Geoxus
        Notiomys

*These three genera account for 80% of the species.

Genus *Akodon* Meyen, 1833

Description

A diagnosis for the "tribe" Akodontini is included above. Members of the genus *Akodon* are distrib-uted throughout South America. The molars tend to be somewhat simplified or tetralophodont, and the M₃ tends to be reduced in size (see fig. 11.18; Reig 1987). The head and body length of adults ranges from 72 to 141 mm; the tail is 50 to 100 mm. The baculum is complex in structure (except in "*Abo-thrix*"), the tail is shorter than the head and body, the hind feet are relatively short, and ears are rela-tively small. The dorsum varies from gray to brown, and the venter varies from white to gray. These ro-dents remind a North Temperate Zo..e worker of voles, but their dietary habits are quite different.

Distribution

The genus *Akodon* as here defined is confined to South America from Colombia to Argentina. Species range from sea level to 5,000 m.

Life History and Ecology

Depending on the authority, some twenty-six to forty-one species may be included in this genus. The confusion depends in part on the subgroupings, some of which may be raised to generic status (see comment on the definition of the Akodontini and table 11.3). Natural history data that are available suggest that the species of *Akodon* are omnivorous and feed on fruits, seeds, insects, and some herba-ceous vegetation.

Akodon albiventer Thomas, 1897
Ratón Ventriblanco

Measurements

	Mean	Min.	Max.	N	Loc.	Source[a]
TL	159.7	153.0	163.0	6	C	1
	159.6	153.0	170.0	8	A	2
HB	93.2	87.0	97.0	6	C	1
T	66.5	65.0	68.0			
	71.1	66.0	78.0	8	A	2
HF	21.5	21.0	22.0	6	C	1
E	13.5	13.0	14.0			
Wta	22.2	18.0	32.0	5		
	21.4	16.5	29.5	8	A	2

[a](1) Santiago; (2) Mares, Ojeda, and Kosco 1981.

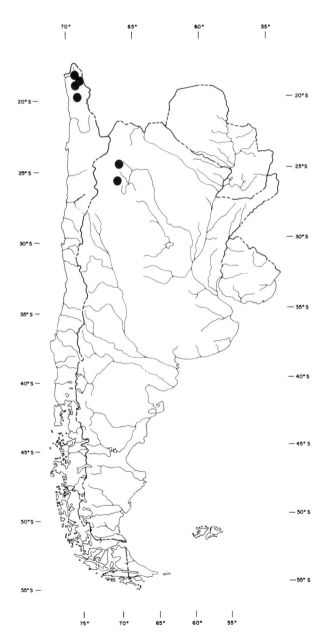

Map 11.47. Distribution of *Akodon albiventer*.

Description

The fur is dense, and the short ears are well haired. These mice are dark gray through olive brown to light brown dorsally, with the throat, venter, and feet a strongly contrasting white to dirty white. There is sometimes a line of buff separating the venter from the dorsum. The sparsely haired tail is darker on top and white on the bottom (Mann 1978; pers. obs.).

Distribution

Akodon albiventer is found in southern and western Bolivia, northwestern Argentina, northern Chile, and southeastern Peru (Honacki, Kinman, and Koeppl 1982; Tamayo and Frassinetti 1980) (map 11.47).

Ecology

In Chile this species of *Akodon* is found only above 3,000 m, and in Salta province, Argentina, it has been trapped above 400 m. It is typically an animal of open habitats, having been found in meadows with dense grass and rock walls, near small marshes, and on high, dry slopes. It seems to prefer more mesic areas but is not confined to them. In Peru densities ranged from 3.53/ha in quenua habitat to 0.14/ha in tola habitat. *A. albiventer* is diurnal and has long claws that it reportedly uses for digging tunnels and finding insect larvae (Koford 1955; Mann 1978; Mares, Ojeda, and Kosco 1981; Pearson and Ralph 1978; Pine, Miller, and Schamberger 1979).

Comment

Akodon albiventer includes *A. berlepschii.*

Akodon andinus (Philippi, 1858)
Ratón Andino

Measurements

	Mean	Min.	Max.	N	Loc.	Source[a]
TL	154.0	144.0	163.0	11	A	1
	155.5	141.0	177.0	8	C	2
T	58.9	52.0	65.0	11	A	1
	58.3	46.0	67.0	8	C	2
HF	20.6	19.8	21.8	11	A	1
	21.9	18.0	25.0	8	C	2
E	14.4	13.0	15.5	11	A	1
	13.1	14.0	16.0	8	C	2
Wta	21.5	18.0	23.8	7	A	1
	26.2	23.0	31.0	8	C	2

[a](1) Contreras and Rosi 1981b; (2) Santiago.

Description

This small *Akodon* has very dense, soft fur, light brown or buffy to agouti gray dorsally with a gray or gray white venter. Behind the ears is a pale patch, and in some populations the ears themselves are covered with white hairs. The lips and chin are also frequently white (Contreras and Rosi 1981b; Mann 1978; Osgood 1943; Pearson 1951; pers. obs.).

Distribution

Akodon andinus is distributed along the higher part of the Andes from Peru south to Chile and Argentina, though in places it is found as low as 950 m. In Chile it occurs south to about 34°, and in Argentina it is found south to Mendoza province (Honacki, Kinman, and Koeppl 1982; Iriarte and Simonetti 1986; Mann 1978; Roig 1965) (map 11.48).

Life History

In Salta province, Argentina, a lactating female was caught in March, and in Peru a female with six embryos was captured (Ojeda and Mares 1989; MVZ).

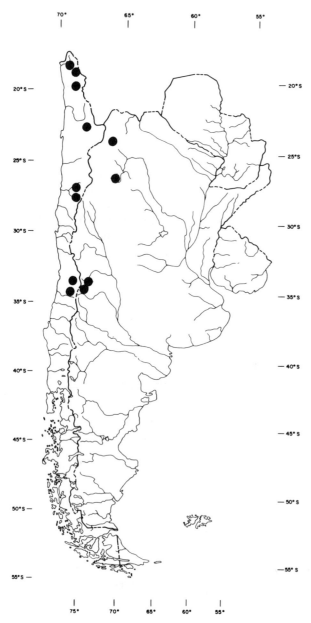

Map 11.48. Distribution of *Akodon andinus*.

Ecology

This species has been reported to be a high-altitude *Akodon*, apparently distributed from about 2,500 m to above 4,500 m, but recent work near Santiago, Chile, has found it as an occasional visitor in the matorral at 950 m. It inhabits sparsely vegetated rocky slopes and is a good digger that constructs a system of galleries about 5 cm deep in the soil, winding among the rocks. These mice can be very common and may undergo dramatic population increases. Diurnal at least at some times during the year, *A. andinus* shows a marked perference for animal material in its diet, though some authors report only plant material. The food eaten may well vary seasonally (Contreras and Rosi 1981b; Fonollat 1984; Iriarte and Simonetti 1986; Mann 1978; Mares, Ojeda, and Kosco 1981; Pearson 1951; Scrocchi, Fonollat, and Salas 1986; Simonetti, Fuentes, and Otaiza 1985).

Akodon azarae (Fisher, 1829)
Ratón de Azarae

Measurements

	Mean	Min.	Max.	N	Loc.	Source[a]
TL	160.5	139.0	180.0	21 m	A	1
	152.0	133.0	174.0	26 f		
	178.8	159.0	199.0	35	U	2
T	69.7	55.0	85.0	21 m	A	1
	67.4	53.0	77.0	26 f		
	77.4	66.0	94.0	35	U	2
HF	19.0	18.0	21.0	21 m		
	18.4	16.0	21.0	29 f		
	21.4	19.0	27.0	35	U	2
E	13.7	11.0	15.0	21 m	A	1
	12.7	11.0	14.5	29 f		
	14.5	12.0	16.0	35	U	2
Wta	35.0	27.0	41.0	5 m	A	1
	27.0	22.0	36.0	12 f		
	30.1	20.3	46.0	35	U	2
	32.5	17.0	45.0	258 m	A	3
	24.2	10.0	40.0	227 f		

[a](1) Fornes and Massoia 1965; (2) Barlow 1965; (3) Dalby 1975.

Description

This typically volelike *Akodon* is of moderate size and has a harsh olive brownish pelage washed with yellowish on the sides and more strongly yellowish and grayish on the venter. In some populations the venter is sharply demarcated from the dorsum. The hind feet are tan, and there is a faint wash of reddish brown on the nose and shoulders and a faint eye-ring. In Paraguay *A. azarae* is easily distinguished from the sympatric *A. varius* by its smaller size (Barlow 1965; Myers and Wetzel 1979; pers. obs.).

Distribution

Akodon azarae is found from southern Brazil south to central Argentina. It occurs in both eastern and western Paraguay, throughout Uruguay, and in Argentina along the eastern part of the country south to Buenos Aires and La Pampa provinces (Barlow 1965; Contreras and Justo 1974; Honacki, Kinman, and Koeppl 1982; Massoia 1971a; UM) (map 11.49).

Life History

In Uruguay this species breeds from October to May, and females have three to seven embryos (mean 4; *n* = 12). In the Argentine pampas *A. azarae* breeds from October to April and the average number of embryos is 4.6 to 5.7 (range 3–10). The gestation period is 24.5 days, and neonates weigh 2.2 g.

In the lab males reach sexual maturity at eighty-four days and females at seventy-five days, whereas in the wild maturity can be reached at sixty days. In the wild 50% of females reach sexual maturity by 22 g, though some can begin reproducing at 12 g. Longevity in the wild is between ten and twelve months for those born in the fall and between seven and eight months for those born in the spring. Young born near the beginning of the season reproduced when only two months old; those born near the end of the season did not breed until they were seven months old.

Akodon azarae is very like *Microtus* in its reproductive habits and can undergo similar population eruptions (Barlow 1969; Crespo 1966; Crespo et al. 1970; Dalby 1975; Pearson 1967; Villafañe 1981a).

Ecology

In Paraguay *A. azarae* is strongly associated with marshes and wet grassland along the Río Paraguay, and in Uruguay it occurs in a variety of habitats including open thorn woodland, tall grass adjacent to riparian woodland, and upland stands of tall grass. These mice are animals of open vegetation formations. During periods of high density they can occur at over 200/ha, though averages of about 50/ha are more common. Where it occurs *A. azarae* is usually the dominant small rodent.

These mice live in shallow holes and will occasionally burrow. They can be nocturnal or diurnal, and in Uruguay the stomachs of eleven individuals contained 25% plant material and 75% invertebrate material, most commonly Coleoptera, Orthoptera, and Hymenoptera (Barlow 1969; Crespo 1966; Crespo et al. 1970; Dalby 1975; Fornes and Massoia 1965; Myers 1982; Villafañe et al. 1973; UM).

Akodon boliviensis Meyen, 1833
Ratón Plomizo

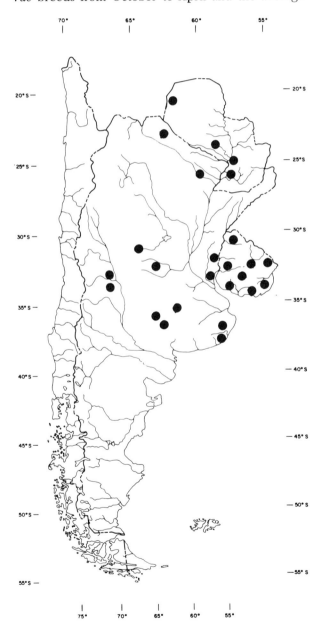

Map 11.49. Distribution of *Akodon azarae*.

Measurements

	Mean	Min.	Max.	N	Loc.	Source[a]
TL	162.9	153.0	183.0	10	A	1
HB	107.3	101.0	120.0	7	A	2
T	76.3	71.0	83.0	10	A	1
	73.2	65.0	82.0	7	A	2
HF	22.3	20.3	24.0	10	A	1
	21.3	19.0	22.6	7	A	2
E	17.4	14.3	25.5	10	A	1
	16.4	14.4	18.3	7	A	2
Wta	19.4	14.3	25.5	10	A	1
	32.5	27.5	42.0	7	A	2

[a](1) Mares, Ojeda, and Kosco 1981; (2) Bárquez et al. 1980.

Description

In this small *Akodon* the dorsum ranges from agouti brown to darker shades of brown. The sides are often washed with tan, and the venter is grayish tan to buffy gray. The ears are lightly furred and the nose and cheeks have a slight orangish cast in some populations. The tail is bicolored (Mares, Ojeda, and Kosco 1981; Pearson 1951; pers obs.).
Chromosome number: $2n = 40$ (Bárquez et al. 1980).

Distribution

Akodon boliviensis is found in southern Peru, northwestern Argentina south to at least Catamarca province, and Bolivia (Bárquez et al. 1980; Honacki, Kinman, and Koeppl 1982; Mares, Ojeda, and Kosco 1981; Olrog 1979; Olrog and Lucero 1981) (map 11.50).

Life History

In captivity the litter size is three or four. Nests have been found under trunks and in holes in the ground (Fonollat 1984).

Ecology

In Salta province, Argentina, this field mouse is found throughout the central, low-elevation moist forested regions. It is particularly common in second-growth areas in mesic forest, where it was taken under rocks and logs, along streams and road cuts,

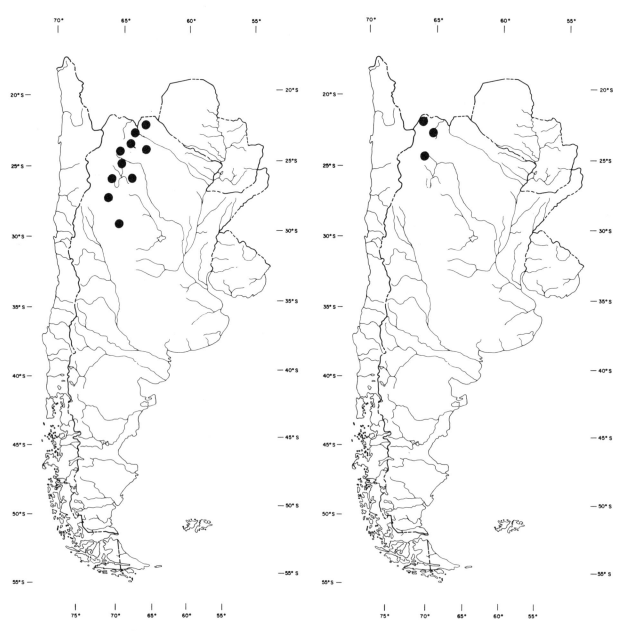

Map 11.50. Distribution of *Akodon boliviensis*.

Map 11.51. Distribution of *Akodon budini*.

and commonly in agricultural areas. In southern Peru up to 4,000 m it is the common mouse of grassy places, and some were caught in homes. In Tucumán Province, Argentina, they show a marked preference for areas with high vegetational cover and are commonly found in cultivated fields.

This terrestrial species is active both day and night. Its diet consists primarily of invertebrates, particularly coleopteran larvae. Twenty-six stomachs from thirteen sites in Peru contained 8% grass, 6% forbs, 8% seeds, and 78% insects (Dorst 1971, 1972; Fonollat 1984; Mares, Ojeda, and Kosco 1981; Ojeda 1979; Pearson 1951; Pizzimenti and de Salle 1980).

Akodon budini (Thomas, 1918)
Ratón de Calilegua

Measurements

	Mean	Min.	Max.	N	Loc.	Source[a]
TL	177.8	157	193.0	13	A	1
HB	98.4	89	108.0			
T	79.4	68	89.0			
HF	24.6	21	26.7			
E	19.5	16	21.6			
Wta	26.9	20	34.0	11		

[a](1) Thomas 1918a; BA, CM.

Description
The general dorsal color of *A. budini* is dark olivaceous brown, becoming browner on the rump. The venter is dark to brownish gray, lightly washed with whitish or buffy. The ears are sometimes darker than the head, the forefeet and hind feet are grayish, and there is a distinct white spot on the chin. The claws are comparatively long, and the tail is bicolored, and blackish above and whitish below (Thomas 1918a; pers. obs.).
Chromosome number: $2n = 38$; $FN = 42$ (Kajon et al. 1984).

Distribution
Akodon budini is found only in the mountains of northwestern Argentina in the provinces of Jujuy, Tucumán, and probably Salta. It is known from only a few localities (Honacki, Kinman, and Koeppl 1982; Olrog 1979; Thomas 1918a; BA) (map 11.51).

Ecology
In Jujuy province this species is common up to 2,500 m and undergoes population irruptions (Olrog 1979).

Akodon caenosus Thomas, 1918
Ratón Unicolor

Measurements

	Mean	Min.	Max.	N	Loc.	Source[a]
TL	152.7	135.0	166.0	19	A	1
HB	89.0	75.0	103.0			

T	63.7	55.0	71.0	
HF	19.4	18.0	20.3	
E	14.4	12.0	16.0	13
Wta	21.3	13.2	28.0	4

[a](1) Gyldenstolpe 1932; BA, CM, FM.

Description
The dorsum of this small *Akodon* is brownish olive to light reddish brown; the venter is slightly lighter brown or washed with pale yellow not sharply demarcated from the dorsum. The tail is weakly bicolored (Gyldenstolpe 1932; BA; pers. obs.).
Chromosome number: $2n = 34$. *A. caenosus* has a karyotype very similar to that of *A. puer* and may be included in this species (Bárquez et al. 1980; Kajon et al. 1984).

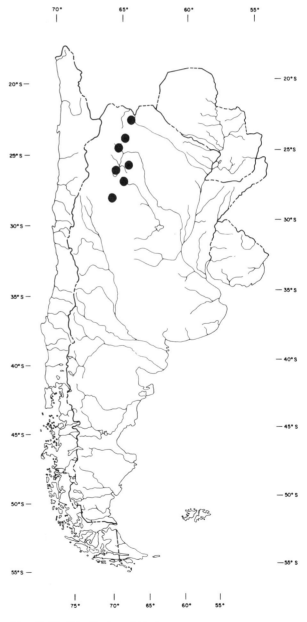

Map 11.52. Distribution of *Akodon caenosus*.

Distribution

Akodon caenosus is found in Bolivia and in northwestern Argentina, in the provinces of Jujuy and Tucumán (Bárquez et al. 1980; Honacki, Kinman, and Koeppl 1982) (map 11.52).

Ecology

This species has been taken in second-growth forest in northeastern Argentina (Bárquez et al. 1980).

Akodon cursor (Winge, 1887)

Measurements

	Mean	Min.	Max.	N	Loc.	Source[a]
TL	200.1	175	230	26	A, P, U	1
HB	110.2	79	128	25		
T	88.6	74	105			
HF	24.8	22	28	24		

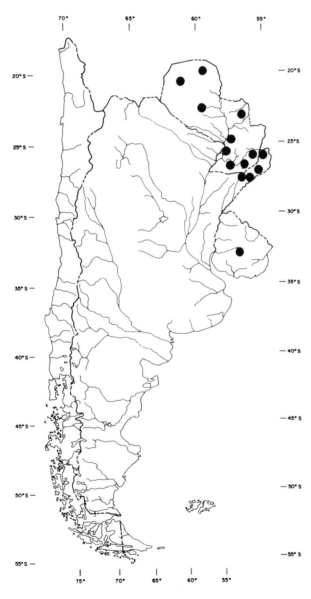

Map 11.53. Distribution of *Akodon cursor*.

	Mean	Min.	Max.	N
E	18.5	16	21	25
Wta	39.7	24	61	23

[a](1) Ximénez and Langguth 1970; BA, PCorps, UM.

Description

Akodon cursor is a medium-sized mouse, reddish brown to olive brown dorsally and grading to more tan on the sides, gradually becoming a reddish tan to gray washed with orange on the venter. The tail is sparsely haired and weakly bicolored, the feet are tan, and the face shows some blackish hairs (pers. obs.).

Chromosome number: $2n = 24$; FN = 42 (Liascovich and Reig 1989).

Distribution

The species is distributed in southeastern and central Brazil, Uruguay, Paraguay, and northeastern Argentina (Crespo 1982b; Honacki, Kinman, and Koeppl 1982; Massoia 1980a; UM) (map 11.53).

Life History

In northeastern Argentina the litter size is three and the breeding season is September to March (Crespo 1982b).

Ecology

In eastern Paraguay this is one of the most common species in the forests and forest-grassland ecotones. In Misiones province, Argentina, it is found in most habitats but prefers flatter, drier areas. *A. cursor* undergoes large population cycles, and stomachs of this species have been found to contain plant material, seeds, and adult and larval coleopterans, lepidopterans, and dipterans (Crespo 1982b; Myers 1982).

Akodon dolores Thomas, 1916
Ratón Cordobés

Measurements

	Mean	Min.	Max.	N	Loc.	Source[a]
TL	196.5	182.0	207.0	4	A	1
HB	113.8	105.0	124.0			
T	81.8	77.0	89.0			
HF	21.8	20.0	23.0			
E	18.0	17.0	19.0	3		
Wta	50.5	39.8	60.7			

[a](1) Gyldenstolpe 1932; BA.

Description

This rather large *Akodon* has soft, very dense dorsal pelage. The dorsum is light brown with an olive cast, blending imperceptibly to gray on the sides and venter. On some specimens the venter is washed with white, and there is sometimes a reddish tinge to the sides and venter. The tail is bicolored and haired for its entire length (Gyldenstolpe 1932; pers. obs.).

Distribution

Akodon dolores is apparently confined to the Sierra de Córdoba in the central Argentine province of Córdoba (Honacki, Kinman, and Koeppl 1982; Bianchi, Merani, and Lizarralde 1979) (map 11.54).

Life History

In the lab age of first reproduction is three months, and in the wild this species reproduces from September to March (Roldán et al. 1984).

Comment

This species exhibits heterozygosity, with chromosome numbers ranging from 34 to 37. In the lab fertile hybrids have been produced between *A. dolores* and *A. molinae* (Bianchi, Merani, and Lizarralde 1979; Merani et al. 1978).

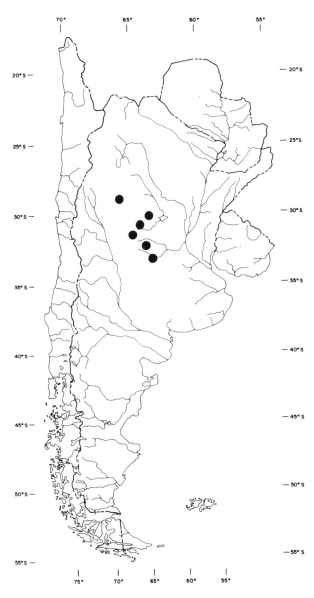

Map 11.54. Distribution of *Akodon dolores.*

Akodon hershkovitzi Patterson, Gallardo, and Freas, 1984

Description

One specimen measured TL 175; HB 104; T 71; HF 24 (Patterson, Gallardo, and Freas 1984). *A. hershkovitzi*, recently described, is a medium-sized *Akodon*, larger than any other species in the subgenus *Akodon* in southern South America except *A. markhami*. In coloration it is similar to *A. xanthorhinus* in having an agouti dorsum, head, and sides, buffy white chin and venter, rufescent dorsal surfaces of the feet, a bicolored tail, and differentially pigmented sides of the nose. It is easily distinguishable from *A. xanthorhinus* by its larger size (over 30 g) and its relatively longer tail (exceeding 65% of head and body length) (Patterson, Gallardo, and Freas 1984; pers. obs.).

Distribution

This species is found on the outer islands of the Chilean archipelago off the southern tip of South America (Patterson, Gallardo, and Freas 1984) (map 11.55).

Ecology

This mouse has been trapped in the Magellanic steppe and coastal forests of offshore islands (Patterson, Gallardo, and Freas 1984).

Comment

Akodon hershkovitzi appears to be a larger, island form derived from *A. xanthorhinus* (Patterson, Gallardo, and Freas 1984).

Akodon illuteus (Thomas, 1925)
Ratón Grande

Measurements

	Mean	Min.	Max.	N	Loc.	Source[a]
TL	188.8	169	209	4	A	1
HB	102.8	71	120			
T	86.0	77	98			
HF	25.0	24	26	3		
E	19.3	17	20	4		

[a](1) BA.

Description

This species has a gray brown dorsum, a gray venter, reddish brown on the nose and head, and a tail only faintly bicolored (pers. obs.).
Chromosome number: $2n = 52$; $FN = 56$ (Liascovich, Bárquez, and Reig 1989).

Distribution

This species is found only in the northwestern provinces of Jujuy and Tucumán, Argentina (Honacki, Kinman, and Koeppl 1982; Lucero 1983; Olrog 1979) (map 11.56).

Ecology

In Tucumán province this mouse is particularly common in brushy vegetation along streams from 1,000 to 4,000 m (Fonollat 1984; Lucero 1983).

Akodon iniscatus Thomas, 1919
Ratón Patagónico

Measurements

	Mean	Min.	Max.	N	Loc.	Source[a]
TL	149.0	134	167.0	11	A	1
HB	90.6	81	102.0			
T	58.4	52	66.0			
HF	19.6	18	21.6	10		
E	12.4	11	15.0	11		

[a](1) Thomas 1919c; BA, FM.

Description

Akodon iniscatus is similar in size and proportions to *A. xanthorhinus*. The dorsum is dark grizzled olivaceous brown to uniform brown, with a gray venter sharply demarcated from the dorsum. The ears are short, the tail is short, well haired, and strongly bicolored, and there is a conspicuous white spot on the chin (Thomas 1919c; pers. obs.).

Distribution

This species is found in southern and central Argentina from the province of Santa Cruz north to at least Neuquén province (Honacki, Kinman, and Koeppl 1982; BA) (map 11.57).

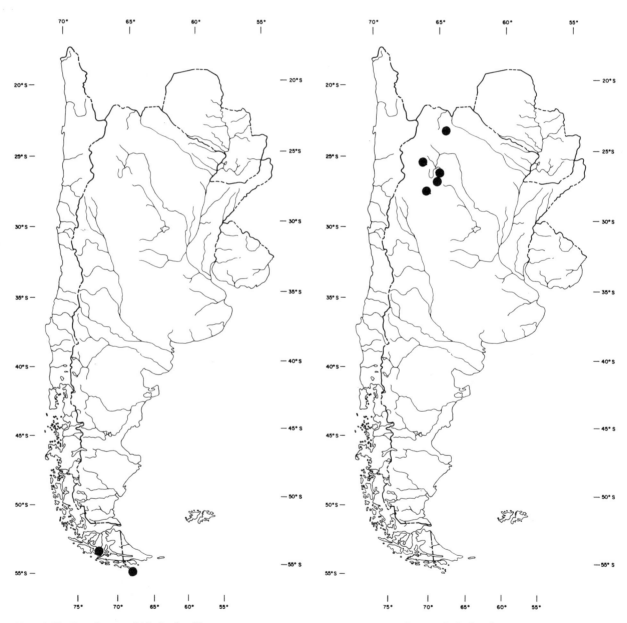

Map 11.55. Distribution of *Akodon hershkovitzi*.

Map 11.56. Distribution of *Akodon illuteus*.

Akodon (Chroeomys) jelskii (Thomas, 1894)
Ratón Tricolor

Measurements

	Mean	Min.	Max.	N	Loc.	Source[a]
TL	178.1	157	191	13	A, P	1
HB	98.7	84	106			
T	79.4	72	88			
HF	23.3	21	25			
E	18.5	18	19			

[a](1) Pearson 1951; Sanborn 1947a; BA.

Description

This is the most striking of the *Akodon* species, and its coloring makes identification easy. The unique features of color alone resulted in its original description as a separate genus. In some populations the nose, face, and ears are red, while in others only the nose is orange. The dorsal pelage is gray brown to chocolate brown, and the venter is sharply contrasting white that may extend onto the sides and the cheeks. In some populations there is a sharply contrasting white patch behind each ear that may connect with the white on the cheeks. The feet are tan, the tail is not bicolored, and the fur is dense and long (Pearson 1951; pers. obs.).

Distribution

Akodon jelskii is distributed from central Peru to Argentina, where it is found in Salta and Jujuy provinces (Honacki, Kinman, and Koeppl 1982; BA) (map 11.58).

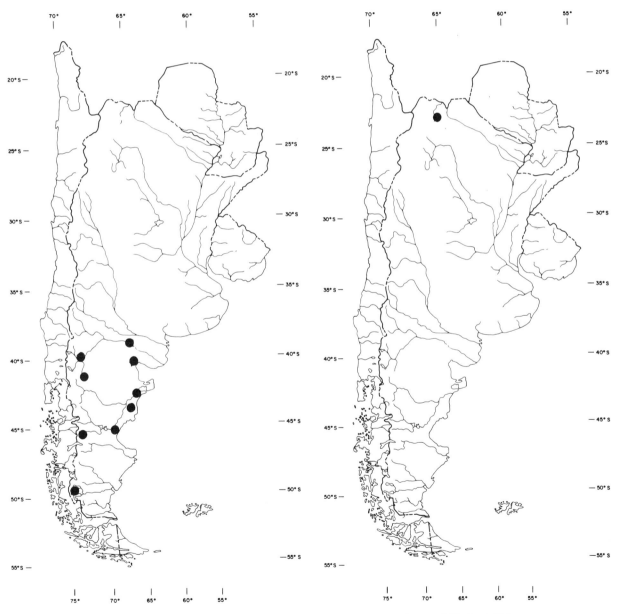

Map 11.57. Distribution of *Akodon iniscatus*.

Map 11.58. Distribution of *Akodon jelskii*.

Life History

Two females from Peru had three embryos each (MVZ).

Ecology

This high-altitude *Akodont* ranges from about 2,800 to 5,600 m. It occupies a variety of habitats in southern Peru, such as grassy places, rocks, vacant huts, and occupied houses. It is active day and night, and seventeen stomachs were found to contain 19% grass, 42% forbs, 4% seeds, and 35% insects (Dorst 1971; Pearson 1951; Pizzimenti and de Salle 1980; Sanborn 1947a).

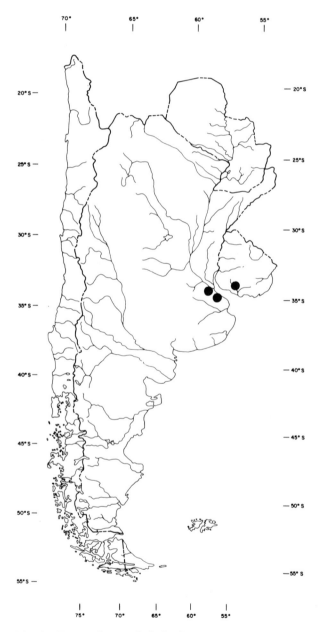

Map 11.59. Distribution of *Akodon kempi*.

Comment

The generic status of this species is in some dispute. Many workers think an elevation to full generic status is in order.

Akodon (Deltamys) kempi (Thomas, 1917)
Ratón del Delta

Measurements

	Mean	Min.	Max.	N	Loc.	Source[a]
TL	178.3	170	195	10	A, U	1
HB	96.3	89	108			
T	82.0	74	87			
HF	20.9	20	22			
E	13.0	12	14			
Wta	26.4			4	U	2

[a](1) Massoia 1964; Thomas 1917; (2) Miller and Anderson 1977.

Description

The eyes are rather small, without eye-rings, and the ears are short and well haired. *A. kempi* has a tail proportionally longer than the sympatric *A. azarae*. The dorsum of *A. kempi* is blackish brown, inconspicuously washed on the head and sides with olivaceous. The venter is dull brownish gray, the feet are tan, and the tail is only faintly bicolored (Massoia 1964; Thomas 1917; pers. obs.).

Distribution

Akodon kempi is known only from the islands of the Río Paraná estuary in both Argentina and Uruguay (Massoia 1963c, 1964) (map 11.59).

Life History

One female with four embryos was captured (MVZ).

Ecology

This is a species of the wet grassy areas of the Río Paraná delta, where it is found with *Holochilus brasiliensis* and *Scapteromys*. It is omnivorous, and one grass nest was found in a log (Massoia 1964).

Akodon lanosus (Thomas, 1897)
Ratón Colorado

Measurements

	Mean	Min.	Max.	N	Loc.	Source[a]
TL	163.4	151.0	168	10	C	1
T	59.0	53.0	65			
HF	21.9	21.5	23			
Wta	25.0	22.0	28	2	A	2

[a](1) Osgood 1943; (2) Pine, Angle, and Bridge 1978.

Description

This small cinnamon brown to olive brown mouse has short, thinly haired ears. The venter is heavily washed with fulvous or white, and it has white feet and a bicolored tail (Osgood 1943; Pine, Angle, and Bridge 1978).

Distribution

Akodon lanosus is found in southern Argentina and southern Chile. In Argentina it occurs in Santa Cruz and Tierra del Fuego, and in Chile it occurs from the extreme south of Ultima Esperanca province south to Tierra del Fuego (Honacki, Kinman, and Koeppl 1982; Tamayo and Frasinetti 1980) (map 11.60).

Ecology

In Chile this species prefers cool, humid forests and has been found in *Bubo* owl pellets (Osgood 1943; Rau, Yáñez, and Jaksić 1978).

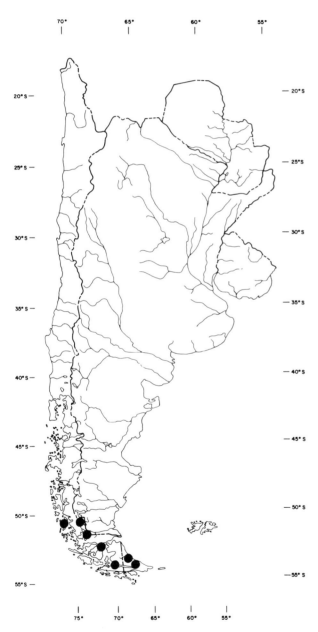

Map 11.60. Distribution of *Akodon lanosus*.

Akodon longipilis (Waterhouse, 1837)
Ratón de Pelos Largos

Measurements

	Mean	S.D.	Min.	Max.	N	Loc.	Source[a]
TL	104.9		91	122	106 m	A	1
	105.4		88	121	82 f		
HB	129.2	2.5	115	145	39	C	2
T	88.1	2.2					
HF	27.7	0.5					
E	20.0	0.7					
Wta	38.2		15	51	121 m	A	1
	36.9		23	60	87 f		

[a](1) Pearson 1983; (2) Jaksić and Yáñez 1979; Yáñez et al. 1978.

Description

This rather large, heavy-bodied mouse has small, thinly haired ears, long, loose pelage, and a dark dorsum. Both color and size vary geographically, but the dorsum ranges from gray with rosy tints to coffee color with rich sepia tones. The venter is gray to grayish white, and the tail is bicolored, though only slightly in some subspecies (Mann 1978; Osgood 1943; pers. obs.).

Distribution

Akodon longipilis ranges from near the Chilean city of Santiago south to Tierra del Fuego, and in Argentina from Tierra del Fuego north to at least the southern part of Mendoza province (Honacki, Kinman, and Koeppl 1982; Olrog and Lucero 1981; Roig 1965; Tamayo and Frassinetti 1980) (map 11.61).

Life History

In the Chilean province of Malleco the average number of embryos was 3.7 (range 2–5; $n = 9$); similarly, in southwestern Argentina the average was 3.78 (range 2–5; $n = 51$). The central Chilean subspecies is reported to have two or three litters a year with six to eight young per litter. In Argentina the males begin breeding in November and December and continue until the autumn, and females undergo a postpartum estrus. In the northern Chilean scrub females had embryo counts of six and four, and no individual reached sexual maturity before five months (Meserve and Le Boulengé 1987; Greer 1966; Mann 1978; Pearson 1983).

Ecology

This volelike mouse is found in many vegetation types ranging from cloud forest to brushy areas and marshes, but it is most common in moister areas with a high proportion of shrub and litter cover. In many habitats it is the most commonly caught small mammal. In central Chile it is generally found in or near rotting logs, in grassy areas at the base of shrubs, and in rocky areas with dense undergrowth. In Patagonia *A. longipilis* can reach densities of 10.8/ha.

In Chile densities ranged from 7.1 to 12 per hectare. The home ranges of females did not overlap, though the home ranges of males overlapped those of other males and of females. In central Chile the average home-range size was 175 m² in a forest of *Nothofagus* and *Chusquea*.

   Akodon longipilis is a good burrower but has also been caught 2 m off the ground. It is active both diurnally and nocturnally. The diet of this species varies geographically, though basically it is frugivorous / omnivorous. In Argentina stomachs were found to contain berries, seeds, fern spores, insects, fungi, worms, and slugs. In the semiarid shrubland of central Chile *A. longipilis* is largely insectivorous, differing from the sympatric *A. olivaceus*, which

eats many more seeds. The consumption of insects increases between spring and summer (Fulk 1975; Contreras and Rosi 1981a; Glanz 1977a,b; Greer 1966; Iriarte, Contreras, and Jaksić 1989; Mann 1978; Meserve 1981a,b; Murúa and González 1979; Murúa, González, and Jofre 1982; Pearson 1983; Pearson and Pearson 1982; Schamberger and Fulk 1974).

Comment

Pearson (1984) has stated that *Chelemys angustus* is a synonym of *A. longipilis*.

Akodon mansoensis De Santis and Justo, 1980

Meausurements

	Mean	Min.	Max.	N	Loc.	Source[a]
TL	184.0	162	198.0	18	A	1
T	83.5	73	92.0			
HF	23.9	23	26.0			
E	16.6	16	17.2			

[a](1) De Santis and Justo 1980.

Description

This medium-sized *Akodon* has a dark chestnut dorsum and a lighter venter. The tail is rosy chestnut dorsally and lighter underneath (De Santis and Justo 1980).

Distribution

Akodon mansoensis is known only from the Argentine province of Río Negro (De Santis and Justo 1980) (map 11.62).

Comment

This species may be a synonym of *A. olivaceus* (O. P. Pearson, pers. comm.).

Akodon markhami Pine, 1973

Measurements

	Mean	Min.	Max.	N	Loc.	Source[a]
TL	184.0	180.0	187.0	4	C	1
HB	104.0	101.0	107.0			
T	80.0	77.0	84.0			
HF	24.9	24.7	25.3			
E	14.6	13.5	15.4			
Wta	29.5	24.0	32.0			

[a](1) Pine 1973.

Description

This medium-sized to large *Akodon* is externally very similar to *A. olivaceus* but has longer hair (Pine 1973).

Distribution

This is one of only two Chilean mammals found on islands but not on the mainland. It occurs only on Isla Wellington in the Chilean province of Magallanes (Pine 1973; Tamayo and Frassinetti 1980) (map 11.63).

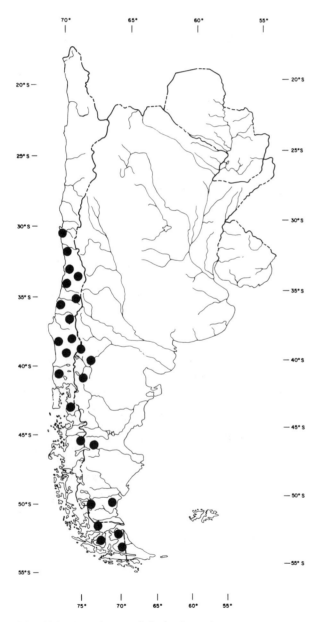

Map 11.61. Distribution of *Akodon longipilis*.

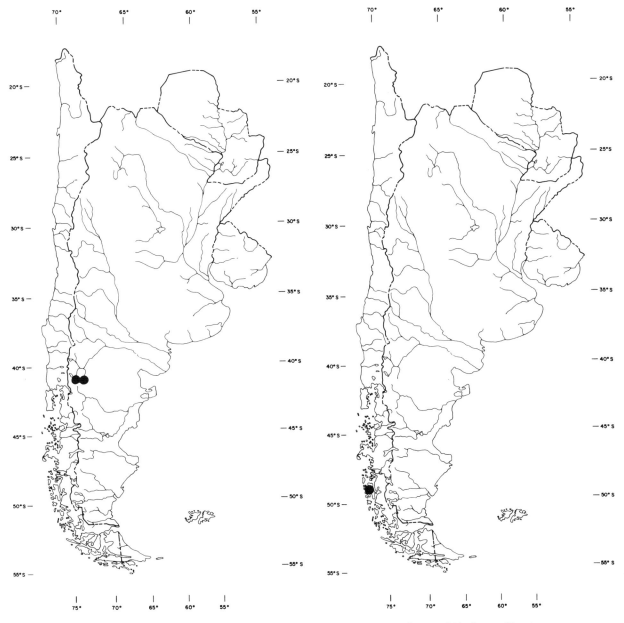

Map 11.62. Distribution of *Akodon mansoensis*.

Map 11.63. Distribution of *Akodon markhami*.

Akodon molinae Contreras, 1968
Ratón Pajizo

Measurements

	Mean	Min.	Max.	N	Loc.	Source[a]
HB	100.8	89.0	114.0	18	A	1
T	78.4	68.0	90.0			
HF	24.2	22.0	25.1			
E	17.2	15.0	18.5			
Wta	31.8	20.5	45.0			

[a](1) Contreras and Rosi 1980d.

Description
This good-sized *Akodon* has long vibrissae and long, dense fur, straw-colored on the dorsum and grizzled gray on the stomach. The lightly haired tail is dark on the top and white along the bottom for the terminal third (Contreras 1968).

Chromosome number: Individuals of this species had 42, 43, or 44 chromosomes (Bianchi, Merani, and Lizarralde 1979).

Distribution
Akodon molinae is confined to central Argentina, where it is found in the provinces of Mendoza, Buenos Aires, Río Negro, and La Pampa (Contreras 1968; Contreras and Rosi 1980d) (map 11.64).

Life History
Animals from a laboratory colony reproduced for the first time at three months and continued re-

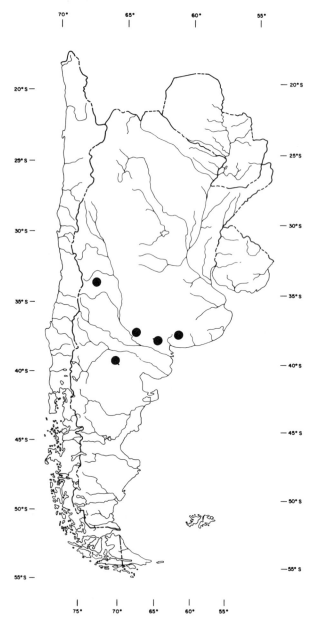

Map 11.64. Distribution of *Akodon molinae*.

HF	18.5	16	20	
E	12.2	11	14	11
Wta	19.9	13	29	

[a](1) BA, UM.

Description

This small, dark *Akodon* is distinguished from the sympatric *A. cursor* by its smaller size and a tail half the length of *A. cursor's*. It is glistening olive brown to reddish brown, finely grizzled with ochraceous on the dorsum and dull ochraceous washed with yellowish to gray brown on the venter. The tail is not bicolored (Gyldenstolpe 1932; Myers and Wetzel 1979; pers. obs.).

Distribution

The species is found in eastern Paraguay, southeastern Brazil, and northeastern Argentina (Massoia 1963a; Myers 1982) (map 11.65).

Life History

In Brazil females have four pairs of mammae, and three females were found to contain three, four, and five embryos (Davis 1947).

Ecology

Akodon nigrita is apparently confined to moist tropical forests, following the Paraná rain forests from southeastern Brazil into eastern Paraguay and Misiones province, Argentina. In Paraguay these mice were trapped in wet tropical forest in runways through grass or near fallen logs, though they were also captured in second growth. In Brazil they were found under logs and tree roots. This species will make tunnels in the leaf litter. It is aggressive, strongly terrestrial and diurnal (Davis 1947; Myers 1982; Myers and Wetzel 1979; UM).

Comment

Akodon nigrita has been included in the genus *Thaptomys* by some authors (Honacki, Kinman, and Koeppl 1982; but see table 11.3).

producing until at least thirteen months of age. In the wild they breed from November to March (Roldán et al. 1984).

Ecology

Akodon molinae has been trapped in clumps of grass near watercourses (Contreras 1968).

Akodon nigrita (Lichtenstein, 1829)
Ratón Subterráneo

Measurements

	Mean	Min.	Max.	N	Loc.	Source[a]
TL	139.3	125	154	12	A, P	1
HB	92.3	83	107			
T	46.9	40	56			

Akodon olivaceus (Waterhouse, 1837)
Ratón Oliváceo

Measurements

	Mean	Min.	Max.	N	Loc.	Source[a]
TL	169.8	160.0	188.0	10	C	1
	95.6	86.0	103.0	36 m	A	2
	93.2	84.0	108.0	22 f		
HB	103.3	93.0	116.0	10	C	1
T	66.5	60.0	72.0			
HF	22.6	22.0	24.0			
E	16.3	15.0	18.0			
Wta	30.1	24.2	40.2			
	28.2	20.0	36.0	41 m	A	2
	25.8	18.5	35.0	23 f		

[a](1) Santiago; (2) Pearson 1983.

Description

This small mouse is grayish brown dorsally, sometimes grizzled with yellow, and has whitish, grayish, or brownish underparts. The hair color is variable, influenced by time of year and the color of the substrate the mice are living on. Size decreases with latitude (Greer 1966; Mann 1978; Osgood 1943; Yáñez, Valencia, and Jaksić 1979).

Chromosome number: $2n = 52$ (Spotorno and Fernández 1976).

Distribution

Akodon olivaceus is found from northernmost Chile south to Tierra del Fuego, and in Argentina along the southern extension of the Andes at least in Río Negro and Neuquén provinces and probably in Chubut as well (Mann 1978; Olrog and Lucero 1981; Pearson 1983; Tamayo and Frassinetti 1980) (map 11.66).

Life History

Near Santiago (Chile) evidence of breeding was found from at least September to March. In Fray Jorge Park, Chile, most reproduction took place in November and December, and the average number of embryos per female was 5.5 (± 0.3; $n = 7$). Other embryo counts range from 5.5 (range 3–8; $n = 189$, Chile) to 5.1 (range 4–8; $n = 12$, Argentina). In Chile *A. olivaceus* is reported to produce two or three litters a year, and the age of first reproduction is about two months (Fulk 1975; Greer 1966; Mann 1978; Meserve and Le Boulengé 1987; Pearson 1983).

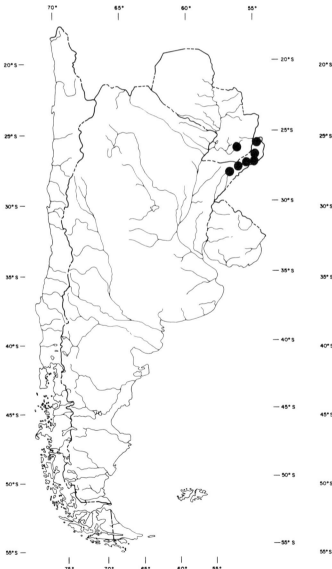

Map 11.65. Distribution of *Akodon nigrita*.

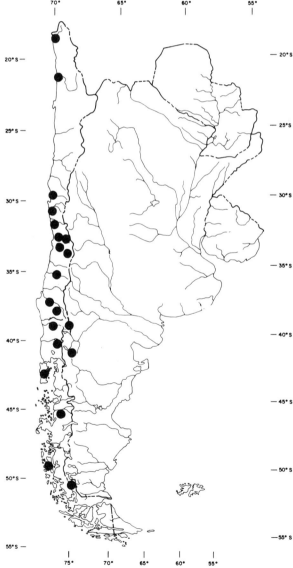

Map 11.66. Distribution of *Akodon olivaceus*.

Ecology

Akodon olivaceus is a ubiquitous cricetine in grassy and brushy areas throughout its range. It is also caught in woods and pastures but is always associated with vegetation that provides good cover, whether bushes or dense grass. The density of this species can vary at one locality from 25 to 100 individuals per hectare. In the Patagonian forests of Argentina it has been recorded at 17.9/ha, and in Fray Jorge Park in Chile a much higher density of 67.2/ha was recorded. Populations of this species reportedly cycle with the flowering of bamboo. In one study the average home range in a marshy area of Malleco province, Chile, was 12.5 m².

In Chile the annual population cycle in shrubland/grassland peaked in the fall with 37.7/ha, whereas in woodlands the population maximum was reached in winter at 37.9/ha. Males living in the woods had home ranges ranging from 0.07 to 0.23 ha, and in woodlands home ranges ranged from 0.12 to 0.25 ha.

These mice are usually nocturnal but in some sites they have been caught mostly during the day. They climb well, can dig, and in Chile have been found living in burrows of other rodents such as *Spalacopus*. In dense grass they create runways, and in rocky areas they tunnel under rocks. They build simple nests of grass, underground or sheltered in roots or rocks. One study concluded that they apparently live in small groups including females, juveniles, and a single male.

A sample of fifty-one stomachs from the forests of southern Chile, taken throughout the year, contained (by volume) 39.1% seeds, 13.9% insects, and 42.1% fungus, plus other minor food items. Another study found that the year-round diet consisted of shrub and herbaceous seeds; and 39.7% vegetative tissue. *A. olivaceus* is apparently a poor conserver of metabolic water (Fulk 1975; González, Murúa, and Feito 1982; Greer 1966; Mann 1978; Meserve 1978, 1981a; Murúa and González 1981; Murúa, González, and Jofre 1982; Pearson 1983; Pearson and Pearson 1982; Schamberger and Fulk 1974).

Akodon puer Thomas, 1902

Measurements

	Mean	Min.	Max.	N	Loc.	Source[a]
TL	154.5	145	165	6	A, B	1
HB	83.0	75	90			
T	71.5	70	75			
HF	18.3	15	20			
E	15.2	15	16			

[a](1) Thomas 1918a; CM, FM.

Description

Dorsally this mouse is dark olivaceous grizzled with tan; the sides are a little more buffy, and the venter is gray washed with yellow (Thomas 1918a; pers. obs.).

Chromosome number: $2n = 34$; FN = 42 (Kajon et al. 1984).

Distribution

Akodon puer is found in western Bolivia, southern and central Peru, and apparently in the Argentine province of Jujuy (Honacki, Kinman, and Koeppl 1982; Thomas 1918a).

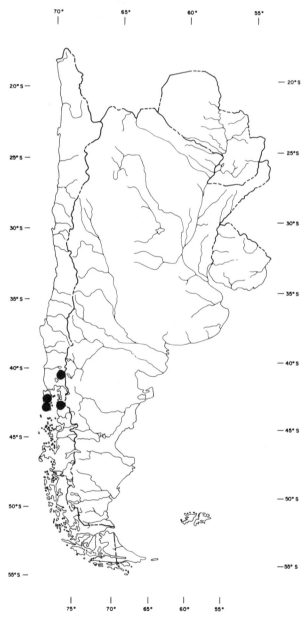

Map 11.67. Distribution of *Akodon sanborni*.

Akodon sanborni Osgood, 1843
Ratón Negruzco

Measurements

	Mean	Min.	Max.	N	Loc.	Source[a]
TL	180.0	169.0	200.0	10	C	1
T	75.8	69.0	85.0			
HF	24.1	23.5	25.5			
E	14.9	13.5	16.0	4	C	2
Wta	25.9	21.0	30.0			

[a](1) Osgood 1943; (2) Santiago.

Description
This small dark mouse is generally a uniform blackish brown, including the tail and feet, though some individuals can be much lighter. In some populations the dorsum is iron gray, not contrasting sharply with the venter. The fur is thick, and the tail is dark and unicolored (Osgood 1943; pers. obs.).

Distribution
Akodon sanborni is found in southern Chile in the region of Los Lagos (Valdivia, Llanquihue, and Chiloé) and adjacent areas of Argentina (Honacki, Kinman, and Koeppl 1982; Olrog and Lucero 1981; Tamayo and Frassinetti 1980) (map 11.67).

Ecology
This *Akodon* is a forest dweller, active both day and night. Its diet includes a high proportion of fungi with moderate amounts of mature arthropods and larvae (15%–30% of diet) (Osgood 1943; Meserve, Lang, and Patterson 1988; Murúa, González, and Jofre 1982).

Comment
Some investigators think this form does not constitute a valid species (O. P. Pearson, pers. comm.).

Akodon serrensis Thomas, 1902

Measurements

	Mean	Min.	Max.	N	Loc.	Source[a]
TL	177.0	166	185	3	A, Br	1
HB	94.5	88	101	2		
T	81.0	78	84			
HF	24.0	24	24			
E	18.0	18	18			

[a](1) Justo and De Santis 1977; Thomas 1902.

Description
The fur of this *Akodon* may be thick and wooly, a grizzled olivaceous above, darker along the spine, paling on the sides and ochraceous below. The venter is not sharply demarcated from the dorsum; the anal region is strikingly ochraceous. The ears are dark brown, slightly darker than the body, and the tail is very thinly haired, almost naked (Justo and De Santis 1977; Thomas 1902; pers. obs.).

Chromosome number: $2n = 44$; FN = 44 (Liascovich and Reig 1989).

Distribution
Akodon serrensis is known from southeastern Brazil and northern Argentina and perhaps into eastern Paraguay (Honacki, Kinman, and Koeppl 1982) (map 11.68).

Ecology
In Argentina an individual of this species was taken in an *Araucaria* forest (Justo and De Santis 1977).

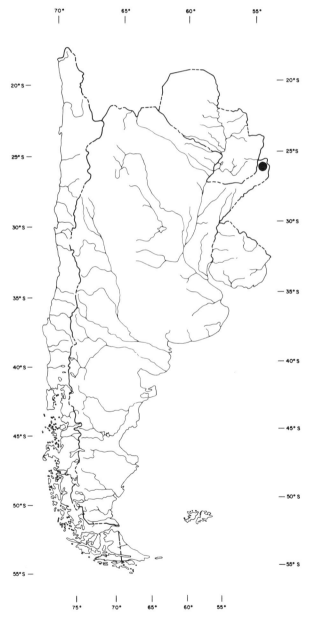

Map 11.68. Distribution of *Akodon serrensis*.

Akodon varius Thomas, 1902
Ratón Variado

Measurements

	Mean	Min.	Max.	N	Loc.	Source[a]
TL	188.0	137.0	206.0	23	P	1
HB	108.2	80.0	125.0	70	A	2
	105.0	84.0	118.0	22	P	1
T	84.1	69.0	101.0	68	A	2
	85.3	72.0	100.0	22	P	1
HF	24.6	22.5	27.2	73	A	2
	24.5	22.0	31.0	23	P	1
E	20.5	17.0	27.2	73	A	2
	16.8	14.0	20.0	23	P	1
Wta	35.7	21.7	62.0	69	A	2
	38.9	32.0	48.0	16	P	2

[a](1) Bárquez et al. 1980; (2) PCorps;
Note: See Comment at end of species account.

Description

As portrayed here the data are probably derived from a composite (see Comment). This large mouse is heavy bodied, with a dark gray to light chestnut dorsum grading laterally to gray; the venter is lighter gray. The feet are white or gray, and some individuals have a distinctive white chin spot. There can be markedly different color in different subspecies; one is entirely gray. The tail is bicolored (Bárquez et al. 1980; Gyldenstolpe 1932; Olrog and Lucero 1981; pers. comm.).
Chromosome number: $2n = 41$ (Bárquez et al. 1980).

Distribution

The species is found in Bolivia, eastern and western Paraguay, and Argentina along the sub-Andean district south to at least Neuquén province (Honacki, Kinman, and Koeppl 1982; Myers 1982; UM) (map 11.69).

Life History

In Tucumán province, Argentina, litter size ranges from two to four, and reproduction starts at the end of September. This species will nest in holes in the ground (Fonollat 1984; Fonollat and Décima 1979).

Ecology

Akodon varius is a mouse of moist areas and has been caught in forested areas, riverbanks, stream banks, sugarcane fields, old fields, orchards, and grasslands. In many areas it is numerically dominant, and in Mendoza province, Argentina, it was found at a density of 21.5/ha. At this study site the average home range was 212.5 m².

In one study a quarter of the captures of this species were made during the day, though in another study it was caught only at night. Animals were often caught from burrows under logs or rocks or in forest litter. Stomachs from one study contained only invertebrate remains, mostly insect larvae; in another study plant material was found with insect larvae (Contreras and Rosi 1980a; Mares 1973; Mares, Ojeda, and Kosco 1981; Ojeda 1979).

Comment

The ecological data probably include material for *Akodon simulator.*

In a recent study several of the subspecies of *Akodon varius* were separated out as species. These included *A. neocenus,* found in central and southern Argentina below 1,000 m; *A. simulator,* found in Salta, Santiago del Estero, and Catamarca provinces, Argentina; and *A. toba,* found in the Chaco of

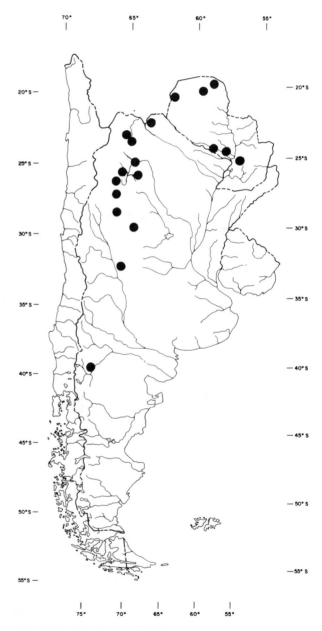

Map 11.69. Distribution of *Akodon varius.*

western Paraguay, eastern Bolivia, and northwestern Argentina. Under the new revision *A. varius* would be found mainly in the Bolivian Andes between 2,000 and 3,000 m (Bárquez et al. 1980; Myers 1989).

Myers (1989) has attempted a monumental revision of the *varius* group within the genus *Akodon*. Some of the species he treats range into Bolivia, an area not covered in this volume. He considers the following species: *Akodon dayi, A. dolores, A. molinae, A. neocensus, A. simulator, A. toba*, and *A. varius*. He concludes that *A. glaucinus* and *A. tartereus* can be considered subspecies of *A. simulator*.

Akodon xanthorhinus (Waterhouse, 1837)
Ratón Hocico Bayo

Measurements

	Mean	S.D.	N	Loc.	Source[a]
TL	92.3	8.1	6 m	C	1
	103.0	3.5	7 f		
T	53.2	1.9	6 m		
	52.6	5.0	7 f		
HF	20.3	0.8	6 m		
	20.3	0.6	7 f		
E	14.4		29	C	2
Wta	26.5		2	A, C	3

[a](1) Patterson, Gallardo, and Freas 1984; (2) Yáñez, Valencia, and Jaksić; (3) BA, Santiago.

Description

This small *Akodon* is distinguished from *A. hershkovitzi* by its smaller size (less than 30 g) and relatively short tail (shorter than 65% of head and body length). The fur is long and dense. Dorsally *A. xanthorhinus* varies from grayish brown to brown suffused with rufous. In one subspecies (*A. x. canescens*) the dorsum and feet are lighter, the forefeet are tan, the hind feet are darker and more rufescent, and the yellow orange color of the nose is more or less defined on all specimens. In the other subspecies (*A. x. xanthorhinus*) the dorsum and feet are darker, the forefeet have traces of orange, the hind feet are predominantly rufescent, and the nose is proportionally darker (Osgood 1943; Patterson, Gallardo, and Freas 1984; pers. obs.).
Chromosome number: $2n = 52$ (Spotorno and Fernández 1976).

Distribution

Akodon xanthorhinus is found in Patagonia and Tierra del Fuego (Argentina and Chile) (Patterson, Gallardo, and Freas 1984) (map 11.70).

Ecology

This akodontine inhabits a wide range of habitats from moist forests to open grasslands and marshes, but it is apparently most common in swampy areas and grasslands surrounded by forest. It is found in nearly all patches of suitable habitat, indicating a well-developed colonization ability. In general the subspecies *A. x. xanthorhinus* is found in forests of *Nothofagus* and *Drimys*, while the subspecies *A. x. canescens* is found more in the pampas habitats (Mann 1978; Osgood 1943; Patterson, Gallardo, and Freas 1984; Pearson and Pearson 1982; Pine, Angle, and Bridge 1978; Pine, Miller, and Schamberger 1979; Reise and Vengeas 1974).

Comment

Akodon llanoi is included in this species (Patterson, Gallardo, and Freas 1984).

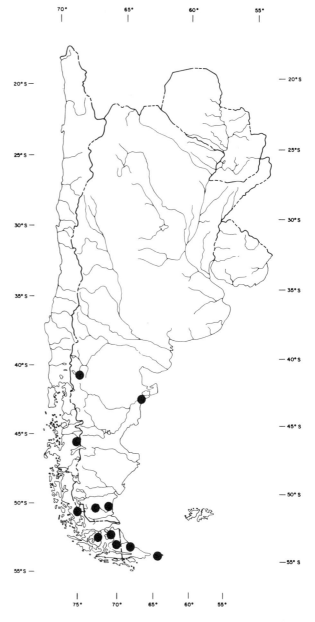

Map 11.70. Distribution of *Akodon xanthorhinus*.

Genus *Bolomys* Thomas 1916

Description

Bolomys is clearly an akodont rodent and thus exhibits the major characters of that tribe (see pp. 000–00). According to Reig (1987, 352–54) the key features of *Bolomys* when constrasted with *Akodon* include: Braincase broad and deep; occipital region short; rostrum rather short and markedly tapering forward in lateral view; upper profile of skull gradually sloping forward from the middle of parietals; nasals short, with anterior borders well posterior to the level of the anterior border of incisors; frontals long, always longer than nasals; parietals short, less than half the length of frontals and extending forward anteriolaterally by means of narrow spines penetrating between frontals and temporals; interparietal noticeably reduced anteroposteriorly and transversely; occiput short and truncated; interorbital area with well-formed, anteriorly convergent borders; posterior palate moderately long and wide, the median posterior border of palatines behind the posterior border of M³; zygomatic plate broad and strong with anterior border straight or slightly concave, perpendicular to diastema; upper incisors orthodont or proodont; molars mesodont, terraced with moderate wear, broad and robust; upper molars with lophs almost completely transverse, and mesoloph usually completely coalesced with paraloph; procingulum of M¹ simple with anteromedian flexus absent or only slightly developed; lower molars with lingual cusps somewhat anterior to the labial ones, with mesolophid remnants and mesostylids usually absent.

Although many workers in the past have been tempted to place *Bolomys* as a subgenus of *Akodon*, Anderson and Olds (1989) make a strong case for the generic status of *Bolomys* and clarify distributional problems on the frontier of Argentina and Bolivia.

Distribution

The genus includes five species distributed in Peru, Brazil, Paraguay, Bolivia, Uruguay, and Argentina.

Bolomys lactens (Thomas, 1918)
Ratón Ventrirufo

Description

One specimen measured HB 101; T 67; HF 22; E 16. The dorsum is mixed blackish and buffy, with the head grayer and the rump more buffy. The sides and belly are distinctly more buffy, and the chin is white. The ears are about the color of the head, and the claws are rather long (Thomas 1918a).

Distribution

Bolomys lactens is known from the northwestern Argentine provinces of Jujuy and Tucumán (Honacki, Kinman, and Koeppl 1982; Lucero 1983; Thomas 1918a) (map 11.71).

Bolomys lasiurus (Lund, 1814)
Ratón Selvático

Measurements

	Mean	Min.	Max.	N	Loc.	Source[a]
TL	177.7	158	212	12	A, P	1
HB	102.5	85	132			
T	75.2	62	90			
HF	24.6	22	27			
E	16.3	14	19			
Wta	35.4	20	58	13		

[a](1) PCorps, UM.

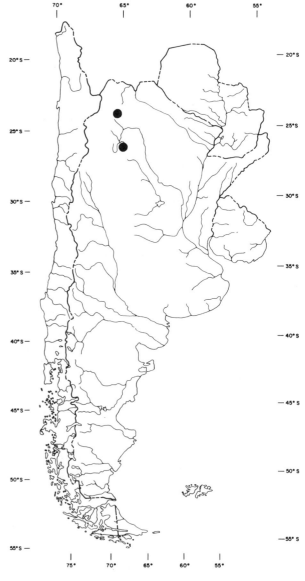

Map 11.71. Distribution of *Bolomys lactens*.

Description

This olivaceous gray mouse has a gray to grayish white venter. It has no distinctive facial markings. The tail is much shorter than the head and body, lightly haired, and unicolored (Macedo and Mares 1987; pers. obs.).

Distribution

The species is found in eastern Brazil and Paraguay and into the Argentine province of Misiones. It occurs patchily throughout eastern and western Paraguay (Honacki, Kinman, and Koeppl 1982; Myers 1982; BA) (map 11.72).

Life History

In central Brazil six females had an average of 4.2 embryos each (Dietz 1983).

Ecology

Bolomys lasiurus is a species of grassland and cerrado, though it is occasionally trapped in forest. In the Brazilian cerrado it is almost exclusively terrestrial, mostly diurnal, and trapped at the forest edge and in bamboo near gallery forest. It is seasonally rare, becoming the most common species during the rainy season. It has been found at an average density of 11.8/ha. In the Brazilian caatinga it is found only in cultivated and abandoned fields and appears to be dependent on these areas. It builds nests of grass and leaves and lives in burrows with two to five openings. *B. lasiurus* depends on free water in its food. In central Brazil its diet consists primarily of seeds (82%; $n = 32$), although in some localities it also eats significant quantities of invertebrates (Alho and de Souza 1982; Borchert and Hansen 1983;

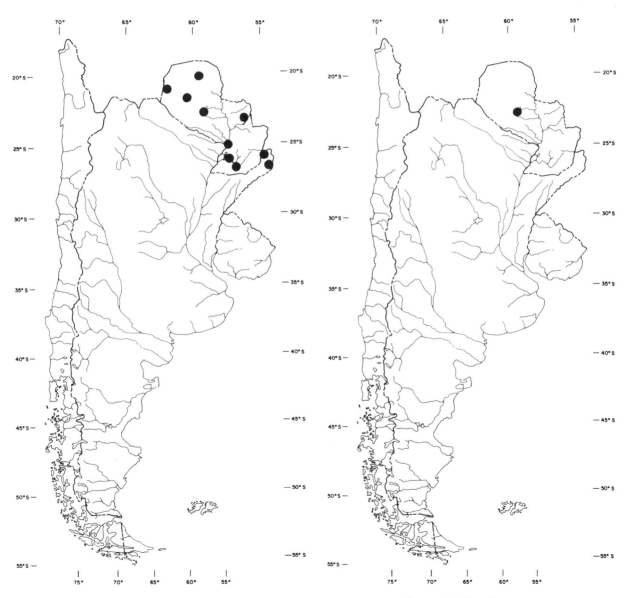

Map 11.72. Distribution of *Bolomys lasiurus*.

Map 11.73. Distribution of *Bolomys lenguarum*.

Dietz 1983; Macedo and Mares 1987; Nitikman and Mares 1987; Myers 1982; Streilein 1982a,b; UM).

Comment

Bolomys lasiurus includes *B. lasiotus*, *Akodon lasiurus*, *A. arviculoides*, *A. lenguarum*, and *Zygodontomys lasiurus* (Honacki, Kinman, and Koeppl 1982; Myers 1982).

Bolomys (Cabreramys) lenguarum (Thomas, 1898)
Ratón Bayo

Description

Two specimens had the following measurements: TL 181–91; HB 110–15; T 71–76; HF 21; E 14.7 (Massoia and Fornes 1967b).

Distribution

The identity and distribution of this species are controversial, but they have been dealt with by Anderson and Olds (1989). Honacki, Kinman, and Koeppl (1982) contend it is found in Bolivia, southwestern Brazil, Paraguay, and Argentina, but Massoia and Fornes say it is known only from Bolivia and the Paraguayan Chaco, and Lucero reports it from Tucumán Province. Olds and Anderson (1989) report that in southern South America this species occurs only in Paraguay (map 11.73).

Comment

This taxon includes *Akodon tapirapoanus* (Honacki, Kinman, and Koeppl 1982).

Bolomys (Cabreramys) obscurus (Waterhouse, 1837)
Ratón Oscuro

Measurements

	Mean	Min.	Max.	N	Loc.	Source[a]
TL	178.3	163.0	199	14 m	A	1
	168.0	153.0	193	8 f		
	177.7	160.0	203	19	U	2
T	68.3	57.0	78	14 m	A	1
	63.1	56.0	78	8 f		
	82.0	68.0	93	19	U	2
HF	22.8	21.0	24	14 m	A	1
	21.4	20.0	23	8 f		
	22.2	22.0	23	19	U	2
E	15.1	14.0	16	14 m	A	1
	14.5	13.0	16	8 f		
	13.6	12.0	16	19	U	2
Wta	52.0	40.0	60	8 m	A	1
	44.0	35.0	53	5 f		
	26.1	19.6	38	19	U	2

[a](1) Fornes and Massoia 1965; (2) Barlow 1965.

Description

This volelike mouse has a very dark brown dorsum tinged with reddish in juveniles and olivaceous in adults, and the venter is gray washed with yellowish. The tail is strongly bicolored and the nose is light brown. The nails are very well developed (Barlow 1965; Massoia and Fornes 1967).

Distribution

This species is found in Uruguay and northeastern and east-central Argentina at least as far south as Buenos Aires province, as well as in one site in the Paraguayan Chaco (Barlow 1965; Honacki, Kinman, and Koeppl 1982; Massoia and Fornes 1967b; UConn) (map 11.74).

Life History

In Uruguay this species has a protracted breeding season (Barlow 1965).

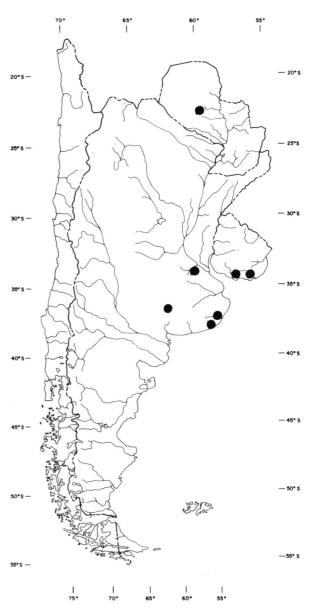

Map 11.74. Distribution of *Bolomys obscurus*.

Ecology

Bolomys obscurus is an animal of moist habitats and can reach high densities. In Buenos Aires province, Argentina, it was found at a density of 147.9/ha. In Uruguay and eastern Argentina it occurs in low, wet areas and in grassy fields adjoining such areas. In some places it has been caught in traps it had to swim to. This diurnal, insectivorous form prefers coleopterans and orthopterans. It is capable of digging (Barlow 1969; Crespo et al. 1970; Fornes and Massoia 1965; Reig 1964; Villafañe et al. 1973).

Comment

This species has been included in the genus *Akodon* and also in *Cabreramys* (Honacki, Kinman, and Koeppl 1982).

Bolomys (Cabreramys) temchuki Massoia, 1980

Measurements

	Mean	Min.	Max.	N	Loc.	Source[a]
TL	203.0	190	220.0	3	A	1
HB	126.7	118	138.0			
	105.1	85	129.0	10	A	2
T	76.3	72	82.0	3	A	1
	74.1	60	94.0	10	A	2
HF	24.0	24	24.0	2	A	1
	21.3	19	23.0	10	A	2
E	16.5	16	17.0	3	A	1
	14.7	13	16.0	10	A	2
Wta	56.8	52	65.3	3	A	1
	37.6	22	52.0	10	A	2

[a](1) Contreras 1982b; (2) Massoia 1982.

Description

The dorsum is dark brownish gray with marked agouti, the flanks are the same color, and the venter is chestnut with gray tints (Massoia 1982).

Distribution

Bolomys temchuki is distributed in the Argentine provinces of Misiones, Corrientes, Chaco, and Formosa (Contreras 1982b) (map 11.75).

Genus *Oxymycterus* Waterhouse, 1837

Description

Systematists generally concede that the genus *Oxymycterus* is affiliated with *Akodon* (Reig 1987). Some ten species have been recorded in the literature; all have the following characters: the baculum of the male is complex in structure; the hind feet are relatively short, as is the tail; the external ears are relatively short; the third molar is small, and the molars exhibit a simplified (tetralophodont) pattern; the rostrum of the skull is long and delicate (fig. 11.7); and the claws on the forefeet are long. The general appearance is of an animal adapted for a semifossorial existence (plate 14). The head and body length ranges from 91 to 170 mm and the tail is from 63 to 153 mm. The dorsal pelage is usually some shade of brown, and the venter is buff to grayish white.

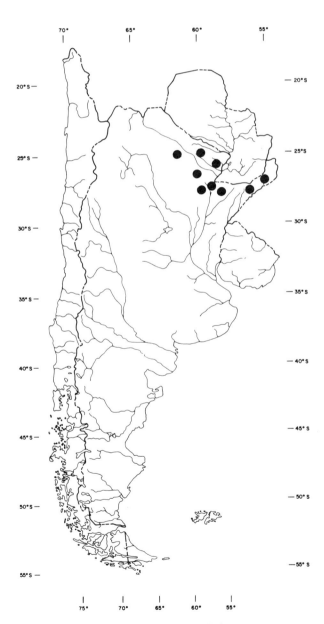

Map 11.75. Distribution of *Bolomys temchuki*.

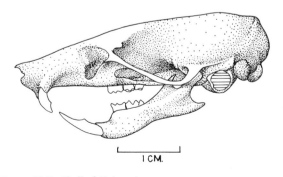

Figure 11.7. Skull of *Oxymycterus* sp.

Distribution

South American in its distribution, the genus has been recorded from Brazil, Bolivia, Argentina, Paraguay, and Uruguay.

Natural History

Although the species accounts are far more informative, in general these mice are semifossorial and highly insectivorous (Redford 1984). In a sense they are an evolutionary bridge between the generalist *Akodon* and the specialist *Blarinomys*.

Oxymycterus akodontius Thomas, 1921
Hocicudo Negro

Description

One specimen measured HB 116; T 79; HF 26; E 18.5. The dorsum is dark blackish brown, and the venter is dark slaty brown washed with buffy. The comparatively short tail is blackish brown. Unlike other species of *Oxymycterus*, *O. akodontius* does not have any rufous coloring (Thomas 1921a). Chromosome number: $2n = 54$; FN $= 64$ (Kajon et al. 1984).

Distribution

The species is known only from the Argentine province of Jujuy (Honacki, Kinman, and Koeppl 1982) (map 11.76).

Ecology

Moist areas in thick woods are where this species has been trapped. It lives in burrows (Thomas 1921a).

Comment

Oxymycterus akodontius may be conspecific with *O. paramensis* (Honacki, Kinman, and Koeppl 1982).

Oxymycterus delator Thomas, 1903

Measurements

	Mean	Min.	Max.	N	Loc.	Source[a]
TL	257.1	238	277	12	P	1
HB	159.4	146	178			
T	98.4	91	110			
HF	29.0	27	31	13		
E	18.7	18	20	12		
Wta	81.5	59	98			

[a](1) UM.

Description

This large, stout-bodied *Oxymycterus* is very dark above: black with olive flecking varying from slight to heavy, lightening on the sides to slightly more orangish. The venter, not sharply demarcated from the sides, is orangish gray (pers. obs.).

Distribution

Oxymycterus delator is found only in the low areas of eastern Paraguay (Myers 1982) (map 11.77).

Ecology

This is a species of low, marshy areas (Myers 1982).

Oxymycterus hispidus Pictet, 1843
Hocicudo Selvático

Measurements

	Mean	Min.	Max.	N	Loc.	Source[a]
TL	264.6	225	317	5	A	1
HB	143.5	108	174	4		
T	119.8	104	143			
HF	33.2	29	36	5		
E	22.2	20	25			

[a](1) Sanborn 1931; BA.

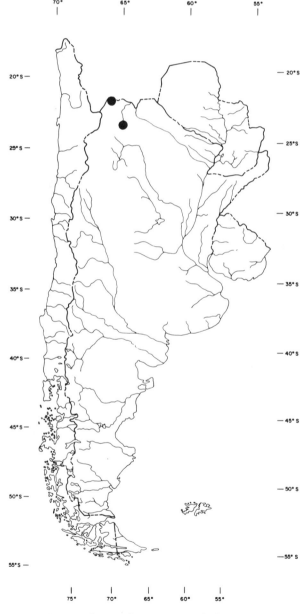

Map 11.76. Distribution of *Oxymycterus akodontius*.

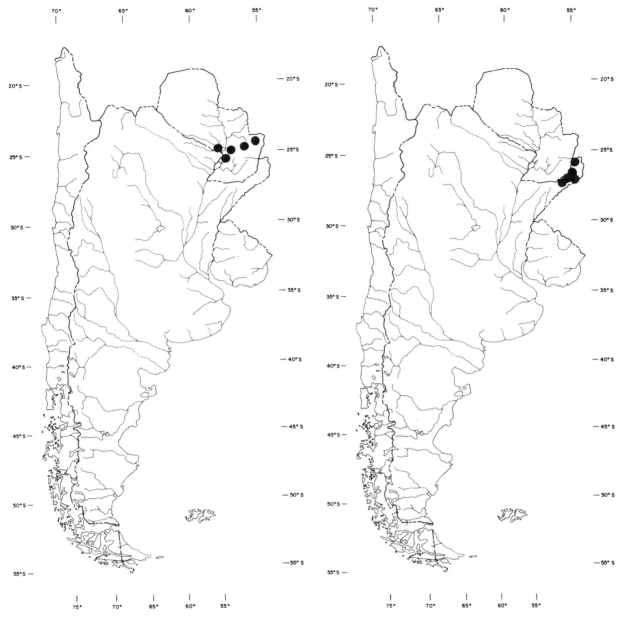

Map 11.77. Distribution of *Oxymycterus delator*.

Map 11.78. Distribution of *Oxymycterus hispidus*.

Description

The general dorsal color of this large *Oxymycterus* is reddish to orangish brown, shading into warm yellow brown on the sides and to almost russet on the rump; ventrally it is buffy gray to orangish gray (Sanborn 1931; pers. obs.).

Distribution

Oxymycterus hispidus is distributed only in eastern Brazil and the northeastern Argentine province of Misiones. It may be found to occur in eastern Paraguay as well (Honacki, Kinman, and Koeppl 1982; Massoia 1980a) (map 11.78).

Oxymycterus iheringi Thomas, 1896
Hocicudo Chico

Measurements

	Mean	Min.	Max.	N	Loc.	Source[a]
HB	103.2	93	111	7	A	1
T	86.2	83	90			
HF	22.4	21	24			
E	17.8	16	19			
Wta	43.0	40	45	3		

[a](1) Massoia and Fornes 1969.

Description

Oxymycterus iheringi is a small, slender mouse that is uniform grizzled brown, only slightly paler below. The brown ears are fairly large and thinly haired. This species resembles *Akodon arviculoides*, with which it is sympatric (Massoia and Fornes 1969; Thomas 1898d).

Distribution

The species is found in Brazil and the northeastern Argentine province of Misiones (Honacki, Kinman, and Koeppl 1982; Massoia and Fornes 1969) (map 11.79).

Ecology

This species has been caught in forests (Massoia 1963b).

Oxymycterus paramensis Thomas, 1902
Hocicudo Parameno

Measurements

	Mean	Min.	Max.	N	Loc.	Source[a]
TL	235.0	218.0	245.0	8	Peru	1
	215.3	192.0	243.0	3	A	2
T	90.0	80.0	96.0	8	Peru	1
	100.0	87.0	113.0	3	A	2
HF	30.0	29.0	31.0	8	Peru	1
	29.0	28.4	29.6	3	A	2
E	20.8	20.6	21.0			
Wta	42.0	36.7	52.0			
	65.0			1		3

[a](1) Osgood 1944; (2) Mares, Ojeda, and Kosco; (3) Bárquez 1976.

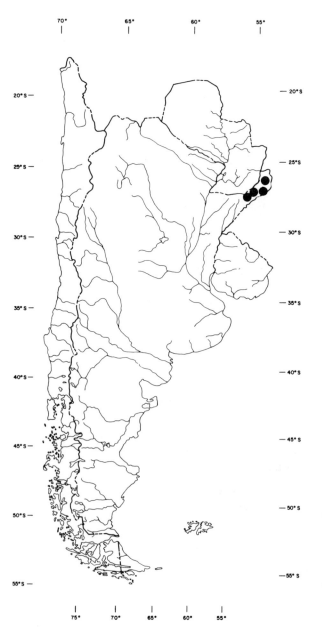

Map 11.79. Distribution of *Oxymycterus iheringi*.

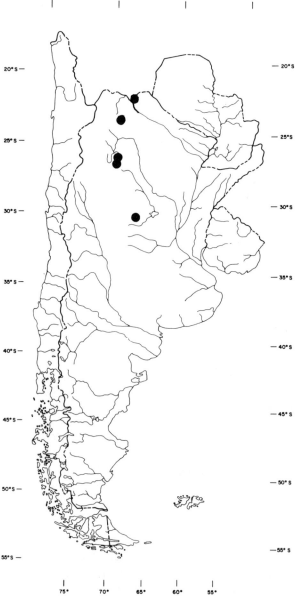

Map 11.80. Distribution of *Oxymycterus paramensis*.

Description

In this species the dorsum and sides are a rich reddish brown and the venter is ochraceous tawny. A sharply marked blackish brown frontal spot extends about 10 mm back from the rhinarium (Osgood 1944; pers. obs.).

Distribution

This species is found from eastern Bolivia south to the Argentine provinces of Salta, Jujuy, Córdoba, and Tucumán (Bárquez 1976; Honacki, Kinman, and Koeppl 1982) (map 11.80).

Ecology

Oxymycterus paramensis is found at intermediate and high altitudes, having been taken at from 1,500 to 4,300 m. In Salta province it inhabits the forest floor of the northern wet forests and probably the mesic forests to the south. In Tucumán province it has been trapped in brushy areas near watercourses and in forests (Bárquez 1976; Lucero 1983; Mares, Ojeda, and Kosco 1981; Osgood 1944; Thomas 1920b).

Oxymycterus rutilans (Olfers, 1818)
Hocicudo Común

Measurements

	Mean	Min.	Max.	N	Loc.	Source[a]
TL	221.7	206.0	243.0	13 m	U	1
	214.5	204.0	227.0	16 f		
	240.3	224.0	251.0	7 m	A	2
	224.1	215.0	249.0	7 f		
T	89.8	67.0	102.0	13 m	U	1
	78.7	74.0	93.0	16 f		
	98.1	94.0	102.0	7 m	A	2
	89.0	81.0	103.0	7 f		
HF	27.6	26.0	29.0	13 m	U	1
	26.5	26.0	28.0	16 f		
	26.2	25.0	27.5	6 m	A	2
	25.1	14.0	16.0	7 f		
E	17.5	15.0	20.0	13 m	U	1
	18.1	16.0	20.0	16 f		
	17.1	15.0	18.5	6 m	A	2
	16.6	16.0	18.0	7 f		
Wta	68.1	54.3	86.0	13 m	U	1
	63.9	49.7	81.0	16 f		
	86.0	75.0	97.0	6 m	A	2
	66.0	60.0	78.0	6 f		
	92.4	62.0	125.0	82 m	A	3
	76.2	46.0	110.0	39 f		

[a](1) Barlow 1965; (2) Fornes and Massoia 1965; (3) Dalby 1975.
Note: Oxymycterus rutilans includes *O. rufus.*

Description

Adult dorsal pelage is a grizzled reddish to yellowish black, darker along the spine, becoming ochraceous on the sides, and the venter is bright ochraceous mixed with gray. The tail is dark, well scaled, and sparsely haired (Barlow 1965; pers. obs.).

Chromosome number: $2n = 54$ (Kajon et al. 1984).

Distribution

This species is distributed from Brazil south through Paraguay and Uruguay and along eastern Argentina to Buenos Aires province (Honacki, Kinman, and Koeppl 1982; Crespo 1964) (map 11.81).

Life History

In Uruguay the breeding season is protracted, and females had an average of 2.1 embryos (range 1–4; $n = 7$). In Buenos Aires province, Argentina, females are found in breeding condition throughout the year, with most activity from September to late

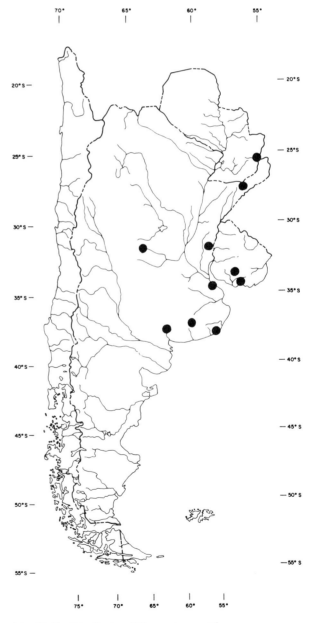

Map 11.81. Distribution of *Oxymycterus rutilans*.

May. There the average litter size is 3.1, and young are weaned at fourteen days (Barlow 1969; Dalby 1975; Kravetz 1972).

Ecology

In Uruguay *O. rutilans* is found in wet meadows with stands of bunchgrass, in tall grass adjacent to streams and rivers, and in drier parts of marshes. In Buenos Aires province, Argentina, it occurs in moist grassy areas and also in rocky hills, though it is most common where there is good cover. It can reach densities of 40/ha in suitable habitat.

In some areas *O. rutilans* is diurnal, and in others it is nocturnal. It is strictly terrestrial, does not swim well, and is primarily an invertebrate eater with a specialized stomach similar to that of the North American grasshopper mouse (*Onychomys*). In Uruguay stomachs had a 100% occurrence of Coleoptera, though Formicidae, Diptera, and Hemiptera were also common. In Buenos Aires province earthworms constituted 20% of the diet, arthropods 60%, plant material 15%, and other animal material 5%. It is also reported to eat small rodents (Barlow 1969; Dalby 1975; Echave Llanos and Vilchez 1964; Fornes and Massoia 1965; Kravetz 1972; Reig 1964; Vaz-Ferreira 1958).

THE "MOLE MICE" OF SOUTH AMERICA

The following three genera are often assigned to the tribe Akodontini. Pearson (1984) noted that some sigmodontine rodents had occupied semifossorial niches and have begun to adapt to the apparently vacant semifossorial insectivore and herbivore niches. These forms have reduced pinnae, reduced eyes, long claws, short tails, and short but dense pelage. Resemblances in morphology may reflect ancestral affinity or convergence. Consider the following comments from Pearson (1984):

Notiomys edwardsi is an insectivorous rodent and within its range can share tunnels with *Ctenomys* sp., the latter being an herbivore. The soil type that is preferred might be characterized as steppe soils and this species appears to be confined to open country. This contrasts with the genus *Geoxus* which prefers moist, forest soils although it is also insectivorous.

Geoxus, a semifossorial rodent highly adapted for insectivory, may be found with the largely herbivorous rodent genera *Aeconomys* and *Chelemys* in Chile and Argentina.

Chelemys is a fossorial rodent that is apparently herbivorous. It prefers moist soils in association with forests. It can co-occur with the semifossorial, insectivore rodent genus *Geoxus*.

Genus *Notiomys* Thomas 1890
Notiomys delfini (Cabrera, 1905)

Description

Measurements: HB 106; T 63; HF 22; E 11. The fur of this short-tailed mouse is long and soft, blackish brown dorsally and only slightly paler on the venter (Gyldenstolpe 1932; Mann 1978).

Distribution

Notiomys delfini is found in southern Argentina and the Chilean province of Magallanes (Honacki, Kinman, and Koeppl 1982; Osgood 1943).

Comment

O. P. Pearson (pers. comm.) believes that *delfini* cannot be assigned to *Notiomys*.

Notiomys edwardsii (Thomas, 1890)
Ratón Topo Chico

Measurements

	Mean	Min.	Max.	N	Loc.	Source[a]
TL	125.4	115.0	135.0	5	A	1
HB	84.8	78.0	92.0			
T	40.6	35.0	46.0			
HF	19.2	17.4	20.5	4		
E	7.6	6.5	8.0			
Wta	21.3	18.5	25.0			

[a](1) Pearson 1984; Thomas 1929.

Description

This short-tailed mouse has long front claws and a very well developed fringe of hairs as long as 3 mm on the margins of the hind feet as well as less conspicuous fringes on the front feet. Its ears are very small and extremely thin, especially at the margin, where they are covered with long, silky white hairs. The nose is tipped with a dark, leathery button; the tiny teeth are almost cuspless. It is grayish fawn above and white below, with a rufous nose and a bright lateral line. Unlike similar genera (*Chelemys* and *Geoxus*), *Notiomys* does not have molelike fur and drab coloration (Osgood 1925; Pearson 1984; Thomas 1929).

Distribution

Notiomys edwardsii is found in southern Argentina in the provinces of Río Negro and Santa Cruz (Allen 1903; Pearson 1984) (map 11.82).

Ecology

The distribution of this species is very similar to that of the bush *Sapium*. These bushes support diverse populations of the insects these rodents eat. *N. edwardsii* was trapped in brushy abandoned pastures and in a *Ctenomys* burrow. It is insectivorous but also eats plant material (Pearson 1984).

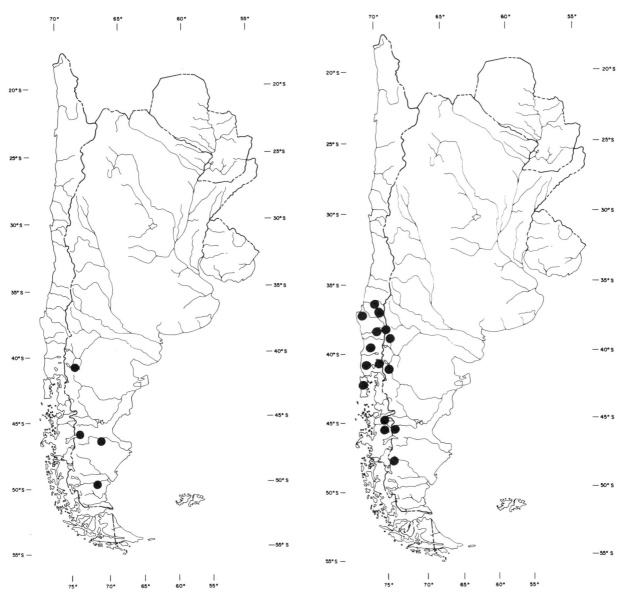

Map 11.82. Distribution of *Notiomys edwardsii*.

Map 11.83. Distribution of *Geoxus valdivianus*.

Genus *Geoxus* Thomas, 1919
Geoxus valdivianus (Philippi, 1858)
Ratón Topo Pardo

Measurements

	Mean	Min.	Max.	N	Loc.	Source[a]
TL	142.2	122.0	165.0	13	C	1
HB	101.7	95.0	106.0	15 m	A	2
	101.7	93.0	108.0	9 f		
T	41.6	33.0	54.0	13	C	1
HF	20.5	19.0	22.0			
E	11.6	10.0	13.0	9		
Wta	26.2	19.8	30.5	8		
	31.7	25.5	38.0	15 m	A	2
	31.3	26.0	41.0	8 f		

[a](1) Osgood 1925, 1943; Pine, Miller, and Schamberger 1979;
FM; Santiago; (2) Pearson 1983.

Description

This small mouse resembles a short-tailed shrew
(*Blarina*). It has small, simple teeth, small eyes, and
very short ears (plate 14). The pelage is dense and
short. The tail is short, and the front claws are elongated. Its body color is rich, uniform olive brown to
blackish washed with reddish, with underparts only
slightly lighter (Osgood 1925, 1943; Pearson 1983;
pers. obs.).
Chromosome number: $2n = 53$ (Pearson 1984).

Distribution

This species is found in southern and central
Chile, including Chilóe Island, and also in southwestern Argentina. In Chile it has been obtained
in the following provinces: Osorno, Ñuble, Bío-Bío,

Valdivia, Malleco, Concepción, and Magallanes; in Argentina it has been taken in Río Negro province (Honacki, Kinman, and Koeppl 1982; Markham 1971; Osgood 1925; Pearson 1984; Pine, Miller, and Schamberger 1979; Schneider 1946) (map 11.83).

Life History
In Argentina four captured females had an average of 3.5 embryos (range 3–4) (Pearson 1983).

Ecology
Geoxus valdivianus frequents the southern rain forests on moist soils. In Argentine Patagonia the animals can occur in pure stands of any one of three *Nothofagus* species, in mixed stands, and in moist habitats such as marshes and meadows. In forests they reach densities of 0.5/ha. In Chile individuals have been taken at timberline in tussock grass and *Nothofagus antarctica*, in bamboo (*Saxegothaea*) and *Nothofagus dombeyi* forests, and in areas of heavy undergrowth of cane and barberry (*Berberis*).

In soft soils this species may make its own burrows or may use those of other animals. Individuals are frequently caught in runways under the overhang of large, decaying logs, in patches of meadow in or near forest, and occasionally in *Aconaemys* burrows.

Stomachs have been found to contain earthworms, slugs, beetle larvae, spiders, and other arthropods. In Chile stomachs of this species ($n = 18$) contained mature arthropods and larvae (31.6%) as well as other animal material including annelids (24.3%), with a substantial proportion of vegetation and fungi (42.0%). This species is active both diurnally and nocturnally (Greer 1966; Meserve, Lang, and Patterson 1988; Murúa, González, and Jofre 1982; Pearson 1983, 1984; Pearson and Pearson 1982; Pine, Miller, and Schamberger 1979; Reise and Venegas 1974).

Comment
Geoxus valdivianus used to be contained within the genus *Notiomys* (see Pearson 1984).

Genus *Chelemys* Thomas, 1903
Chelemys macronyx (Thomas, 1894)
Ratón Topo Grande

Measurements

	Mean	Min.	Max.	N	Loc.	Source[a]
TL	182.8	165	196.0	9	A	1
HB	129.9	116	146.0			
	130.0	109	139.0	14 m		2
	130.8	113	146.0	11 f		
T	52.9	47	59.0	9		1
HF	25.6	23	27.0			
E	15.7	11	17.1			
Wta	74.6	50	96.0	14 m		2
	72.0	46	90.0	11 f		

[a](1) Osgood 1925, 1943; BA; (2) Pearson 1983.

Description
Chelemys macronyx is separable from *C. megalonyx* because the former occurs in Chile and has pale proodont incisors whereas *C. macronyx* is found in both Argentina and Chile and has incisors not nearly as proodont and much more pigmented. *C. macronyx* is a medium-sized, stout-bodied, short-tailed mouse with elongated front claws and dense, soft pelage nearly concealing the small ears. The coloration varies geographically, with the dorsum varying from black grizzled with olivaceous to coffee colored and the venter from gray white to white. The ventral color often extends to the sides, where it is sharply demarcated from the dorsal color. The tail

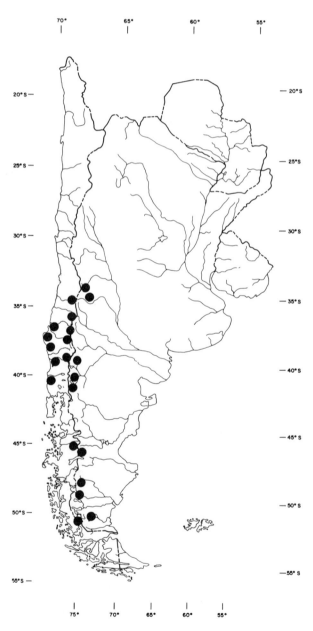

Map 11.84. Distribution of *Chelemys macronyx*.

is always bicolored (Mann 1978; Osgood 1943; Pearson 1984; pers. obs.).

Distribution

The species is distributed in southern and western Chile and eastern and southern Argentina. In Argentina it is found in the provinces of Mendoza, Río Negro, Neuquén, Santa Cruz, and Chubut, while in Chile it has been trapped in Magallanes, Ñuble, Malleco, and Bío-Bío provinces (Contreras 1983; Greer 1966; Honacki, Kinman, and Koeppl 1982; Markham 1971; Pearson 1984; Pine, Miller, and Schamberger 1979; Yepes 1935; BA) (map 11.84).

Life History

In southern Argentina two trapped females had four and five embryos (Pearson 1983).

Ecology

This species typically inhabits *Nothofagus pumilio* forests from timberline to the edges of the precordilleran steppe, where it occurs in moist habitats or under shrubs. In the Patagonian forests it reaches densities of 14.7/ha. In Malleco province, Chile, it prefers deep soils in elevated lowlands. *C. macronyx* is semifossorial and builds networks of tunnels, but it may live in rocks too. In Argentina, after the winter snows melt, castings from subsurface burrows of this species are found lying on the surface of the ground. One excavated burrow was 9 m long. Individuals have been caught during both day and night. Stomachs have been found to contain grass seeds, fruits, fungi, arthropods, and earthworms (Greer 1966; Osgood 1943; Pearson 1983, 1984; Pearson and Pearson 1982; Reise and Venegas 1974).

Comment

This species used to be contained within the genus *Notiomys* (see Pearson 1984).

Chelemys megalonyx (Waterhouse, 1845)

Measurements

	Mean	Min.	Max.	N	Loc.	Source[a]
TL	171.0	157	178	5	C	1
HB	117.6	108	127			
	121.0	114	127	5		2
T	53.4	49	57	5		1
HF	25.4	24	28			
E	18.7	18	19			

[a](1) Osgood 1943; Wolffsohn 1923; (2) Jaksić and Yáñez 1979.

Description

This medium-sized molelike mouse has elongated front claws, thick pelage, and a tail much shorter than the head and body. The dorsum is brown, and the venter is grayish white to brown, sometimes washed with yellow (Osgood 1925; pers. obs.).

Distribution

The species is found in central Chile in Coquimbo, Valparaíso, and Concepción provinces (Honacki, Kinman, and Koeppl 1982; Koford 1955; Osgood 1925) (map 11.85).

Ecology

Chelemys megalonyx is semifossorial, digs tunnels, preferably in humid soils, and has been trapped at 300 m among shrubs and cacti (Koford 1955; Schneider 1946).

Comment

This species used to be contained within the genus *Notiomys* (see Pearson 1984).

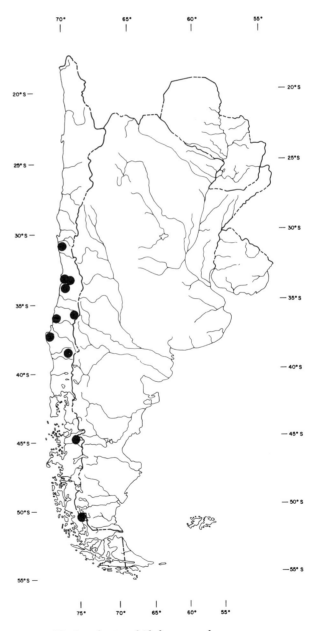

Map 11.85. Distribution of *Chelemys megalonyx*.

TRIBE SIGMODONTINI

Description

Students of the group should be warned that the common character is an S-shaped molar pattern (see fig. 11.18; see also the tribe Phyllotini p. 000). This may not be a natural grouping, and much more research is necessary to establish its phylogenetic integrity. The group includes both dry grassland forms (*Sigmodon* and *Sigmomys*) and semiaquatic forms (*Holochilus*). The last genus is important in the southern cone.

Distribution

As currently recognized, this "tribe" extends from the southeastern United States to northern Argentina.

Genus *Holochilus* Brandt, 1835
Holochilus brasiliensis (Desmarest, 1819)
South American Water Rats, Rata Nutria o Colorada

Measurements

	Mean	Min.	Max.	N	Loc.	Source[a]
TL	405.1	350.0	425.0	12	U	1
	329.8	284.0	381.0	10	A	2
T	201.3	183.0	220.0	12	U	1
	161.2	140.0	192.0	10	A	2
HF	56.9	53.0	58.0	12	U	1
	39.6	35.2	42.7	10	A	2
E	22.9	21.0	26.0	12	U	1
	21.8	20.8	23.0	10	A	2
Wta	146.9	81.0	230.0			
	326.0	275.0	455.0	11	A	3

[a](1) Barlow 1965; (2) Mares, Ojeda, and Kosco 1981; (3) Massoia 1976a.

Description

This species is similar to the North American muskrat (*Ondatra*) (see plates 15 and 18). The fur is soft, dense, and shiny. The webbed hind feet are large, and the tail is sparsely haired with scales clearly visible. There is a good deal of variation in size. The dorsal pelage is ochraceous tawny mixed with blackish. The sides are paler and more orange, and the venter is white, occasionally orangish (Barlow 1965; Fornes and Massoia 1965; Massoia 1976a; pers. obs.).

Distribution

Holochilus brasiliensis is found from eastern Brazil into Uruguay, Paraguay, and Argentina. In southern South America it is found throughout eastern Paraguay and in gallery forests into the Chaco, throughout Uruguay, and in Argentina (see Massoia 1976a) (Honacki, Kinman, and Koeppl 1982; Vaz-Ferreira 1958; UConn, UM) (map 11.86).

Life History

Females with three and four embryos have been examined. Rainfall seems to influence reproductive pattern in this species, and it can undergo population irruptions (Barlow 1969; Mares, Ojeda, and Kosco 1981; Veiga-Borgeaud, Lemos, and Bastos 1987).

Ecology

This semiaquatic rodent is distributed throughout southern and eastern South America in low, marshy areas. It constructs spherical nests of leaves, sticks, and grasses that average 22 to 27 cm in diameter. These nests are often in tangles of vines and terminal

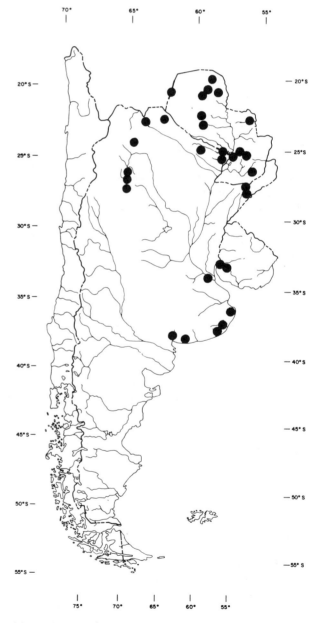

Map 11.86. Distribution of *Holochilus brasiliensis*.

forks of tree branches, but the animals also dig burrows with up to three entrances.

H. brasiliensis primarily eats the more tender parts of aquatic, riparian, and domestic plants. It can be a pest in sugarcane and rice fields (Barlow 1969; Mares, Ojeda, and Kosco 1981; Massoia 1971b,c, 1976a).

Comment

This species may be a composite containing *Holochilus balnearum* and *H. amazonicus* (Reig 1986).

Holochilus chacarius Thomas, 1906

Measurements

	Mean	Min.	Max.	N	Loc.	Source[a]
HB	177	162	201	10	A	1
T	167	152	183			
HF	39	36	42			
E	19	17	20			
Wta	186	152	262			

[a](1) Massoia 1976a.

Description

In this smallest species of *Holochilus*, the hind foot (without claws) never exceeds 45 mm. The interdigital membranes are rudimentary. The dorsum is a comparatively light orange chestnut, sometimes reddish with a few dark hairs mixed in. The venter is pure white to grayish white (Massoia 1976a).

Distribution

Holochilus chacarius is found in Paraguay and northeastern Argentina (Honacki, Kinman, and Koeppl 1982; Massoia 1976a) (map 11.87).

Ecology

This semiaquatic rodent swims, dives, and climbs well. It makes grass nests measuring 20–30 cm in rice fields, banana plantations, and long river courses. It is strictly herbivorous and can cause extensive damage to rice, bananas, sugarcane, and other crops. In places it is very common (Massoia 1976a).

Holochilus magnus Hershkovitz, 1955

Measurements

	Mean	Min.	Max.	N	Loc.	Source[a]
TL	436.0			1	U	1
HB	228.0	205	242	4	A	2
T	250.0			1	U	1
	267.0	246	289	4	A	2
HF	62.0			1	U	1
	62.0	60	65	4	A	2
E	25.0			1	U	1
	25.0	25	27	4	A	2
Wta	238.5	227	250	2	A, U	1, 2

[a](1) Barlow 1965; (2) Massoia 1976a.

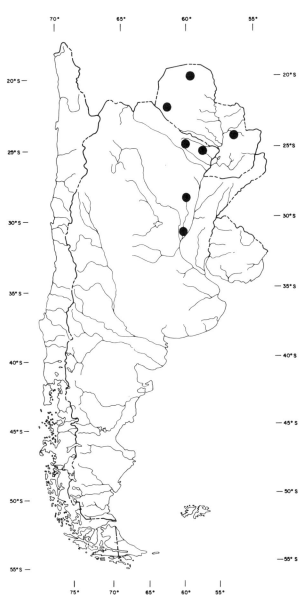

Map 11.87. Distribution of *Holochilus chacarius*.

Description

Holochilus magnus is one of the largest South American cricetines, with very large feet and a long, strong tail. The dorsum is ochraceous to tawny mixed with grayish or blackish, especially along the midline. The venter is tawny to ochraceous, and the feet are whitish with reasonably well developed interdigital membranes (Barlow 1965; Massoia 1976a; pers. obs.).

Distribution

The species is found in Uruguay, adjacent Argentina, and southeastern Brazil (Barlow 1965; Honacki, Kinman, and Koeppl 1982; Vaz-Ferreira 1958) (map 11.88).

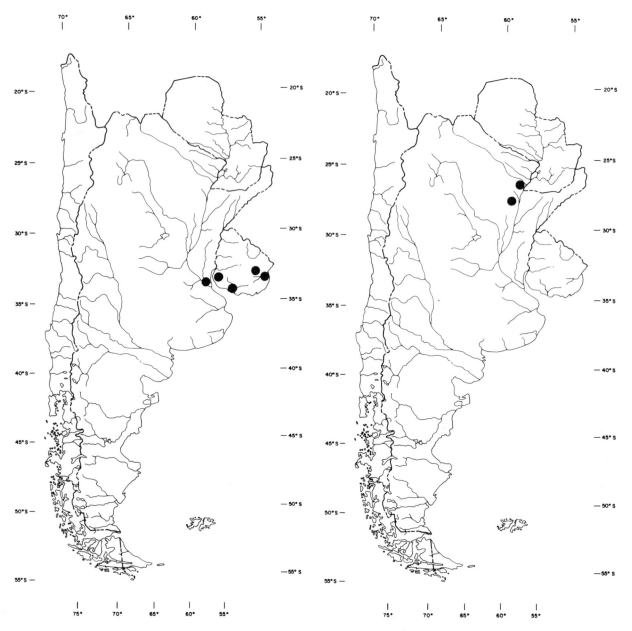

Map 11.88. Distribution of *Holochilus magnus*.

Map 11.89. Distribution of *Bibimys chacoensis*.

Ecology

In Uruguay this semiaquatic species has been taken along streams in tall bunchgrass and reeds and in marshy areas. Its diet is 95% plant material (Barlow 1969).

TRIBE SCAPTEROMYINI

Description

This vexing group was reviewed by Hershkovitz (1966). Basically these are semiaquatic rodents with considerable fossorial ability: the ears are short, the eyes are small, the claws on the forefeet tend to be long, and the tail is shorter than the head and body. The skull is robust, and the molars are complex in structure. In common with other South American genera, the baculum is complex. The color pattern is variable (see species accounts). Massoia (1979) has proposed that the genus *Bibimys* may be allied to the classical genera that Hershkovitz includes within this group.

Distribution

The tribe as we recognize it is confined to the wetlands of southeastern Brazil and to eastern Paraguay, adjacent portions of Uruguay, and northeastern Argentina.

Genus *Bibimys* Massoia, 1979

Description

Massoia (1979) suggested that *Bibimys* is closely related to the genera *Scapteromys* and *Kunsia*. All three genera are clearly adapted for a fossorial life in wetlands. The ear and tail are reduced in length, and the claws are enlarged. The skull is robust, and certain dental characters reinforce the inclusion of

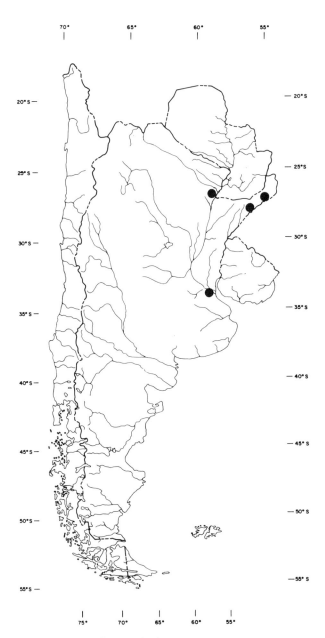

Map 11.90. Distribution of *Bibimys torresi*.

the genus *Bibimys* in the Scapteromyini. Head and body length ranges from 84 to 97 mm. The dorsum is a variable shade of brown, and the venter is usually gray or buff. *B. torresi* has a reddish nose.

Distribution

The genus is recorded from southeastern Brazil and adjacent parts of northeastern Argentina.

Bibimys chacoensis (Shamel, 1931)

Description

One specimen measured TL 160; T 66; HF 22.5. This mouse is olivaceous with some buff about the eyes and sides of head and along the sides of the body and dark along the back. The venter is whitish with a slight buffy tinge (Shamel 1931).

Distribution

The species is known only from a few localities in north-central Argentina (Myers 1982; Shamel 1931) (map 11.89).

Ecology

Bibymys chacoensis has been caught in marsh grass in open savanna (Shamel 1931).

Comment

Massoia moved this species from *Akodon* to *Bibimys* (Massoia 1980b).

Bibimys torresi Massoia, 1979

Measurements

	Mean	Min.	Max.	N	Loc.	Source[a]
TL	164.0	149	175	3	A	1
HB	92.3	84	97			
T	71.7	65	78			
HF	20.7	20	22			
E	16.0	15	17			
Wta	28.0	23	34			

[a](1) Massoia 1979.

Description

This small robust mouse has a short tail with a ring of transverse scales, well-developed front and hind claws, and small eyes. It has soft fur, dark chestnut on the back, with a gray venter, gray feet, and a bicolored tail. The distinguishing character is the rose-colored nose, which distinguishes it from the sympatric *Akodon azarae* (Massoia 1979).

Distribution

The distribution of *B. torresi* is poorly known, but it appears to be confined to northeastern and eastern Argentina (Honacki, Kinman, and Koeppl 1982; Massoia 1979, 1980a) (map 11.90).

Genus *Kunsia* Hershkovitz, 1966
Kunsia fronto (Winge, 1887)
Rata Acuática Grande

Measurements

	Mean	Min.	Max.	N	Loc.	Source[a]
HB		160.0	205	24	Br	1
T		75.0	108			
HF		25.0	38			
E		17.8	21			

[a](1) Hershkovitz 1966.

Description

This species is similar in size to *Rattus norvegicus*, but the tail is proportionally shorter (plates 15 and 18). It has small, densely haired ears. The subspecies *K. fronto* from central Brazil is black or blackish above, with sides speckled with white, and grayish white ventrally (Avila-Pires 1972; Hershkovitz 1966).

Distribution

Kunsia fronto is known from Chaco province, Argentina, and from the central Brazilian plateau. Its precise distribution is poorly understood (Hershkovitz 1966; Honacki, Kinman, and Koeppl 1982) (map 11.91).

Ecology

In central Brazil specimens of this species were captured in a marsh (Avila-Pires 1972; pers. obs.).

Comment

Formerly placed in *Scapteromys*, *Kunsia fronto* includes *K. chacoensis*.

Genus *Scapteromys* Waterhouse 1837
Scapteromys tumidus Waterhouse, 1837
Rata Acuática

Measurements

	Mean	Min.	Max.	N	Loc.	Source[a]
TL	336.7	326.0	357.0	31	U	1
HB	180.7	160.0	198.0	36 m	A	2
	170.0	152.0	190.0	26 f		
T	154.0	140.0	168.0	31	U	1
	147.4	132.0	158.0	36 m	A	2
	144.1	132.0	159.0	26 f		
HF	41.2	39.0	43.0	31	U	1
	36.4	35.0	38.0	37 m	A	2
	35.2	34.0	37.0	27 f		
E	24.9	23.0	29.0	31	U	1
	22.0	19.0	24.0	37 m	A	2
	21.5	20.0	24.0	27 f		
Wta	146.0	107.9	195.3	22	U	1

[a](1) Barlow 1965; (2) Massoia and Fornes 1964.

Description

Scapteromys is similar to *Rattus* in appearance but with a relatively larger head, stouter body, larger feet, and relatively longer tail, which often has a distinct fringe of stiff hairs (plates 15 and 18). The forefeet are well developed, with fairly long claws. The pollex, as well as all the other digits, is equipped with a claw. The dorsal pelage is long and glossy, ranging from medium to dark brown with a grayish wash in some individuals. The sides have a yellowish cast, and the venter is a dirty white. Juveniles are dark grayish on the dorsum and grayish white to light gray on the venter (Barlow 1965; Hershkovitz 1966; pers. obs.).

Distribution

The species is distributed in Uruguay, adjacent Brazil, eastern Argentina, and central Paraguay along the Río Paraguay (Honacki, Kinman, and Koeppl 1982; Massoia 1964; Myers 1982) (map 11.92).

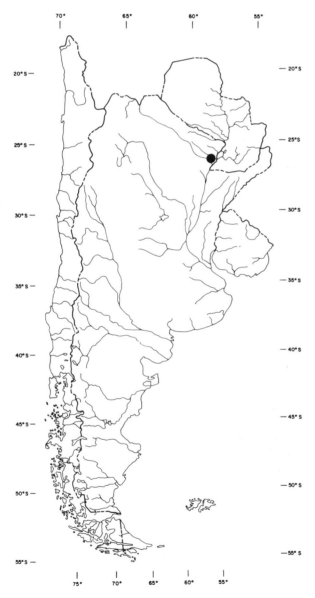

Map 11.91. Distribution of *Kunsia fronto*.

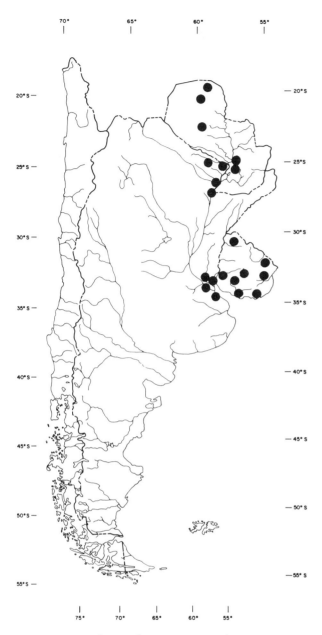

Map 11.92. Distribution of *Scapteromys tumidus*.

Life History

In Uruguay the average number of embryos or young was four, with a range of three to five; females have eight nipples. Breeding males were found all year round (Barlow 1969).

Ecology

Scapteromys tumidus is a semiaquatic species of low, flooded grasslands, including salt marshes. Throughout its range it is found only in or near areas with standing water. A good swimmer, using its tail with its "swimming fringe," *S. tumidus* has also been observed to dive. It digs for its food and is trapped throughout the day and night. Its diet con-

sists of arthropods (85%), mostly beetles, and earthworms. There is no indication that this species digs burrows, and young have been found in shallow depressions beneath matted grass (Barlow 1969; Massoia and Fornes 1964; Myers 1982).

Comment

Scapteromys tumidus is considered to include *S. aquaticus*, but Reig (1986) reports $2n = 32$ for *S. aquaticus* and $2n = 24$ for *S. tumidus*.

FAMILY MURIDAE
SUBFAMILY MURINAE
Old World Rats and Mice

Diagnosis

These rodents have a dental formula of I 1/1, C 0/0, P 0/0, M 3/3 and typically have the molar cusps aligned in three longitudinal rows (fig. 11.18), differing from the varied sigmodontine forms. They are superficially very similar to cricetine rodents in body form and proportions.

Distribution

These rodents are native to the Old World, but species of the subfamily Murinae have been widely introduced around the world by humans. Three species occur near settlements in the area covered by this volume: *Rattus rattus*, *Rattus norvegicus*, and *Mus musculus*.

Genus *Rattus* Fischer, 1803
Rattus rattus (Linnaeus, 1758)
Black Rat

Description

Total lenth ranges from 327 to 430 mm; the tail is 160 to 220 mm, and the hind foot average 35.5 mm. A large male may weigh 200 g. This rat may be distinguished from *R. norvegicus* by its longer tail. The dorsum is grayish black to brown, and the venter may be gray or almost black.

Distribution

As a species introduced by humans, this rat tends to be associated with dwellings. It is highly arboreal and may extend into forest regions far from points of original introduction. Generally it occurs near coastal settlements.

Life History and Ecology

This rat is highly adaptable and an excellent climber. It is omnivorous but readily eats fruits and nuts when occupying natural forests. Three to five young are produced after twenty-three days' gestation. The behavior of black rats has been described by Ewer (1971).

Rattus norvegicus (Berkenhout, 1769)
Brown Rat

Description
Total length ranges from 320 to 480 mm; the tail is 153 to 218 mm, and the hind foot 37 to 44 mm. The tail is shorter than the head and body, and this rat is more robust than *R. rattus*. The dorsal pelage is brown interspersed with black hairs, and the underparts are pale gray.

Distribution
This terrestrial species is associated with large urban centers. It does not penetrate into undisturbed habitats in Patagonia.

Ecology
This species is much more terrestrial than *R. rattus*. It tunnels actively, and its presence is usually first detected when its burrows are noticed. Up to seven young are born after twenty-three days' gestation. The brown rat is an omnivore and is often a serious pest when it exploits stored cereal grains. Its ecology and behavior have been summarized by Calhoun (1962).

Genus *Mus* Linnaeus, 1766
Mus musculus Linnaeus, 1766
House Mouse

Description
Total length ranges from 148 to 205 mm; the tail is 69 to 85 mm, the hind foot 16 to 20 mm. The dorsum is gray brown to brown; the venter is lighter and often buffy.

Distribution
This species is closely associated with human dwellings and farms. It has been widely introduced in coastal cities.

Life History and Ecology
This mouse is omnivorous and mainly terrestrial. It is almost always associated with humans as a commensal. It climbs well but does not nest arboreally. Five to seven young are born after twenty to twenty-one days' gestation. The species' natural history in Australia has been summarized by Newsome (1971) and that in England by Chitty and Southern (1954).

HYSTRICOGNATH RODENTS

Introduction
As indicated above under "History and Classification" (pp. 253–54), the earliest rodents on the continent of South America are hystricognathous. The classification employed here follows Woods (1982), who groups the South American hystricognath rodents into four superfamilies, which occur within the range covered by this volume and include the Erethizontoidea, the Cavioidea, the Chinchilloidea, and the Octodontoidea.

The hystricognath rodents exhibit a number of unique reproductive characters. Litter size tends to be small (one to three), and the gestation period is long. The young are precocial. The biology of hystricognath rodents is admirably summarized in the volume edited by Rowlands and Weir (1974). Reproduction has been summarized by Weir (1974b) and by Kleiman, Eisenberg, and Maliniak (1979). Behavior patterns were reviewed by Kleiman (1974), Eisenberg (1974), and Eisenberg and Kleiman (1977), and distribution and ecology by Mares and Ojeda (1982).

SUPERFAMILY ERETHIZONTOIDEA
FAMILY ERETHIZONTIDAE
Porcupines, Puercos Espinosos

Diagnosis
The dental formula is I 1/1, C 0/0, P 1/1, M 3/3. The genera of this family are adapted for an arboreal life; the thumb is replaced by a broad movable pad, and the sole of the hind foot is wide. The dorsal pelage is modified so that many hairs have become spinelike (plate 16), and their tips have minute barbs. The spines are easily detached and embed themselves in potential predators.

Distribution
Species are distributed from northern Canada and Alaska south to northern Argentina. The northern species belong to the genus *Erethizon*. The species covered by this volume are in the genera *Sphiggurus* and *Coendou*.

Genus *Coendou* Lacépède, 1799
• *Coendou prehensilis* (Linnaeus, 1758)
Coendú Grande

Measurements

	Mean	Min.	Max.	N	Loc.	Source[a]
TL	1,044.2	920	1,190	6	B, Br, P	1
HB	524.2	455	590			
T	520.0	400	600			
HF	82.7	65	103			
E	15.0	10	25	3		
Wta	4.89 kg			4 f	Captivity	2
	4.51 kg			5 m		

[a](1) Crespo 1974a; Vieira 1945; PCorps; (2) Roberts, Brand, and Maliniak 1985.

Description
In this rather large porcupine, the quills are conspicuous and not covered by dorsal hairs. The tail almost equals the head and body in length and is fully prehensile (see plate 16). The shape of the skull is distinctive (figure 11.8).

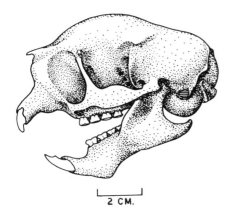

Figure 11.8. Skull of *Coendou prehensilis*.

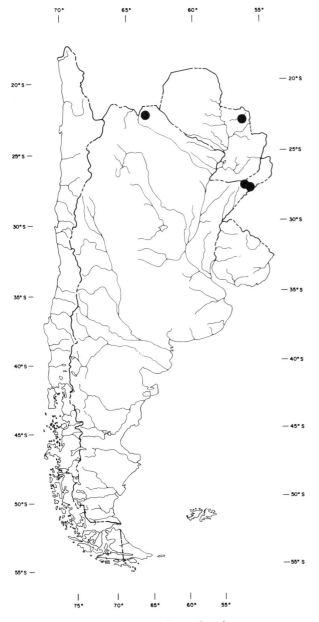

Map 11.93. Distribution of *Coendou prehensilis*.

Distribution

Coendou prehensilis is broadly distributed from Venezuela south to northern Argentina. In southern South America it has been collected in Uruguay, eastern Paraguay, and the Argentine provinces of Salta and Misiones (Honacki, Kinman, and Koeppl 1982; Olrog 1976; Tálice 1969; Yepes 1938; PCorps) (map 11.93).

Life History

In captivity gestation has been determined to last between 195 and 210 days, with a postpartum estrus and a minimum interbirth interval of 203 days. There is no seasonality in reproduction, but rainfall can cause birth peaks in selected areas. The litter size is one, and the average neonatal weight is 415 g (range 364–447; $n = 8$). Young are born with long red fur that hides the spines. Lactation lasts approximately fifteen weeks. The young is left alone in a sheltered place and nursed at least once a day. Female age of first reproduction is about nineteenth months, and animals can reproduce for more than twelve years (Roberts, Brand, and Maliniak 1985).

Ecology

In Venezuela *C. prehensilis* is nocturnal and rests in trees during the day, often sheltering in cavities or remaining inconspicuous high in a tree crown. The home range of this solitary species may be very large depending on the availability of suitable feeding trees. In the Venezuelan llanos animals range over an area of 15–20 ha.

In Suriname *C. prehensilis* was exclusively herbivorous, with a large proportion of the diet consisting of buds, bark, fruits, and unripe seeds of large fruit. In captivity these porcupines sleep aboveground and are nocturnal. Adults are primarily solitary but are socially tolerant. Both sexes anal rub, and males mark females with urine. Nine vocalization types are produced, including a long moan employed as a contact call between isolated individuals (Charles-Dominique et al. 1981; Montgomery and Lubin 1978; Roberts, Brand, and Maliniak 1985).

Genus *Sphiggurus* Husson, 1978

Sphiggurus spinosus (Lichtenstein, 1818)
Coendú Chico

Measurements

	Mean	Min.	Max.	N	Loc.	Source[a]
TL	580.7	510	650	12	A, U	1
HB	329.4	310	390			
T	251.0	200	280			
HF	48.8	40	62			
E	13.8	102	311			

[a](1) Barlow 1965; Crespo 1974a; BA.

Description

Sphiggurus spinosus has gray brown hair; black is always prominent on the distal portion of the body hairs. Quills are scarcely visible anywhere on the body, being obscured by the longer hairs (plate 16), and the tail is relatively short with no quills on the tip. Young are covered with reddish hair. This species is easily distinguished from *Coendou prehensilis* by its much smaller size and longer hair with no quills showing through (Miranda-Ribeiro 1936; pers. obs.).

Distribution

The species is found in southern and eastern Brazil, eastern Paraguay, throughout Uruguay, and in the Argentine province of Misiones (Crespo 1974a;

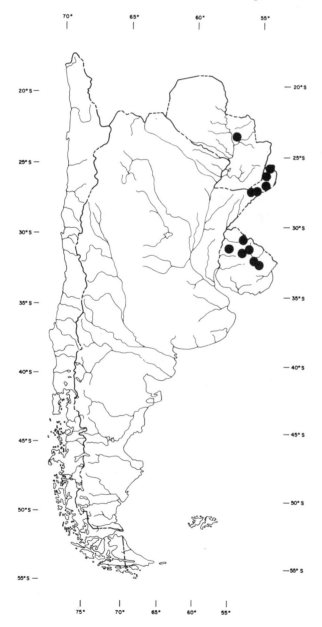

Map 11.94. Distribution of *Sphiggurus spinosus*.

Honacki, Kinman, and Koeppl 1982; Ximénez, Langguth, and Praderi 1972; UM) (map 11.94).

Ecology

In Suriname this genus is reported to feed on fruit, ant pupae, vegetables, and roots (Husson 1978).

SUPERFAMILY CAVIOIDEA
FAMILY CAVIIDAE

Diagnosis

The dental formula is I 1/1, C 0/0, P 1/1, M 3/3. The cheek teeth are ever growing and have two enamel prisms with sharp folds and angular projections. The body is stout and the tail is extremely reduced. Reduction in the number of toes is common; four toes are retained on the forefeet and three on the hind feet (fig. 11.9; plate 16).

Distribution

Disjunct populations characterize the northern distribution, but the genus occurs over most of South America except the easternmost parts of Brazil and the extremely arid portions of southern Peru and Chile.

Genus *Cavia* Pallas, 1766

Description

Most characters are as noted for the family; however, in the genus *Cavia* the tail is virtually absent (see species accounts and plate 16).

• *Cavia aperea* Erxleben, 1777
Cavy, Guinea Pig, Cuis Selvatico, Cuis Campestre

Measurements

	Mean	Min.	Max.	N	Loc.	Source[a]
TL	274.7	196	320	7	A, P	1
HB	271.9	191	320			
	306.0	290	325	11	A	2
T	2.4	0	7	7	A, P	1
HF	46.7	44	50			
	46.0	43	50	11	A	2
E	26.4	22	29	7	A, P	1
	28.0	27	31	11	A	2
Wta	637.0	520	795	9		
	461.0	450	475	3	Br	3

[a](1) Massoia and Fornes 1967b; PCorps; (2) Massoia 1973a; (3) Ximénez 1980.

Description

This species closely resembles the domestic guinea pig but is less robust. It has an extremely short tail or none. The dorsum is dark olive mixed with brownish and blackish, and the venter is paler, usually whitish or yellowish gray (plate 16) (Barlow 1965, 1969; pers. obs.).

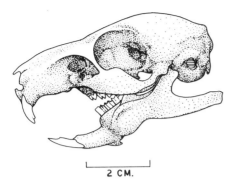

Figure 11.9. Skull of *Cavia aperea*.

Chromosome number: $2n = 64$; FN $= 116$ (Maia 1984).

Distribution

The species is found from Colombia south through Brazil and into Argentina. In southern South America it is found in Uruguay, eastern and northern Paraguay, and northeastern and east-central Argentina south to Buenos Aires province (Contreras 1972b, 1980; Honacki, Kinman, and Koeppl 1982; Massoia 1973a; Ximénez, Langguth, and Praderi 1972; UM) (map 11.95).

Life History

In captivity the gestation period is sixty-two days, the litter size averages 2.3 (range 1–5), and neonates average 57.6 g. In Uruguay this species has a protracted breeding season beginning in September and extending until April or May. In Buenos Aires province, Argentina, young are born throughout the year; two neonates weighed 56.6 g and 58.7 g, the litter size averages 2.1, and females can have up to five litters a year. The minimum age of first reproduction is thirty days (Barlow 1969; Kleiman, Eisenberg, and Maliniak 1979; Rood 1972).

Ecology

Cavia aperea is a grassland herbivore, ecologically analogous to some of the North American microtine rodents. It does not excavate burrows but constructs a maze of interconnecting surface tunnels and runs from 8 to 12 cm in diameter. Piles of its characteristic bean-shaped fecal pellets and plant cuttings are found at intervals along these runs. It may forage in the open but remains close to shelter.

These animals are grazers but will eat grass inflorescences as well, and they feed during evening and early morning hours. They can occur in high densities, and in Buenos Aires province, Argentina, they reached a maximum of 38.7/ha. In this habitat male home ranges averaged 1,387 m² and female home ranges 1,173 m². This species can depress the biomass of other grazing rodents (Barlow 1969; Dalby 1975; Rood 1972).

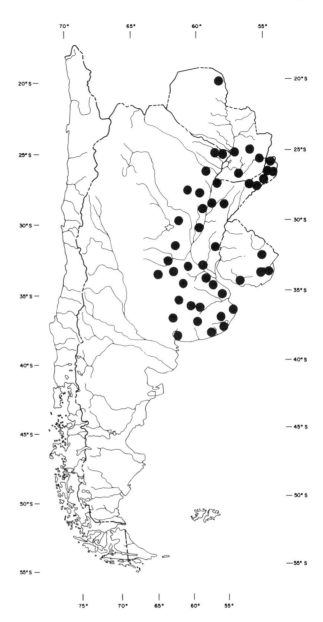

Map 11.95. Distribution of *Cavia aperea*.

Comment

Cavia aperea includes *C. pamparum*.

Cavia tschudii Fitzinger, 1857
Cuis Serrano

Measurements

	Mean	Min.	Max.	N	Loc.	Source[a]
HB	247.4	218.0	270	8	A, B, C, Peru	1
T	3.6	0.0	10	5		
HF	43.1	36.5	49	8	A, C	1
E	27.8	22.0	35	6	A, C	1

[a](1) Sanborn 1949; Thomas 1926; CM, Santiago.

Description

The color of this species varies throughout its range. In Peru specimens were very dark, almost reddish brown, heavily mixed with black and with dark buffy gray underparts. In Chile specimens were pale agouti brown dorsally, and in Bolivia they are agouti olive and brown with a white or cream venter (Sanborn 1949; pers. obs.).

Distribution

This species is distributed from Peru south to the northern Chilean province of Tarapacá (Arica) and in northwesternmost Argentina in the province of Tucumán and probably also Jujuy; archaeological remains have been found in Salta province as well (Honacki, Kinman, and Koeppl 1982; Massoia 1973a; Pine, Miller, and Schamberger 1979; Tonni 1984; Thomas 1926) (map 11.96).

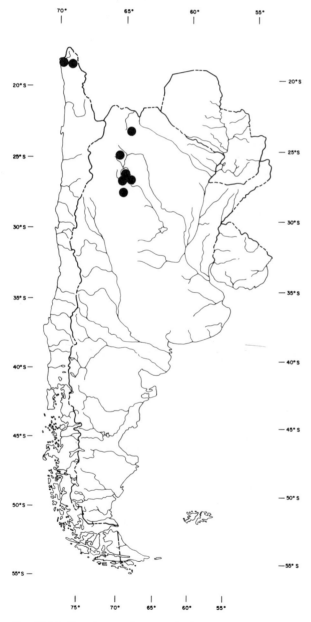

Map 11.96. Distribution of *Cavia tschudii*.

Life History

In captivity the gestation period is 63.3 days, litters average 1.9 (range 1–4), and the average age of first reproduction is two months (Weir 1974b).

Ecology

Cavia tschudii typically prefers moist habitats with scattered rocks between 2,000 and 3,800 m, and it has been taken in dense riparian habitats where extensive runways were evident. In Peru the animals live in thick grass, in which they make distinct runways. In Argentina this species has been reported to live in burrows with multiple entrances (Lucero 1983; Olrog 1979; Pearson 1957; Pine, Miller, and Schamberger 1979; Tamayo and Frassinetti 1980; Thomas 1926).

Genus *Dolichotis* Desmarest, 1820
Patagonian Cavy, Mara

Description

The description includes *Pediolagus* Marelli, 1927. These rather large rodents have a head and body length ranging from 430 to 780 mm. The limbs are long, and the feet are digitigrade (plate 16). The tail is extremely short and the ears are of modest length (see species accounts). The body form resembles that of a small artiodactyl, and they can run rather fast. They cannot be confused with any other rodent in the southern cone of South America.

Distribution

The current range of the two species includes the scrub and grassland areas of southern Bolivia, Paraguay, and Argentina.

Dolichotis patagonum (Zimmerman, 1780)
Mara, Mara Patagónica, Liebre Patagónica

Measurements

	Mean	Min.	Max.	N	Loc.	Source[a]
TL	707.2	610	810	10	A	1
HB	677.0	605	735	7		
T	34.7	25	40			
HF	140.5	130	160	6		
E	97.0	90	103	7		

[a](1) BA.

Description

This large, harelike animal has a square muzzle and short, rich fur. The dorsum is agouti gray with a distinct black area on the rump. The sides are lighter, grading to an orangish cream belly. The very short, dark tail is contained in a narrow band of striking white across the rump. The chin is orange (FM).

Distribution

Dolichotis patagonum is found in Argentina from approximately 28° to 50° S (Honacki, Kinman, and Koeppl 1982; BA) (map 11.97).

Life History

Females in a group have synchronized births and a postpartum estrus. In captivity pairs reproduce year round with up to four litters a year and a litter size averaging two (range 1–3). In Patagonia, however, maras produced only one litter a year. The age of first reproduction is eight months. In captivity and in the wild young leave the burrow during the first six weeks only to nurse. Lactation lasts about eleven weeks (Dubost and Genest 1974; Taber and Macdonald 1984).

Ecology

In northwestern Argentina this species is most common along gully forests and in *Larrea* flats. The

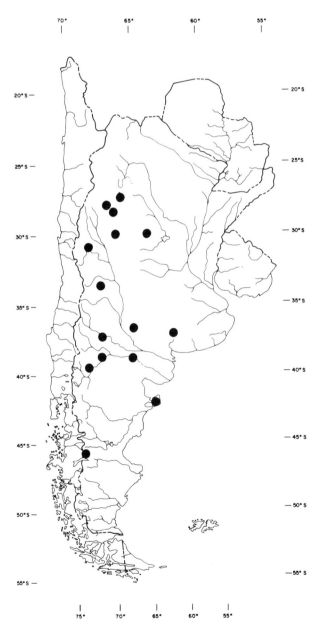

Map 11.97. Distribution of *Dolichotis patagonum*.

burrows, used only by the young, are in *Larrea* flats, washes, or vertical riverbanks. In southern Argentina *D. patagonum* prefers open brushy areas, frequently with sandy soil. In this area maras generally form perennial pairs, or sometimes trios (one male and two females), and occupy widely overlapping home ranges. Despite the overlap, pairs avoid contact with other pairs, occupying essentially floating territories. The members of a pair always travel in close proximity, and up to twenty-two pairs may deposit their young in a communal burrow. The members of a pair maintain contact through acoustic signals, and there is also extensive scent marking.

The usual activity period in northern Argentina is from sunrise to noon and from midafternoon to shortly after sunset. *D. patagonum* feeds on green vegetation, *Acacia* seeds, and cactus. It is heavily hunted and can run up to 80 kmph when pursued. In behavior and morphology maras resemble small artiodactyls (Daciuk 1974; Dubost and Genest 1974; Mares 1973; Taber and Macdonald 1984).

Dolichotis (Pediolagus) salinicola
(Burmeister, 1876)
Mara Chico, Conejo del Palo

Measurements

	Mean	Min.	Max.	N	Loc.	Source[a]
TL	460.4	380.0	512.0	13	P	1
	443.3	420.0	470.0	5	A	2
HB	435.3	359.0	491.0	13	P	1
T	25.1	11.0	38.0			
	23.6	19.0	30.0	5	A	2
HF	100.1	91.0	106.0	12	P	1
	97.1	91.0	100.2	5	A	2
E	60.7	52.0	70.0	13	P	1
	60.5	56.7	64.6	5	A	2
Wta	2.13 kg	1.0 kg	2.7 kg	9	P	1
	1.89 kg	1.5 kg	2.2 kg	5	A	2

[a](1) PCorps, UConn; (2) Mares, Ojeda, and Kosco 1981.

Description

This harelike animal has very thin legs and pointed ears that are moderately long and broad. The vibrissac are long and rather stout. The dorsum is agouti brown, extending onto the throat, the sides are lighter, and there is white on the venter and chin, as well as white patches in front of and behind the eyes. The cheeks are washed with tan (pers. obs.).

Distribution

The species is found in extreme southern Bolivia, in northwestern Argentina as far south as the province of Córdoba, and in the Paraguayan Chaco (Honacki, Kinman, and Koeppel 1982; BA, UConn) (map 11.98).

Life History

Females with one and two embryos have been caught (*n* = 3). In captivity the gestation period is seventy-seven days, litter size averages 1.5 (range 1–3), and neonates averaged 199 g (Kleiman, Eisenberg, and Maliniak 1979; Mares, Ojeda, and Kosco 1981; UConn).

Ecology

Dolichotis salinicola is an animal of the arid Chaco of Bolivia, Paraguay, and Argentina and is typically found in dry, low, flat thorn scrub, where it digs large burrows, made conspicuous by the extensive piles of dirt outside the entrances (Mares, Ojeda, and Kosco 1981; pers. obs.).

As in *D. patagonica*, the young are integrated into the social unit by various forms of chemical marking and "playful social interactions" (see Wilson and Kleiman 1974).

Genus *Galea* Meyen, 1832
Galea musteloides Meyen, 1832
Cuis Común

Measurements

	Mean	Min.	Max.	N	Loc.	Source[a]
TL	204.3	170	240	12	A, P	1
HF	38.3	36	44			
E	18.6	10	25			
Wta	225.7	187	283	9		

[a](1) Lord 1964; Mares, Ojeda, and Kosco 1981; CM, PCorps, UM.

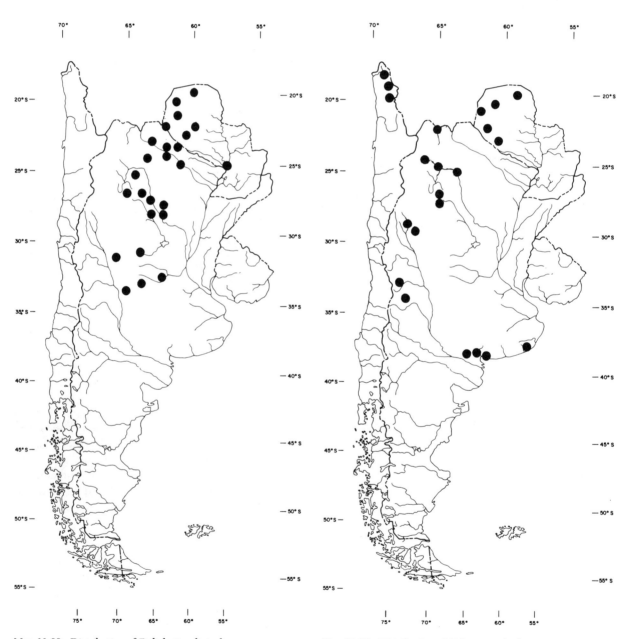

Map 11.98. Distribution of *Dolichotis salinicola*.

Map 11.99. Distribution of *Galea musteloides*.

Description

This large, hamster-sized rodent is very similar to *Cavia aperea* in coloration, but it has a slightly whiter belly and is less robust and considerably smaller (plate 16). It ranges from dark agouti flecked with black dorsally, slightly darker on the rump and head, to much lighter agouti brown. The dorsum is sharply demarcated from the white belly (pers. obs.).

Distribution

The species is found from southern Peru south into Chile and Argentina. In Chile it is confined to the high altitudes of Parinacota and Iquique; in Argentina it is found across most of the northern and central part of the country except for the northeastern corner; and in Paraguay it is found throughout the Chaco (Honacki, Kinman, and Koeppl 1982; Olrog and Lucero 1981; Tamayo and Frassinetti 1980; UConn) (map 11.99).

Life History

In captivity the gestation period is fifty-three days, litter size is 2.7 (range 1–5), and neonates average 37 g. Lactation lasts about three weeks, and the age of first reproduction is two to three months. In Paraguay females have been examined with one and two embryos. In Buenos Aires province, Argentina, young are born throughout the year; females may have up to seven litters a year; male neonates average 36.4 g; and the minimum age of first reproduction is twenty-eight days (Kleiman, Eisenberg, and Maliniak 1979; Rood 1972; Weir 1974a; UConn).

Ecology

Galea musteloides is found through an enormous altitudinal range: in the Andes it occurs up to 5,000 m, and in Paraguay it is found in the low Chaco. In Salta province, Argentina, *G. musteloides* is most common in moist areas such as stream edges and croplands. The home range of a female in Buenos Aires province, Argentina, was 4,275 m². It can be a dominant species: on one grid in Salta province, Argentina, it exceeded a biomass of 8,000 g/ha. It is a diurnal and often found in association with *Microcavia australis;* in fact, it shares runways and engages in social grooming with this species. During courtship males mark females using chin-gland secretions (Cabrera, 1953; Kleiman 1974; Mann 1978; Mares, Ojeda, and Kosco 1981; Ojeda 1979; Ojeda and Mares 1989; Rood 1970, 1972).

Genus *Microcavia* Gervais and Ameghino, 1880

Description

As noted for the family, these small, tailless rodents have short ears and short but stout claws.

Head and body length ranges from 200 to 220 mm, and they are appreciably smaller than the closely allied genus *Cavia*. The eye is offset by a rim of light hairs, yielding the effect of a ring (plate 16). The adults lack a submandibular gland, thus distinguishing them from the similar-sized *Galea*.

Distribution

The genus has been recorded from the montane regions of Bolivia, southeastern Peru, and Chile. *M. australis* occurs in lowland Argentina and southern Chile (see species accounts).

Microcavia australis (I. Geoffrey and d'Orbigny, 1833)
Cuis Chico

Measurements

	Mean	Min.	Max.	N	Loc.	Source[a]
TL	217.5	210.0	230.0	8 m	A	1
	218.8	210.0	230.0	8 f		
	209.8	197.0	225.0	5		2
	190.1			10	?	3
HF	50.0	49.0	52.0	8 f		1
	43.8	40.0	47.7	5		2
	41.5			10	?	3
E	16.3	12.0	22.9	5		2
	16.7			10	?	3
Wta	286.1	248.2	326.0	4		2

[a](1) Allen 1903; (2) Daciuk 1974; Mares 1973; Mares, Ojeda, and Kosco 1981; (3) Cabrera 1953.

Description

This species has small rounded ears and no external tail. The dorsum ranges from yellowish gray through very light agouti brown to gray. It is lighter on the sides, and the venter is gray to white washed with yellow (Allen 1903; Osgood 1943; pers. obs.).

Distribution

Microcavia australis is distributed in southern Chile and in Argentina between Jujuy and Santa Cruz provinces. It may also occur in extreme southern Bolivia. This species has also been recovered from archaeological sites in Magallanes province, Chile (Honacki, Kinman, and Koeppl 1982; Sielfeld 1979) (map 11.100).

Life History

In Buenos Aires province, Argentina, this species has a birth season lasting over nine months and an average litter size of 2.8 ($n = 16$). It is born after a gestation period of fifty-three to fifty-five days. Females have a minimum age of first reproduction of eighty-five days and exhibit postpartum estrus. Neonates average 30 g in weight, and lactation lasts three to four weeks. Females can have up to four litters a year (Kleiman, Eisenberg, and Maliniak 1979; Rood 1970, 1972).

Ecology

In Salta province, Argentina, this species prefers riparian situations or forested areas along gullies or in sandy, forested flats. It is particularly abundant along cultivated fields, under rock walls, and in rock piles. In the forested areas burrows are found under shrubs with overhanging branches or under fallen trees. The shallow tunnels have many openings. An individual may forage more than 50 m from its burrow but will race back when frightened. Groups may have one primary homesite or may move between several.

In Buenos Aires province, Argentina, *M. australis* clears paths between clumps of thorn scrub. Only occasionally will it build burrows, though it frequently uses holes dug by armadillos and viscachas. Individuals sleep and rest in shallow depressions dug under bushes. In this open habitat this species occurs at a density of 24.4/ha. Individuals are nonterritorial, but there is male-male aggression and a linear dominance hierarchy. The grison (*Galictis*) is the main predator.

Though not a graceful climber, *M. australis* frequently climbs bushes and trees and will jump down when startled. The diet consists of grass, leaves, seed heads, flowers, and berries. The animals are diurnal and appear not to need free water (Daciuk 1974; Mares 1973; Mares, Ojeda, and Kosco 1981; Rood 1970, 1972).

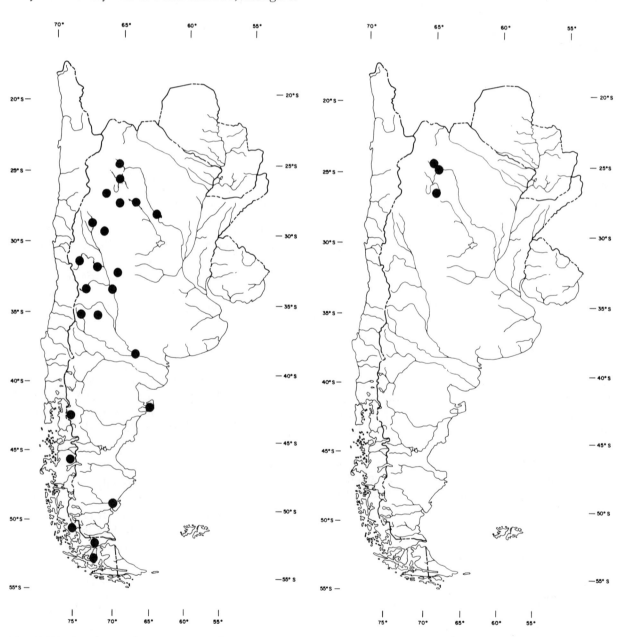

Map 11.100. Distribution of *Microcavia australis*.

Map 11.101. Distribution of *Microcavia shiptoni*.

Microcavia shiptoni (Thomas, 1925)
Cuis Andino

Measurements

	Mean	Min.	Max.	N	Loc.	Source[a]
TL	185.9	170.0	200.0	7	A	1
HF	36.8	36.0	38.1	6		
E	17.1	16.5	17.5			

[a](1) BA.

Description

This species has a light brown to olivaceous brown dorsum and a tan venter. The ventral tan may be reduced to a band in the midventer only. There is tan behind the ears (pers. obs.).

Distribution

Microcavia shiptoni is found only in the mountains of Tucumán, Catamarca, and Salta provinces, Argentina (Honacki, Kinman, and Koeppl 1982; Lucero 1983) (map 11.101).

Ecology

In Tucumán province, Argentina, *M. shiptoni* is found in brush and steppe areas from 3,000 to 4,000 m (Lucero 1983).

FAMILY HYDROCHAERIDAE
Genus *Hydrochaeris* Brunnich, 1772
• *Hydrochaeris hydrochaeris* (Linnaeus, 1766)
Capybara, Carpincho, Chigüire

Measurements

	Mean	Min.	Max.	N	Loc.	Source[a]
HB	1,176.3	960	1,280.0	12	A, U	1
HF	223.0	198	260.0			
E	62.5	59	70.0	11		
Wta	63.2 kg	48 kg	105.4 kg	6		
	43.0 kg	34 kg	57.0 kg	12 m	Br	2
	40.0 kg	30 kg	58.0 kg	9 f		

[a](1) Barlow 1965; Crespo 1974a; González 1973; Massoia 1976a; (2) Schaller and Crawshaw 1981.

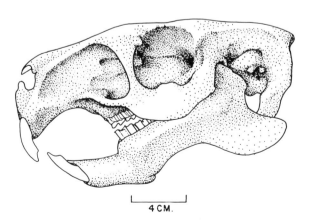

Figure 11.10. Skull of *Hydrochaeris hydrochaeris*.

4 CM.

Figure 11.11. The capybara *H. hydrochaeris*.

Description

The dental formula is I 1/1, C 0/0, P 1/1, M 3/3. The molars consist of transverse lamellae joined with cementum, and the third molar is longer than the other three, so that the teeth are distinctive (fig. 11.10). This is the largest living rodent and cannot be confused with any other form. Height at the shoulder averages 500 mm. With its massive skull and vestigial tail, the capybara gives the appearance of an enormous guinea pig (fig. 11.11). There are four semiwebbed toes on the forefeet and three on the hind feet. In the sexually mature male a bare raised glandular area on top of the snout is conspicuous. The upperparts are reddish brown and the underparts are lighter, usually some shade of yellow brown (Barlow 1965; Massoia 1976a; Schaller and Crawshaw 1981; pers. obs.).

Distribution

Hydrochaeris hydrochaeris has an extensive distribution from Panama south into northern Argentina. In southern South America it is found in eastern Paraguay, in Uruguay, and in Argentina in the northern and northeastern provinces south to Buenos Aires province. The distribution of capybaras has been severely affected by hunting, with local populations eliminated; their original range may have been more extensive (Barlow 1969; Honacki, Kinman, and Koeppl 1982; Massoia 1976a; Rabinovich et al. 1987; UM) (map 11.102).

Life History

In Venezuela capybaras reproduce throughout the year; there is usually one breeding cycle a year, though two are possible in excellent habitat. In Uruguay the breeding season is from early spring to late summer; in the Brazilian Pantanal reproduction is seen year round, with the birth peak in mid-February and a range in litter size of one to seven and an average of 3.5; and in northeastern Argentina reproduction is from September to March with a litter size of four to seven. The gestation period is 120 days, neonates average 1,400 g, and lactation lasts approximately ten weeks (Alho et al. 1989; Barlow 1969; Crespo 1982b; Ojasti 1973; Kleiman 1974; Kleiman, Eisenberg, and Maliniak 1979; Schaller and Crawshaw 1981).

Ecology

Capybaras are confined to areas with permanent standing or running water and can occur in marshes or estuaries and along rivers and streams. Depending on habitat and hunting pressure, capybaras can be found singly or in groups. In the Venezuelan llanos the median group size was 10.5. The normal herd is composed of a dominant male, several adult females with young, and subordinate males as peripheral members. In the Brazilian Pantanal the largest group was thirty-seven, with the mean group size varying between 3.6 and 5.8 depending on the

season; densities could reach 12.5/ha. In Uruguay herds range from seven to fifteen individuals.

During the dry season groups are larger as animals concentrate around water holes. Within a single population it is possible to see several different social structures. Males mark using the large sebaceous gland on the muzzle and have been known to kill young animals.

Capybaras can be diurnal or nocturnal depending on season and hunting pressure. When alarmed an animal gives a characteristic "woof" alarm bark and flees into the water. Capybaras are excellent swimmers and can remain submerged for several minutes. They are grazers, eating grass as well as aquatic vegetation.

Capybara leather is valued in southern South America, and between 1976 and 1979 almost 80,000 skins were exported from Argentina. There is also a sizable internal market for these skins (Barlow 1969; Kleiman 1974; Mares and Ojeda 1984; Massoia 1976a; Ojasti 1973; Ojasti and Burgos 1985; Schaller and Crawshaw 1981; pers. obs.).

FAMILY AGOUTIDAE

Genus *Agouti* (= *Cuniculus*) Lacépède, 1799

Description

The dental formula is I 1/1, C 0/0, P 1/1, M 3/3. Head and body length averages 600 to 795 mm, and the vestigial tail barely exceeds 20 mm; the hind foot averages 188 mm and the ear 45 mm. Weight can reach 10 kg. The adult male is about 15% larger than the adult female. The zygomatic arch is massive and distinctively sculpted (fig. 11.12); a small slit on its underside is visible externally, opening to a blind cavity on either side of the head. There are four toes on the forefeet and four on the hind feet. The dorsal

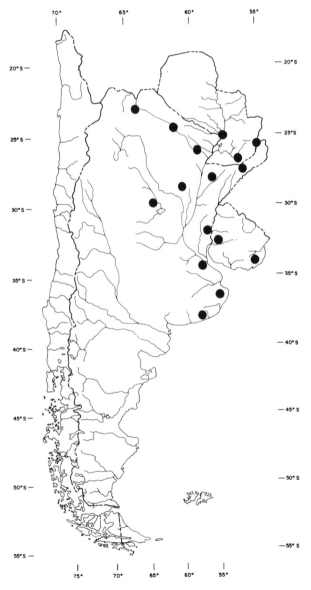

Map 11.102. Distribution of *Hydrochaeris hydrochaeris*.

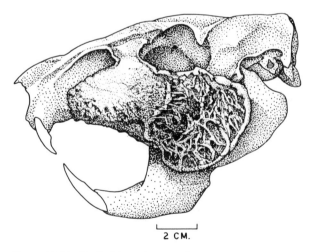

Figure 11.12. Skull of *Agouti paca* male.

pelage is brown to almost black, with four lines of white dots on each side of the body, and the venter tends to be white to buff (plate 16).

Distribution
Members of the genus are distributed from southern Mexico to northern Argentina in suitable lowland habitats.

• *Agouti paca* (Linnaeus, 1766)
Paca, Lapa

Measurements

	Mean	Min.	Max.	N	Loc.	Source[a]
TL	636.1	602.0	698.0	9	Br	1
T	18.8	10.0	24.0	8		
HF	113.1	104.0	120.0			
Wta	7.49 kg	4.3 kg	10.5 kg	47	P	2

[a](1) AMNH; (2) K. Hill, unpubl. data (only animals over 4 kg included).

Description
This large rodent has short legs and a moderately robust body. The skull is unmistakable, with rugose, greatly enlarged zygomatic arches, more pronounced in males. The eyes shine brilliant yellow when spotlighted. The square muzzle has large lips, elaborate nostrils, large eyes, and stiff vibrissae, as well as a prominent externally opening, fur-lined cheek pouch. The coloration is distinctive: the dorsum varies from reddish brown to dark chocolate, with two to seven prominent irregular rows of white spots on the flanks. The rows are more numerous in young animals and become progressively indistinct with age. The ventral coloration is white to cream (Collett 1981; Smythe 1970).

Distribution
Agouti paca is widely distributed in mesic habitats from Mexico south through eastern Paraguay and into the northeastern Argentine province of Misiones (Crespo 1974a; Honacki, Kinman, and Koeppl 1982; UM) (map 11.103).

Life History
In captivity, animals from Panama have one or two single births a year. Young are born weighing 650 g and nurse for approximately ninety days. The gestation period is 116 days. In Venezuela animals give birth throughout the year, whereas in Argentina reproduction takes place from August to January (Collett 1981; Crespo 1982b; Matamoros 1982; Mondolfi 1972).

Ecology
Pacas are restricted to forested habitats, but they occupy a wide range of forest types including mangrove swamps, narrow gallery forests, and dense up-

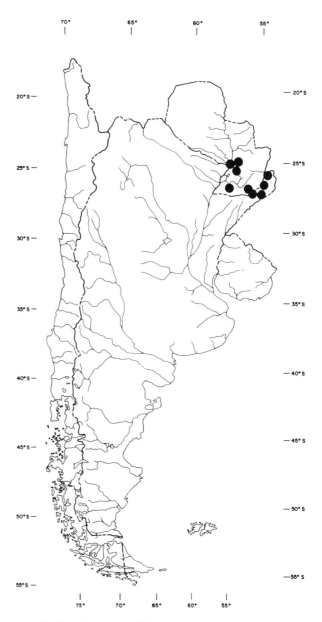

Map 11.103. Distribution of *Agouti paca*.

land scrub. They are strongly associated with moist areas and are frequently encountered near permanent water such as rivers. They are strictly terrestrial, spending the day in a burrow either of their own construction or built by another animal such as an armadillo. They are excellent swimmers and often take to the water when pursued. The entrance to the burrow is frequently concealed with leaf litter. In many areas pacas form an important part of the mammalian community. In Colombia in favorable habitat they occurred at densities ranging from eighty-four to ninety-three per square kilometer. Territories are defended by mated pairs.

Pacas are primarily frugivorous, though they also browse. They frequently take fruit to a "midden" to eat it, and this species may be an important seed disperser. Pacas are important game animals through-out their range and are frequently the preferred bush meat (Collett 1981; Crespo 1982b; Mondolfi 1972; Redford and Robinson 1987; Smythe 1970; pers. obs.).

FAMILY DASYPROCTIDAE

Diagnosis

The dental formula is I 1/1, C 0/0, P 1/1, M 3/3 (fig. 11.13). These large cursorial rodents are similar to the pacas but lack the spotted pattern and are more slender in build (plate 16).

Distribution

The family ranges from Veracruz in Mexico south to northern Argentina and Uruguay in appropriate moist habitats.

Genus *Dasyprocta* Illiger, 1811

Comment

The genus *Dasyprocta* is in need of taxonomic revision. The names employed here are in accordance with Honacki, Kinman, and Koeppl (1982).

Dasyprocta azarae Lichtenstein, 1823
Agouti, Agutí Bayo

Measurements

	Mean	Min.	Max.	N	Loc.	Source[a]
TL	516.4	475	550	7	A, Br	1
HB	494.3	475	530	4		
T	24.3	20	32	6		
HF	113.3	105	120	4		
E	37.0	35	40	4		
Wta	2.7 kg	2 kg	4 kg	8	P	2

[a](1) BA, FM; K. Hill, unpubl. data (only individuals above 2 kg included).

Description

This species is light to medium agouti brown with a pronounced speckled pattern. It can be very yellow on the sides and venter (pers. obs.).

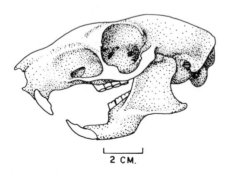

Figure 11.13. Skull of *Dasyprocta* sp.

Distribution

Dasyprocta azarae is found from east-central and southern Brazil into eastern Paraguay and the north-eastern Argentine provinces (Honacki, Kinman, and Koeppl 1982; Massoia 1980a; UConn) (map 11.104).

• *Dasyprocta punctata* Gray, 1842
Agouti, Agutí Rojizo

Description

This species is similar in size to *D. azarae;* two specimens weighed 3.1 and 3.6 kg (AMNH). Animals from Colombia weighed 3.5 to 4 kg. This agouti is a pronounced reddish agouti brown with much

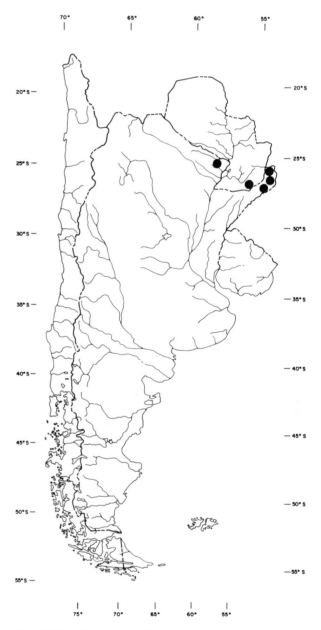

Map 11.104. Distribution of *Dasyprocta azarae.*

less pronounced speckling than *D. azarae*. When aroused or fleeing, it erects its long rump hairs (Kleiman 1974; Vergara 1982; pers. obs.).

Distribution
Dasyprocta punctata is distributed from southern Mexico south to the northwestern Argentine provinces of Jujuy and Salta, and apparently into the Paraguayan Chaco (Honacki, Kinman, and Koeppl 1982; BA) (map 11.105).

Life History
In captivity this species breeds throughout the year, with an interbirth interval of 127 days and a somewhat shorter gestation period. Neonates weigh 22.7 g (range 200–256; $n = 17$); litter size is two, and two litters can be produced within a year. Animals can breed for the first time by at least sixteen months. In Venezuela the normal litter size is two (range 1–3), and animals apparently breed year round.

In Panama young are generally born near some burrow or crevice to which they retreat for shelter except when called by the female (Meritt 1983; Ojasti 1972; Smythe 1978; Vergara 1982).

Ecology
These cursorial, diurnal rodents occur in broadleaf forests up to 1,500 m in Argentina, though they are most common at lower elevations. In Panama mated pairs occupy an area of 2–3 ha that they defend against conspecifics. Adults dig burrows and build nests where the young are born. In a dry forest in Guatemala they occur at a density no greater than 0.1/ha, while in the Brazilian Pantanal they occur at densities ranging from 0.9 to 3.2 per square kilometer, with the highest densities in gallery forest. In Costa Rica one female had a home range of 3.9 ha, larger than the 1–2 ha ranges reported from Panama.

Agoutis are frugivores and bury fruit that they do not immediately eat. Because of this habit of "scatter hoarding," they are important agents for the dispersal of seeds. When fleeing, animals emit a sharp alarm bark (Cant 1977; Eisenberg 1974; Olrog 1979; Rodríguez 1985; Schaller 1983; Smythe 1970, 1978).

Comment
Many workers would assign the name *Dasyprocta variegata* to the form in northwestern Argentina.

SUPERFAMILY CHINCHILLOIDEA
FAMILY CHINCHILLIDAE

Diagnosis
The dental formula is I 1/1, C 0/0, P 1/1, M 3/3. The lacrimal bone is large, and the lacrimal canal opens on the side of the rostrum; the angular process is delicate and not strongly hystricognathous; there is no massateric crest on the dentary; and the auditory canal is deflected dorsally and has an accessory foramen at the base (Woods 1984). Weight ranges from 0.5 kg to over 9 kg. These stout rodents have fully furred tails that are shorter than the head and body; they look like rabbits with short ears. On the hind foot the toes are reduced, to a minimum of three in *Lagostomus*. The color is highly variable, from a blue gray dorsum with cream underparts (*Chinchilla*) to a brown dorsum and a bold black band across the face in *Lagostomus* (see plate 16).

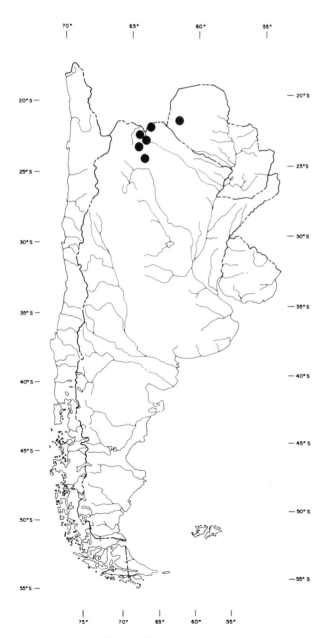

Map 11.105. Distribution of *Dasyprocta punctata*.

Distribution

The family occurs from the highlands of Peru south through the Andes to Chile. *Lagostomus* occurs in the lowlands of Argentina. By and large, this is a Temperate Zone family that has colonized both montane and lowland habitats.

Natural History

Chinchilla is colonial and prefers montane habitats. Animals have been taken above 4,500 m. Formerly it occurred in Chile, Peru, Boliva, and Argentina, but now wild colonies have been reduced to remnants. *Lagidium* is also colonial and occupies montane habitats extending above 5,000 m in the north, but in the southerly distributions in temperate portions of Argentina and Chile it may occur at much lower elevations. *Lagostomus* is colonial as well and is confined to elevations below 500 m. All species appear to be entirely herbivorous, eating a variety of herbaceous vegetation. In common with other hystricognath rodents, they have relatively long gestation periods and give birth to exceedingly precocial young. There is size dimorphism between the sexes. In *Lagostomus* the male is much larger than the female, whereas in *Chinchilla* the female is larger. Further discussion will be deferred to the species accounts.

Genus *Chinchilla* Bennett, 1829

Description

The tympanic bullae are inflated (fig. 11.14; see plate 16 and species accounts). This stout, rabbitlike rodent weighs about 0.5 kg but has a short, bushy tail.

Chinchilla brevicaudata Waterhouse, 1848

Description

This species is basically brownish gray and is distinguished from *C. lanigera* by its shorter tail and smaller ears (Mann 1978).

Distribution

The species is found only in the Andes, from southern Peru to northern Chile and northwestern Argentina. Severe human persecution has undoubtedly greatly reduced its original range, and no animals have been recorded from Peru for over fifty years (Honacki, Kinman, and Koeppl 1982; Ojeda 1985; O. P. Pearson, pers. comm.; Tamayo and Frassinetti 1980; BA) (map 11.106).

Ecology

Chinchilla brevicaudata lives in mountain grassland and scrub areas, very similar to those occupied by the viscacha, *Lagidium* (Mann 1978).

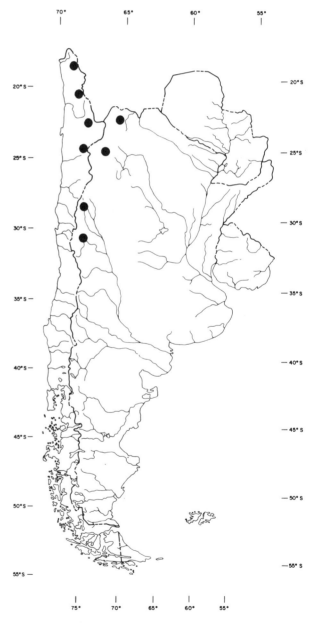

Map 11.106. Distribution of *Chinchilla brevicaudata*.

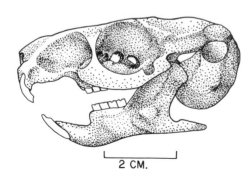

2 CM.

Figure 11.14. Skull of *Chinchilla laniger*.

Comment

Some authors include *Chinchilla brevicaudata* in *C. lanigera* (Honacki, Kinman, and Koeppl 1982).

Chinchilla lanigera (Molina, 1782)
Chinchilla

Measurements

	Mean	Min.	Max.	N	Loc.	Source[a]
TL	365.0	350	384	4	C	1
T	140.5	130	156			
HF	56.0	54	58	2		
E	46.7	45	48	3		
Wta	435.5	390	500	25		2

[a](1) Santiago; (2) Mohlis 1983.

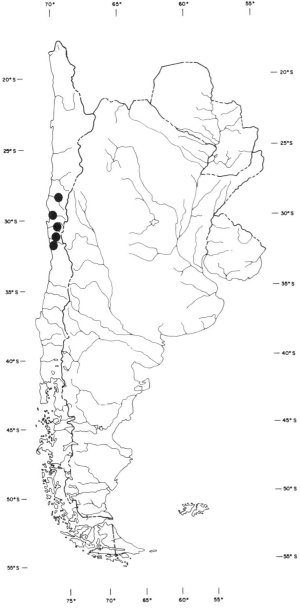

Map 11.107. Distribution of *Chinchilla lanigera*.

Description

This medium-sized rodent has a silvery gray pelt and is distinguished from *C. brevicaudata* by its longer tail and larger ears. Adults weigh about 500 g. It has large, rounded ears and eyes, five front toes, and three well-developed hind toes plus a rudimentary fourth. The tail has long gray and black hairs on the dorsal surface. The vibrissae are very long and can reach 110 mm (Kleiman, Eisenberg, and Maliniak 1979; Mann 1978; Mohlis 1983; Osgood 1943).

Distribution

Chinchilla lanigera is found in the foothills of the Andes and in the coastal mountains of Chile but is almost extinct in the wild (Honacki, Kinman, and Koeppl 1982; Mohlis 1983) (map 11.107).

Life History

The gestation period is 111 days, litter size is two, and neonates weigh 35 g. Lactation lasts between six and eight weeks, and the age of first reproduction is about eight months. In Chile lactating females have been taken from October to April, with births recorded from September to February (Kleiman, Eisenberg, and Maliniak 1979; Mohlis 1983; Weir 1974b).

Ecology

This species is found in rocky areas and is strictly herbivorous (Mohlis 1983).

Genus *Lagidium* Meyen, 1833
Viscacha, Vizcacha

Description

These rabbit-sized rodents have long ears and rather long tails. They resemble a larger version of *Chinchilla* (see plate 16). Head and body lengths range from 300 to 450 mm, and the tail is 200 to 400 mm. The thick fur of the dorsum ranges from gray to brown, contrasting with the cream to white venter.

Distribution

This genus is montane in its distribution, contrasting with the closely related plains denizen, *Lagostomus*. *Lagidium* is found from the highlands of Peru and Bolivia south to central Chile and western Argentina.

Lagidium peruanum Meyen, 1833
Montane Viscacha

Measurements

	Mean	Min.	Max.	N	Loc.	Source[a]
TL	639.4	615.0	660.0	5	P	1
T	267.2	245.0	295.0			
HF	91.0	87.0	93.0			
E	65.2	61.0	70.0			
Wta	1.29 kg	1.22 kg	1.36 kg	2		

[a](1) MVZ.

Description

The tail is medium in length with a long black pencil. The rabbitlike ears are fairly long, and the vibrissae are very long and stout. This species has short, soft, dense fur of an ash brown, browner on the shoulders. The sides are washed with yellow and the belly is washed with white.

Distribution

The species is found in the Andes of central and southern Peru. One specimen has been collected from Parinacota, Chile (Santiago) (map 11.108).

Life History

The gestation period is 140 days; the usual litter size is one; neonates weigh 180 g, and lactation lasts

about eight weeks. In Peru mating takes place from October through November (Pearson 1948; Weir 1974b).

Ecology

These animals shelter in rocky crevices, usually with one entrance. In Peru they occur at 3,000 to 5,000 m. They are colonial and sit on rocky perches when not feeding, and they emit a high-pitched alarm whistle on sighting a potential predator. Colonies can occur at a crude density of eleven individuals per square kilometer. Colonies range from four to seventy-five. Larger colonies are subdivided into smaller units of two to five animals, presumably family groups (Pearson 1948). They eat cacti as well as other vegetation (Dávila et al. 1982).

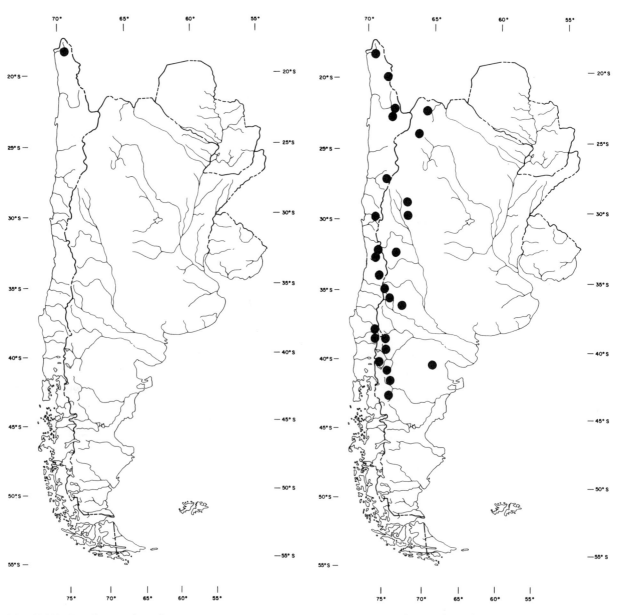

Map 11.108. Distribution of *Lagidium peruanum*.

Map 11.109. Distribution of *Lagidium viscacia*.

Lagidium viscacia (Molina, 1782)
Chinchillón, Vizcacha Serrana

Measurements

	Mean	Min.	Max.	N	Loc.	Source[a]
TL	677.0	510.0	840.0	14	A, C	1
HB	395.9	295.0	464.0	16		
T	287.1	215.0	376.0	14		
HF	97.4	82.0	113.0			
E	70.3	61.0	82.0	12		
Wta	1.54 kg	0.75 kg	2.1 kg	8		

[a](1) Crespo 1963; Greer 1966; Jaksić and Yáñez 1979; Osgood 1943; Pine, Miller, and Schamberger 1979; Wolffsohn 1923; Santiago.

Description

This large rodent has very dense, soft, slightly woolly fur, elongated ears, and a long, heavily penicillate tail covered with long hairs on the dorsal surface and haired to the tip. The tail is held half curled or fully curled when the animal is at rest but is carried fully extended when moving. There is a great deal of individual variation in pelage color, since the fur is molted throughout the year and there are pronounced age-related changes in color, but the basic coloration is agouti gray and brown on the dorsum, with a range of cream to black, and the venter pale yellow to tan (see plate 16) (Osgood 1943; Mann 1978; Rowlands 1974; pers. obs.).

Distribution

Lagidium viscacia is distributed from southern Peru along the Andes to south-central Chile and Argentina (Honacki, Kinman, and Koeppl 1982; Olrog and Lucero 1981; Tamayo and Frassinetti 1980) (map 11.109).

Ecology

This gregarious species lives in colonies in rocky outcroppings up to 4,000 m. It is diurnal and apparently does not need free water (Mann 1978; Rowlands 1974).

Lagidium wolffsohni (Thomas, 1907)
Chinchilla Anaranjada

Measurements

	Mean	Min.	Max.	N	Loc.	Source[a]
TL	777.0	775	781	3	C	1
HB	471.7	470	475			
T	305.3	305	306			
HF	110.7	107	113			
E	65.7	62	70			

[a](1) Wolffsohn 1923.

Description

This species is readily distinguished from other species of *Lagidium* by its larger size, rich color, long fur, immensely bushy tail, and short, black ears (Osgood 1943).

Distribution

Lagidium wolffsohni is found in southwestern Argentina and adjacent Chile (Honacki, Kinman, and Koeppl 1982) (map 11.110).

Genus *Lagostomus* Brooks, 1828
Lagostomus maximus (Desmarest, 1817)
Plains Viscacha, Vizcacha

Measurements

	Mean	Min.	Max.	N	Loc.	Source[a]
TL	753.2	705.0	820.0	55 m	A	1
	640.9	568.0	735.0	55 f		
T	181.8	165.0	205.0	34 m		
	160.2	135.0	173.0	21 f		
HF	126.8	118.0	136.0	30 m		
	112.8	104.0	118.0	19 f		
E	58.6	55.0	65.0	32 m		
	56.4	52.0	59.0	21 f		
Wta	6.37 kg	5.06 kg	8.84 kg	55 m		
	4.01 kg	3.52 kg	5.02 kg	55 f		

[a](1) Llanos and Crespo 1952.

Description

The head is large and massive, and the skull is distinctive (see fig. 11.15). Males have considerably larger heads and heavier bodies than the more gracile females. The vibrissae are very long and stout, and females have a set of stout bristles on the cheeks below the eyes. The face is broad and prominently marked by two parallel black stripes, one passing between the eyes and the other over the nose, separated by a white band. There is additional white on the cheeks, above the eyes, and at the base of the moderately haired ears. The dorsal fur is gray brown, lightening on the sides to a pure white venter (pers. obs.).

Distribution

Lagostomus maximus is found in extreme western Paraguay, probably in southeastern Bolivia, and into Argentina, then south across the country to at least Río Negro province, with at least one specimen recorded from Uruguay. It may be expanding its range

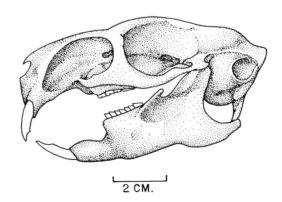

2 CM.

Figure 11.15. Skull of *Lagostomus maximus*.

toward the north and west into the Monte because of human persecution in the preferred grassland habitat. On the other hand, it has been eliminated from large parts of its range owing to intense eradication programs (Honacki, Kinman, and Koeppl 1982; Rabinovich et al. 1987; Weir 1974b; BA, Peace Corps, UConn) (map 11.111).

Life History

In central Argentina the birth peak of plains viscachas is from July to September, with one litter a year and two young per litter. In captivity litters of five have been born after a gestation period of 154 days. The birthweight is 200 g, and the age of first reproduction is eight months. Females have four nipples. In northeastern Argentina this species produces two litters a year, one in July to August and the other between February and April.

In La Pampa province, Argentina, the breeding season is from May to June; there may be two breeding peaks, and there is usually only one litter a year. Lactation lasts less than one month, and if two young are born usually only one is raised (Crespo 1982a; Hudson 1872; Kleiman, Eisenberg, and Maliniak 1979; Llanos and Crespo 1952; Maik-Siembida 1974; Rabinovich et al. 1987).

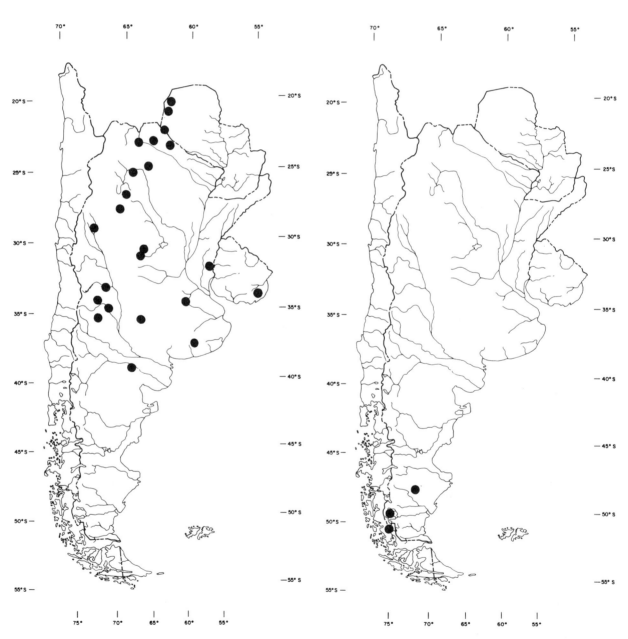

Map 11.110. Distribution of *Lagidium wolffsohni*.

Map 11.111. Distribution of *Lagostomus maximus*.

Ecology

Lagostomus maximus is characteristically found in open grassy or brushy areas, frequently with little rain. They are colonial, burrowing animals, and their presence is easy to determine because of the extensive burrow systems. By preference these burrows are placed near the bases of bushes or trees, though they may also occur in cracks in limestone. The average number of openings per burrow is 3.6 (range 1–11; $n = 35$) some burrow systems have been in use for at least eight years. The species is colonial and gregarious, with a complex social system incorporating dominant breeding males. There are several animals per burrow, and the burrow entrances are often clumped. Viscachas collect bones, sticks, and stones and pile them at the burrow entrances. Many other animals also use these burrows.

In central Argentina densities range from 2.0 to 7.5 animals per hectare. Plains viscachas are strictly grazing animals, and the areas around burrow entrances are often heavily grazed. They feed nocturnally and appear not to need surface water. They are very vocal and have a rich repertoire.

With the increase in cattle grazing, the overall range of *L. maximus* has undoubtedly increased, though local populations have been eliminated because they damage pasture and provide attractive meat. One author and gourmet commented that the flesh of young animals was "rather insipid; the old males are tough, but the mature females are excellent." Viscacha skins are exported from Argentina in large numbers: between 1976 and 1979 over 370,000 skins left the port of Buenos Aires (Hudson 1872; Llanos and Crespo 1952; Mares, Ojeda, and Kosco 1981; Mares and Ojeda 1984; Rabinovich et al. 1987).

SUPERFAMILY OCTODONTOIDEA
FAMILY MYOCASTORIDAE

Diagnosis

The dental formula is I 1/1, C 0/0, P 1/1, M 3/3; deciduous premolars are retained throughout life. The body is heavy; head and body length may reach 700 mm, and the tail 450 mm. The sexes differ in body size, with males 2–3 kg heavier than females. Ear size is reduced, the tail is long and sparsely haired, and the hind feet are webbed (fig. 11.16). The fur is long, and the dense underfur makes the pelt commercially valuable.

Distribution

Members of this group are found in the temperate portions of Chile and Argentina, northward to Bolivia and southern Brazil. They have been introduced to parts of Europe and the United States.

Figure 11.16. The coypu, *Myocastor.*

Natural History

These animals are highly adapted for an aquatic way of life; they feed on roots and aquatic grasses and are also known to take snails and freshwater mussels. They construct a simple burrow containing a nest chamber, where they feed heavily. Platform nests may be constructed in the portions of the marsh that have been eaten out.

Genus *Myocastor* Kerr, 1792
Myocastor coypus (Molina, 1782)
Nutria, Coypu, Coipo

Measurements

	Mean	Min.	Max.	N	Loc.	Source[a]
TL	862.3	610.0	995.0	10	A, C, U	1
HB	498.8	370.0	563.0			
T	363.5	240.0	435.0			
HF	132.6	118.0	153.0	8		
E	30.3	24.0	35.0	10		
Wta	4.99 kg	3.5 kg	5.8 kg	3		
	3.95 kg			101 m	A	2
	3.64 kg			98 f		

[a](1) Barlow 1965; Greer 1966; Massoia 1976a; Santiago;
(2) Crespo 1974b.

Description

This large rodent has a rounded head, short neck, short snout, and large, stout vibrissae. The hind feet are large, and the middle toes are connected by a basal swimming membrane. The mammae are high on the sides rather than on the abdomen. It has glossy pelage with long guard hairs and a long, rounded, tapering, thinly haired tail. The color, uniform over the body, is highly variable, ranging from light brown to reddish black (Barlow 1965; Mann 1978; Massoia 1976a; Osgood 1943).

Distribution

Myocastor coypus is found in southern Brazil, Paraguay, Uruguay, Bolivia, Argentina, and Chile. In southern South America it occurs in eastern Paraguay; throughout Uruguay; in Chile from Elqui department, Coquimbo province, to at least Malleco province and from Chilóe to the Strait of Magellan; and in Argentina in the northern and central provinces south to about 47° (Honacki, Kinman, and Koeppl 1982; Massoia 1976a; Tamayo and Frassinetti 1980; UM) (map 11.112).

Life History

In northern Argentina nutrias reach sexual maturity at four months or at a weight of 2–3 kg. Adult females cycle every twenty-three to twenty-six days throughout the year, and the average number of embryos per female is 4.9 (range 2–7). In Chile females

give birth in spring and summer and can have up to two litters a year, with litters ranging in size from two to eleven. The gestation period is 123–50 days, neonates weigh 225 g, and the female has four or five pairs of nipples (Crespo 1974b; Kleiman, Eisenberg, and Maliniak 1979; Mann 1978).

Ecology

Myocastor coypus is a semiaquatic rodent confined to areas with permanent water, from swamps and marshes to streams and drainage ditches, but almost always with succulent vegetation in or near the water. Waterways with steep banks seem to be preferred as burrow sites, but if these are not available nest platforms are built aboveground among tall marsh vegetation. Nutrias are excellent swimmers and divers and can remain submerged up to seven minutes. They eat aquatic vegetation. The species is a very important one in the skin trade; over nine million skins were exported from Argentina between 1976 and 1979 (Barlow 1969; Crespo 1974b; Greer 1966; Mares and Ojeda 1984; Massoia 1976a).

FAMILY OCTODONTIDAE

Diagnosis

The dental formula is I 1/1, C 0/0, P 1/1, M 3/3; the cheek teeth have a simplified occlusal pattern in the shape of a figure eight. The jugal process is variable, and the lacrimal does not open on the side of the rostrum; the paraoccipital process is short and fused to the bulla; and the coronoid process of the dentary is very pronounced (Woods 1984). Weight ranges from 200 to 300 g. These rather small, stout-bodied rodents usually have the tail shorter than the head and body. All are adapted for burrowing; the most extreme adaptation is shown in the genus *Spalacopus*. The fur is soft, and the dorsum ranges from gray to brown, with the venter usually cream or white.

Distribution

Octodonts are confined to southern South America, extending from southwestern Peru through the Andes of Bolivia, Chile, and Argentina. Forms such as *Octodontomys* may occur at high elevations. Others, such as *Spalacopus*, are confined to low-lying plains. The biogeography of this group has been reviewed by Contreras, Torres-Mura, and Yáñez (1987).

Natural History

These rodents are all to some extent adapted for digging. Most live in burrows of their own construction. They appear to be herbivorous, eating a wide variety of herbaceous vegetation. One can recognize

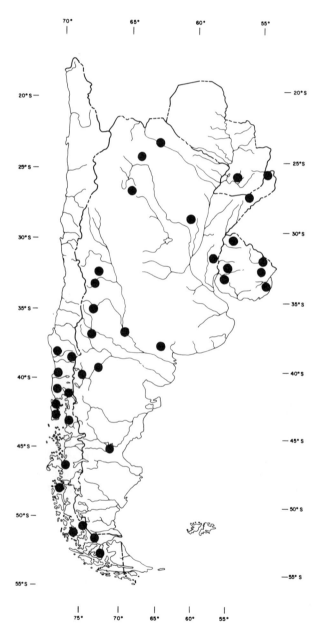

Map 11.112. Distribution of *Myocastor coypus*.

a continuum from strongly fossorial forms, such as *Spalacopus*, to forms that live in burrows but forage extensively on the surface, such as *Octomys*. Within *Octomys*, the postulated subgenus *Tympanoctomys* has tympanic bullae that are inflated, suggesting a convergence toward aridadapted rodents such as *Dipodomys* and *Salpingotus*. In this volume *Tympanoctomys* is considered a genus. All members of the family have extended gestation periods relative to their body size and produce precocial young.

Genus *Aconaemys* Ameghino, 1891
Aconaemys fuscus (Waterhouse, 1842)
Rata de los Pinares

Measurements

	Mean	Min.	Max.	N	Loc.	Source[a]
TL	225.7	206	250	9	C	1
HB	155.6	135	172			
	178.0	172	187	8	A	2
T	70.1	58	78	9	C	1
	78.0	74	80	8	A	2
HF	30.8	29	34	9	C	1
	35.3	34	37	8	A	2
E	18.5	17	21	8	C	1
	20.9	20	22	8	A	2
Wta	123.1	100	143	5	C	1
	134.0	121	143	8	A	2

[a](1) Greer 1966; Santiago; (2) Pearson 1984.

Description
This short-tailed octodont rodent is the same size and shape as some species of the genus *Ctenomys* but less specialized for burrowing: the ears are not as reduced; the front claws are not as large; and the fringe of hairs on the hind feet, though present, is not as well developed. It is rich dark brown dorsally and bright rufous to lighter brown or whitish ventrally (plate 17) (Ipinza, Tamayo, and Rottman 1971; Mann 1978; Osgood 1943; Pearson 1984).

Distribution
The species is found only in the high Andes of Argentina and Chile, between about 33° and 41° south (Contreras, Torres-Mura, and Yáñez 1987; Honacki, Kinman, and Koeppl 1982) (map 11.113).

Life History
One female with five embryos was examined (Greer 1966).

Ecology
Aconaemys fuscus shows a wide habitat tolerance: in Chile it is found in *Nothofagus* and *Araucaria* forests up to and above the tree line, in close association with bunchgrass; in southern Argentina it is found only in dense forests of *Nothofagus dombeyi* with thick bamboo understory.

This semifossorial rodent constructs a network of interconnected tunnels close to the surface, sometimes becoming runways. Entrances to this burrow system can occur every meter. Its burrows can be confused with those of *Ctenomys* or *Spalacopus*. Burrows are sometimes shared with *Akodon longipilis* and *Notiomys valdivianus. A. fuscus* appears to live in colonies. It is active under the snow during winter and is mainly nocturnal, but it is also active during the day. This species is herbivorous; in Chile *A. fuscus* stores food and seems to have a diet based on the seeds and roots of *Araucaria* (Contreras, Torres-Mura, and Yáñez 1987; Greer 1966; Mann 1978; Osgood 1943; Pearson 1983, 1984; Pearson and Pearson 1982).

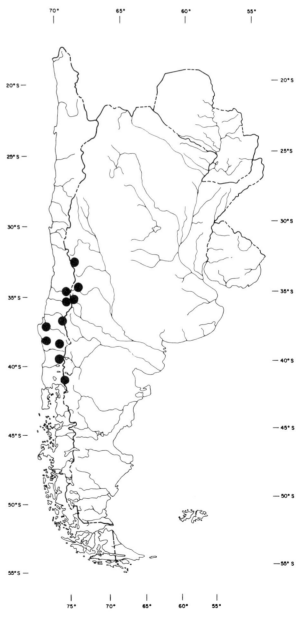

Map 11.113. Distribution of *Aconaemys fuscus*.

Aconaemys sagei Pearson, 1984

Measurements

	Mean	Min.	Max.	N	Loc.	Source[a]
TL	226.9	215	247	7	C	1
HB	165.0	151	174			
T	61.9	51	73			
HF	30.9	30	32			
E		17	19	5	A	2
Wta		83	110			

[a](1) FM; (2) Pearson 1984.

Description

This species is rich agouti brown dorsally, slightly lighter tawny ventrally, and has a lightly furred, slightly bicolored tail. The fur is short and soft. It has shorter front claws than *A. fuscus* and is much smaller and much darker and duller colored, without the bright rufous venter of the sympatric *A. fuscus* (Pearson 1984; pers. obs.).

Distribution

This newly described species is known from Neuquén province, Argentina, and Malleco province, Chile; some of the previously reported specimens of *Aconaemys* may belong to this species (Pearson 1984; FM) (map 11.114).

Ecology

This diurnal herbivore is especially abundant in ungrazed bunchgrass but is also found in second-growth *Nothofagus* forest with bamboo understory. It uses aboveground runways and burrows (Pearson 1984).

Genus *Octodon* Bennett, 1832
Degu

Description

The dental formula is I 1/1, C 0/0, P 1/1, M 3/3. The degus are stout-bodied, rat-sized rodents with the tail shorter than the head and body. The tail is well haired and often penicillate.

Distribution

The various species occur in the foothills and up to 1,200 m in the Andes of Chile and Argentina (see species accounts).

Octodon bridgesi Waterhouse, 1845

Measurements

	Mean	Min.	Max.	N	Loc.	Source[a]
TL	323.0	250	370	4	C	1
HB	184.8	148	203			
	197.0	190	203	3		2
T	138.3	102	167	4		1
HF	38.5	34	40			
E	22.0	20	23	3		
Wta	92.5			1		

[a](1) Greer 1966; Osgood 1943; Wolffsohn 1923; Santiago; (2) Jaksić and Yáñez 1979.

Description

This dark grayish to agouti brown rat with a gray to tan venter has a rounded, slightly penicillate tail shorter than the head and body. The forefeet have a rudimentary fifth toe, and the hind feet have granulated soles. Behind the "armpits" and in the groin region there are always well-delineated white spots. Compared with *O. degus*, *O. bridgesi* has larger eyes, proportionally smaller ears, softer fur, and a grayer pelage, and its tail has a much less pronounced pencil composed of shorter hairs (Ipaniza, Tamayo, and Rottman 1971; Mann 1978; Osgood 1943, pers. obs.).

Distribution

The species is found in the Chilean Andes from about 34°15′ to at least 40° S. Formerly it occurred

Map 11.114. Distribution of *Aconaemys sagei*.

as far north as O'Higgins and Curicó provinces, but in the past five years its range has diminished, and now it seems to occur only in three discrete populations that probably were connected before pre- and postconquest agricultural practices isolated them (Contreras, Torres-Mura, and Yáñez 1987; Miller 1980) (map 11.115).

Ecology

Octodon bridgesi was apparently once a common rodent, but now it is rare. It occurs between sea level and about 1,200 m in the forest zones of Mediterranean and south-central Chile. It has been trapped in streamside thickets of bamboo (*Chusquea*). It is diurnal and herbivorous, eating mostly leaves, seeds, and grass. Unlike other octodontids, it is not a digger. Nests are made in bushes or thick

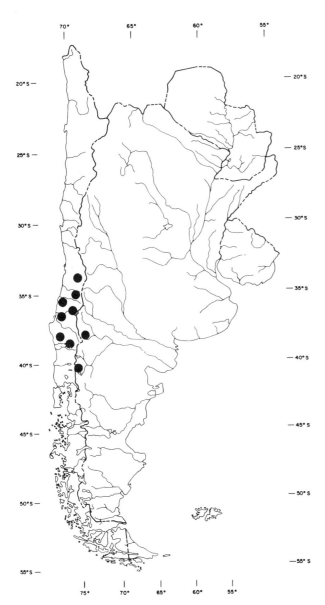

Map 11.115. Distribution of *Octodon bridgesi*.

grass (Glanz 1977a; Greer 1966; Ipinza, Tamayo, and Rottman 1971; Miller 1980; Tamayo and Frassinetti 1980).

Octodon degus (Molina, 1782)
Degu

Measurements

	Mean	Min.	Max.	N	Loc.	Source[a]
TL	266.5	200	307	75	C	1
HB	182.6	169	212	40		2
T	111.4	81	138	64		1
HF	24.7	19	31	75		
E	35.5	31	40	75		
Wta	215.0	170	260	?		

[a](1) Yáñez and Jaksić 1978; (2) Jaksić and Jáñez 1979.

Description

This species, the smallest of the *Octodon* group, has large eyes, a moderately long, thinly haired, black-tipped, penicillate tail, and soft fur. There are four well-developed toes and a poorly developed fifth toe; the forefeet bear nails instead of claws, and long, stiff, comblike bristles project over the claws of the hind feet. The dorsum is yellow brown and the venter is creamy yellow (see plate 17). There is an area of pale yellow above and below the eye and often a pale band around the neck. The feet are pale gray to white (Osgood 1943; Mann 1978; Woods and Boraker 1975; pers. obs.).

Distribution

The species is found at middle altitudes of the Chilean Andes between Vallenar and Curicó (Contreras, Torres-Mura, and Yáñez 1987; Honacki, Kinman, and Koeppl 1982) (map 11.116).

Life History

Octodon degus is flexible in the timing of its breeding, and there is evidence that precipitation directly triggers reproduction. It is a seasonal breeder with a birth season between September and December. During this time, in some areas, two litters can be produced. In some populations females exhibit a postpartum estrus. In other areas births peak in September, the last month of the rainy season and the time of greatest herbaceous growth. The gestation period is eighty-seven to ninety days, females have an average of 5.3 embryos (range 3–8), young are born weighing 14 g, and the age of first reproduction is about ten months. In captivity lactation lasts five to six weeks (Kleiman 1974; Meserve and Le Boulengé 1987; Meserve, Martin, and Rodríguez 1983, 1984).

Ecology

The species is typical of the central Chilean Mediterranean semiarid scrub (matorral) between sea level and 2,000 m. It is unusual for an octodontid in being diurnal and living in arid and semiarid habitats. Its requirements include low bushes or rocky outcrops, good herbaceous cover, and soil suitable for digging burrows. It appears to benefit from clearing of matorral. In areas where thickets are more common, *O. degus* appears to be replaced by *O. lunatus*. The degu can occur at densities ranging from 49–73/ha at one site to a maximum 192–259/ha at another site. Density values ranging from 2.5 to 1,033 per hectare have been reported. It can reach pest status in agricultural areas.

Degus dig their own burrows under bushes or rocks. They probably used to dig out in the open,

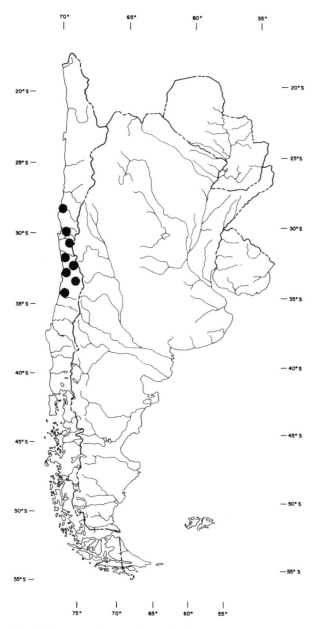

Map 11.116. Distribution of *Octodon degus*.

but trampling by stock has caused them to place burrows in protected areas. The burrows are elaborate, with several chambers and several openings. In each one lives a family group consisting of one male, two or three females with young, and four to six juvenile females. In the fall juveniles disperse to form new colonies. *O. degus* has a rich repertoire of vocalizations and sounds alarm calls at the approach of predators. The animals are common prey for raptors and foxes.

Degus frequently dust bathe. Adults carry grass into the burrows to make nests. They are diurnal herbivore-folivores. In one study the diet consisted of 25.6% by volume seeds and 60.0% shrub and herbaceous foliage; in another study grass composed 41.7%, herbs 15.4%, flowers 8.4%, seeds 9.6%, fruit 7.8%, bush bark and leaves 16%, and adult insects 0.8%. They obtain food by climbing in bushes and low trees as well as foraging on the ground (Contreras and Bustos-Obregón 1977; Contreras, Torres-Mura, and Yáñez 1987; Fulk 1976b; Glanz 1977a; Iriarte, Contreras, and Jaksić 1989; Wilson and Kleiman 1974; Le Boulengé and Fuentes 1978; Meserve, Martin, and Rodríguez 1983, 1984; Miller 1980; Pine, Miller, and Schamberger 1979; Rosenmann 1977; Schamberger and Fulk 1974; Tamayo and Frassinetti 1980; Woods and Boraker 1975; Yáñez and Jaksić 1978; Zunino and Vivar 1986).

Octodon lunatus Osgood, 1943

Measurements

	Mean	Min.	Max.	N	Loc.	Source[a]
TL	360.0	328	382	4	C	1
HB	187.8	167	221	6		2
T	157.0	152	161	4		1
HF	40.7	40	42			
E	28.0					3
Wta	233.0					

[a](1) Osgood 1943; (2) Jaksić and Yáñez 1979; (3) Glanz 1977a.

Description

Octodon lunatus closely resembles *O. bridgesi*, but its dorsum is more chestnut and ventrally the tail is at least half black. Both species have white spots in the "armpit" and groin areas. The major difference separating these two species is the configuration of the last maxillary molar, though there appears to be intrapopulational variability in this trait (Contreras, Torres-Mura, and Yáñez 1987; Ipinza, Tamayo, and Rottman 1971).

Distribution

The species is confined to the coastal range from about 35° S in Aconcagua province, Chile, north to Illapel (31°30′ S) (Contreras, Torres-Mura, and Yáñez 1987) (map 11.117).

Ecology

Octodon lunatus prefers thick, thorny scrub and eats shrubby vegetation, forbs, and grass (Glanz 1977b).

Genus *Octodontomys* Palmer, 1903
Octodontomys gliroides (Gervais and d'Orbigny, 1844)
Rata Cola Pincel

Measurements

	Mean	Min.	Max.	N	Loc.	Source[a]
TL	313.4	297.0	335	10	A, C	1
HB	170.9	161.0	182			
T	142.5	130.0	160			
HF	36.0	33.0	45			
E	28.5	25.8	32			
Wta	158					1

[a](1) Pine, Miller, and Schamberger 1979; BA.

Description

The tympanic bullae are large, but not as extreme as in *Tympanoctomys*. This large rat has large ears, a long bicolored tail with a pencil, and soft, dense fur with longer hairs along the spine. The dorsum is light brown, but the dark tips of the dorsal hairs make the coat look darker than that of *Octomys* (plate 17). The venter and feet are white (Ipinza, Tamayo, and Rottman 1971; Mann 1978; pers. obs.).

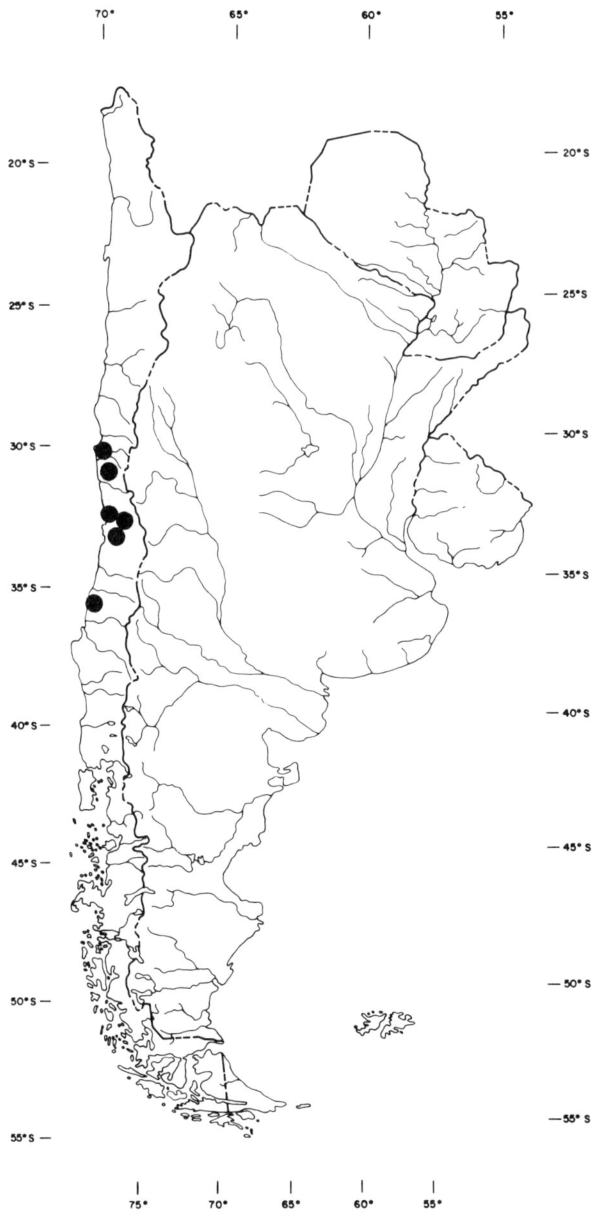

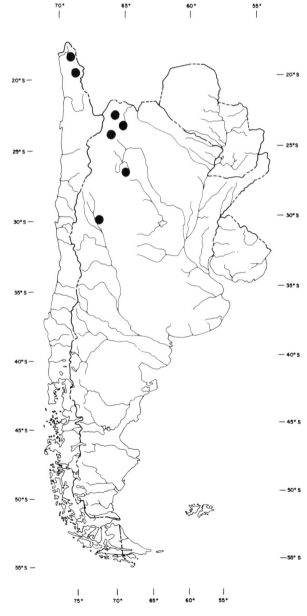

Map 11.117. Distribution of *Octodon lunatus*.

Map 11.118. Distribution of *Octodontomys gliroides*.

Distribution

The species is found in the high Andes of southwestern Bolivia, northernmost Chile, and the northwestern Argentine provinces from Jujuy to La Rioja (Contreras, Torres-Mura, and Yáñez 1987; Honacki, Kinman, and Koeppl 1982; Lucero 1983; Mann 1945; BA) (map 11.118).

Life History

In captivity gestation lasts 99–104 days, litter size averages two (range 1–4), young weigh 15 g at birth, and lactation lasts five to six weeks (Kleiman 1974; Kleiman, Eisenberg, and Maliniak 1979; Weir 1974b).

These forms dust bathe and have elaborate vocal repertoires. The behavior of the young during maturation has been documented by Wilson and Kleiman (1974).

Ecology

Octodontomys gliroides is an animal of the arid high Andes. It occurs between 2,000 and 5,000 m in open, dry habitats with bushes, cacti, and rocks. It is nocturnal and lives in small galleries in rocks and cacti roots, with superficial trails between burrow mouths. It does not dig and is a good rock climber (Ipinza, Tamayo, and Rottman 1971; Mann 1978; Tamayo and Frassinetti 1980).

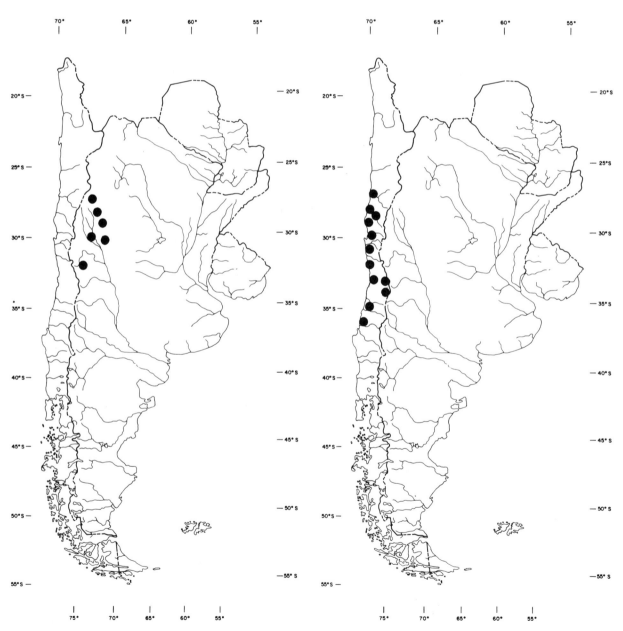

Map 11.119. Distribution of *Octomys mimax*.

Map 11.120. Distribution of *Spalacopus cyanus*.

Genus *Octomys* Thomas, 1920
Octomys mimax Thomas, 1920
Rata Cola Peluda

Measurements

	Mean	Min.	Max.	N	Loc.	Source[a]
TL	314.3	300.0	330.0	6	A	1
HB	151.8	140.0	165.0			
T	162.5	151.0	178.0	1		
HF	36.0	34.9	37.0	7		1
E	24.9	21.0	27.4	8		1
Wta	131.0	121.0	144.0	4		

[a](1) BA, CM.

Description

This species resembles *Octodontomys* but has a shorter tail that is less penicillate, and it has less developed tympanic bullae. The ears are relatively hairless and have guard hairs in front of the meatus. The dorsum is light brown, and the venter is cream. The tail is well haired to the tip and bicolored, with a conspicuous terminal tuft for about 50% of its length. The feet are tan (pers. obs.).

Distribution

The species is found in the western Argentine provinces of La Rioja, Catamarca, Mendoza, and San Juan (Honacki, Kinman, and Koeppl 1982; Mares 1973) (map 11.119).

Ecology

Octomys mimax appears to be very similar ecologically to the North American wood rat *Neotoma albigula*. It is found in high-altitude arid areas, where it burrows under large boulders and along dirt embankments. It apparently feeds on cacti, seeds, and other plant material (Mares 1973).

Genus *Spalacopus* Wagler, 1832
Spalacopus cyanus (Molina, 1782)

Measurements

	Mean	Min.	Max.	N	Loc.	Source[a]
TL	188.4	142	219.0	20	C	1
HB	143.9	115	165.0	18		
T	48.6	40	57.0			
HF	29.4	21	37.0	20		
E	11.0	10	12.0			
Wta	94.0	50	149.5			
	118.3			14 m	2,500 m	2
	112.5			14 f		
	99.3			14 m	70 m	
	80.2			14 f		

[a](1) Reig 1970; Wolffsohn 1923; Santiago; (2) Contreras 1983.

Description

Spalacopus cyanus is a medium-sized rat showing considerable modification for a fossorial life. It has extremely reduced ears, a short tail, and large front feet with large, strong claws (plate 17). It is a uni-form dark brown, though some specimens from higher elevations have a small white patch on the nape of the neck. The mountain population is phenotypically different from the coastal ones in skull morphology as well as being larger and showing different metabolic rates (Contreras 1983; Mann 1978; Reig, Spotorno, and Fernández 1972; pers. obs.). Chromosome number: $2n = 58$; there is very little intra- and interpopulational variation (Reig, Spotorno, and Fernández 1972).

Distribution

This species is confined to Chile between about 27° and 36° S. The range of this species used to be larger but it has been affected by human activity (Contreras, Torres-Mura, and Yáñez 1987; Miller 1980) (map 11.120).

Life History

Females give birth once or twice a year, between November and March. Litter size is two or three (Mann 1978).

Ecology

This fossorial species inhabits areas with shrub cover less than 60%, allowing the development of an herb layer on which it depends for food. It is particularly numerous in moist, sandy soils and is commonly found in wet flats along streams. Burrows are at depths of 10–12 cm and are 5–7 cm in diameter. The openings are not plugged. Digging is done with the incisors and the forefeet.

Spalacopus cyanus lives in small colonies composed of several males and females. It was thought that these colonies moved in search of food, but more recent research has shown that home ranges are quite stable. The diet is apparently composed exclusively of underground tubers and roots. When alarmed these animals emit a musical trill (Contreras 1983; Contreras, Torres-Mura, and Yáñez 1987; Eisenberg 1974; Glanz 1977a; Mann 1978; Miller 1980; Pine, Miller, and Schamberger 1979; Reig 1970; Reig, Spotorno, and Fernández 1972).

Genus *Tympanoctomys* Yepes, 1940
Tympanoctomys barrerae (Lawrence, 1941)
Rata Vizcacha Colorada

Description

This species resembles a small *Octodontomys*. The tympanic bullae are greatly inflated, so the intact animal appears to have a large head. One specimen from Argentina measured HB 121; T 143; HF 29; E 15; another specimen weighed 82 g. The dorsum is very pale tan to pinkish buff, and the venter is tan. The distal half of the heavily penicillate tail is dark brown dorsally, but ventrally only the distal third is so colored. It has short white hind feet (Lawrence 1941; Ojeda et al. 1989; pers. obs.).

Distribution
Tympanoctomys barrerae is found in the Argentine province of Mendoza and perhaps adjacent provinces (Honacki, Kinman, and Koeppl 1982; Roig 1965) (map 11.121).

Comment
Woods (1982) includes *Tympanoctomys barrerae* in *Octomys*.

Ecology
This species has morphological features typical of a desert rodent. It is a strict herbivore (feces from twenty-three specimens), feeding mainly on leaves and stems of *Heterostachys ritteriana* (Chenopodia-

ceae). It has been trapped only in places with halophytic vegetation and appears to be specialized for salt-rich desert plants, with a distribution that may reflect Pleistocene refugia (Mares 1973; Ojeda et al. 1989; Torres-Mura, Lemus, and Contreras 1989).

FAMILY CTENOMYIDAE
Tuco-tucos

Diagnosis
This family includes the single genus *Ctenomys*. The dental formula is I 1/1, C 0/0, P 1/1, M 3/3 (fig. 11.17). Skull characters are similar to those of the Octodontidae, but the cheek teeth are kidney shaped rather than like a figure eight. A jugal process is always present, and the zygoma is very broad. Average weight varies from 100 to 700 g. The tail is extremely short, and overall the animals are clearly adapted for a fossorial life (plate 17); the eyes are reduced in size, as are the external ears, and the foreclaws are very long. The thick fur is usually some shade of brown.

Distribution
The family is found from the western lowlands at Peru south to Tierra del Fuego. The distribution extends into eastern Brazil but is discontinuous.

Natural History
These animals spend most of their life underground. Some species appear to lead a solitary existence with one adult per burrow; others seem more colonial. The long, complex tunnel systems contain chambers for food storage as well as a nesting chamber. They feed on roots and grasses and are convergent in habits and behavior with the gophers (Geomyidae) of North America and the mole rats (Spalacidae) of Eurasia. Their vernacular name, tuco-tuco, refers to the call they produce; it may be made at the entrance to burrows and also underground.

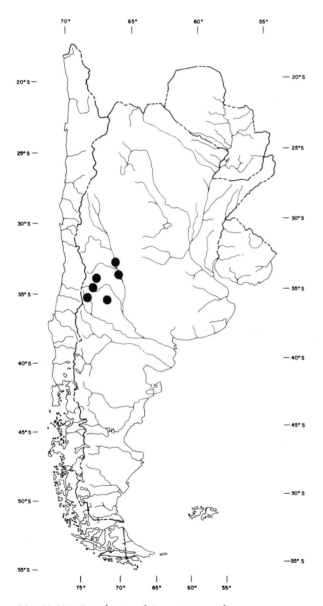

Map 11.121. Distribution of *Tympanoctomys barrerae*.

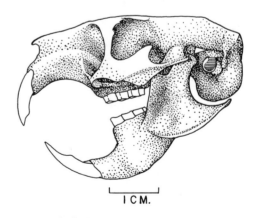

I CM.

Figure 11.17. Skull of *Ctenomys torquatus*.

Comment

This vexing group may have many lessons to teach us concerning the rates and modes of organic evolution. Although there appear to be many species of similar morphology, fossorial rodents are prone to speciation because they have very limited means of aboveground movement. Hence localized populations not too far apart may exhibit limited exchange of genetic material. Localized populations may also be subjected to extreme environmental selection resulting from profound differences in soil texture, depth, mean low temperatures, and so on

(Nevo 1979; Patton and Yang 1977). Clearly, evolutionary divergence among populations of *Ctenomys* is in a very dynamic state. Some putative species may show extreme genetic divergence, whereas others may exhibit much less divergence from a presumptive common ancestral form. The application of advanced genetic analysis to these groups is in its infancy but offers great promise (Novello and Lessa 1986; Sage et al. 1986).

Some recently described species of *Ctenomys* were omitted from the species accounts because the publications were not seen in time. Thus the num-

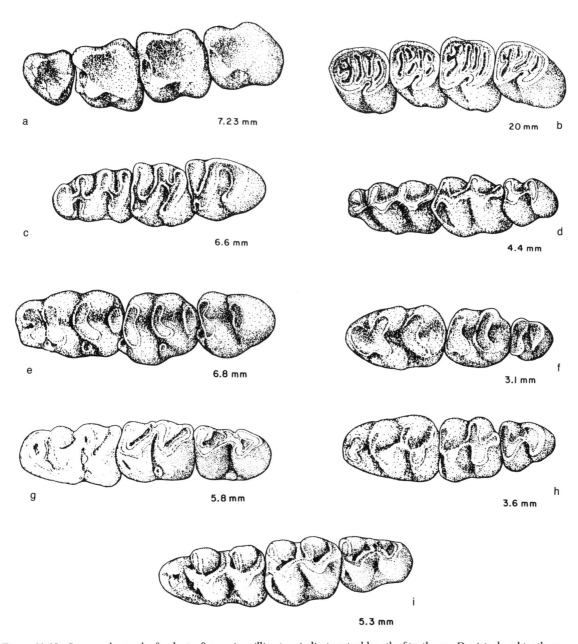

Figure 11.18. Some molar teeth of rodents: figures in millimeters indicate actual length of tooth row. Depicts dorsal tooth row, anterior left, posterior right: (*a*) *Sciurus aestuans*; (*b*) *Dasyprocta* sp.; (*c*) *Sigmodon hispidus*; (*d*) *Akodon* sp.; (*e*) *Rattus rattus*; (*f*) *Mus musculus*; (*g*) *Oryzomys capito*; (*h*) *Calomys laucha*; (*i*) *Rhipidomys leucodactylus*.

ber of described and currently recognized species covered below is a minimum. Also considered should be *C. sociabilis* Pearson and Christie 1985.

Genus *Ctenomys* Blainville, 1826
Ctenomys argentinus Contreras and Berry, 1982

Measurements

	Mean	Min.	Max.	N	Loc.	Source[a]
HB	177.9	173.0	183	8	A	1
T	81.1	75.0	87			
HF	36.9	34.3	39			
Wta	221.3	164.5	270			

[a](1) Contreras and Berry 1982b.

Description
This recently described medium-sized *Ctenomys* has a dorsal black band, a light collar, and a light belly that contrasts with the dorsum (Contreras and Berry 1982b).

Distribution
This species is known only from the north-central region of the Argentine province of Chaco (Contreras and Berry 1982b) (map 11.122).

Ctenomys australis Rusconi, 1934
Tuco-tuco de los Médanos

Measurements

	Mean	Min.	Max.	N	Loc.	Source[a]
HB	212.3	188.0	246.0	16	A	1
T	95.8	90.0	104.0			
HF	40.9	38.5	43.5			
Wta	349.0	248.0	500.0			

[a](1) Contreras and Reig 1965.

Description
This is a very large, pale species.

Chromosome number: $2n = 46$ ($n = 1$); *C. australis* differs from *C. porteousi*, with which it has sometimes been combined (Reig and Kiblisky 1969).

Distribution
Ctenomys australis is found in eastern Argentina in the province of Buenos Aires (Contreras and Reig 1965; Reig and Kiblisky 1969) (map 11.123).

Ecology
This species is found only along the Atlantic coast and inhabits vegetated dunes. Its burrows are 10–12 cm in diameter, up to 60 cm deep, and up to 75 m long. Food is frequently stored in the burrows. Never has more than one individual been found inhabiting a burrow. This species is much less vocal than *C. talarum* or *C. torquatus* and is usually heard calling only during the hottest part of the day (Contreras and Reig 1965).

Ctenomys azarae Thomas, 1903
Tuco-tuco Pampeano

Description
Ctenomys azarae is similar in appearance and size to *C. australis*.

Chromosome number: $2n = 48$ (n = 1); *C. azarae* is sereologically different from *C. mendocinus* and more closely allied to *C. australis talarum* and *C. porteousi* (Reig and Kiblisky 1969).

Distribution
The species is known only from La Pampa province, Argentina (Honacki, Kinman, and Koeppl 1982; Contreras 1972b) (map 11.124).

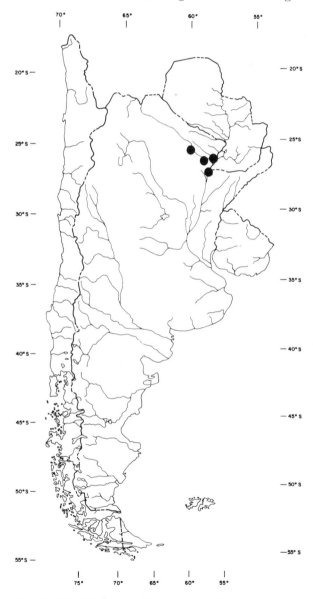

Map 11.122. Distribution of *Ctenomys argentinus*.

Ctenomys boliviensis Waterhouse, 1848
Tuco-tuco Boliviano

Measurements

	Mean	Min.	Max.	N	Loc.	Source[a]
TL	314.6	288	356	9	B	1
T	81.8	76	90			
HF	42.7	38	46			
E	9.3	8	12			
Wta	420.5	300	650	8		

[a](1) AMNH.

Description

This handsome *Ctenomys* demonstrates considerable variation in color, but all specimens have a dark middorsal stripe grading into lighter brown on the flanks and cream to chestnut on the venter. The venter also has varying amounts of white. There is a distinct light stripe extending from the base of the ear to the chin (pers. obs.).

Distribution

The species is described from central Bolivia, recorded in western Paraguay, and reported to occur in Formosa province, Argentina (Honacki, Kinman, and Koeppl 1982; Olrog and Lucero 1981; PCorps) (map 11.125).

Ctenomys bonettoi Contreras and Berry, 1982

Measurements

	Mean	Min.	Max.	N	Loc.	Source[a]
TL	246.0	232.0	260.0	2	A	1
HB	177.0	171.0	183.0			
T	69.0	61.0	77.0			
HF	34.8	32.8	36.8			
E	8.8	7.2	10.5			
Wta	202.3	184.5	220.0			

[a](1) Contreras and Berry 1982a.

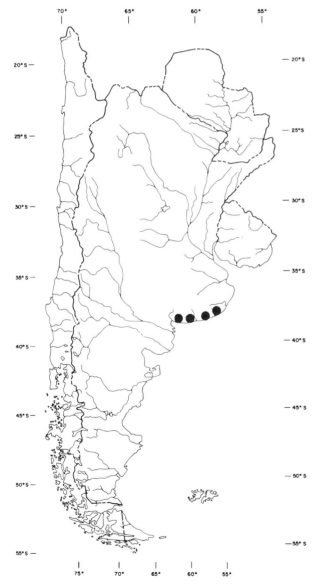

Map 11.123. Distribution of *Ctenomys australis*.

Map 11.124. Distribution of *Ctenomys azarae*.

Description

This medium-sized *Ctenomys* is light gray brown flecked with dark hairs dorsally, with faint dark bands on the back. The flanks are lighter, and the venter is yellowish (Contreras and Berry 1982a).

Distribution

Ctenomys bonettoi is known only from Chaco province, Argentina (Contreras and Berry 1982a) (map 11.126).

Comment

This species appears to be related to *Ctenomys tucumanus* and *C. argentinus*, from which it is separated by at least 30 km (Contreras and Berry 1982a).

Ctenomys colburni J. A. Allen, 1903
Tuco-tuco Ventriblanco

Measurements

	Mean	Min.	Max.	N	Loc.	Source[a]
TL	224.5	210	250	15 m	A	1
	213.0	200	225	17 f		
T	69.0	60	80	15 m		
	62.2	60	65	17 f		
HF	30.0	28	33	15 m		
	29.5	29	31	17 f		

[a](1) Allen 1903.

Description

Ctenomys colburni is similar to *C. sericeus* but larger, much more strongly suffused with fulvous,

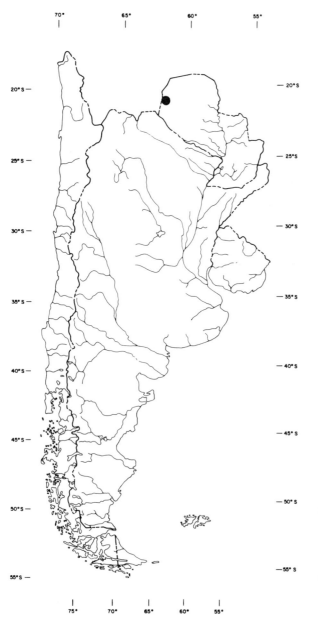

Map 11.125. Distribution of *Ctenomys boliviensis*.

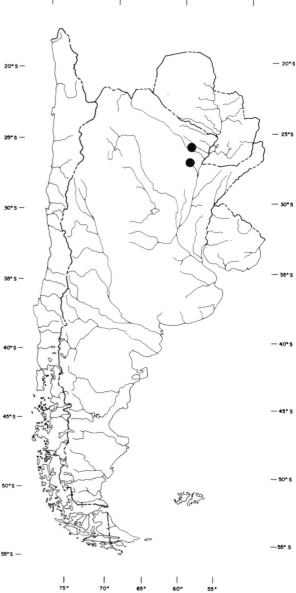

Map 11.126. Distribution of *Ctenomys bonettoi*.

and less varied with black. Younger animals are grayer (Allen 1903).

Distribution

This species is known only from the province of Santa Cruz, Argentina (Honacki, Kinman, and Koeppl 1982) (map 11.127).

Ctenomys conoveri Osgood, 1946

Measurements

	Mean	Min.	Max.	N	Loc.	Source[a]
TL	403.3	338	442	9	P	1
HB	293.3	245	328			
T	110.0	93	137			
HF	58.0	52	62			
E	13.4	10	16			
Wta	900.0			1		

[a](1) Osgood 1946; PCorps, UConn, UM.

Description

The proodont incisors are very broad and heavy; each of their anterior surfaces has an inner and an outer wide lateral groove and three shallow, narrow median grooves, the last little more than striations and not always fully continuous. This is a very large, stout *Ctenomys*. The dorsum is tan with a faint mid-dorsal line. The top of the head is gray brown, lightening to flecking on the back of the head, the cheeks are brownish yellow, and there are touches of white around the mouth and on the chin. The venter is a lighter yellowish (Osgood 1946; pers. obs.).

Distribution

Ctenomys conoveri is distributed in the Paraguayan Chaco and across the border into Argentina (Honacki, Kinman, and Koeppl 1982; Olrog and Lucero 1981) (map 11.128).

Ctenomys dorsalis Thomas, 1900

Measurements

	Mean	Min.	Max.	N	Loc.	Source[a]
TL	222.4	202	238	11	P	1
HB	161.8	152	178			
T	60.5	46	66			
HF	31.6	30	34			
E	30.2					

[a](1) Thomas 1900a; FM.

Description

The fur of this species is soft and fine. The dorsum is buffy fawn with a thick black dorsal stripe running from the tip of the nose to the rump; on the head this line may be sharply defined. There are no dark lateral face markings around the eyes or ears. The belly is lighter tan to white with a pale buffy line down the center. Some individuals have white blotches. The tail hairs are mixed with black and white (Thomas 1900a; pers. obs.).

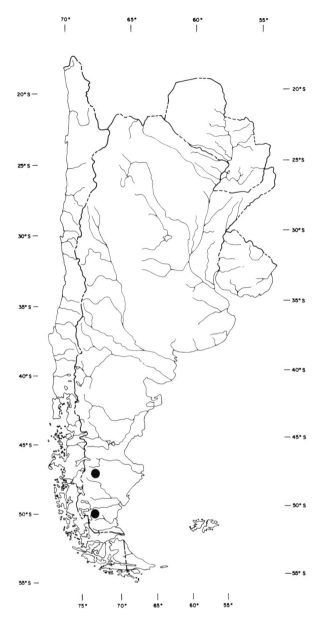

Map 11.127. Distribution of *Ctenomys colburni*.

Distribution

Ctenomys dorsalis is found only in the northern Paraguayan Chaco (Honacki, Kinman, and Koeppl 1982) (map 11.129).

Ctenomys emilianus Thomas and St. Leger, 1926
Tuco-tuco de las Dunas

Measurements

	Mean	Min.	Max.	N	Loc.	Source[a]
TL	385.6	264	302	3	A	1
HB	197.3	189	211			
T	86.3	75	93			
HF	32.7	22	39			

[a](1) Thomas and St. Leger 1926; BA, FM.

Description

This species is a pale glossy fawn to light gray brown above; most specimens have an almost pinkish cast. They are colored uniformly on the head and back, with no black markings. The sides and belly are whitish, passing onto the hips and thighs, where the color contrasts with the brown of the rump. The hind feet, forefeet, and tail are buffy whitish. The tail has practically no black on its terminal crest (Pearson 1984; Thomas and St. Leger 1926; pers. obs.).

Distribution

Ctenomys emilianus is found in Neuquén province, Argentina (Honacki, Kinman, and Koeppl 1982) (map 11.130).

Ecology

This species appears to be confined to sand dune habitats at about 800 m (Pearson 1984; Thomas and St. Leger 1926).

Ctenomys frater Thomas, 1902
Tuco-tuco Colorado

Measurements

	Mean	Min.	Max.	N	Loc.	Source[a]
TL	255.4	234.0	285.0	14	A	1
HB	175.6	157.0	210.0			
T	79.7	67.0	94.0			
HF	35.3	32.7	38.5			
E	8.28	6.8	11.3	10		
Wta	172.9	134.5	234.0	9		

[a](1) Mares, Ojeda, and Kosco 1981; BA, CM.

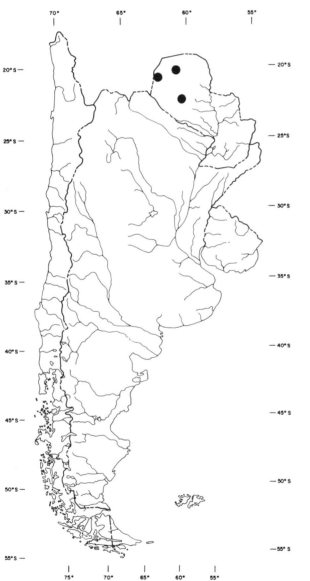

Map 11.128. Distribution of *Ctenomys conoveri*.

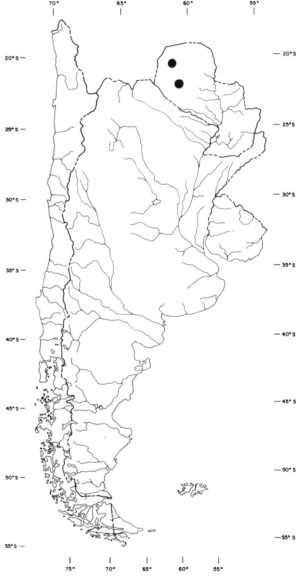

Map 11.129. Distribution of *Ctenomys dorsalis*.

Description

Individuals of this species exhibit a great range in dorsal color, ranging from rich dark to light brown, almost tan. The venter is gray brown to light brown, in most cases not contrasting sharply with the dorsum. In some populations there is white on the venter and along the hind legs and forelegs (pers. obs.).

Distribution

The species is distributed in the Andes from southwestern Bolivia into the Argentine provinces of Jujuy and Salta (Honacki, Kinman, and Koeppl 1982) (map 11.131).

Ecology

Ctenomys frater has been taken in moist forest, in grassy meadows at 2,000 m, and at 4,500 m in Ju-juy province, Argentina. In Salta province, Argentina, it inhabits flat areas with deep soil, often near creeks, does not appear to burrow extensively, and is not as vocal as desert *Ctenomys* (Mares, Ojeda, and Kosco 1981; Olrog 1979; BA).

Comment

Ctenomys frater includes *C. budini* and *C. sylvanus.*

Ctenomys fulvus Philippi, 1860
Tuco-tuco Coludo

Measurements

	Mean	Min.	Max.	N	Loc.	Source[a]
TL	316.6	296	341	5	A, C	1
HB	220.0	199	240	5	A, C	1
	165.0			7	A	2
T	99.6	92	104	5	A, C	1
	75.0	70	80	5 f		
	82.0			7		2
HF	45.8	36	51	4	A, C	1
	37.0	35	40	5 f		
	34.0			7		2
E	7.5	7	8	2	A, C	1
Wta	360.3	300	401	7	A, C	1

[a](1) Pine, Miller, and Schamberger 1979; Thomas 1921c; Santiago; (2) Mares 1973.

Description

This large species of *Ctenomys* is uniformly pale buffy, some individuals being varied with blackish. The short ears barely project above the fur, and the head often has some blackish. The venter is slightly lighter, and the tail is bicolored and penicillate. In at least some populations the females are significantly smaller than the males (Allen 1903; Osgood 1943; pers. obs.).

Distribution

Ctenomys fulvus robustus is found only in a very small area in Pica Oasis in Tarapacá province, Chile. *C. f. fulvus* is found in the Atacama Desert, Chile, and into Argentina in the provinces of San Juan, Mendoza, Catamarca, and Salta (Honacki, Kinman, and Koeppl 1982; Mann 1978) (map 11.132).

Ecology

These are animals of well drained, very sandy soils at reasonably high altitudes. In northern Chile they inhabit isolated oases surrounded by deserts barren of vegetation. In Salta province, Argentina, they are found in various habitats, though most commonly in *Larrea* flats with taller shrubs and sandy soils. They make burrow systems usually 25 cm or more below the surface, with the mouth of the burrow frequently in dense vegetation. The burrow temperature remains between 19 and 25° C when the outside temperature ranges from 6 to 62°C. The burrow is plugged with earth most of the time but is

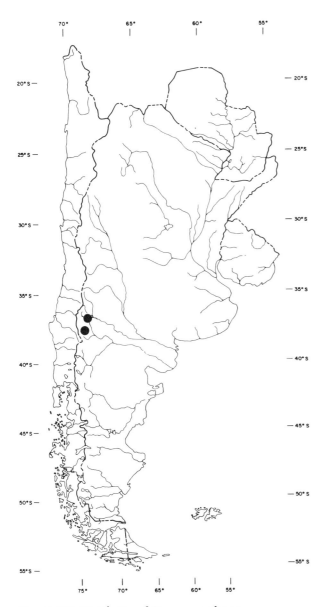

Map 11.130. Distribution of *Ctenomys emilianus.*

opened between 5:00 and 7:00 A.M. to throw out soil and feces.

Animals are rarely observed aboveground or at any distance from the burrow mouth, and then they are seen only at night. One individual inhabits a burrow system. Animals call only from their burrows, and a single call triggers a wave of other calls. Feeding is done from regular burrow openings or from special feeding holes. In Salta province, Argentina, animals appear to be active aboveground only during the driest part of the year. *Larrea* leaves up to 20 cm off the ground are eaten (Mann 1978; Mares 1973; Miller et al. 1983; Rosenmann 1959).

Comment

In *Ctenomys fulvus fulvus* in Chile, $2n = 26$ and $FN = 52$, while in *C. f. robustus* $2n = 25$ and $FN =$

52. Unlike all other Chilean *Ctenomys*, the sperm of the two forms is symmetric (Gallardo 1979).

Ctenomys haigi Thomas, 1917
Tuco-tuco Patagónico

Measurements

	Mean	Min.	Max.	N	Loc.	Source[a]
TL	230	225	235	2	A	1
HB	160	155	165			
T	70	70	70			
HF	29	28	30			
E	6			1		
Wta	164			7 m		2

[a](1) Thomas 1919c; (2) Pearson 1984.

Description

This fairly small *Ctenomys* has soft, fine, silky fur, agouti gray brown above with no darker markings on

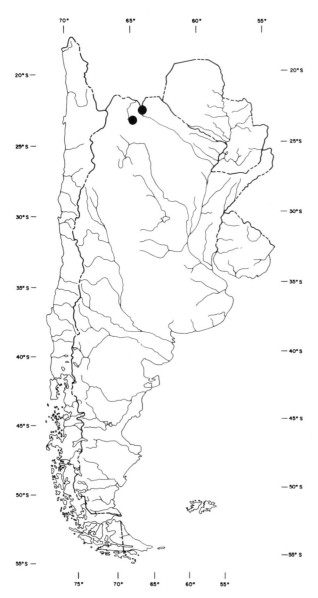

Map 11.131. Distribution of *Ctenomys frater*.

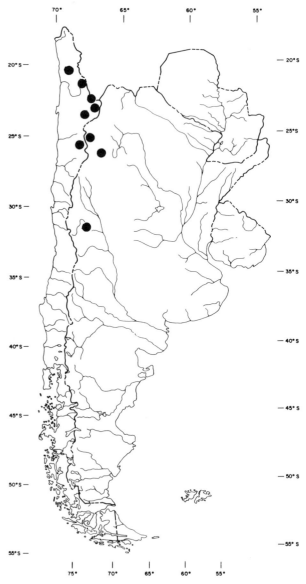

Map 11.132. Distribution of *Ctenomys fulvus*.

the rump or crown except for the top of the nose, which is dark. The sides may be lighter gray, and the lower flanks are rather sharply buffy as a continuation of the orangish wash on the venter. The tail is gray or orangish on the sides and beneath and blackish above (Thomas 1919c; pers. obs.).

Chromosome number: The karyotype is 50, unlike any other described *Ctenomys* species (Pearson 1984).

Distribution

Ctenomys haigi is known from the Argentine provinces of Chubut and Río Negro (Pearson 1984; Thomas 1919c) (map 11.133).

Life History

The average litter size is 2.6 (range 2–4) (Pearson 1984).

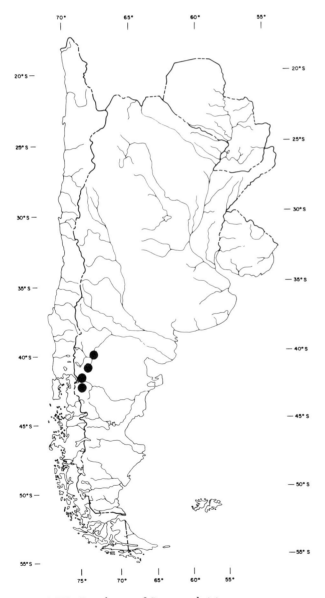

Map 11.133. Distribution of *Ctenomys haigi*.

Ecology

This species inhabits the steppe and precordillera but not the forests. Compared with other species of *Ctenomys* it does not dig extensively, but it is rarely seen aboveground. Its call, a short "tuc-tuc" uttered once each second with as many as thirty calls per bout, is heard day and night (Pearson 1984).

Comment

Ctenomys haigi has been combined with *C. mendocinus* but seems to be a good species (Pearson 1984).

Ctenomys knighti Thomas, 1919
Tuco-tuco Catamarqueño

Measurements

	Mean	Min.	Max.	N	Loc.	Source[a]
TL	288.6	237	324.0	7	A	1
HB	205.7	168	234.0			
T	82.9	69	94.0			
HF	38.8	30	47.9			
E	9.4	7	12.0	5		
Wta	471.7	410	565.0	3		

[a](1) Thomas 1919a; CM, FM.

Description

This account includes *Ctenomys viperinus*. This very large, robust *Ctenomys* has a reddish brown dorsum, mottled with gray as the color of the hair bases shows through. Laterally it is more orange, and the venter is mottled gray and orange. The ventral white coloring extends in a patch on the nose and between the eyes and at the base of each ear. The extent of this white varies between individuals (pers. obs.).

Distribution

Ctenomys knighti is found in La Rioja province and north to Tucumán province, Argentina (Thomas 1919a) (map 11.134).

Ctenomys latro Thomas, 1918

Measurements

	Mean	Min.	Max.	N	Loc.	Source[a]
TL	217.2	188	241.0	9	A	1
HB	153.0	136	170.0			
T	64.2	51	71.0			
HF	26.7	20	30.5			

[a](1) Thomas 1918b; BA, FM.

Description

This small tuco-tuco has a light to medium brown dorsum and a darker face and crown. The venter is noticeably paler. Behind each ear there is a light buffy patch. The tail is dark brown on top and pale buffy on the sides and below. *C. latro* is much lighter than the similar *C. tucumanus* (Reig and Kiblisky 1969; Thomas 1918b; pers.obs.).

Chromosome number: *Ctenomys latro* exhibits autosomal polymorphism; $2n = 42$ (Reig and Kiblisky 1968, 1969).

Distribution

The species is found in the northwestern Argentine provinces of Tucumán and Salta (Reig and Kiblisky 1969; BA) (map 11.135).

Ecology

This species lives in soft, sandy soils that are yellowish, very similar to its dorsal color. It is found in dry areas at about 600 m with chaco, xeric vegetation (Reig and Kiblisky 1969; Thomas 1918b).

Comment

This species was synonymized with *C. tucumanus*, but it is clearly separable (Reig and Kiblisky 1968, 1969).

Ctenomys magellanicus Bennett, 1836
Tuco-tuco Magellánico

Measurements

	Mean	Min.	Max.	N	Loc.	Source[a]
TL	274.6	244	304	15	C	1
HB	193.2	172	222			
T	81.4	68	92			
HF	37.2	32	41			

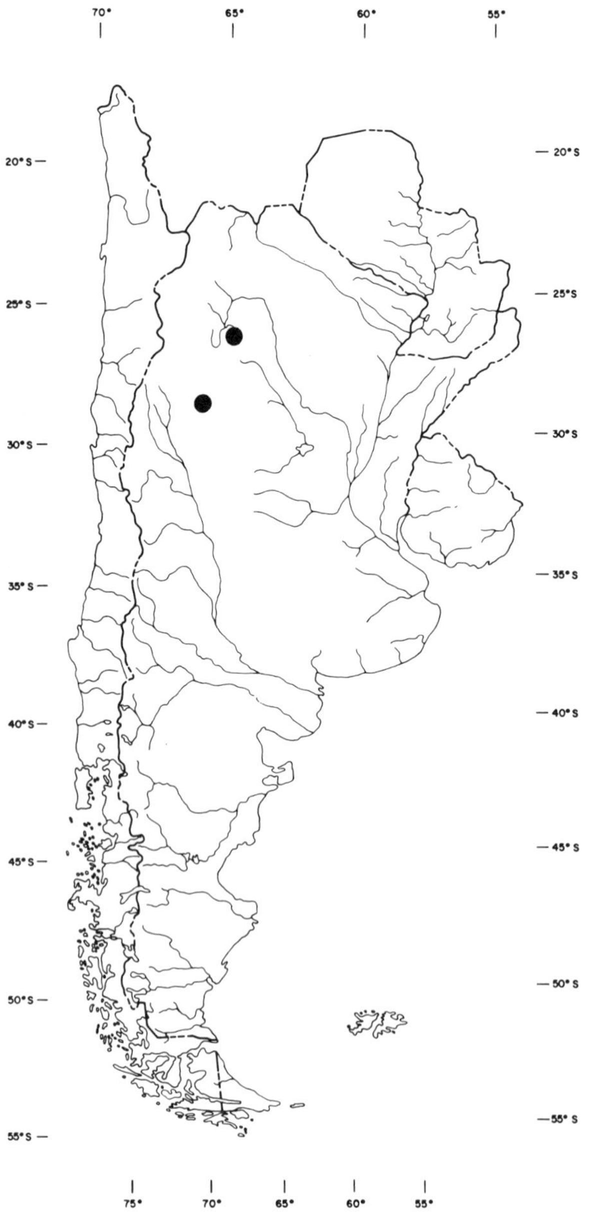

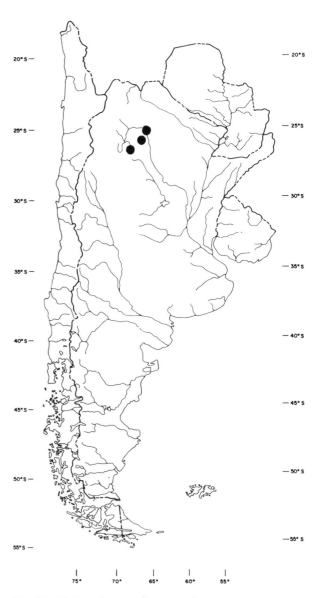

Map 11.134. Distribution of *Ctenomys knighti*.

Map 11.135. Distribution of *Ctenomys latro*.

E	8.2	7	9	11	
Wta	246.7	165	317		
	271.0	165	370	12	2

[a](1) Osgood 1943; Texera 1975; Santiago; (2) Atalah, Sielfeld, and Venegas 1980.

Description

Different geographical races of this species exhibit different colors, ranging from light colored with pale grizzled grayish buff and cinnamon buff below to blackish and buffy gray above and below. Some individuals are fulvous above and orange on the venter (Allen 1903; Osgood 1943; pers. obs.). Chromosome number: $2n = 34$ or 36; FN = 68 (Gallardo 1979; Reig and Kiblisky 1969).

Distribution

Ctenomys magellanicus is found in extreme southern Chile and southern Argentina. The ranges of many of the subspecies have been greatly reduced because of sheep grazing (Honacki, Kinman, and Koeppl 1982; Miller et al. 1983; Tamayo and Frassinetti 1980) (map 11.136).

Ecology

This species inhabits the Patagonian steppe, where it lives in open meadows with dense grass cover, often dotted with low bushes. Its tunnels are at least 30 cm below the surface, and it eats the roots of grasses and shrubs. The subspecies *C. m. fuegi-*

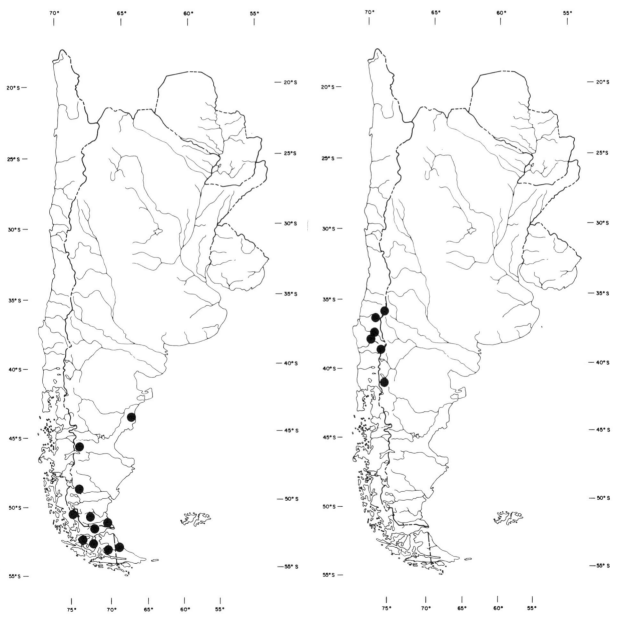

Map 11.136. Distribution of *Ctenomys magellanicus*.

Map 11.137. Distribution of *Ctenomys maulinus*.

nus was an important dietary item for the aboriginal Ona (Mann 1978; Miller et al. 1983).

Ctenomys maulinus Philippi, 1872

Measurements

	Mean	Min.	Max.	N	Loc.	Source[a]
TL	291.3	275	305	7	C	1
HB	204.7	192	210			
T	86.6	78	95			
HF	39.4	37	42			
Wta	164.0			3 m	A	2

[a](1) Osgood 1943; (2) Pearson 1984.

Description

There is a good deal of intra- and interpopulational variation in color in this species, ranging from very light brown through olivaceous brown to dark brown. In part this variation tracks changes in soil color. The venter is often only slightly lighter than the dorsum. The tail has a white pencil (Greer 1966; Mann 1978; Osgood 1943; Pearson 1984).
Chromosome number: $2n = 26$; FN $= 50$ (Gallardo 1979; Venegas 1973; Venegas and Smith 1974).

Distribution

This species is found in the Chilean provinces of Talca, Ñuble, Malleco, and Cautín, and in Argentina it occurs in the province of Neuquén (Pearson 1984; Tamayo and Frassinetti 1980) (map 11.137).

Ecology

Ctenomys maulinus has been caught in habitats ranging from open volcanic sands to *Nothofagus-Araucaria* forests, from 900 to 2,000 m. Two burrows measured 49 and 14 m in length and were found to contain stored plant material (Contreras 1983; Greer 1966; Tamayo and Frassinetti 1980).

Ctenomys mendocinus Philippi, 1869
Tuco-tuco Mendocino

Measurements

	Mean	Min.	Max.	N	Loc.	Source[a]
TL	223.4	200.0	250	18	A	1
HB	156.7	135.0	180			
T	66.7	58.0	80			
HF	29.2	24.6	32			
E	8.0			1		

[a](1) BA, FM.

Description

This small species shows considerable variation in dorsal color, from gray to gray brown and buffy. The blackish line on the tail is sometimes strongly developed but is sometimes absent. The venter is gray to yellowish gray (pers. obs.).

Distribution

The species is found on the eastern slope of the Andes from Santa Cruz north to Mendoza province and probably into Salta province, Argentina (Mares, Ojeda, and Kosco 1981; Roig 1965; BA) (map 11.138).

Comment

Ctenomys mendocinus includes *C. juris*, *C. fachi*, *C. bergi*, and *C. pundti* (Cabrera 1960).

Ctenomys minutus Nehring, 1887

Measurements

	Mean	Min.	Max.	N	Loc.	Source[a]
HB	174.8	155	210	31	A	1
T	76.9	62	94	30		
HF	32.2	28	35	31		
E	7.0			1		

[a](1) Reig, Contreras, and Piantanida 1965; FM.

Description

Ctenomys minutus exhibits color variation; some specimens are black on the dorsum and venter with no markings, while others are glossy tan, with yellow on the sides and venter and a dark stripe down the tail and darker on the head (Langguth and Abella 1970; AMNH).
Chromosome number: $2n = 50$ (Reig and Kiblisky 1969).

Distribution

This species is found in southern Brazil, Uruguay, and east-central Argentina in the province of Entre Ríos (Honacki, Kinman, and Koeppl 1982; Langguth and Abella 1970) (map 11.139).

Comment

A subspecies of *Ctenomys minutus*, *C. m. rionegrensis*, is regarded by Altuna and Lessa (1985) as a full species.

Ctenomys occultus Thomas, 1920
Tuco-tuco Montaraz

Measurements

	Mean	Min.	Max.	N	Loc.	Source[a]
TL	274.8	264.0	295.0	4	A	1
T	83.5	83.0	85.0			
HF	36.2	35.5	36.8			

[a](1) BA.

Description

This species is smaller than *Ctenomys latro* but is very similar in color and skull morphology. Dorsally it is pale brown to cinnamon brown, with darkening on the crown of some specimens. The venter is not strongly lighter, though it is washed with a paler shade (Thomas 1920a; pers. obs.).

Chromosome number: The 2*n* number is very low, only 22. *Ctenomys occultus* is similar to *C. latro* (Reig and Kiblisky 1968, 1969).

Distribution

The species is found in northwestern Argentina in Tucumán province and neighboring provinces (Reig and Kiblisky 1969; Thomas 1920a; BA) (map 11.140).

Ecology

This tuco-tuco lives in areas with xeric vegetation (Lucero 1983; Reig and Kiblisky 1969).

Ctenomys opimus Wagner, 1848
Tuco-tuco Tujo

Measurements

	Mean	Min.	Max.	N	Loc.	Source[a]
TL	314.0	288	330	14 m	Peru	1
	286.0	252	320	32 f		
T	91.0	78	103	14 m		
	84.0	68	95	32 f		
HF	44.3	42	47	14 m		
	40.1	35	44	32 f		
E	7.0	7	8	4 f		2
Wta	439.0	370	530	14 m		1
	284.0	233	370	32 f		

[a](1) Pearson 1959; (2) Pearson 1951.

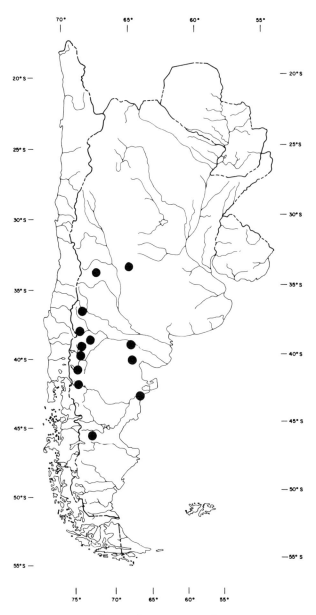

Map 11.138. Distribution of *Ctenomys mendocinus*.

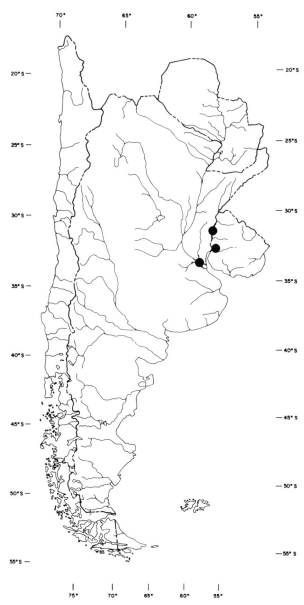

Map 11.139. Distribution of *Ctenomys minutus*.

Description

The anterior surface of the incisors is a bright orange yellow. Males are much larger than females; this is one of the few mammals of the Peruvian puna to accumulate fat. In contrast to other Chilean *Ctenomys*, this species has long, silky pelage. Dorsally it is light brown to buffy gray, lighter on the sides and venter, with a blackish head and sometimes a faint dorsal line (Mann 1978; Osgood 1943; Pearson 1951; Thomas 1900a; pers. obs).

Chromosome number: $2n = 26$ (Gallardo 1979; Reig and Kiblisky 1969).

Distribution

This species is found from southern Peru south to northern Chile and northwestern Argentina (Ho-nacki, Kinman, and Koeppl 1982; Mares, Ojeda, and Kosco 1981; Tamayo and Frassinetti 1980; Thomas 1900a) (map 11.141).

Life History

In southern Peru most births are between October and March; females have three pairs of nipples and average 1.6 embryos (range 1–2; $n = 10$). Animals are capable of breeding in the first year (Pearson 1959).

Ecology

Ctenomys opimus is a tuco-tuco of the altiplano, occurring between 2,500 and 5,000 m. It is abundant where vegetation is sparse and the soil is loose. It is found on sandy, gravely, or cindery soils, usually on slopes.

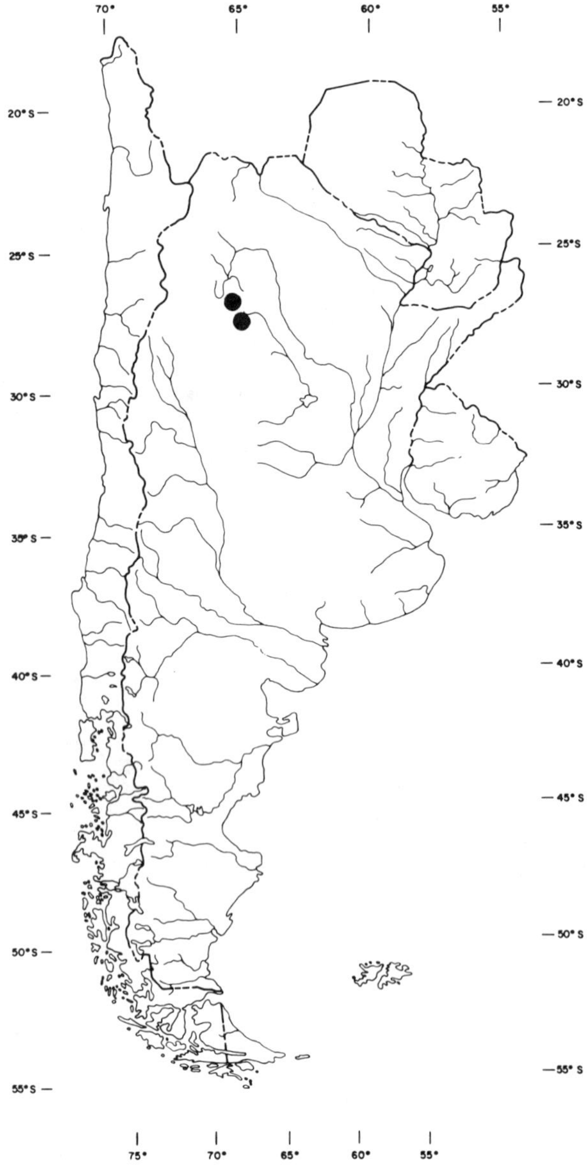

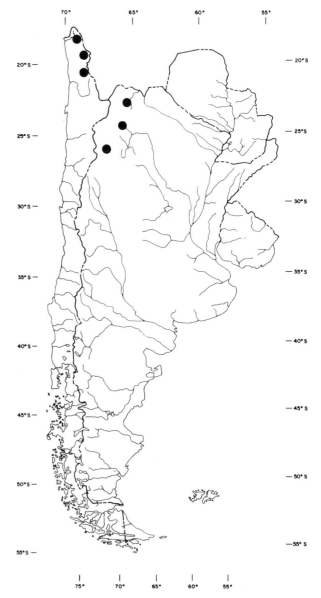

Map 11.140. Distribution of *Ctenomys occultus*.

Map 11.141. Distribution of *Ctenomys opimus*.

Burrows are dug by loosening earth with the fore-feet and sweeping it out of tunnels with the hind feet. Burrow systems usually consist of a single main tunnel, from which short lateral branches diverge every few meters, and include one or more chambers with stored vegetation and other chambers with nests. Only one animal is found per burrow system. The burrows are expanded as vegetation is eaten, and *C. opimus* reveals its presence by stripping large areas of vegetation. In southern Peru *C. opimus* can occur at densities of one to seventeen individuals per acre.

Feeding and digging usually occur between 6:30 and 11:00 A.M.; animals are not seen aboveground very often. The food consists of roots, stems, or leaves of most of the available plants. The burrow is dug to the food source, and the tuco-tuco forages no more than two or three body lengths outside its burrow, bringing the food back to the hole to eat (Mares, Ojeda, and Kosco 1981; Pearson 1951, 1959; Pine, Miller, and Schamberger 1979; Tamayo and Frassinetti 1980).

Ctenomys pearsoni Lessa and Langguth, 1983

Measurements

	Mean	Min.	Max.	N	Loc.	Source[a]
TL	262.4	245.0	277	11	U	1
T	77.3	72.0	82			
HF	35.7	32.5	37			
E	8.3	8.0	9			
Wta	212.0	165.0	250			

[a](1) Lessa and Langguth 1983.

Description

This *Ctenomys* is tannish brown dorsally grading to creamish tan on the venter, which has white inguinal and axillary patches. There is a pronounced lighter band extending from the base of each ear to the chin (pers. obs.).
Chromosome number: $2n = 56$ and 70 (Novello and Lessa 1986).

Distribution

The species is known from Soriano, San José, and Colonia departments in Uruguay (Lessa and Langguth 1983) (map 11.142).

Ecology

This species demonstrates great variation in its burrow construction, but usually there is one principal gallery with a constant depth that forms the major axis. Tunnels have an average of thirteen openings with a range of five to twenty-four, and vary in length from 7.1 to 49.1 m. They range in diameter from 70 to 130 cm. There are one or two nests of dried grass per tunnel (Altuna 1983).

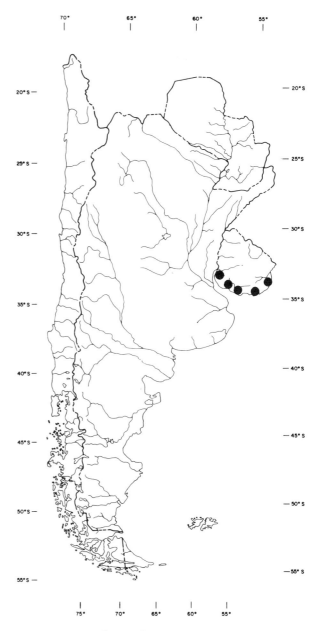

Map 11.142. Distribution of *Ctenomys pearsoni*.

Comment

The skull of *Ctenomys pearsoni* is not as broad or as high as that of *C. torquatus*, and it has a longer muzzle (Lessa and Langguth 1983).

Ctenomys perrensis Thomas, 1898
Tuco-tuco Misionero

Description

Ctenomys perrensis is similar in size to *C. torquatus*. One specimen from Argentina measured HB 200; T 67; HF 31. The dorsum is dark buff or clay colored, heavily mixed with black along the median line of the face and back. The venter is slightly paler: a rich buff with pure white patches on the axillae and groin. From eye to ear and below the ear there is a lighter patch. The upper surfaces of the forefeet and hind feet are white (Thomas 1898d; pers. obs.).

Distribution
This species is found in Entre Ríos, Misiones, and Corrientes provinces, Argentina (Honacki, Kinman, and Koeppl 1982; Thomas 1898d; BA) (map 11.143).

Ctenomys pontifex Thomas, 1918
Tuco-tuco Marrón

Description
Measurements from one skin were HB 183; T 77; HF 34. *C. pontifex* is a medium-sized *Ctenomys*, a

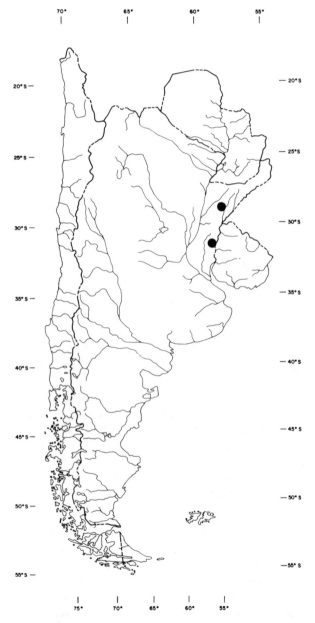

Map 11.143. Distribution of *Ctenomys perrensis*.

uniform drab brown dorsally without darker markings. Ventrally it is paler and more buffy; the tail is brown above and whitish below (Thomas 1918b).

Distribution
This species is found in at least Mendoza and San Luis provinces, Argentina (Honacki, Kinman, and Koeppl 1982; Roig 1965; Thomas 1918b) (map 11.144).

Ctenomys porteousi Thomas, 1916
Tuco-tuco Acanelado

Measurements

	Mean	Min.	Max.	N	Loc.	Source[a]
HB	186.3	174.0	205.0	10	A	1
T	77.1	68.0	89.0			
HF	30.4	28.4	33.5			
Wta	192.4	115.0	240.0			

[a](1) Contreras and Reig·1965.

Description
Dorsally this species is cinnamon brown lined with blackish. The middle of the back is dark in some specimens, but the black is not sharply defined. The tops of the muzzle and crown are also blackish. The venter, forefeet, and hind feet are lighter, and the tail is light brown with a darker pencil (Thomas 1916b).
Chromosome number: $2n = 48$ (Reig and Kiblisky 1969).

Distribution
Ctenomys porteousi is found in Buenos Aires province and possibly adjoining parts of La Pampa province, Argentina (Reig and Kiblisky 1969) (map 11.145).

Ctenomys saltarius Thomas, 1912
Tuco-tuco Salteño

Description
One specimen from Argentina measured HB 203; T 80; HF 37.5; E 20.3; Wta 230 (Mares, Ojeda, and Kosco 1981).

Distribution
The species is found in the Argentine provinces of Salta and Jujuy (Honacki, Kinman, and Koeppl 1982) (map 11.146).

Ecology
Ctenomys saltarius inhabits the hillsides and valleys of the northern Monte Desert and is especially common in *Larrea* flats and in areas with *Prosopis* on soft soils. It feeds on creosote and other small shrubs by opening its burrows at the base of the plants. It vocalizes readily (Mares, Ojeda, and Kosco 1981).

Ctenomys sericeus J. A. Allen, 1903
Tuco-tuco Enano

Measurements

	Mean	Min.	Max.	N	Loc.	Source[a]
TL	200.0	195	208	5 m	A	1
	210.0			1 f		
T	56.6	51	62	5 m		
	60.0			1 f		
HF	26.2	25	28	5 m		
	27.0			1 f		

[a](1) Allen 1903.

Description
This species has short, soft, silky fur. The yellowish gray dorsum is strongly varied with black; the flanks and venter are buff; the sides of the nose are yellowish brown; the upper surfaces of the feet are dingy gray with a yellowish cast; and the tail is pale yellowish with a median dusky stripe along half of the upper surface (Allen 1903).

Distribution
Ctenomys sericeus is found in Santa Cruz, Chubut, and Río Negro provinces, Argentina (Honacki, Kinman, and Koeppl 1982; Thomas 1929) (map 11.147).

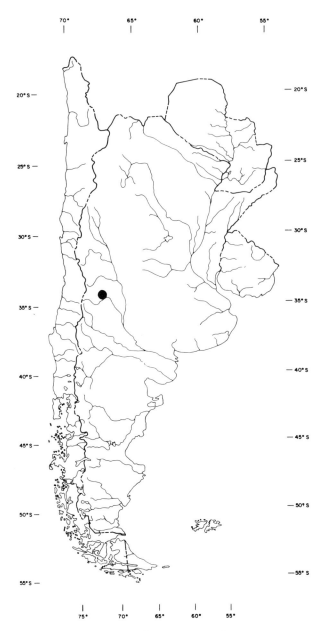

Map 11.144. Distribution of *Ctenomys pontifex*.

Ctenomys talarum Thomas, 1898
Tuco-tuco de los Talares

Measurements

	Mean	Min.	Max.	N	Loc.	Source[a]
HB	162.3	146.7	177.0	45	A	1
T	61.0	56.5	66.0			
HF	27.6	25.5	29.5			
Wta	133.0	92.0	193.0			
	186.0			40 m		2
	128.0			36 f		

[a](1) Contreras and Reig 1965; (2) Pearson et al. 1968.

Description
This small tuco-tuco has a very dark chestnut gray to black dorsum. The head is often quite black. There is a small whitish patch at the base of each ear, and the inguinal and axillary areas are white. The venter is cream to yellow, the tail is bicolored, and the incisors are deep yellow orange (Thomas 1898c; pers. obs.).

Chromosome number: $2n = 48$; the species exhibits autosomal polymorphism (Ortells et al. 1984; Reig and Kiblisky 1969).

Distribution
Ctenomys talarum is found along the coast of Buenos Aires province, Argentina, possibly extending into Santa Fe province (Honacki, Kinman, and Koeppl 1982; Reig and Kiblisky 1969) (map 11.148).

Life History
In captivity the gestation period is 102 days, litter size is five (range 1–7), lactation lasts five weeks, and age of first reproduction is eight months. Females come into estrus twice a year (Contreras and Reig 1965; Weir 1974a).

Ecology
This species prefers well-vegetated, firm soils, in contrast to *C. australis*, which prefers sandy soils closer to the coast. In grazed pasture it has been found at a density of 207/ha. Its burrows seem to be shorter than 25 m, measure 6--7 cm in diameter, and are usually no less than 30 cm deep. Up to three nests have been found in one burrow. Like *C. tor-*

quatus, C. talarum opens and closes the entrances to its tunnels according to wind direction.

Some investigators report frequently finding a male and a female in the same tunnel, though others say only one animal occurs per burrow system. It is probable that animals are found together only during the breeding season. This species vocalizes often: the song is apparently sung only by the male; it consists of thirty to seventy-five "tucs" and lasts ten to twenty-five seconds. This song may serve to space males. Males seem easier to trap than females. Burrowing owls are reported to eat young *C. talarum* but never adults (Contreras and Reig 1965; Pearson et al. 1968; Weir 1971).

Ctenomys torquatus Lichtenstein, 1830
Tuco-tuco de Collar

Measurements

	Mean	Min.	Max.	N	Loc.	Source[a]
TL	266.8	235.0	289.0	19 m	U	1
	262.7	230.0	283.0	38 f		
T	76.4	61.0	86.0	19 m		
	77.4	65.0	93.0	38 f		
HF	37.5	35.0	40.0	19 m		
	36.4	34.0	39.0	38 f		
E	7.3	6.0	9.0	19 m		
	6.7	5.0	9.0	38 f		
Wta	228.9	178.3	303.7	19 m		
	190.1	156.0	236.0	38 f		

[a](1) Barlow 1965.

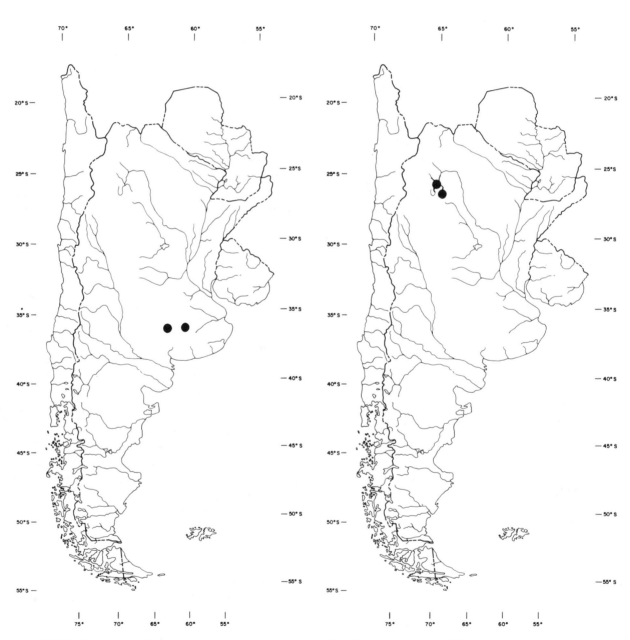

Map 11.145. Distribution of *Ctenomys porteousi*.

Map 11.146. Distribution of *Ctenomys saltarius*.

Description

This tuco-tuco has a dark yellow to mahogany dorsum, sometimes with a darker line along the spine. The flanks and venter are yellowish white, and most individuals have a whitish or yellowish collar; in the inguinal and axillary regions many have light patches. The tail is usually a uniform dark brown. There is great inter- and intrapopulational variation in color (Barlow 1965; Fernández 1965; Freitas and Lessa 1984; Langguth and Abella 1970; pers. obs.).

Chromosome number: $2n = 68$, one of the highest recorded for the genus. However, more recent work has shown that $2n = 56-70$ and that *C. torquatus* is probably a superspecies (Altuna and Lessa 1985; Lessa and Altuna 1984; Reig and Kiblisky 1969).

Distribution

Ctenomys torquatus is found in extreme southern Brazil through Uruguay and into Argentina in the provinces of Entre Ríos and Corrientes (Honacki, Kinman, and Koeppl 1982; Reig and Kiblisky 1969) (map 11.149).

Life History

Females are monoestrous, breeding from June to October. Young are born in September to December after a gestation period of about 105 days. Females

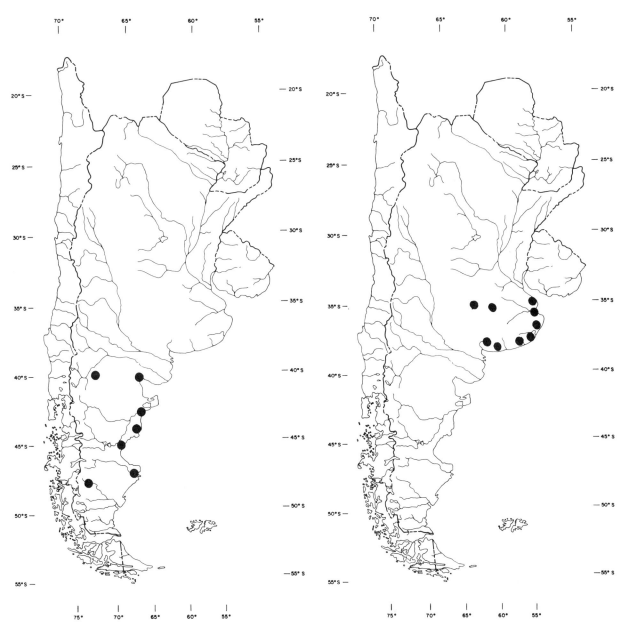

Map 11.147. Distribution of *Ctenomys sericeus*.

Map 11.148. Distribution of *Ctenomys talarum*.

have two or three embryos (Barlow 1969; Tálice, Mosera, and Sprechmann 1972).

Ecology

In Argentina this species prefers high dry areas with sandy soil and is common in open palm savannas. In Uruguay its distribution is probably limited by soil composition, since *C. torquatus* prefers sandy, rock-free soil and disappears in areas under cultivation. The density of this species depends on the density of the food plants. The animals live singly except during the reproductive season, when they are found in pairs. Twice as many females as males were captured in one study (*n* = 180). In six study sites the male-to-female ratio varied from 1 : 2.5 to 1 : 14.5. This species has good vision and is aggressive toward conspecifics.

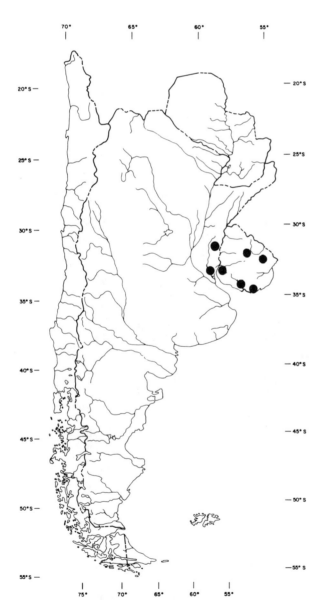

Map 11.149. Distribution of *Ctenomys torquatus*.

Burrows nowhere exceed 60 cm in depth, and they average 4.5 m long and 6–10 cm in diameter. Food is stored in side galleries. *C. torquatus* is diurnal in its surface activity. Animals have never been seen more than two-thirds of the way out of a burrow entrance when foraging or resting. They are strictly herbivorous, feeding primarily at the surface and eating entire plants in the vicinity of the burrow mouth (Barlow 1969; Crespo 1982a; Fernández 1965; Freitas and Lessa 1984; Tálice and Momigliano 1954; Tálice, Caprio, and Momigliano 1954; Tálice, Mosera, and Sprechmann 1972).

Ctenomys tuconax Thomas, 1925
Tuco-tuco Robusto

Measurements

	Mean	Min.	Max.	N	Loc.	Source[a]
TL	255.0	230	285	6	A	1
HB	178.3	150	200			
T	76.7	70	85			
HF	37.5	32	42			
E	5.0	5	5	5		

[a](1) Thomas 1925; FM.

Description

This large chestnut *Ctenomys* has a median line of black, and the venter is similar to the dorsum but washed with yellow. The forefeet and hind feet are thinly haired and dull whitish, and the tail is pale brown. Some individuals have white axillary patches (Thomas 1925).

Chromosome number: $2n = 61$ (Reig and Kiblisky 1968, 1969).

Distribution

Ctenomys tuconax is found in Tucumán province, Argentina (Honacki, Kinman, and Koeppl, 1982) (map 11.150).

Ecology

In Tucumán province *C. tuconax* is found in the humid plains to 3,000 m (Lucero 1983; Reig and Kiblisky 1969).

Ctenomys tucumanus Thomas, 1900
Tuco-tuco Tucumano

Measurements

	Mean	Min.	Max.	N	Loc.	Source[a]
TL	243.8	238.0	249	5	A	1
HB	173.2	170.0	177			
T	70.6	68.0	72			
HF	30.9	30.1	32			

[a](1) Thomas 1900b; BA, FM.

Description

This medium to fairly small *Ctenomys* has a brownish fawn to dark brown dorsum gradually shading to light brownish yellow on the sides and venter.

The forehead is frequently very dark. The venter is sometimes washed with white. The midline of the venter and the axillary and inguinal regions may be almost white. The tail is definitely bicolored. (Thomas 1900b; pers. obs.).

Chromosome number: $2n = 28$ (Reig and Kiblisky 1969).

Distribution

The species is found only in the Argentine province of Tucumán (Honacki, Kinman, and Koeppl 1982) (map 11.151).

Ecology

Ctenomys tucumanus inhabits the humid areas of the plains of central Tucumán province (Reig and Kiblisky 1969).

Ctenomys validus Contreras, Roig and Suzarte, 1977
Tuco-tuco Válido

Measurements

	Mean	Min.	Max.	N	Loc.	Source[a]
HB	204.0	186.0	218.0	8 m	A	1
	187.0	179.0	194.0	6 f		
T	85.0	80.0	96.0	8 m		
	82.0	79.0	84.0	6 f		
HF	39.6	36.8	43.5	8 m		
	36.5	33.0	38.8	6 f		
E	8.7	8.0	9.2	8 m		
	8.3	7.7	8.8	6 f		
Wta	266.0	186.7	371.0	8 m		
	199.2	176.2	220.0	6 f		

[a](1) Contreras, Roig, and Suzarte 1977.

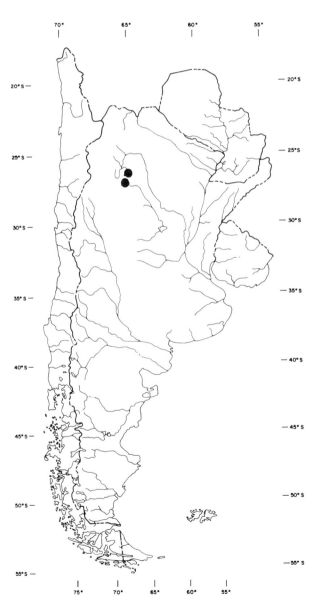

Map 11.150. Distribution of *Ctenomys tuconax*.

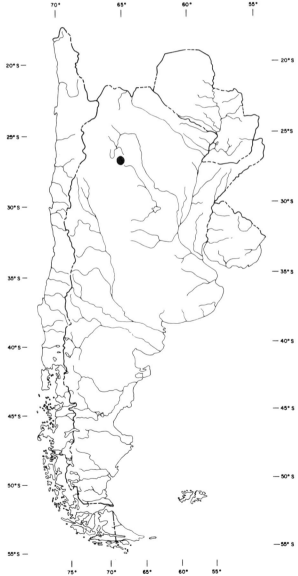

Map 11.151. Distribution of *Ctenomys tucumanus*.

Description

This relatively large *Ctenomys* is generally gray with a yellowish wash on the dorsum and is lighter on the venter. Behind the ear and on the throat there is a conspicuously lighter area. The nose has a dark stripe. The relatively long tail is bicolored and has a dark pencil (Contreras, Roig, and Suzarte 1977).

Distribution

Ctenomys validus is known only from Mendoza province, Argentina (Honacki, Kinman, and Koeppl 1982) (map 11.152).

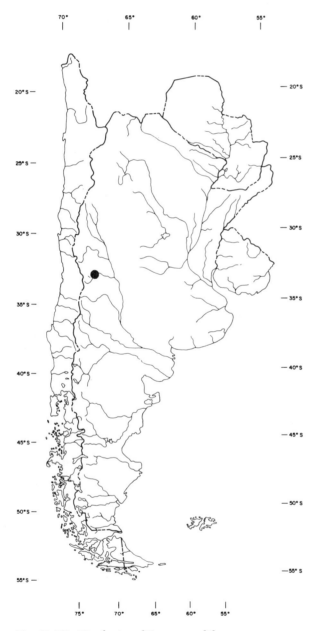

Map 11.152. Distribution of *Ctenomys validus*.

FAMILY ABROCOMIDAE

Diagnosis

The dental formula is I 1/1, C 0/0, P 1/1, M 3/3. Members of this family differ from the Octodontidae in that the lacrimal canal opening on the rostrum is near the dorsal root of the zygoma. The palatal foramina are long and narrow. These rodents are approximately the size of a large rat; head and body length ranges from 150 to 250 mm. They have short limbs, a rather pointed snout, and large ears (Woods 1984).

Distribution

The family is found from southern Peru to Chile and northwestern Argentina. *Abrocoma bennetti* is found in lowland hills of Chile up to 3,700 m; *A. cinerea* occurs at higher elevations, usually in cold, dry habitats.

Natural History

These rodents are vegetarians and live in burrows. *Abrocoma bennetti* is unique in that it uses the burrows of *Octodon degus*. Little is known concerning the natural history of *A. cinerea*, but it has been studied in captivity. In common with other hystricognath rodents, it gives birth to precocial young after a lengthy gestation period.

Genus *Abrocoma* Waterhouse, 1837
Abrocoma bennetti Waterhouse, 1837

Measurements

	Mean	Min.	Max.	N	Loc.	Source[a]
TL	355.8	300	405	12	C	1
HB	218.0	179	260			
T	137.8	121	168			
HF	34.3	30	37			
E	32.5	26	36			
Wta	250.5	220	307	6		

[a](1) Pine, Miller, and Schamberger 1979; Wolffsohn 1923; Santiago.

Description

Abrocoma bennetti can be separated from *A. cinerea* by its longer tail and larger ears. The ears are sparsely haired and the tail is well haired. The feet have only four digits, and the soles are granulated. It is brownish gray with a gray to gray brown venter (plate 17) (Mann 1978; Osgood 1943; pers. obs.).

Distribution

The species is found in Chile from Copiapó to the area of the Río Bío-Bío (Honacki, Kinman, and Koeppl 1982) (map 11.153).

Life History

Females give birth twice a year to litters that range from one to six (Mann 1978).

Ecology

Abrocoma bennetti is typical of the Mediterranean area of Chile but extends into the arid regions to the north. It prefers rocky areas with dense brush from sea level to 2,000 m. This is a highly commensal species and is usually found living in burrows made by *Octodon degus*, *Aconaemys fuscus*, and *Chinchilla lanigera*. It will also build its own tunnels, often under rocks. In one study site near Santiago, Chile, *A. bennetti* reached a density of 1.3 ha, though the population size was highly variable throughout the year.

Abrocoma bennetti is almost strictly nocturnal; it is a good rock climber and also climbs bushes in search of food. It eats shrubs, grasses, forbs, and seeds. In one area its diet was dominated by the foliage of *Chenopodium* (51.9% by volume) and shrub connective tissue (22.8%) (Glanz 1977b; Iriarte, Contreras, and Jaksić 1989; Mann 1978; Meserve 1981a, b; Meserve, Martin, and Rodríguez 1983; Miller 1980; Schamberger and Fulk 1974; Tamayo and Frassinetti 1980).

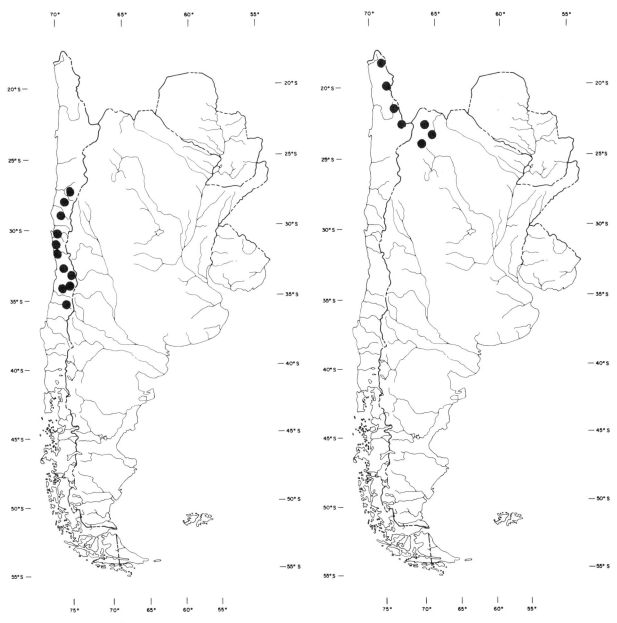

Map 11.153. Distribution of *Abrocoma bennetti*.

Map 11.154. Distribution of *Abrocoma cinerea*.

Abrocoma cinerea Thomas, 1919
Rata Chinchilla

Measurements

	Mean	Min.	Max.	N	Loc.	Source[a]
TL	234.9	207.0	262.0	11	A, Peru	1
HB	173.9	157.0	192.0			
T	61.0	50.0	74.0			
HF	25.9	22.0	35.0			
E	24.3	21.6	29.5			

[a](1) Pearson 1951. Thomas 1919a; BA.

Description

Abrocoma cinerea is smaller than *A. bennetti* and has a much shorter tail. This large rat has large, rounded ears, delicate incisors, and granular soles on its hind feet. It is silver gray with a paler venter (Pearson 1951; pers. obs.).

Distribution

The species is found from southwestern Peru south to the Chilean departments of Parinacota and El Loa and into Argentina along the Andes from Salta province and perhaps as far south as Mendoza province (Honacki, Kinman, and Koeppl 1982; Roig 1965; Tamayo and Frassinetti 1980; BA) (map 11.154).

Life History

In captivity the gestation period is 116 days, litter size averages 2.2 (range 1–3), and neonates weigh 22 g (Kleiman, Eisenberg, and Maliniak 1979).

Ecology

This is a species of arid, rocky zones at elevations up to 5,000 m. In tola habitat in Peru it was found at a density of 0.17 ha. It lives in small colonies and digs burrows about 5 cm in diameter at the bases of boulders and shrubs. *A. cinerea* is not as arboreal as *A. bennetti* (Koford 1955; Lucero 1983; Mann 1978; Pearson and Ralph 1978; Tamayo and Frassinetti 1980).

FAMILY ECHIMYIDAE

Spiny Rats, Ratas Espinosas, Casiraguas

Description

The dental formula is I 1/1, C 0/0, P 1/1, M 3/3. These rat-sized hystricognath rodents usually have long tails; as the common name implies, many species, but not all, have spinescent hairs interspersed in their dorsal pelage. The dorsal pelage is highly variable in color but is usually some shade of brown to almost black. The venter often contrasts sharply with the dorsum, being white or yellowish.

Distribution

Echimyid rodents are distributed from Honduras south through the Isthmus over much of South America to northern Argentina.

Natural History

The best-studied echimyid rodents are members of the genus *Proechimys*. These terrestrial forms use hollow logs, natural crevices, or burrows they construct to cache food, bear their young, and seek refuge. Some terrestrial genera of southern South America are poorly known, including *Clyomys* and *Euryzygomatomys*. *Proechimys* may be one of the

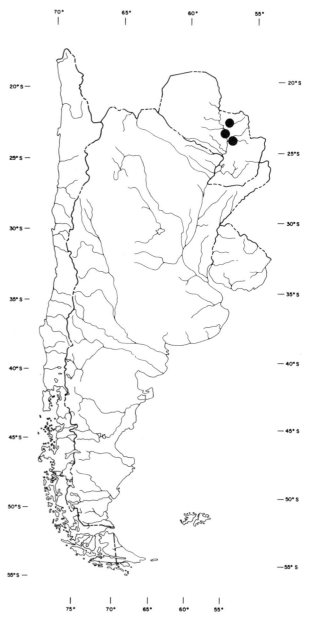

Map 11.155. Distribution of *Clyomys laticeps*.

most abundant rodents in the lowland tropical rain forest of the northern Neotropics (Emmons 1982). Other spiny rat genera are strongly arboreal, including *Dactylomys*, *Kannabateomys*, *Echimys*, *Diplomys*, *Isothrix*, *Mesomys*, and *Thrinacodus*. The arboreal genera have not been studied in as much detail as the terrestrial forms.

Genus *Clyomys* Thomas, 1916
Clyomys laticeps (Thomas, 1909)

Measurements

	Mean	Min.	Max.	N	Loc.	Source[a]
HB	187	107	230	16	Br, P	1
T	73	48	80	12		
HF	32	21	39	16		

[a](1) Bishop 1974.

Description

In this medium-sized spiny echimyid the claws are well developed and the ears and tail are short. It is agouti buffy gray brown or grizzled rufous black, blacker on the back and becoming strongly rufous on the rump and head. The venter is dull whitish or buffy, and the tail is dark brown (Bishop 1974; Moojen 1952b; pers. obs.).

Distribution

Clyomys laticeps is found in southern Brazil and into eastern Paraguay (Honacki, Kinman, and Koeppl 1982; Moojen 1952b) (map 11.155).

Life History

In Mato Grosso state, Brazil, three females obtained from June to December each had a single embryo (Bishop 1974).

Ecology

This highly fossorial species is one of the few South American rodents apparently confined to savanna. In the Brazilian Pantanal it is a common member of the small-mammal fauna, inhabiting edge habitats and frequently caught in gardens. It is colonial (Alho et al. 1987; Bishop 1974).

Genus *Euryzygomatomys* Goeldi, 1901
Euryzygomatomys spinosus (G. Fischer, 1814)
Rata Guira

Measurements

	Mean	Min.	Max.	N	Loc.	Source[a]
TL	245.3	215	270	6	Br	1
HB	195.0	167	224			
T	50.3	46	55			
HF	35.3	34	36			
E	16.8	16	17	4		
Wta	187.5	165	210	2		2

[a](1) FM; (2) Stallings 1988a.

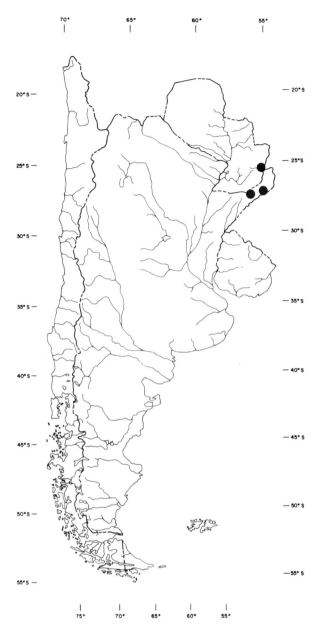

Map 11.156. Distribution of *Euryzygomatomys spinosus*.

Description

This species is somewhat less spiny than *Clyomys*, with a shorter tail, a longer tooth row, and broad incisors; though the bullae are somewhat inflated, they are less so than in *Clyomys*. This stout rat has short ears. It is agouti brown and black; the venter is more tan than the dorsum and is marked with distinctive pure white hourglass-shaped blotches. The chin and throat are often heavily washed with tan (plate 17) (Bishop 1974; pers. obs.).

Distribution

Euryzygomatomys spinosus is found in southern and eastern Brazil, in northeastern Argentina, and in Paraguay (Honacki, Kinman, and Koeppl 1982) (map 11.156).

Ecology

In Minas Gerais state, Brazil, this species was caught in wet meadow habitat (Stallings 1988b).

Genus *Kannabateomys* Jentink, 1891
Kannabateomys amblyonyx (Wagner, 1845)
Rata Tacuarera

Measurements

	Mean	Min.	Max.	N	Loc.	Source[a]
TL	568.9	527	620	9	A	1
HB	247.2	225	290			
T	321.7	301	340			
HF	22.7	20	25	7		
Wta	475.0			1		

[a](1) Crespo 1950, 1982b.

Description

This large rodent is unmistakable because of its size and its very long tail, which is well haired along the basal 6 cm, becoming less well haired and ending in a pronounced pencil (plate 17). The dorsum is light agouti brown along the midline, grading to reddish brown on the sides. The venter and the sides of the face are reddish white (pers. obs.).

Distribution

Kannabateomys amblyonyx is found in southern Brazil, into eastern Paraguay, and in the northeastern province of Misiones, Argentina (Honacki, Kinman, and Koeppl 1982) (map 11.157).

Ecology

This nocturnal, arboreal rodent appears to be confined to moist tropical forest, where it lives in thickets of bamboo, especially *Guadua angustifolia* when it grows near water. When alarmed this species makes loud squeaks or shrieks. A variant of these calls may serve a spacing function. In its ecology it seems similar to *Dactylomys* (see Emmons 1981; Crespo 1950, 1982b).

Genus *Proechimys* J. A. Allen, 1899
Proechimys longicaudatus (Rengger, 1830)

Measurements

	Mean	Min.	Max.	N	Loc.	Source[a]
TL	374.3	331	450	7	Br, P	1
HB	221.9	187	250			
T	152.4	121	200			
HF	47.1	41	55			
E	21.0			1		

[a](1) Allen 1916a,b; Moojen 1948; Vieira 1945; PCorps.

Description

This large rat has a glossy chestnut dorsum, more orange on the sides, and a sharply demarcated pure white belly and chin (plate 17). The fur is not heav-ily bristled. The relatively short tail is bicolored and lightly haired and is frequently broken off (pers. obs.).

Chromosome number: In Peru, $2n = 28$; FN $= 50$ (Patton and Gardner 1972).

Distribution

The species ranges from lowland eastern Peru through Brazil and into Paraguay (Honacki, Kinman, and Koeppl 1982; PCorps) (map 11.158).

Life History

Proechimys longicaudatus is an Amazonian basin form not closely related to *P. setosus*, found in Minas Gerais state, Brazil. In the region covered by this volume *Proechimys* barely extends its range into

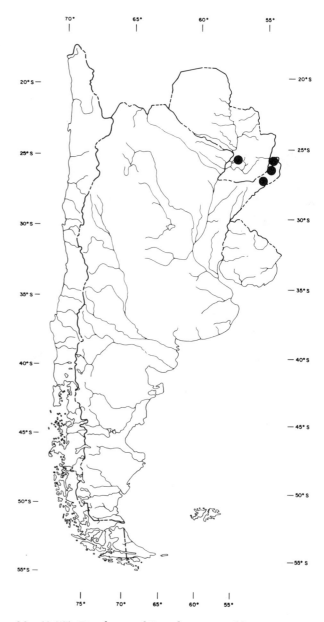

Map 11.157. Distribution of *Kannabateomys amblyonyx*.

northeastern Paraguay. The natural history of several species of *Proechimys* is reasonably well known, including *P. cuvieri*, *P. guayanensis*, *P. semispinosus*, and *P. setosus* (see Eisenberg 1989; Emmons 1982; and Fonseca and Kierulff 1989 for further discussion). Most species of *Proechimys* studied to date are terrestrial herbivores sheltering in natural cavities or burrows of their own construction. The female gives birth to a litter of three to five precocial young after a two-month gestation period. They tend to be nocturnal and may be important contributors to the ecology of the communities they occupy, since they feed on fungi, fruits, and seeds. Much more research is necessary to determine the role of this species within its ecosystem.

Genus *Thrichomys* Trouessart, 1880
Thrichomys aperoides (Lund, 1839)

Measurements

	Mean	Min.	Max.	N	Loc.	Source[a]
TL	440.3	408	468	12	P	1
HB	236.9	225	256			
T	203.4	182	226			
HF	47.2	43	53	13		
E	24.6	23	27			
Wta	338.7	247	500	11		

[a](1) PCorps, UM.

Description

In this large, stocky rat the tail is long, well pencilled, strongly bicolored, and very well furred. The ears are relatively small. It has remarkably long vi-

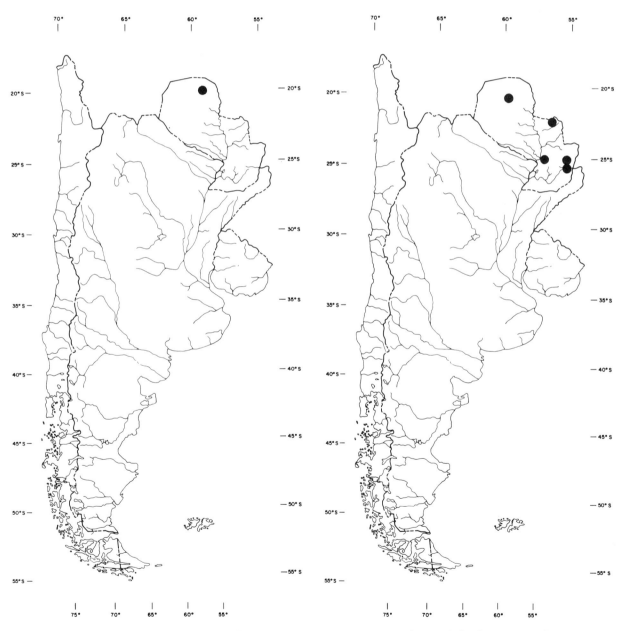

Map 11.158. Distribution of *Proechimys longicaudatus*.

Map 11.159. Distribution of *Thrichomys apereoides*.

brissae and long vibrissae between the ears and eyes. The dorsum is agouti greenish brown, with more tan on the sides and a grayish white venter. There are often white spots above the eyes (pers. obs.).

Distribution

The species is found in southern Brazil south to Paraguay (Alho et al. 1987; Honacki, Kinman, and Koeppl 1982; UM) (map 11.159).

Life History

In the Brazilian caatinga there is some reproductive activity throughout the year, and two or three litters are produced each year; litters average 3.1 (range 1–6). The gestation period is eighty-nine days. The young are very precocial, weighing about 29 g, and start eating solid food on the first day. In captivity lactation lasts fifty to sixty days. The age of first reproduction is between seven and nine months (Kleiman, Eisenberg, and Maliniak 1979; Roberts, Thompson, and Cranford 1988; Strelein 1982a,c).

Ecology

Thrichomys aperoides is usually associated with rocky outcroppings, though a recent study in the Brazilian Pantanal found it abundant in patches of cerrado with no rocks. Its nests are typically under large rocks, and when scared this species hides in the cracks in rocks. It is crepuscular and moderately good at conserving water. Like *Proechimys*, there is a zone of fracture so that the tail is easily detached from the body when grabbed (Alho et al. 1987; Myers 1982; Strelein 1982a,b).

References

Alho, C. J. R., Z. M. Campos, and H. C. Gonçalves. 1989. Ecology, social behavior and management of the capybara (*Hydrochaeris hydrochaeris*) in the Pantanal of Brazil. In *Advances in Neotropical mammalogy*, ed. K. H. Redford and J. F. Eisenberg, 163–94. Gainesville, Fla.: Sandhill Crane Press.

Alho, C. J. R., T. E. Lacher, Jr., Z. M. S. Campos, and H. C. Gonçalves. 1987. Mamíferos da Fazenda Nhumirim, sub-regiâo de Nhecolândia, Pantanal do Mato Grosso do Sul: 1. Levantamento preliminar de espécies. *Rev. Brasil. Zool., São Paulo* 4(2): 151–64.

Alho, C. J. R., and M. J. de Souza. 1982. Home range and use of space in *Zygodontomys lasiurus* (Cricetidae, Rodentia) in the cerrado of central Brazil. *Ann. Carnegie Mus.* 51:127–32.

Allen, J. A. 1901. New South American Muridae and a new *Metachirus*. *Bull. Amer. Mus. Nat. Hist.* 14:405–12.

———. 1903. Descriptions of new rodents from southern Patagonia, with a note on the genus *Euneomys coues,* and an addendum to article 4, on Siberian mammals. *Bull. Amer. Mus. Nat. Hist.* 19:185–96.

———. 1916a. Mammals collected on the Roosevelt Brazilian expedition, with field notes by Leo E. Miller. *Bull. Amer. Mus. Nat. Hist.* 34:559–610.

———. 1916b. New mammals collected on the Roosevelt Brazilian expedition. *Bull. Amer. Mus. Nat. Hist.* 35:523–30.

Altuna, C. A. 1983. Sobre la estructura de las construcciones de *Ctenomys pearsoni* Lessa y Langguth, 1983 (Rodentia, Octodontidae). *Res. Com. J. C. Nat., Montevideo* 3:70–72.

Altuna, C. A., and E. P. Lessa. 1985. Penial morphology in Uruguayan species of *Ctenomys* (Rodentia: Octodontidae). *J. Mammal.* 66:483–88.

Anderson, S., and N. Olds. 1989. Notes on Bolivian mammals. 5. Taxonomy and distribution of *Bolomys* (Muridae, Rodentia). *Amer. Mus. Novitat.* 2935:1–22.

Atalah G., A., W. Sielfeld K., and C. Venegas C. 1980. Antecedentes sobre el nicho trófico de *Canis g. griseus* Gray 1836 en Tierra del Fuego. *Anal. Inst. Pat. Punta Arenas* (Chile) 11:259–71.

Avila-Pires, F. D. de. 1972. A new subspecies of *Kunsia fronto* (Winge, 1888) from Brazil (Rodentia, Cricetidae). *Rev. Brasil. Biol.* 32(3): 419–22.

Barlow, J. C. 1965. Land mammals from Uruguay: Ecology and zoogeography. Ph.D. diss., University of Kansas.

———. 1969. Observations on the biology of rodents in Uruguay. *Life Sci. Contrib., Roy. Ontario Mus.* 75:1–59.

Bárquez, R. M. 1976. Nuevo registro de distribución de *Oxymycterus paramensis* (Mammalia, Rodentia, Cricetidae). *Neotrópica* 22(68): 115.

———. 1983. La distribución de *Neotomys ebriosus* Thomas en la Argentina y su presencia en la provincia de San Juan (Mammalia, Rodentia, Cricetidae). *Hist. Nat.* 3(22): 189–91.

Bárquez, R. M., D. F. Williams, M. A. Mares, and H. H. Genoways. 1980. Karyology and morphometrics of three species of *Akodon* (Mammalia: Muridae) from northwestern Argentina. *Ann. Carnegie Mus.* 49:279–403.

Baskin, J. A. 1978. *Bensonomys, Calomys,* and the origin of the phyllotine group of Neotropical cricetines (Rodentia: Cricetidae). *J. Mammal.* 59: 125–35.

Bianchi, N. O., S. Merani, and M. Lizarralde. 1979. Cytogenetics of the South-American *Akodon* rodents (Cricetidae). 6. Polymorphism in *Akodon dolores* (Thomas). *Genetica* 50(2): 99–104.

Bishop, I. R. 1974. An annotated list of caviomorph rodents collected in north-eastern Mato Grosso, Brazil. *Mammalia* 38(3): 489–502.

Borchert, M., and R. L. Hansen. 1983. Effects of flooding and wildfire on valley side wet campo rodents in central Brazil. *Rev. Brasil. Biol.* 43(3): 229–40.

Cabrera, A. L. 1953. *Los roedores argentinos de la familia Caviidae*. Publicación 6. Buenos Aires: Facultad de Agronomía y Veterinaria, Escuela de Veterinaria.

———. 1960. Catálogo de los mamíferos de América del Sur. 2. Sirenia-Perissodactyla-Artiodactyla-Lagomorpha-Rodentia-Cetacea. *Rev. Mus. Argent. Cienc. Nat. "Bernardino Rivadavia," Zool.* 4(2): 308–732.

Calhoun, J. B. 1962. *The ecology and behavior of the Norway rat*. Washington, D.C.: National Institutes of Health.

Cant, J. G. H. 1977. A census of the agouti (*Dasyprocta punctata*) in seasonally dry forest at Tikal, Guatemala, with some comments on strip censusing. *J. Mammal.* 58:688–90.

Carleton, M. D. 1980. Phylogenetic relationships in neotomine-peromyscine rodents (Muroidea) and a reappraisal of the dichotomy with New World Cricetinae. *Misc. Publ. Mus. Zool. Univ. Michigan* 157:1–146.

Carleton, M. D., and G. G. Musser. 1984. Muroid rodents. In *Orders and families of Recent mammals of the world*, ed. S. Anderson and J. Knox Jones, Jr., 289–379. New York: John Wiley.

Charles-Dominique, P., M. Atramentowicz, M. Charles-Dominique, H. Gérard, A. Hladik, C. M. Hladik, and M. F. Prévost. 1981. Les mammifères frugivores arboricoles nocturnes d'une forêt guyanaise: Interrelations plantes-animaux. *Rev. Ecol. (Terre et Vie)* 35:342–435.

Chitty, D., and H. N. Southern, ed. 1954. *Control of rats and mice*. 3 vols. Oxford: Clarendon Press.

Collett, S. F. 1981. Population characteristics of *Agouti paca* (Rodentia) in Colombia. *Publ. Mus. Michigan State Univ., Biol. Ser.* 5:487–601.

Contreras, J. R. 1968. *Akodon molinae* una nueva especie de ratón de campo del sur de la provincia de Buenos Aires. *Zool. Platense* 1(2): 9–12.

———. 1972a. El home range en una población de *Oryzomys longicaudatus* Philippi (Landbeck) (Rodentia, Cricetidae). *Physis* 31(83): 353–61.

———. 1972b. Nuevos datos acerca de la distribución de algunos roedores en las provincias de Buenos Aires, La Pampa, Entre Ríos, Santa Fe y Chaco. *Neotrópica* 18(55): 27–30.

———. 1980. Sobre el limite occidental de la distribución geográfica del cuis grande *Cavia aperea pamparum* en la Argentina. *Hist. Nat.* 1(11): 73–74.

———. 1982a. Mamíferos de Corrientes. 1. Nota preliminar sobre la distribución de algunas especies. *Hist. Nat.* 2(10): 71–72.

———. 1982b. Nota acerca de *Bolomys temchuki* (Massoia, 1982) en el noreste argentino con la descripción de dos nuevas subspecies (Rodentia, Cricetidae). *Hist. Nat.* 2(20): 174–76.

———. 1982c. *Graomys griseoflavus* (Waterhouse, 1837) en la provincia del Chaco, república Argentina (Rodentia, Cricetidae). *Hist. Nat.* 2(27): 252.

Contreras J. R., and L. M. Berry. 1982a. *Ctenomys bonettoi*, una nueva especie de tucu-tucu procedente de la provincia del Chaco, república Argentina (Rodentia, Octodontidae): Diagnosis preliminar. *Hist. Nat.* 2(14): 123–24.

———. 1982b. *Ctenomys argentinus*, una nueva especie de tucu-tucu procedente de la provincia del Chaco, república Argentina (Rodentia, Octodontidae). *Hist. Nat.* 2(20): 165–73.

———. 1983. Notas acerca de los roedores del género *Oligoryzomys* de la provincia del Chaco, república Argentina (Rodentia, Cricetidae). *Hist. Nat.* 3(15): 145–48.

Contreras, J. R., and A. N. C. de Contreras. 1984. Diagnosis preliminar de una nueva especie de "auguya-tutu" (género *Ctenomys*) para la provincia de Corrientes, Argentina (Mammalia: Rodentia). *Hist. Nat.* 4(13): 131–32.

Contreras, J. R., and E. R. Justo. 1974. Aportes a la mastozoología pampeana. 1. Nuevas localidades para roedores Cricetidae (Mammalia, Rodentia). *Neotrópica* 20(62): 91–96.

Contreras, J. R., and O. A. Reig. 1965. Datos sobre la distribución del género *Ctenomys* (Rodentia, Octodontidae) en la zona costera de la provincia de Buenos Aires comprendida entre Necochea y Bahía Blanca. *Physis* 25 (69): 169–86.

Contreras, J. R., V. G. Roig, and C. M. Suzarte. 1977. *Ctenomys validus*, una nueva especie de "tunduque" de la provincia de Mendoza (Rodentia; Octodontidae). *Physis*, sec. C, 36(92): 159–62.

Contreras, J. R., and M. I. Rosi. 1980a. Comportamiento territorial y fidelidad al hábitat en una población de roedores del centro de la provincia de Mendoza. *Ecología* 5:17–29.

———. 1980b. El ratón de campo *Calomys musculinus cordovensis* (Thomas) en la provincia de Mendoza. 1. Consideraciones taxonómicas. *Hist. Nat.* 1(5): 17–28.

———. 1980c. Una nueva subespecie del ratón colilargo para la provincia de Mendoza: *Oligoryzomys flavescens occidentalis* (Mammilia, Rodentia, Cricetidae). *Hist. Nat.* 1(22): 157–60.

————. 1980d. Acerca de la presencia en la provincia de Mendoza del ratón de campo *Akodon molinae* Contreras, 1968 (Rodentia: Cricetidea). *Hist. Nat.* 1(26): 181–84.

————. 1981a. Notas sobre los Akodontini argentinos (Rodentia, Cricetidae). 1. *Abrothrix longipilis moerens* Thomas, 1919, en el Parque Nacional Nahuel Huapi. *Hist. Nat.* 1(30): 209–12.

————. 1981b. Notas sobre los Akodontini argentinos (Rodentia, Cricetidae). 2. *Akodon andinus andinus* (Philippi, 1868) en la provincia de Mendoza. *Hist. Nat.* 1(32): 233–36.

Contreras, L. C. 1983. Physiological ecology of fossorial mammals: A comparative study. Ph.D. diss., University of Florida.

Contreras, L. C., and E. Bustos-Obregón. 1977. Ciclo reproductivo anual en *Octodon degus* (Molina) Macho. *Med. Amb.* 3(1): 83–90.

Contreras, L. C., J. C. Torres-Mura, and J. L. Yáñez. 1987. Biogeography of octodontid rodents: An eco-evolutionary hypothesis. *Fieldiana: Zool.*, n.s., 39:401–11.

Crespo, J. A. 1950. Nota sobre mamíferos de Misiones nuevos para Argentina. *Comun. Inst. Nac. Invest. Cienc. Nat., Mus. Argent. Cienc. Nat. "Bernardino Rivadavia," Zool.* 1(14): 1–14.

————. 1963. Dispersión del chinchillón, *Lagidium viscacia* (Molina) en el noreste de Patagonia y descripción de una nueva subespecie (Mammalia; Rodentia). *Neotrópica* 9(29): 61–63.

————. 1964. Cita de mamíferos para el sudoeste de la provincia de Buenos Aires. *Neotrópica* 10(33): 102.

————. 1966. Ecología de una comunidad de roedores silvestres en el Partido de Rojas, provincia de Buenos Aires. *Rev. Mus. Argent. Cienc. Nat. "Bernardino Rivadavia," Ecol.* 1(3): 79–134.

————. 1974a. Comentarios sobre nuevas localidades para mamíferos de Argentina y de Bolivia. *Rev. Mus. Argent. Cienc. Nat. "Bernardino Rivadavia," Zool.* 11(1): 1–31.

————. 1974b. Observaciones sobre la reproducción de la nutria en estado silvestre. *Primer Congreso Argentino de Producción Nutriera,* fasc. 1:60–73.

————. 1982a. Introducción a la ecología de los mamíferos del Parque Nacional el Palmar, Entre Ríos. *Anal. Parques Nac.* 15:1–33.

————. 1982b. Ecología de la comunidad de mamíferos del Parque Nacional Iguazú, Misiones. *Rev. Mus. Argent. Cienc. Nat. "Bernardino Rivadavia," Ecol.* 3(2): 45–162.

Crespo, J. A., M. S. Sabattini, M. J. Piantanida, and G. de Villafañe. 1970. *Estudios ecológicos sobre roedores silvestres, observaciones sobre densidad, reproducción y estructura de comunicades de roedores silvestres en el sur de Córdoba.* Buenos Aires: Ministerio de Bienestar Social.

Daciuk, J. 1974. Notas faunísticas y bioecológicas de Península Valdés y Patagonia. 12. Mamíferos colectados y observados en la Península Valdés y zona litoral de los Golfos San José y Nuevo (provincia de Chubut, república Argentina). *Physis,* sec. C, 33(86): 23–39.

Dalby, P. L. 1975. Biology of pampa rodents, Balcarce area, Argentina. *Publ. Mus., Michigan State Univ. (Biol. Ser.)* 5(3): 153–271.

Dalby, P. L., and M. A. Mares. 1974. Notes on the distribution of the coney rat, *Reithrodon auritus,* in northwestern Argentina. *Amer. Midl. Nat.* 92(1): 205–6.

Dávila, J., E. López, G. Mamani, and P. Jiménez. 1982. Consideraciones ecológicas de algunas poblaciones de vizcachas en Arequipa-Perú. In *Zoología neotropical,* vol. 2, ed. P. J. Salinas, 949–58. Mérida, Venezuela: Congreso Latinamericano de Zoología.

Davis, D. E. 1947. Notes on the life histories of some Brazilian mammals. *Bol. Mus. Nac., Zool.* 76:1–8.

De Santis, L. J. M., and E. R. Justo. 1980. *Akodon (Abrothrix) mansoensis* sp. nov. un nuevo "ratón lanoso" de la provincia de Río Negro, Argentina (Rodentia, Cricetidae). *Neotrópica* 26(75): 121–27.

Dietz, J. M. 1983. Notes on the natural history of some small mammals in central Brazil. *J. Mammal.* 64:521–23.

Dorst, J. 1971. Nouvelles recherches sur l'écologie des rongeurs des hauts plateaux Peruviens. *Mammalia* 35:515–47.

————. 1972. Morphologie de l'estomac et régime alimentaire de quelques rongeurs des hautes Andes du Pérou. *Mammalia* 36:647–56.

Dubost, G., and H. Genest. 1974. Le comportement social d'une colonie de maras *Dolichotis patagonum* Z. dans le Parc de Branféré. *Z. Tierpsychol.* 35:225–302.

Echave Llanos, J. M., and C. A. Vilchez. 1964. Anatomía microscópica del estómago del ratón hocicudo (*Oxymycterus ritulans*). *Rev. Soc. Argent. Biol.* 40:187–92.

Eisenberg, J. F. 1974. The function and motivational basis of hystricomorph vocalizations. In *The biology of hystricomorph rodents,* ed. I. W. Rowlands and B. J. Weir, 211–44. London: Academic Press.

————. 1984. New World rats and mice. In *The encyclopaedia of mammals,* ed. D. Macdonald, 640–49. New York: Facts on File.

———. 1989. *Mammals of the Neotropics*. Vol. 1. *Mammals of the northern Neotropics: Panama, Colombia, Venezuela, Guyana, Suriname, French Guiana*. Chicago: University of Chicago Press.

Eisenberg, J. F., and D. G. Kleiman. 1977. Communication in lagomorphs and rodents. In *How animals communicate*, ed. T. Sebeok, 634–54. Bloomington: Indiana University Press.

Emmons, L. H. 1981. Morphological, ecological, and behavioral adaptations for arboreal browsing in *Dactylomys dactylinus* (Rodentia, Echimyidae). *J. Mammal.* 62:183–89.

———. 1982. Ecology of *Proechimys* (Rodentia, Echimyidae) in southeastern Peru. *Trop. Ecol.* 23(2): 280–90.

Ernest, K. A. 1986. *Nectomys squamipes*. *Mammal. Species* 265:1–5.

Ewer, R. F. 1971. The biology and behavior of a free-living population of black rats (*Rattus rattus*). *Anim. Behav. Monogr.* 4(3): 127–74.

Fernández, R. G. 1965. Tuco-tuco—Um roedor nocivo. *Rev. Fac. Agron. Vet. Porto Alegre* 7: 253–57.

Fonollat, A. M. P. de. 1984. Cricétidos de la provincia de Tucumán (Argentina). *Acta Zool. Lilloana* 37(2): 219–25.

Fonollat, A. M. P. de, and M. S. de Décima. 1979. Comportamiento en cautividad de *Oryzomys longicaudatus* (Bennett, 1832) y *Akodon varius* (Thomas, 1902) (Rodentia-Chordata). *Acta Zool. Lilloana* 33(2): 91–94.

Fonseca, G. A. B. 1988. Patterns of small mammal species diversity in the Brazilian Atlantic forest. Ph.D. diss., University of Florida.

Fonseca, G. A. B., and M. C. M. Kierulff. 1989. Biology and natural history of Brazilian Atlantic forest small mammals. *Bull. Florida State Mus.* 34(1): 99–152.

Fornes, A., and E. Massoia. 1965. Micromamíferos (Marsupialia y Rodentia) recolectados en la localidad Bonaerense de Miramar. *Physis* 25(69): 99–108.

Freitas, T. R. O., and E. P. Lessa. 1984. Cytogenetics and morphology of *Ctenomys torquatus* (Rodentia: Octodontidae) *J. Mammal.* 65:637–42.

Fulk, G. W. 1975. Population ecology of rodents in the semiarid shrublands of Chile. *Occas. Pap. Mus., Texas Tech Univ.* 33:1–40.

———. 1976a. Owl predation and rodent mortality: A case study. *Mammalia* 40:423–27.

———. 1976b. Notes on the activity, reproduction, and social behavior of *Octodon degus*. *J. Mammal.* 57:495–505.

Gallardo, M. 1979. Las especies chilenas de *Cteno-mys* (Rodentia, Octodontidae). 1. Estabilidad cariotípica. *Arch. Biol. Med. Exp.* 12:71–82.

Gardner, A. L., and J. L. Patton. 1976. Karyotypic variation in oryzomyine rodents (Cricetinae) with comments on chromosomal evolution in the Neotropical cricetine complex. *Occas. Pap. Mus. Zool., Louisiana State Univ.* 49:1–48.

Glanz, W. E. 1977a. Comparative ecology of small mammal communities in California and Chile. Ph.D. diss., University of California, Berkeley.

———. 1977b. Small mammals. In *Chile-California Mediterranean scrub atlas: A comparative analysis*, ed. N. J. W. Thrower and D. E. Bradbury, 232–37. Stroudsburg, Pa.: Dowden, Hutchinson and Ross.

González, J. C. 1973. Observaciones sobre algunos mamíferos de Bopicuá (Dpto. de Río Negro, Uruguay). *Comun. Mus. Mun. Hist. Nat. Río Negro, Uruguay* 1(1): 1–14.

González, L. A., R. Murúa, and R. Feito. 1982. Densidad poblacional y patrones de actividad espacial de *Akodon olivaceus* (Rodentia, Cricetidae) en hábitats diferentes. In *Zoología neotropical*, vol. 2, ed. P. J. Salinas, 935–47. Mérida, Venezuela: Congreso Latinoamericano de Zoología.

Greer, J. K. 1966. Mammals of Malleco province, Chile. *Publ. Mus., Michigan State Univ., Biol. Ser.* 3(2): 49–152.

Guillotin, M. 1982. Rythmes d'activité et régimes alimentaires de *Proechimys cuvieri* et d'*Oryzomys capito velutinus* (Rodentia) en forêt guyanaise. *Rev. Ecol. (Terre et Vie)* 36:337–71.

Gyldenstolpe, N. 1932. A manual of Neotropical sigmodont rodents. *Kungl. Svenska Vetenskapsakad. Hand.* 11(3): 1–164.

Hershkovitz, P. 1944. A systematic review of the Neotropical water rats of the genus *Nectomys* (Cricetinae). *Misc. Publ. Mus. Zool. Univ. Mich.* 58:1–88.

———. 1959. Two new genera of South American rodents (Cricetinae). *Proc. Biol. Soc. Washington* 72:5–10.

———. 1962. Evolution of Neotropical cricetine rodents (Muridae) with special reference to the phyllotine group. *Fieldiana: Zool.* 46:1–524.

———. 1966. South American swamp and fossorial rats of the scapteromyine group (Cricetinae, Muridae) with comments on the glans penis in murid taxonomy. *Z. Säugetierk.* 31:81–149.

Hodara, V. L., A. E. Kajón, C. Quintans, L. Montoro, and M. S. Merani. 1984. Parámetros métricos y reproductivos de *Calomys musculinus* (Thomas, 1913) y *Calomys callidus*, Thomas, 1916 (Rodentia, Cricetidae). *Rev. Mus. Argent. Cienc.*

Nat. "Bernardino Rivadavia," Zool. 13(1–60): 453–59.

Honacki, J. H., K. E. Kinman, and J. W. Koeppl, eds. 1982. *Mammal species of the world.* Lawrence, Kans.: Allen Press and Association of Systematics Collections.

Hooper, E. T., and G. G. Musser. 1964. The glans penis in Neotropical cricetines, with comments on the classification of murid rodents. *Misc. Publ. Mus. Zool. Univ. Michigan* 123:1–57.

Hudson, W. H. 1872. On the habits of the vizcacha (*Lagostomus trichodatylus*). *Proc. Zool. Soc. London* 1872:822–33.

Husson, A. M. 1978. *The mammals of Suriname.* Leiden: E. J. Brill.

Ipinza R., J., M. Tamayo H., and J. Rottmann S. 1971. Octodontidae en Chile. *Not. Mens.* 16(183): 3–10.

Iriarte, J. A., and J. A. Simonetti. 1986. *Akodon andinus* (Philippi, 1858): Visitante ocasional del matorral esclerófilo centro-Chileno. *Not. Mens., Mus. Nac. Hist. Nat.* 311:6–7.

Iriarte, J. A., L. C. Contreras, and F. M. Jaksić. 1989. A long-term study of a small-mammal assemblage in the central Chilean mattoral. *J. Mammal.* 70:79–87.

Jaksić, F., and J. Yáñez. 1979. Tamaño corporal de los roedores del distrito mastozoológico santiaguino. *Not. Mens.* 23(271): 3–4.

Justo, E. R., and L. de Santis. 1977. *Akodon serrensis serrensis* Thomas en la Argentina (Rodentia, Cricetidae). *Neotrópica* 23(69): 47–48.

Kajon, A. E., O. A. Scaglia, C. Horgan, C. Velázquez, M. S. Merani, and O. A. Reig. 1984. Tres nuevos cariotipos de la tribu Akodontini (Rodentia, Cricetidae). *Rev. Mus. Argent. Cienc. Nat. "Bernardino Rivadavia," Zool.* 13(1–60): 461–69.

Kleiman, D. G. 1974. Patterns of behavior in hystricomorph rodents. In *The biology of hystricomorph rodents*, ed. I. W. Rolands and B. Weir, 171–209. London: Academic Press.

Kleiman, D. G., J. F. Eisenberg, and E. Maliniak. 1979. Reproductive parameters and productivity of caviomorph rodents. In *Vertebrate ecology in the northern Neotropics*, ed. J. F. Eisenberg, 173–83. Washington, D.C.: Smithsonian Institution Press.

Koford, C. B. 1955. New rodent records for Chile and for two Chilean provinces. *J. Mammal.* 36: 465–66.

Kravetz, F. O. 1972. Estudio del régimen alimentario, periodos de actividad y otros rasgos ecológicos en una población de "ratón hocicudo" (*Oxymycterus rufus platensis* Thomas) de Punta Lara. *Acta Zool. Lilloana* 29:201–12.

Kravetz, F. O., and de Villafañe, G. 1981. Poblaciones de roedores en cultivo de maíz durante las etapas de madurez y rastrojo. *Hist. Nat.* 1(31): 213–32.

Landry, S. 1970. The Rodentia as omnivores. *Quart. Rev. Biol.* 45:351–72.

Langguth, A. 1963. Las especies Uruguayas del género *Oryzomys* (Rodentia, Cricetidae). *Comun. Zool. Mus. Hist. Nat. Montevideo* 7(99): 1–19.

Langguth, A., and A. Abella. 1970. Las especies Uruguayas del género *Ctenomys* (Rodentia= Octodontidae). *Comun. Zool. Mus. Hist. Nat. Montevideo* 10(129): 1–20.

Lawrence, B. 1941. A new species of *Octomys* from Argentina. *Proc. New England Zool. Club* 18: 43–46.

Le Boulengé, E., and E. R. Fuentes. 1978. Quelques données sur la dynamique de population chez *Octodon degus* (Rongeur, Hystricomorphe) du Chili central. *Terre et Vie* 32:325–41.

Lessa, E. P., and C. A. Altuna. 1984. Estudio comparativo de la morfología del pene en poblaciones uruguayas de *Ctenomys* (Rodentia, Octodontidae). *Rev. Mus. Argent. Cienc. Nat. "Bernardino Rivadavia," Zool.* 13(1–60): 471–78.

Lessa, E. P., and A. Langguth. 1983. *Ctenomys pearsoni*, n. sp. (Rodentia, Octodontidae), del Uruguay. *Res. Com. J. C. Nat., Montevideo* 3: 86–88.

Liascovich, R. C., R. M. Bárquez, and O. A. Reig. 1989. A karyological and morphological reassessment of *Akodon* (*Abrothrix*) *illuteus* Thomas. *J. Mammal.* 70:386–91.

Liascovich, R. C., and O. A. Reig. 1989. Low chromosome number in *Akodon cursor montensis* Thomas and karyologic confirmation of *Akodon serrensis* Thomas in Misiones, Argentina. *J. Mammal.* 70:391–95.

Llanos, A. C., and J. A. Crespo. 1952. Ecología de la vizcacha (*Lagostomus maximus maximus* Blainv.) en el nordeste de la provincia de Entre Ríos. *Rev. Invest. Agric.* 6(3–4): 289–378.

Lord, R. D. 1964. Range extension of the yellow-toothed cavy in the province of Buenos Aires. *J. Mammal.* 45:315–16.

Lucero, M. M. 1983. Lista y distribución de aves y mamíferos de la provincia de Tucumán. Ministero de Cultura y Educación, Fundación Miguel Lillo, *Miscelánea* 75:5–53.

Macedo, R. H., and M. A. Mares. 1987. Geographic variation in the South American cricetine rodent *Bolomys lasiurus. J. Mammal.* 68:278–94.

Maia, V. 1984. Karyotypes of three species of Caviinae (Rodentia, Caviidae). *Experientia* 40: 564–66.

Maik-Siembida, J. 1974. Breeding viscachas *Lagostomus maximus* at Lodz Zoo. *Int. Zoo Yearb.* 14:116–17.

Mann, G. 1944. Dos nuevas especies de roedores. *Biológica* 1:94–129.

———. 1945. Mamíferos de Tarapaca: Observaciones realizadas durante una expedición al alto norte de Chile. *Biológica* 2:23–141.

———. 1978. *Los pequeños mamíferos de Chile.* Gayana: Zoología 40. Chile: Universidad de Concepción.

Mares, M. A. 1973. Climates, mammalian communities and desert rodent adaptations: An investigation into evolutionary convergence. Ph.d. diss., University of Texas at Austin.

———. 1977a. Aspects of the water balance of *Oryzomys longicaudatus* from northwestern Argentina. *Comp. Biochem. Physiol.* 57A:237–38.

———. 1977b. Water economy and salt balance in a South American desert rodent, *Eligmodontia typus. Comp. Biochem. Physiol.* 56A:325–32.

———. 1977c. Water balance and other ecological observations on three species of *Phyllotis* in northwestern Argentina. *J. Mammal.* 58:514–20.

———. 1977d. Water independence in a South American non-desert rodent. *J. Mammal.* 58:653–56.

———. 1988. Reproduction, growth, and development in Argentine gerbil mice, *Eligmodontia typus. J. Mammal.* 69:852–54.

Mares, M. A., and R. A. Ojeda. 1982. Patterns of diversity and adaptation in South American hystricognath rodents. In *Mammalian biology in South America,* ed. M. A. Mares and H. H. Genoways, 393–432. Pymatuning Symposia in Ecology 6. Special Publication Series. Pittsburgh: Pymatuning Laboratory of Ecology, University of Pittsburgh.

———. 1984. Faunal commercialization and conservation in South America. *BioScience* 34(9): 580–84.

Mares, M. A., R. A. Ojeda, and M. P. Kosco. 1981. Observations on the distribution and ecology of the mammals of Salta province, Argentina. *Ann. Carnegie Mus.* 50(6): 151–206.

Markham, B. J. 1971. Catálogo de los anfibios, reptiles, aves y mamíferos de la provincia de Magallanes (Chile). *Publ. Inst. Patagonia, Punta Arenas,* 1–64.

Massoia, E. 1963a. Sobre la posición sistemática y distribución geográfica de *Akodon (Thaptomys) nigrita* (Rodentia-Cricetidae). *Physis* 24(67): 73–80.

———. 1963b. *Oxymycterus iheringi* (Rodentia-Cricetidae) nueva especie para la Argentina. *Physis* 24(67): 129–36.

———. 1963c. Sistemática, distribución geográfica y rasgos etoecológicas de *Akodon (Deltamys) kempi* (Rodentia, Cricetidae). *Physis* 24(67): 240.

———. 1964. Sistemática, distribución geográfica y rasgos etoecológicos de *Akodon (Deltamys) kempi* (Rodentia-Cricetidae). *Physis* 24(68): 299–305.

———. 1971a. Descripción y rasgos bioecológicos de una nueva subespecie de cricétido: *Akodon azarae bibianae* (Mammalia-Rodentia). *Rev. Invest. Agropecuarias, INTA* (Buenos Aires), ser. 4, *Patalog. Anim.* 8(5): 131–40.

———. 1971b. Caracteres ye rasgos bioecológicos de *Holochilus brasiliensis chacarius* Thomas (rata nutria) de la provincia de Formosa y comparaciones con *Holochilus brasiliensis vulpinus* (Brants) (Mammalia-Rodentia-Cricetidae). *Rev. Invest. Agropecuarias, INTA* (Buenos Aires), ser. 1, *Biol. Prod. Anim.* 8(1): 13–40.

———. 1971c. Las ratas nutrias argentinas del género *Holochilus* descritas como *Mus brasiliensis* por Waterhouse (Mammalia-Rodentia-Cricetidae). *Rev. Invest. Agropecuarias, INTA* (Buenos Aires), ser. 4, *Patalog. Anim.* 8(5): 141–48.

———. 1973a. Zoogeografía del género *Cavia* en la Argentina con comentarios bioecológicos y sistemáticos (Mammalia-Rodentia-Caviidae). *Rev. Invest. Agropecuarias, INTA* (Buenos Aires), ser. 1, *Biol. Prod. Anim.* 10(1): 1–11.

———. 1973b. Descripción de *Oryzomys fornesi,* nueva especie y nuevos datos sobre algunas especies y subespecies argentinas del subgénero *Oryzomys (Oligoryzomys)* (Mammalia-Rodentia-Cricetidae). *Rev. Invest. Agropecuarias, INTA* (Buenos Aires), ser. 1, *Biol. Prod. Anim.* 10(1): 21–37.

———. 1976a. Mammalia. In *Fauna de agua dulce de la república Argentina,* vol. 44, *Mammalia.* Buenos Aires: Fundación para la Educación, la Ciencia y la Cultura.

———. 1976b. Datos sobre un cricétido nuevo para la Argentina: *Oryzomys (Oryzomys) capito* intermedius y sus diferencias con *Oryzomys (Oryzomys) legatus* (Mammalia-Rodentia). *Rev. Invest. Agropecuarias, INTA* (Buenos Aires), ser. 5, 11(1): 1–7.

———. 1979. Descripción de un género y especie nuevos: *Bibimys torresi* (Mammalia-Rodentia-Cricetidae-Sigmodontinae-Scapteromyini). *Physis,* sec. C, 38(95): 1–7.

———. 1980a. Mammalia de Argentina. 1. Los mamíferos silvestres de la provincia de Misiones. *Iguazú* 1(1): 15–43.

———. 1980b. El estado sistemático de cuatro especies de cricétidos sudamericanos y comentarios sobre otras congenéricas (Mammalia: Rodentia). *Ameghiana* 17:280–87.

―――. 1982. Diagnosis previa de *Cabreramys temchuki*, nueva especie (Rodentia, Cricetidae). *Hist. Nat.* 2(11): 91–92.

Massoia, E., and A. Fornes. 1964. Notas sobre el género *Scapteromys* (Rodentia-Cricetidae). 1. Systemática, distribución geográfica y rasgos etoecológicos de *Scapteromys tumidus* (Waterhouse). *Physis* 24(68): 279–97.

―――. 1965a. Nuevos datos sobre la morfología, distribución geográfica y etoecología de *Calomys callosus callosus* (Rengger) (Rodentia-Cricetidae). *Physis* 25(70): 325–31.

―――. 1965b. *Oryzomys* (*Oecomys*) Thomas, 1906, Nuevo subgénero de cricétidos para la república Argentina (Rodentia). *Physis* 25(70): 319–24.

―――. 1967a. Roedores recolectados en la Capital Federal (Caviidae, Cricetidae y Muridae). *IDIA* 240:47–53.

―――. 1967b. El estado sistemático, distribución geográfica y datos etoecológicos de algunos mamíferos neotropicales (Marsupialia y Rodentia) con la descripción de *Cabreramys*, género nuevo (Cricetidae). *Acta Zool. Lilloana* 23:407–30.

―――. 1969. Caracteres comunes y distintivos de *Oxymycterus nasutus* (Waterhouse) y *O. iheringi* Thomas (Rodentia, Cricetidae). *Physis* 28(77): 315–21.

Massoia, E., A. Fornes, R. L. Wainberg, and T. G. de Fronza. 1968. Nuevos aportes al conocimiento de las especies bonaerenses del género *Calomys* (Rodentia-Cricetidae). *Rev. Invest. Agropecuarias, INTA* (Buenos Aires), ser. 1, *Biol. Prod. Anim.* 5(4): 63–92.

Matamoros, H., Y. 1982. Notas sobre la biología del tepezcuinte, *Cuniculus paca*, Brisson (Rodentia: Dasyproctidae) en cautiverio. *Brenesia* 19/20: 71–82.

Mello, D. A. 1978a. Some aspects of the biology of *Oryzomys eliurus* (Wagner, 1845) under laboratory conditions (Rodentia, Cricetidae). *Rev. Brasil. Biol.* 38(2):293–95.

―――. 1978b. Biology of *Calomys callosus* (Rengger, 1830) under laboratory conditions (Rodentia, Cricetinae). *Rev. Brasil. Biol.* 38(4): 807–11.

Merani, S., M. Lizarralde, D. Oliveira, and N. Bianchi. 1978. Cytogenetics of the South American akodont rodents (Cricetidae). 4. Interspecific crosses between *Akodon dolores* × *Akodon molinae*. *J. Exp. Zool.* 206:343–46.

Meritt, D. A, Jr. 1983. Preliminary observations on reproduction in the Central American agouti, *Dasyprocta punctata*. *Zoo Biol.* 2:127–31.

Meserve, P. L. 1978. Water dependence in some Chilean arid zone rodents. *J. Mammal.* 59: 217–19.

―――. 1981a. Resource partitioning in a Chilean semi-arid small mammal community. *J. Anim. Ecol.* 40:747–57.

―――. 1981b. Trophic relationships among small mammals in a Chilean semiarid thorn scrub community. *J. Mammal.* 62:304–14.

Meserve, P. L., B. K. Lang, and B. D. Patterson. 1988. Trophic relationships of small mammals in a Chilean temperate rainforest. *J. Mammal.* 69(4): 721–30.

Meserve, P. L., and E. B. Le Boulengé. 1987. Population dynamics and ecology of small mammals in the northern Chilean semiarid region. *Fieldiana: Zool.*, n.s., 39:413–31.

Meserve, P. L., R. E. Martín, and J. Rodríguez. 1983. Feeding ecology of two Chilean caviomorphs in a central Mediterranean savanna. *J. Mammal.* 64:322–25.

―――. 1984. Comparative ecology of the caviomorph rodent *Octodon degus* in two Chilean Mediterranean-type communities. *Rev. Chil. Hist. Nat.* 57:79–89.

Miller, L. M., and S. Anderson. 1977. Bodily proportions of Uruguayan myomorph rodents. *Amer. Mus. Novitat.* 2615:1–10.

Miller, S. D. 1980. Human influences on the distribution and abundance of wild Chilean mammals: Prehistoric–present. Ph.D. diss., University of Washington, Seattle.

Miller, S. D., J. Rottman, K. J. Raedeke, and R. D. Taber. 1983. Endangered mammals of Chile: Status and conservation. *Biol. Conserv.* 25:335–52.

Miranda-Ribeiro, A de. 1936. The new-born of the Brazilian tree-porcupine (*Coendou prehensilis* Linn.) and of the hairy tree-porcupine (*Sphiggurus villosus* F. Cuv.). *Proc. Zool. Soc. London* 1936:971–74.

Mohlis, C. 1983. Información preliminar sobre la conservación y manejo de la *Chinchilla silvestre* en Chile. Boletín Técnico 3, Corporación Nacional Forestal, Chile.

Mondolfi, E. 1972. La lapa o paca. *Defensa Nat.* 2(5): 4–16.

Montgomery, G. G., and Y. D. Lubin. 1978. Movements of *Coendou prehensilis* in the Venezuelan llanos. *J. Mammal.* 59:887–88.

Moojen, J. 1948. Speciation in the Brazilian spiny rats (genus *Proechimys*, family Echimyidae). *Univ. Kansas Publ. Mus. Nat. Hist.* 1(19): 301–406.

―――. 1952a. Os reodores do Brasil. *Bibl. Cient. Brasil.* (Rio de Janeiro), ser. A-2, 1–214.

―――. 1952b. A new *Clyomys* from Paraguay (Rodentia: Echimyidae). *J. Washington Acad. Sci.* 42(3): 102.

Murúa, R., and L. A. González. 1979. Distribución de roedores silvestres con relación a las caracteristicas del hábitat. *Anal. Mus. Hist. Nat. Valparaiso* 12:69–75.

———. 1981. Estudiós de preferencias y hábitos alimentarios en dos especies de roedores cricétidos. *Med. Amb.* 5(1–2): 115–24.

———. 1982. Microhabitat selection in two Chilean cricetid rodents. *Oecologia* (Berlin) 52:12–15.

Murúa, R., L. González, and C. Jofre. 1982. Estudiós ecológicos de roedores silvestres en los bosques templados fríos de Chile. *Mus. Nac. Hist. Nat. Publ. Ocas.* 38:105–16.

Myers, P. 1977. A new phyllotine rodent (genus *Graomys*) from Paraguay. *Occas. Pap. Mus. Zool. Univ. Michigan* 676:1–7.

———. 1982. Origins and affinities of the mammal fauna of Paraguay. In *Mammalian biology in South America*, ed. M. A. Mares and H. H. Genoways, 85–84. Pymatuning Symposia in Ecology 6. Special Publication Series. Pittsburgh: Pymatuning Laboratory of Ecology, University of Pittsburgh.

———. 1989. A preliminary revision of the *varius* group of *Akodon* (*A. dayi, dolores, molinae, neocenus, simulator, toba,* and *varius*). In *Advances in Neotropical mammalogy*, ed. K. H. Redford and J. F. Eisenberg, 5–54. Gainesville, Fla.: Sandhill Crane Press.

Myers, P., and M. D. Carleton. 1981. The species of *Oryzomys* (*Oligoryzomys*) in Paraguay and the identity of Azare's "rat sixième ou rat tarse noir." *Misc. Publ. Mus. Zool. Univ. Michigan* 161:1–41.

Myers, P., and R. M. Wetzel. 1979. New records of mammals from Paraguay. *J. Mammal.* 60: 638–41.

Nevo, E. 1979. Adaptive convergence and divergence of subterranean mammals. *Ann. Rev. Ecol. Syst.* 10:269–308.

Newsome, A. E. 1971. The ecology of house mice in cereal haystacks. *J. Anim. Ecol.* 40:116.

Nitikman, L. Z., and M. A. Mares. 1987. Ecology of small mammals in a gallery forest of central Brazil. *Ann. Carnegie Mus.* 56:75–95.

Novello, A. F., and E. P. Lessa. 1986. G-band homology in two karyomorphs of the *Ctenomys pearsoni* complex (Rodentia: Octodontidae) of Neotropical fossorial rodents. *Z. Säugetierk.* 51: 378–80.

O'Connell, M. A. 1981. Population ecology of small mammals from northern Venezuela. Ph.D. diss., Texas Tech University.

Ojasti, J. 1972. Revisión preliminar de los picures o agutís de Venezuela. *Mem. Soc. Cienc. Nat. La Salle* 32(93): 150–204.

———. 1973. *Estudio biológico del chigüire o capibara*. Caracas: Fundo Nacional de Investigaciones Agropecuarias.

Ojasti, J., and L. M. Sosa Burgos. 1985. Density regulation in populations of capybara. *Acta Zool. Fennica* 173:81–83.

Ojeda, R. A. 1979. Aspectos ecológicos de una comunidad de cricétidos del bosque subtropical de transición. *Neotrópica* 25(73): 27–35.

———. 1985. A biogeographic analysis of the mammals of Salta province, Argentina: Patterns of community assemblage in the Neotropics. Ph.D. diss., University of Pittsburgh.

Ojeda, R. A., and M. A. Mares. 1989. *A biogeographic analysis of the mammals of Salta province, Argentina*. Special Publications of the Museum 27. Lubbock: Texas Tech University Press.

Ojeda, R. A., and V. G. Roig, E. P. de Cristaldo, and C. N. de Moyano. 1989. A new record of *Tympanoctomys* (Octodontidae) from Mendoza province, Argentina. *Texas J. Sci.* 41(3): 333–36.

Olds, N., and S. Anderson. 1987. Notes on Bolivian mammals: Taxonomy and distribution of rice rats of the subgenus *Oligoryzomys*. *Fieldiana: Zool.*, n.s., 39:261–82.

———. 1989. A diagnosis of the tribe Phyllotini (Rodentia, Muridae). In *Advances in Neotropical mammalogy*, ed. K. H. Redford and J. F. Eisenberg, 55–74. Gainesville, Fla.: Sandhill Crane Press.

Olrog, C. C. 1976. Sobre mamíferos del noroeste argentino. *Acta Zool. Lilloana* 32:5–12.

———. 1979. Los mamíferos de la selva húmeda, Cerro Calilegua, Jujuy. *Acta Zool. Lilloana* 33: 9–14.

Olrog, C. C., and M. M. Lucero. 1981. *Guía de los mamíferos argentinos*. Tucumán, Argentina: Ministerio de Cultura y Educación, Fundación Miguel Lillo.

Ortells, M. O., O. A. Reig, R. L. Wainberg, G. E. Hurtado de Catalfo, and T. M. L. Gentile de Granza. 1989. Cytogenetics and karyosystematics of phyllotine rodents (Cricetidae, Sigmodontinae). 2. Chromosome multiformity and autosomal polymorphism in *Eligmodontia*. *Z. Säugetierk.* 54:129–70.

Ortells, M. O., A. Vitullo, M. S. Merani, and O. A. Reig. 1984. Identidad cromosómica entre dos razas geográficas de tuco-tucos (*Ctenomys talarum*). *Rev. Mus. Argent. Cienc. Nat. "Bernardino Rivadavia," Zool.* 13(1–60): 479–84.

Osgood, W. H. 1925. The long-clawed South American rodents of the genus *Notiomys*. *Field Mus. Nat. Hist., Zool. Ser.* 12(9): 113–25.

————. 1933. Two new rodents from Argentina. *Field Mus. Nat. Hist., Zool. Ser.* 20:11–14.

————. 1943. The mammals of Chile. *Field Mus. Nat. Hist., Zool. Ser.* 30:1–268.

————. 1944. Nine new South American rodents. *Field Mus. Nat. Hist., Zool. Ser.* 29(13):191–204.

————. 1946. A new octodont rodent from the Paraguayan Chaco. *Fieldiana: Zool.* 31(6):47–49.

Patterson, B. D., M. H. Gallardo, and K. E. Freas. 1984. Systematics of mice of the subgenus *Akodon* (Rodentia: Cricetidae) in southern South America, with the description of a new species. *Fieldiana: Zool.,* n.s., 23:1–16.

Patton, J. L. 1984. Systematic status of the large squirrels (subgenus *Urosciurus*) of the western Amazon basin. *Stud. Neotrop. Fauna Environ.* 19(2):53–72.

Patton, J. L., and A. L. Gardner. 1972. Notes on the systematics of *Proechimys* (Rodentia: Echimyidae), with emphasis on Peruvian forms. *Occas. Pap. Mus. Zool. Louisiana State Univ.* 44:1–30.

Patton, J. L., and S. Y. Yang. 1977. Genetic variation in *Thomomys bottae* pocket gophers: Macrogeographic patterns. *Evolution* 31:697–720.

Pearson, O. P. 1948. Life history of mountain viscachas in Peru. *J. Mammal.* 29:345–74.

————. 1951. Mammals in the high-lands of southern Peru. *Bull. Mus. Comp. Zool.* 106(3):117–74.

————. 1957. Additions to the mammalian fauna of Peru and notes on some other Peruvian mammals. *Breviora* 73:1–7.

————. 1958. A taxonomic revision of the rodent genus *Phyllotis. Univ. Calif. Publ. Zool.* 56(4):391–496.

————. 1959. Biology of the subterranean rodents, *Ctenomys,* in Peru. *Mem. Mus. Hist. Nat. "Javier Prado"* 9:1–56.

————. 1967. La estructura por edades y la dinámica reproductiva en una población de ratones de campo, *Akodon azarae. Physis* 27(74):53–58.

————. 1975. An outbreak of mice in the coastal desert of Peru. *Mammalia* 39:375–86.

————. 1983. Characteristics of a mammalian fauna from forests in Patagonia, southern Argentina. *J. Mammal.* 64:476–92.

————. 1984. Taxonomy and natural history of some fossorial rodents of Patagonia, southern Argentina. *J. Zool.* (London) 202:225–37.

————. 1987. Mice and the postglacial history of the Traful Valley of Argentina. *J. Mammal.* 68:469–78.

————. 1988. Biology and feeding dynamics of a South American herbivorous rodent, *Reithrodon. Stud. Neotrop. Fauna Environ.* 23(1):25–39.

Pearson, O. P., N. Binsztein, L. Boiry, C. Busch, M. Di Pace, G. Gallopin, P. Penchaszadeh, and M. Piantanida. 1968. Estructura social, distribución espacial y composición por edades de una población de tuco-tucos (*Ctenomys talarum*). *Invest. Zool. Chil.* 13:47–80.

Pearson, O. P., and M. I. Christie. 1985. Los tuco-tucos (género *Ctenomys*) de los Parques Nacionales Lanín y Nahuel Huapi, Argentina. *Hist. Nat.* 5(37):337–43.

Pearson, O. P., S. Martin, and J. Bellati. 1987. Demography and reproduction of the silky desert mouse (*Eligmodontia*) in Argentina. *Fieldiana: Zool.,* n.s., 23:433–46.

Pearson, O. P., and J. L. Patton. 1976. Relationships among South American phyllotine rodents based on chromosome analysis. *J. Mammal.* 57:339–50.

Pearson, O. P., and A. K. Pearson. 1982. Ecology and biogeography of the southern rainforests of Argentina. In *Mammalian biology in South America,* ed. M. A. Mares and H. H. Genoways, 129–42. Pymatuning Symposia in Ecology 6. Special Publication Series. Pittsburgh: Pymatuning Laboratory of Ecology, University of Pittsburgh.

Pearson, O. P., and C. P. Ralph. 1978. The diversity and abundance of vertebrates along an altitudinal gradient in Peru. *Memo. Mus. Hist. Nat. "Javier Prado"* 18:1–97.

Pefaur, J. E., J. L. Yáñez, and F. M. Jaksić. 1979. Biological and environmental aspects of a mouse outbreak in the semi-arid region of Chile. *Mammalia* 43:313–22.

Pereira, C. 1941. Sobre as "ratadas" no sul do Brasil e o ciclo vegetativo das taquaras. *Arq. Inst. Biol. São Paulo* 12:175–96.

Pine, R. H. 1973. Una nueva especie de *Akodon* (Mammalia: Rodentia: Muridae) de la Isla Wellington, Magallanes, Chile. *Anal. Inst. Patagonia* 4(1–3):423–26.

————. 1980. Notes on rodents of the genera *Wiedomys* and *Thomasomys* (including *Wilfredomys*). *Mammalia* 44:195–202.

Pine, R. H., J. P. Angle, and D. Bridge. 1978. Mammals from the sea, mainland and islands at the southern tip of South America. *Mammalia* 42:105–14.

Pine, R. H., S. D. Miller, and M. L. Schamberger. 1979. Contributions to the mammalogy of Chile. *Mammalia* 43:339–76.

Pine, R. H., and R. M. Wetzel. 1975. A new subspecies of *Pseudoryzomys wavrini* (Mammalia: Rodentia: Muridae: Cricetinae) from Bolivia. *Mammalia* 39:649–55.

Pizzimenti, J. J., and R. de Salle. 1980. Dietary and

morphometric variation in some Peruvian rodent communities: The effect of feeding strategy on evolution. *Biol. J. Linnean Soc.* 13:263–85.

Rabinovich, J., A. Capurro, P. Folgarait, T. Kitzberger, G. Kramer, A. Novaro, M. Puppo, and A. Travaini. 1987. Estado del conocimiento de 12 especies de la fauna silvestre Argentina de valor comercial. Documento presentado, para su estudio y discusión, al 2° taller de trabajo: "Elaboración de propuestas de investigación orientada al manejo de la fauna silvestre de valor comercial," Buenos Aires.

Rau, J., J. Yáñez, and F. Jaksić. 1978. Confirmación de *Notiomys macronyx alleni* O. y *Eligmodontia typus typus* C., y primer registro de *Akodon (Abrothrix) lanosus* T. (Rodentia: Cricetidae) en la zona de Ultima Esperanza (XII region, Magallanes). *Anal. Inst. Patagonia, Punta Arenas* (Chile) 9:203–4.

Redford, K. H. 1984. Mammalian predation on termites: Tests with the burrowing mouse (*Oxymycterus roberti*) and its prey. *Oecologia* 65:145–52.

Redford, K. H., and J. G. Robinson. 1987. The game of choice: Patterns of Indian and colonist hunting in the Neotropics. *Amer. Anthropol.* 89:650–67.

Reig, O. A. 1964. Roedores y marsupiales del partido de General Pueyrredón y regiones adyacentes (provincia de Buenos Aires, Argentina) *Publ. Mus. Mun. Cienc. Nat. Mar del Plata* 1(6): 203–24.

———. 1970. Ecological notes on the fossorial octodont rodent *Spalacopus cyanus* (Molina). *J. Mammal.* 51:592–601.

———. 1980. A new fossil genus of South American cricetid rodents allied to *Wiedomys*, with an assessment of the Sigmodontinae. *J. Zool.* (London) 192:257–81.

———. 1984. Significado de los métodos citogenéticos para la distinción y la interpretación de las especies, con especial referencia a los mamíferos. *Rev. Mus. Argent. Cienc. Nat. "Bernardino Rivadavia," Zool.* 13(1–60): 19–44.

———. 1986. Diversity patterns and differentiation of high Andean rodents. In *High altitude tropical biogeography,* ed. F. Vuilleumier and M. Monasterio, 404–42. New York: Oxford University Press.

———. 1987. An assessment of the systematics and evolution of the Akodontini, with the description of new fossil species of *Akodon* (Cricetidae: Sigmodontinae). *Fieldiana: Zool.,* n.s., 39:347–99.

Reig, O. A., J. R. Contreras, and M. J. Piantanida. 1965. Contribución a la elucidación de la sistemática de las entidades del género *Ctenomys* (Rodentia, Octodontidae). 1. Relaciones de parentesco entre muestras de ocho poblaciones de tuco-tucos inferidas del estudio estadístico de variables del fenotipo y su correlación con las características del cariotipo. *Univ. Buenos Aires, Fac. Cienc. Exact. Nat., Contrib. Cient., Ser. Zool.* 1(6): 301–52.

Reig, O. A., and P. Kiblisky. 1968. Chromosomes in four species of rodents of the genus *Ctenomys* (Rodentia, Octodontidae) from Argentina. *Experientia* 24:274–76.

———. 1969. Chromosome multiformity in the genus *Ctenomys* (Rodentia, Octodontidae). *Chromosome* (Berlin) 28:211–44.

Reig, O. A., N. Olivo, and P. Kiblisky. 1971. The idiogram of the Venezuelan vole mouse, *Akodon urichi venezuelensis* Allen (Rodentia, Cricetidae). *Cytogenetics* 10:99–114.

Reig, O. A., A. Spotorno, and R. Fernández. 1972. A preliminary survey of chromosomes in populations of the Chilean burrowing octodont rodent *Spalacopus cyanus* Molina (Caviomorpha, Octodontidae). *Biol. J. Linnean Soc.* 4:29–38.

Reise, D., and W. Venegas. 1974. Observaciones sobre el comportamiento de la fauna de micromamíferos en la región de Puerto Ibáñez (Lago General Carrera), Aysén, Chile. *Bol. Soc. Biol. Concepción* 47:71–85.

Roberts, M., S. Brand, and E. Maliniak. 1985. The biology of captive prehensile-tailed porcupines, *Coendou prehensilis. J. Mammal.* 66:476–82.

Roberts, M. S., K. V. Thompson, and J. Cranford. 1988. Reproduction and growth in captive punare of the Brazilian caatinga with reference to the reproductive strategies of the Echimyidae. *J. Mammal.* 69:542–51.

Rodríguez, J. M. 1985. Ecología de la guatusa (*Dasyprocta punctata punctata* Gray). In *Investigaciones sobre fauna silvestre de Costa Rica,* 9–22. San José, Costa Rica: Ministerio de Agricultura y Ganadería.

Roig, V. G. 1965. Elenco sistemático de los mamíferos y aves de la provincia de Mendoza y notas sobre su distribución geográfica. *Bol. Est. Geogr.* 12(49): 175–222.

Roldán, E. R. S., A. M. Acuña, O. Vercellini, C. Horgan, and M. S. Merani. 1984. Parámetros reproductivos de *Akodon molinae*, Contreras, 1968, *Akodon dolores*, Thomas, 1916 y sus híbridos (Rodentia, Cricetidae). *Rev. Mus. Argent. Cienc. Nat. "Bernardino Rivadavia," Zool.* 13(1–60): 485–90.

Rood, J. P. 1970. Ecology and social behavior of the desert cavy (*Microcavia australis*). *Amer. Midl. Nat.* 83:415–54.

———. 1972. Ecological and behavioral comparisons in three genera of Argentine cavies. *Anim. Behav. Monogr.* 5:67–83.

Rosenmann A., M. 1959. *Ctenomys fulvus* Phil., su

hábitat (Rodentia, Ctenomyidae). *Invest. Zool. Chil.* 5:217–20.

———. 1977. Regulación térmica en *Octodon degus*. *Med. Amb.* 3(11): 127–31.

Rosi, M. I. 1983. Notas sobre la ecología, distribución y sistemática de *Graomys griseoflavus griseoflavus* (Waterhouse, 1837) (Rodentia, Cricetidae) en la provincia de Mendoza. *Hist. Nat.* 3(17): 161–67.

Rowlands, I. W. 1974. Mountain viscacha. *Symp. Zool. Soc. London* 34:131–41.

Rowlands, I. W., and B. J. Weir, eds. 1974. *The biology of hystricomorph rodents*. London: Academic Press.

Sage, R. D., J. R. Contreras, V. G. Roig, and J. L. Patton. 1986. Genetic variation in the South American burrowing rodents of the genus *Ctenomys* (Rodentia: Ctenomyidae). *Z. Säugetierk.* 51: 158–72.

Sanborn, C. C. 1931. A new *Oxymycterus* from Misiones, Argentina. *Proc. Biol. Soc. Washington* 44:1–2.

———. 1947a. Geographical races of the rodent *Akodon jelskii* Thomas. *Fieldiana: Zool.* 31(17): 133–42.

———. 1947b. The South American rodents of the genus *Neotomys*. *Fieldiana: Zool.* 31:51–57.

———. 1949. Cavies of southern Peru. *Proc. Biol. Soc. Washington* 63:133–34.

———. 1950. Small rodents from Peru and Bolivia. *Publ. Mus. Hist. Nat. "Javier Prado,"* ser. A, *Zool.* 5:1–16.

Schaller, G. B. 1983. Mammals and their biomass on a Brazilian ranch. *Arq. Zool. São Paulo* 31(1): 1–36.

Schaller, G. B., and P. G. Crawshaw, Jr. 1981. Social organization in a capybara population. *Säugetierk. Mitt.* 29:3–16.

Schamberger, M., and G. Fulk. 1974. Mamíferos del Parque Nacional Fray Jorge. *Idesia* (Chile) 3:167–79.

Schneider, C. O. 1946. Catálogo de los mamíferos de la provincia de Concepción. *Bol. Soc. Biol. Concepción* 21:67–83.

Scrocchi, G., A. M. P. de Fonollat, and H. H. Salas. 1986. *Akodon andinus dolichonyx* (Philippi), (Rodentia, Cricetidae) en la provincia de Tucumán, Argentina. *Acta Zool. Lilloana* 38(2): 113–18.

Shamel, H. H. 1931. *Akodon chacoensis,* a new cricetine rodent from Argentina. *J. Mammal.* 21: 427–29.

Sielfeld, W. H. 1979. Presencia de *Microcavia australis* (G. y d'O) en Magallanes (Mammalia: Caviidae). *Anal. Inst. Patagonia, Punta Arenas* (Chile) 20:197–99.

Simonetti, J. A., E. R. Fuentes, and R. D. Otaiza. 1985. Habitat use by two rodent species in the high Andes of central Chile. *Mammalia* 49: 19–25.

Simpson, G. G. 1945. The principles of classification and a classification of the mammals. *Bull. Amer. Mus. Nat. Hist.* 85:1–350.

———. 1959. The nature and origin of supraspecific taxa. *Cold Spring Harbor Symp. Quant. Biol.* 24:255–72.

Smythe, N. D. E. 1970. Ecology and behavior of the agouti (*Dasyprocta punctata*) and related species on Barro Colorado Island, Panama. Ph.D. diss., University of Maryland.

———. 1978. The natural history of the Central American agouti (*Dasyprocta punctata*). *Smithsonian Contrib. Zool.* 257:1–52.

Spotorno O., A. 1976. Análisis taxonómico de tres especies altiplánicas del género *Phyllotis* (Rodentia, Cricetidae). *Anal. Mus. Hist. Nat.* 9:141–61.

Spotorno O., A., and R. Fernández. 1976. Chromosome stability in southern *Akodon* (Rodentia, Cricetidae). *Mammal. Chrom. Newsl.* 17:13–14.

Spotorno O., A., and L. Walker 1983. Análisis electroforético y biométrico de dos especies de *Phyllotis* en Chile central y sus híbridos experimentales. *Rev. Chil. Hist. Nat.* 56(1): 15–23.

Stallings, J. R. 1988a. Small mammal communities in an eastern Brazilian park. Ph.D. diss., University of Florida, Gainesville.

———. 1988b. Small mammal inventories in an eastern Brazilian park. *Bull. Florida State Mus.* 34(4): 153–200.

Streilein, K. E. 1982a. The ecology of small mammals in the semiarid Brazilian caatinga. 1. Climate and faunal composition. *Ann. Carnegie Mus.* 51(5): 79–107.

———. 1982b. The ecology of small mammals in the semiarid Brazilian caatinga. 2. Water relations. *Ann. Carnegie Mus.* 51(6): 109–26.

———. 1982c. The ecology of small mammals in the semiarid Brazilian caatinga. 3. Reproductive biology and population ecology. *Ann. Carnegie Mus.* 51(13): 251–69.

Taber, A. B., and D. W. Macdonald. 1984. Scent dispensing papillae and associated behaviour of the mara, *Dolichotis patagonum* (Rodentia: Caviomorpha). *J. Zool.* 203:298–301.

Tálice, R. V. 1969. Mamíferos autóctonos. *Nuestra Tierra* 5:1–68.

Tálice, R. V., R. Caprio, and E. Momigliano. 1954. Distribución geográfica y hábitat de *Ctenomys torquatus*. *Arch. Soc. Biol. Montevideo* 21:133–39.

Tálice, R. V., and E. Momigliano. 1954. Arquitectura y microclima de las tuqueras o moradas de *Ctenomys torquatus*. *Arch. Soc. Biol. Montevideo* 21:126–33.

Tálice, R. V., S. L. de Mosera, and A. M. S. de Sprechmann. 1972. Problemas de captura y sobrevida en cautividad en *Ctenomys torquatus*. *Rev. Biol. Uruguay* 1(2): 121–28.

Tamayo, M., and D. Frassinetti. 1980. Catálogo de los mamíferos fósiles y vivientes de Chile. *Mus. Nac. Hist. Nat. Chile* 37:323–99.

Texera, W. A. 1975. Descripción de una nueva subespecie de *Ctenomys magellanicus* (Mammalia: Rodentia: Ctenomyidae) de Tierra del Fuego, Magallanes, Chile. *Anal. Inst. Patagonia, Punta Arenas* (Chile) 6(1–2): 163–67.

Thomas, O. 1898a. On the small mammals collected by Dr. Borelli in Bolivia and northern Argentina. *Boll. Mus. Zool. Anat. Comp.* 13(315): 1–4.

———. 1898b. On some new mammals from the neighbourhood of Mount Sahama, Bolivia. *Ann. Mag. Nat. Hist.*, ser. 7, 1:277–83.

———. 1898c. Description of two new Argentine rodents. *Ann. Mag. Nat. Hist.*, ser. 7, 1:283–86.

———. 1898d. On new small mammals from the Neotropical region. *Ann. Mag. Nat. Hist.*, ser. 6, 18:301–14.

———. 1900a. Descriptions of new rodents from western South America. *Ann. Mag. Nat. Hist.*, ser. 7, 6:383–87.

———. 1900b. Descriptions of new rodents from western South America. *Ann. Mag. Nat. Hist.*, ser. 7, 6:294–302.

———. 1902. On mammals from the Serra do Mar of Paraná, collected by Mr. Alphonse Robert. *Ann. Mag. Nat. Hist.*, ser. 7, 4(9): 59–64.

———. 1916a. Notes on Argentine, Patagonian, and Cape Horn Muridae. *Ann. Mag. Nat. Hist.*, ser. 8, 17:182–87.

———. 1916b. Two new Argentine rodents, with a new subgenus of *Ctenomys*. *Ann. Mag. Nat. Hist.*, ser. 8, 18:304–6.

———. 1917. On small mammals from the delta of the Paraná. *Ann. Mag. Nat. Hist.* 20:95–100.

———. 1918a. On small mammals from Salta and Jujuy collected by Mr. E. Budin. *Ann. Mag. Nat. Hist.*, ser. 9, 1:186–93.

———. 1918b. Two new tuco-tucos from Argentina. *Ann. Mag. Nat. Hist.*, ser. 9, 1:38–40.

———. 1919a. On small mammals from "Otro Cerro," north-eastern Rioja, collected by Sr. E. Budin. *Ann. Mag. Nat. Hist.*, ser. 9, 3:489–500.

———. 1919b. List of mammals from the highlands of Jujuy, north Argentina, collected by Sr. E. Budin. *Ann. Mag. Nat. Hist.*, ser. 9, 4:128–35.

———. 1919c. On small mammals collected by Sr.

E. Budin in north-western Patagonia. *Ann. Mag. Nat. Hist.*, ser. 9, 3(14): 199–212.

———. 1920a. A new tuco-tuco from Tucumán. *Ann. Mag. Nat. Hist.*, ser. 9, 6:243–44.

———. 1920b. A further collection of mammals from Jujuy. *Ann. Mag. Nat. Hist.*, ser. 9, 5(26): 188–96.

———. 1921a. A new tuco-tuco from Bolivia. *Ann. Mag. Nat. Hist.*, ser. 9, 7(37): 136–37.

———. 1921b. Two new Argentine forms of skunk. *Ann. Mag. Nat. Hist.*, ser. 9, 8(44): 221–22.

———. 1921c. The tuco-tuco of San Juan, Argentina. *Ann. Mag. Nat. Hist.*, ser. 9, 7(42): 523–24.

———. 1925. On some Argentine mammals. *Ann. Mag. Nat. Hist.*, ser. 9, 15:582–86.

———. 1926. The Spedan Lewis South American exploration. 3. On mammals collected by Sr. Budin in the province of Tucumán. *Ann. Mag. Nat. Hist.*, ser. 9, 17(101): 602–8.

———. 1927. On a further collection of mammals made by Sr. E. Budin in Neuquén, Patagonia. *Ann. Mag. Nat. Hist.*, ser. 9, 19(114): 650–58.

———. 1929. The mammals of Señor Budin's Patagonian expedition, 1927–1928. *Ann. Mag. Nat. Hist.*, (4)10: 35–45.

Thomas, O., and J. St. Leger. 1926. The Spedan Lewis South American exploration. 5. Mammals obtained by Señor E. Budin in Neuquén. *Ann. Mag. Nat. Hist.*, ser. 9, 18:635–41.

Tonni, E. P. 1984. The occurrence of *Cavia tschudi* (Rodentia, Caviidae) in the southwest of Salta province, Argentina. *Stud. Neotrop. Fauna Environ.* 19(3): 155–58.

Torres-Mura, J. C., M. L. Lemus, and L. C. Contreras, 1989. Herbivorous specialization of the South American desert rodent *Tympanoctomys barrae*. *J. Mammal.* 70:646–48.

Vallejo, S., and E. Gudynas. 1981. Notas sobre la distribución y ecología de *Calomys laucha* en Uruguay (Rodentia: Cricetidae). *C. E. D. Orione Cont. Biol.* (4)1: 1–16.

Vaz-Ferreira, R. 1958. Nota sobre Cricetinae del Uruguay. *Arch. Soc. Biol. Montevideo* 24:66–75.

Veiga-Borgeaud, T. 1982. Données écologiques sur *Oryzomys nigripes* (Desmarest, 1819) (Rongeurs, Cricétidés) dans le foyer naturel de peste de Barração dos Mendes (état de Rio de Janeiro, Brasil). *Mammalia* 46:335–59.

Veiga-Borgeaud, T., D. R. Lemos, and C. B. Bastos. 1987. Etude de la dynamique de population d'*Holochilus brasiliensis* (Rongeurs, Cricétidés), réservoir sauvage de *Schistosoma mansoni* (Baixada do Maranhão, São Luiz, Brasil). *Mammalia* 51(2): 249–57.

Venegas, W. 1973. El cariotipo de *Ctenomys mauli-*

nus maulinus Philippi (Rodentia, Ctenomyidae). *Bol. Soc. Biol. Concepción* 46:145–54.

Venegas, W., and C. Smith. 1974. Los cromosomas de *Ctenomys maulinus bruneus* Osgood (Rodentia, Ctenomyidae). *Bol. Soc. Biol. Concepción* 48:281–87.

Vergara, S. G. 1982. Cría masiva y explotación del agutí en cautiverio. In *Zoología neotropical*, vol. 2, ed. P. J. Salinas, 1479–96. Mérida, Venezuela: Congreso Latinamericano de Zoología.

Vieira, C. 1945. Sobre uma coleção de mamíferos de Mato Grosso. *Arq. Zool. São Paulo* 4:395–430.

Villafañe, G. de. 1981a. Reproducción y crecimiento de *Akodon azarae azarae* (Fischer, 1829). *Hist. Nat.* 1(28): 193–204.

———. 1981b. Reproducción y crecimiento de *Calomys musculinus murillus* (Thomas, 1916). *Hist. Nat.* 1(33): 237–56.

Villafañe, G., and S. M. Bonaventura. 1987. Ecological studies in crop fields of the endemic area of Argentine hemorrhagic fever: *Calomys musculinus* movements in relation to habitat and abundance. *Mammalia* 51(2): 233–48.

Villafañe, G. de, F. O. Kravetz, M. J. Piantanida, and J. A. Crespo. 1973. Dominancia, densidad e invasión en una comunidad de roedores de la localidad de Pergamino (provincia de Buenos Aires). *Physis*, sec. C, 32(84): 47–59.

Vorontzov, N. N. 1960. The ways of food specialization and evolution of the alimentary system in Muroidea. In *Symposium theriologicum*, ed. J. Kratochvil, 360–77. Brno: Ceskoslovenska Akademic Ved.

Voss, R. S. 1988. Systematics and ecology of ichthyomyine rodents (Muroidea): Patterns of morphological evolution in a small adaptive radiation. *Bull. Amer. Mus. Nat. Hist.* 188(2): 259–493.

Walker, L. I., A. E. Spotorno, and J. Arrau. 1984. Cytogenetic and reproductive studies of two nominal subspecies of *Phyllotis darwini* and their experimental hybrids. *J. Mammal.* 65:220–30.

Weir, B. J. 1971. A trapping technique for tuco-tucos, *Ctenomys talarum. J. Mammal.* 52:836–39.

———. 1974a. The tuco-tuco and plains viscacha. *Symp. Zool. Soc. Lond.* 34:113–30.

———. 1974b. Reproductive characteristics of hystricomorph rodents. In *The biology of hystricomorph rodents*, ed. W. Rowlands and B. J. Weir, 265–99. London: Academic Press.

Williams, D. F., and M. A. Mares. 1978. A new genus and species of phyllotine rodent (Mammalia: Muridae) from northwestern Argentina. *Ann. Carnegie Mus.* 47(9): 193–221.

Wilson, S. C., and D. G. Kleiman. 1974. Eliciting play: A comparative study. *Amer. Zool.* 14: 341–70.

Wolffsohn, J. A. 1923. Medidas máximas y mínimas de algunos mamíferos chilenos colectados entre los años 1896 y 1917. *Rev. Chil. Hist. Nat.* 27: 159–67.

Wood, A. E. 1955. A revised classification of the rodents. *J. Mammal.* 36:165–87.

———. 1974. The evolution of the Old World and New World hystricomorphs. In *The biology of hystricomorph rodents*, ed. W. Rowlands and B. J. Weir, 21–54. London: Academic Press.

Woods, C. A. 1982. The history and classification of South American hystricognath rodents: Reflections on the far away and long ago. In *Mammalian biology in South America*, ed. M. A. Mares and H. H. Genoways, 377–92. Pymatuning Symposia in Ecology 6. Special Publication Series. Pittsburgh: Pymatuning Laboratory of Ecology, University of Pittsburgh.

———. 1984. Hystricognath rodents. In *Orders and families of Recent mammals of the world*, ed. S. Anderson and J. Knox Jones, Jr., 389–446. New York: John Wiley.

Woods, C. A., and D. K. Boraker. 1975. *Octodon degus. Mammal. Species* 67:1–5.

Worth, C. B. 1967. Reproduction, development and behavior of captive *Oryzomys laticeps* and *Zygodontomys brevicauda* in Trinidad. *Lab. Anim. Care* 17(4): 355–61.

Ximénez, A. 1980. Notas sobre el genéro *Cavia* Pallas con la descripción de *Cavia magna* sp. n. (Mammalia-Caviidae). *Rev. Nordest. Biol.* 3 (special issue): 145–79.

Ximénez, A., and A. Langguth. 1970. *Akodon cursor montensis* en el Uruguay (Mammalia-Cricetinae). *Com. Zool. Mus. Hist. Nat. Montevideo* 10(128): 1–5.

Ximénez, A., A. Langguth, and R. Praderi. 1972. Lista sistemática de los mamíferos del Uruguay. *Anal. Mus. Nac. Hist. Nat. Montevideo* 7(5): 1–45.

Yáñez, J., and F. Jaksić. 1978. Historia natural de *Octodon degus* (Molina) (Rodentia, Octodontidae). *Mus. Nac. Hist. Nat., Publ. Ocas.* 27:3–11.

Yáñez, J., W. Sielfeld, J. Valencia, and F. Jaksić. 1978. Relaciones entre la sistemática y la morfometria del subgénero *Abrothrix* (Rodentia: Cricetidae) en Chile. *Anal. Inst. Patagonia, Punta Arenas* (Chile) 9:185–97.

Yáñez, J. L., and J. C Torres-Mura, J. R. Rau, and L. C. Contreras. 1987. New records and current status of *Euneomys* (Cricetidae) in southern South America. *Fieldiana: Zool.*, n.s., 39:283–87.

Yáñez, J., J. Valencia, and F. Jaksić. 1979. Morfometría y sistemática del subgénero *Akodon* (Rodentia) en Chile. *Arch. Biol. Med. Exp.* 12: 197–202.

Yepes, J. 1935. Los mamíferos de Mendoza y sus relaciones con las faunas limítrofes. In *Novena Reunión de la Sociedad Argentina de Patología Regional,* 689–725. Jujuy: Mision de Estudios de patología regional Argentina.

———. 1938. Disquisiciones zoogeográficas referidas a mamíferos: Comunes a las faunas de Brasil y Argentina. *Anal. Soc. Argent. Estud. Geogr.* 6:37–60.

———. 1944. Comentarios sobre cien localidades nuevas para mamíferos sudamerícanos. *Rev. Argent. Zoogeogr.* 4(1–2): 59–71.

Zunino, S., and C. Vivar. 1986. Densité de population d'*Octodon degus* (Rongeurs, Octodontidés) au Chili central. *Mammalia* 50(1): 116–18.

Diagnosis

Lagomorphs are small, digitigrade mammals with the tail either extremely short and well furred or virtually absent. The upper lip is cleft but mobile and can be compressed in the midline. The dental formula for the family is variable: I 2/1, C 0/0, P 3/2, M 2–3/3. The first upper incisor has a longitudinal groove down the midline. The second pair of upper incisors is immediately behind the first, and this character is diagnostic (fig. 12.1).

Distribution

Aside from introduction by humans, members of the Lagomorpha were found on all major continents and islands except Australia, New Zealand, and Antarctica. They were introduced by Europeans in Australia and New Zealand, where they have multiplied to become serious pests.

History and Classification

There are two Recent families, the Ochotonidae and the Leporidae; both originated in the Oligocene. The oldest fossils believed to belong to this order trace from the Paleocene. Lagomorphs radiated and diversified in Eurasia and North America; they did not enter South America until the completion of the Panamanian land bridge at the end of the Pliocene. European hares and rabbits have been widely introduced in southern South America (see chap. 14).

FAMILY LEPORIDAE
Rabbits and Hares

Diagnosis

The dental formula is I 2/1, C 0/0, P 3/2, M 3/3. The ears are rather long. The hind limbs are longer

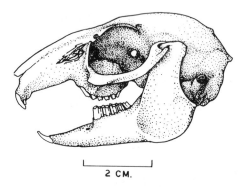

Figure 12.1. Skull of *Sylvilagus brasiliensis*.

than the forelimbs and typify an adaptation for quadrupedal, ricochetal locomotion. There is a strong reduction of the first digit on the forefeet and hind feet. The tail is short but well haired, with a fluffy appearance.

Distribution
See the ordinal account.

Genus *Sylvilagus* Gray, 1867
Cottontail Rabbit, Conejo

Description

These rather small rabbits have a head and body length ranging from 250 to 450 mm; the tail is 25 to 60 mm (fig. 12.2). Weight ranges from 400 to 2,300 g. There are thirteen species. The South American forms were reviewed by Tate (1933).

Distribution

The genus is distributed from southern Canada to Argentina, occupying a wide range of habitats from semiarid, brushy areas to forests. One species occurs in the area covered by this volume, *Sylvilagus brasiliensis*. See chapter 14 for a discussion of introduced forms.

Figure 12.2. *Sylvilagus brasiliensis.*

• *Sylvilagus brasiliensis* (Linnaeus, 1758)
Brazilian Cottontail, Tapití

Measurements

	Mean	Min.	Max.	N	Loc.	Source[a]
TL	340.9	267	405	24	A, Br, P	1
	353.0	338	365	6	P	2
HB	320.0	257	390	23	A, Br, P	2
T	20.9	10	36			
HF	71.1	59	80			
	73.0	71	76	6	P	2
E	53.5	45	61	18	A, Br, P	1
	55.0	52	59	6	P	2
Wta	933.8	575	1,500	4	A, Br, P	1

[a](1) Thomas 1920; Schaller 1983; PCorps; UConn. (2) Hershkovitz 1950.

Description
Its small size, delicate form, short tail, warmly colored rump, and six mammae distinguish S. *brasiliensis* from all other Neotropical rabbits. The dorsum is brown, but the black tips of the dorsal hairs give a speckled appearance. There is a rufous spot at the nape of the neck. The chest is rufous, as is the underside of the tail, and the venter is white. There are pale spots above the eye and on the muzzle (Hershkovitz 1950; pers. obs.).

Distribution
Sylvilagus brasiliensis ranges from southern Mexico to northern Argentina. In southern South America it is found in both eastern and western Paraguay and in Argentina across the northern provinces and at least as far south as Tucumán province (Honacki, Kinman, and Koeppl 1982; Ojeda and Mares 1989; BA, UConn, UM) (map 12.1).

Life History
In Misiones province, Argentina, females reproduce in September. In Paraguay a female with three embryos was collected (Crespo 1982a,b; UConn).

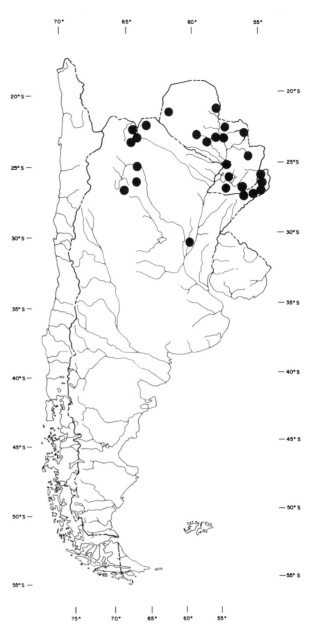

Map 12.1. Distribution of *Sylvilagus brasiliensis.*

Ecology
In Jujuy province, Argentina, this rabbit is common to about 2,500 m, preferring forest edges and brushy areas. It nests in brush heaps, hollow trunks of trees, and at the bases of trees among roots. In Paraguay specimens were collected in forest with bamboo understory, in orange groves, and among bromeliads (Hershkovitz 1950; Olrog 1979; UM).

References
Crespo, J. A. 1982a. Introducción a la ecología de los mamíferos del Parque Nacional el Palmar, Entre Ríos. *Anal. Parques Nac.* 15:1–33.

———. 1982b. Ecología de la comunidad de mamíferos del Parque Nacional Iguazú, Misiones. *Rev. Mus. Argent. Cienc. Nat. "Bernardino Rivadavia," Ecol.* 3(2): 45–162.

Hershkovitz, P. 1950. Mammals of northern Colombia, preliminary report no. 6: Rabbits (Leporidae), with notes on the classification and distribution of the South American forms. *Proc. U.S. Nat. Mus.* 100:327–75.

Honacki, J. H., K. E. Kinman, and J. W. Koeppl, eds. 1982. *Mammal species of the world.* Lawrence, Kans.: Allen Press and Association of Systematics Collections.

Ojeda, R. A., and M. A. Mares. 1989. *A biogeographic analysis of the mammals of Salta province, Argentina.* Special Publication of the Museum 27. Lubbock: Texas Tech University Press.

Olrog, C. C. 1979. Los mamíferos de la selva húmeda, Cerro Calilegua, Jujuy. *Acta Zool. Lilloana* 33:9–14.

Schaller, G. B. 1983. Mammals and their biomass on a Brazilian ranch. *Arq. Zool. São Paulo* 31(1): 1–36.

Tate, G. H. H. 1933. Taxonomic history of the Neotropical hares of the genus *Sylvilagus. Amer. Mus. Novitat.* 661:1–10.

Thomas, O. 1920. A further collection of mammals from Jujuy. *Ann. Mag. Nat. Hist.*, ser. 9, 5(26): 188–96.

Plates

All specimens on a given plate are drawn to the same scale
unless otherwise indicated.

Plate 1: (a) *Philander opossum*; (b) *Caluromys lanatus*; (c) *Metachirus nudicaudatus*; (d) *Lutreolina crassicaudata*; (e) *Chironectes minimus*; (f) *Didelphis albiventris*; (g) face of *Didelphis marsupialis*.

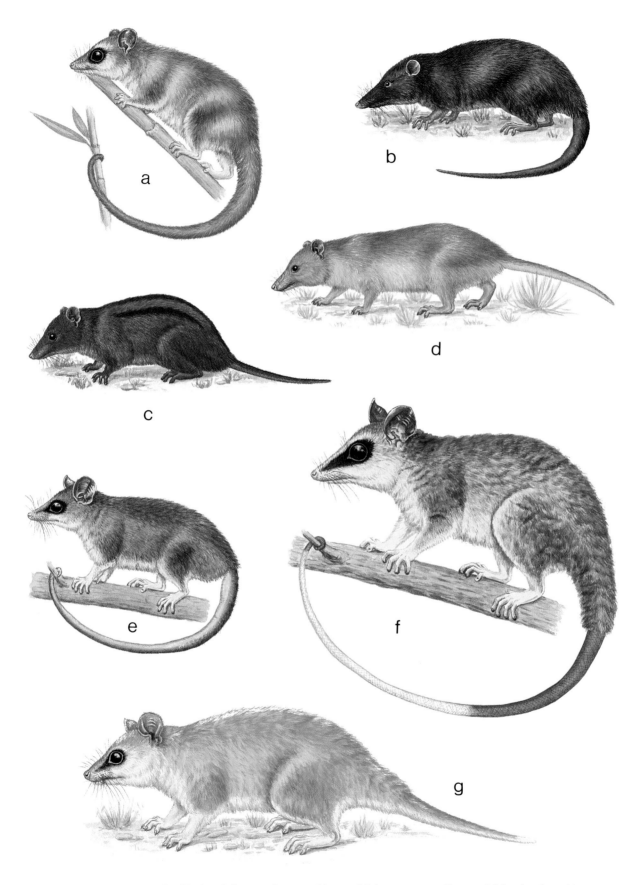

Plate 2: (a) *Dromiciops australis;* (b) *Ryncholestes raphanurus;* (c) *Monodelphis americana;* (d) *Monodelphis dimidiata;* (e) *Marmosa elegans;* (f) *Marmosa cinerea;* (g) *Lestodelphys halli.*

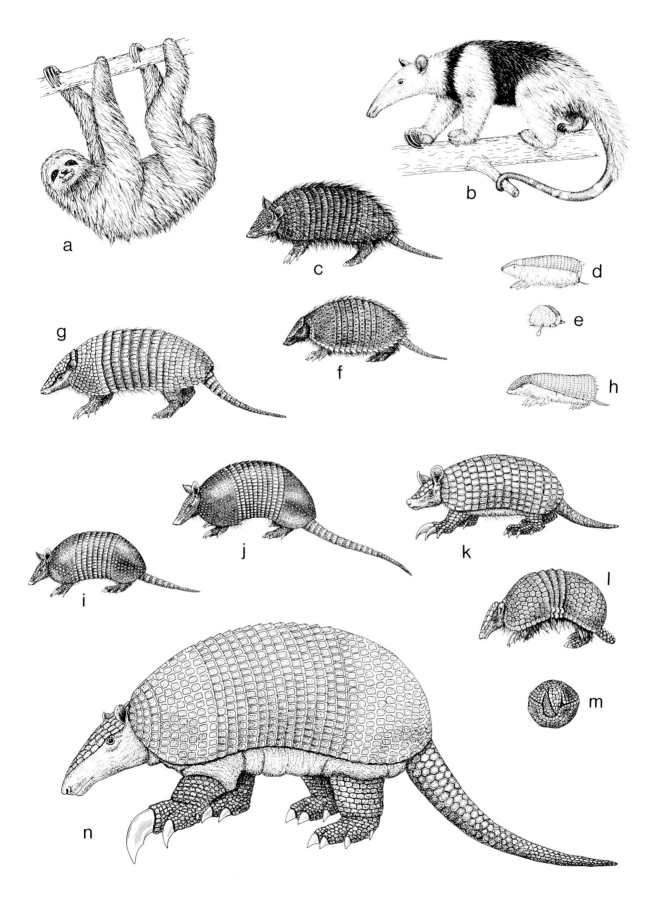

Plate 3: (a) *Bradypus variegatus*; (b) *Tamandua tetradactyla*; (c) *Chaetophractus* sp.; (d) *Chlamyphorus truncatus*; (e) Rump shield of *Chlamyphorus truncatus*; (f) *Zaedyus pichiy*; (g) *Euphractus sexcinctus*; (h) *Chlamyphorus retusus*; (i) *Dasypus hybridus*; (j) *Dasypus novemcinctus*; (k) *Cabassous unicinctus*; (l) *Tolypeutes matacus*; (m) *Tolypeutes* rolled into a ball; (n) *Priodontes maximus*.

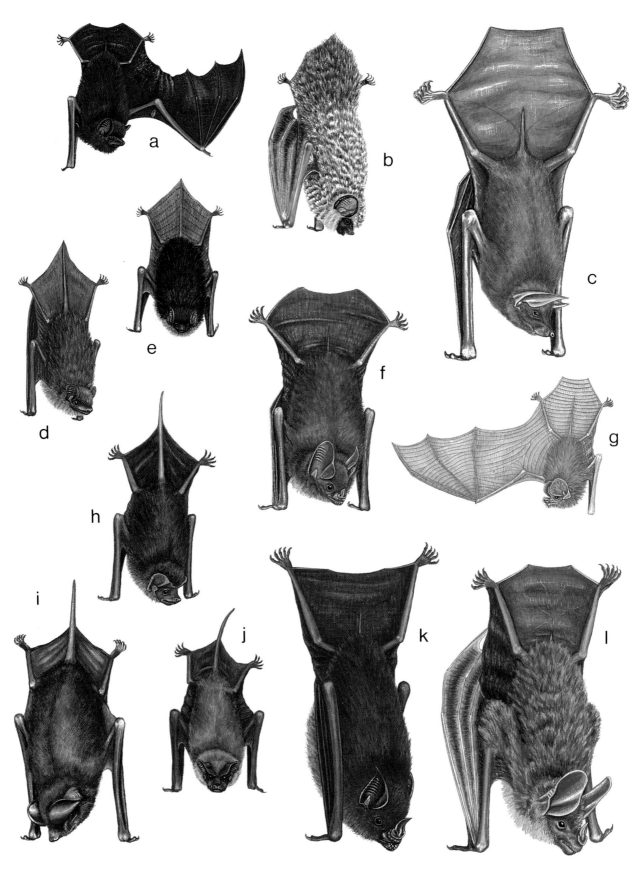

Plate 4: (a) *Peropteryx macrotis*; (b) *Lasiurus cinereus*; (c) *Noctilio leporinus*; (d) *Eptesicus brasiliensis*; (e) *Myotis nigricans*; (f) *Tonatia bidens*; (g) *Amorphochilus schnablii*; (h) *Promops nasutus*; (i) *Eumops glaucinus*; (j) *Tadarida brasiliensis*; (k) *Phyllostomus hastatus*; (l) *Chrotopterus auritus*.

Plate 5: (a) *Aotus azarae*; (b) *Callithrix argentata*; (c) *Cebus apella*; (d) *Calicebus moloch*; (e) *Alouatta fusca* (♂/♀); (f) *Alouatta caraya* (♂/♀).

Plate 6: (a) *Cerdocyon thous*; (b) *Dusicyon griseus*; (c) *Dusicyon culpaeus*; (d) *Dusicyon gymnocerus*; (e) *Speothos venaticus*; (f) *Chrysocyon brachyurus*.

Plate 7: (a) *Eira barbara*; (b) *Lyncodon patagonicus*; (c) *Conepatus chinga*; (d) *Galictis cuja*; (e) *Procyon cancrivorus*; (f) *Nasua nasua*.

L. provocax

L. felina

L. longicaudis

Plate 8: (a) *Lutra provocax* with rhineria of (top to bottom) *Lutra provocax, Lutra felina*, and *Lutra longicaudis*; (b) *Pteronura*; (c) *Panthera onca*; (d) *Puma concolor*.

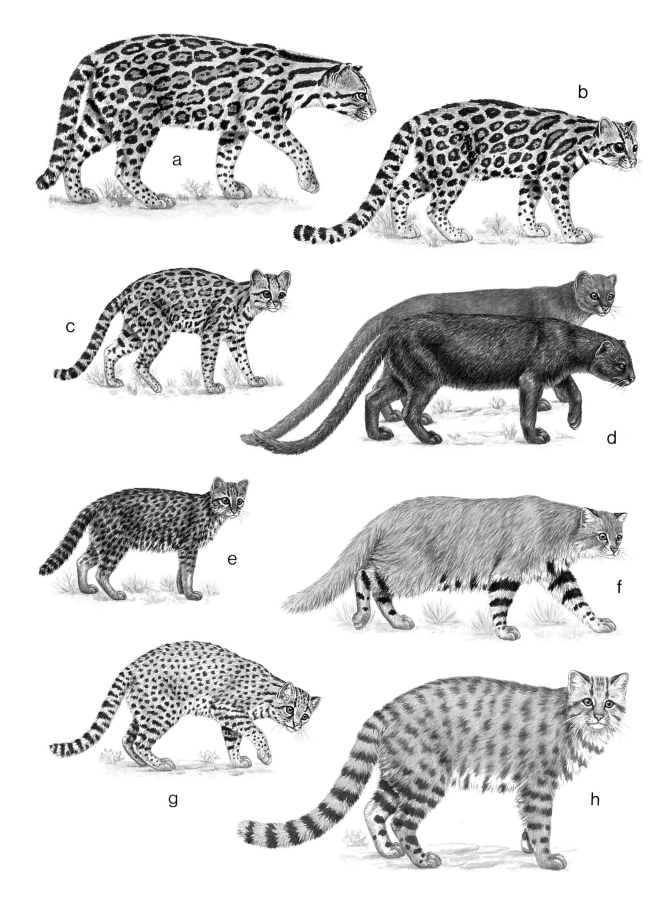

Plate 9: (a) *Felis pardalis*; (b) *Felis wiedii*; (c) *Felis tigrina*; (d) *Felis yagouaroundi* (two color phases); (e) *Felis guigna*; (f) *Felis colocolo*;
(g) *Felis geoffroyi*; (h) *Felis jacobita*.

Plate 10: (a) *Otaria byronia*; (b) *Arctocephalus australis*; (c) *Mirounga leonina*; (d) *Phocoena spinipinnis*; (e) *Lagenorhynchus obscurus*; (f) *Stenella coeruleoalba*; (g) *Cephalorhynchus commersonii*; (h) *Tursiops truncatus*; (i) *Pontoporia blainvillei*. (Pinnipeds not to the scale of cetaceans.)

Plate 11: (a) *Ziphius cavirostris*; (b) *Mesoplodon layardii*; (c) *Orcinus orca* male with dorsal fin of female; (d) *Balaenoptera acutorostrata*; (e) *Megaptera novaeangliae*; (f) *Physeter catodon*; (g) *Kogia breviceps*; (h) *Caperea marginata*; (i) *Balaena glacialis*.

Plate 12: (a) *Pudu puda*; (b) *Hippocamelus bisculus*; (c) *Ozotoceros bezoarticus*; (d) *Blastocerus dichotomus*; (e) *Tapirus terrestris*.

Plate 13: (a) *Mazama americana*; (b) *Mazama gouazoubira*; (c) *Lama vicugna*; (d) *Mazama rufina*; (e) *Catagonus wagneri*; (f) *Lama guanicoe*; (g) *Tayassu pecari*; (h) *Tayassu tajacu*.

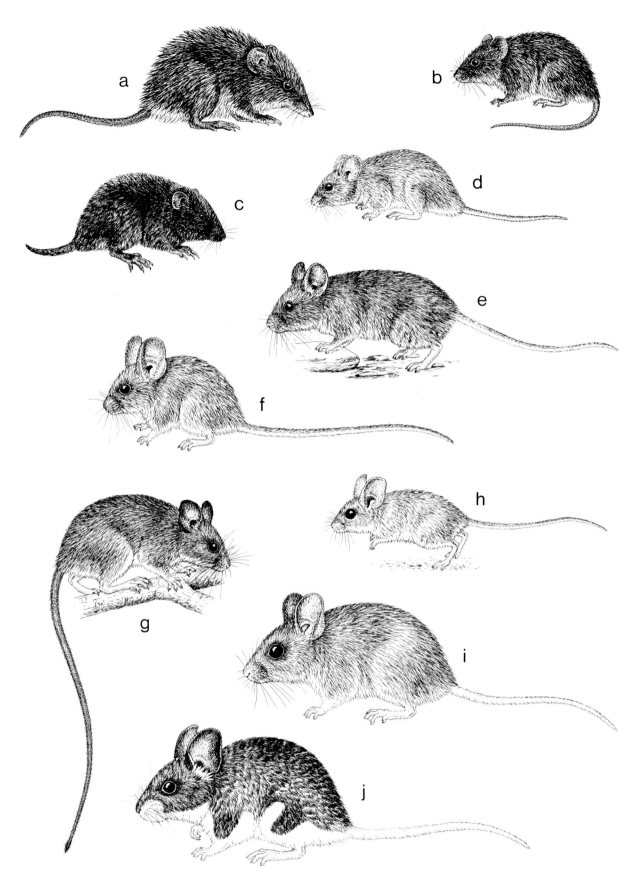

Plate 14: (a) *Oxymycterus rutilans;* (b) *Akodon olivaceus;* (c) *Geoxus valdivianus;* (d) *Calomys laucha;* (e) *Auliscomys micropus;* (f) *Phyllotis darwini;* (g) *Irenomys tarsalis;* (h) *Eligmodontia typus;* (i) *Reithrodon physodes;* (j) *Chinchillula sahamae.*

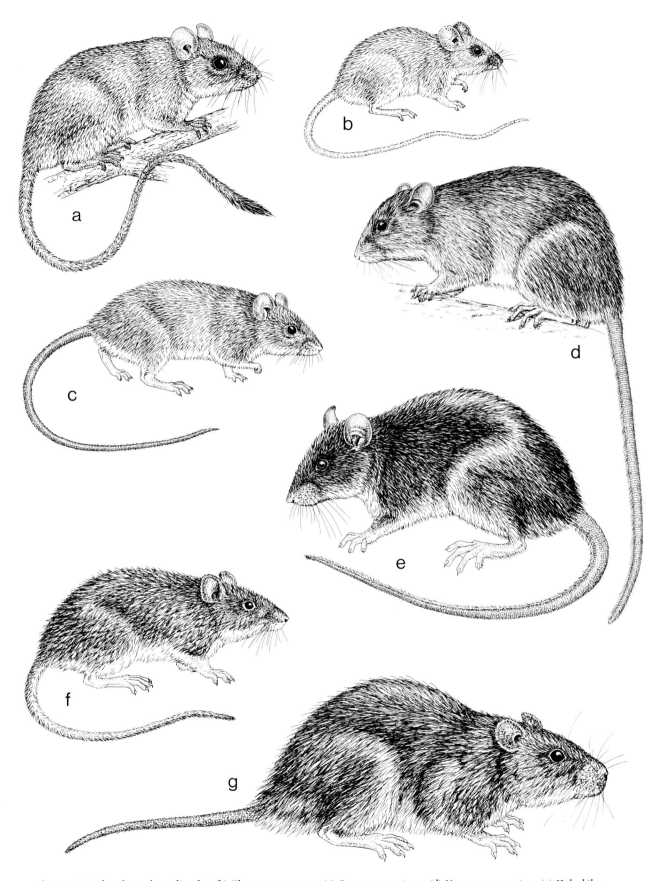

Plate 15: (a) *Rhipidomys leucodactylus*; (b) *Thomasomys oenax*; (c) *Oryzomys ratticeps*; (d) *Nectomys squamipes*; (e) *Holochilus brasiliensis*; (f) *Scapteromys tumidus*; (g) *Kunsia tomentosus*.

Plate 16: (a) *Sphiggurus spinosus*; (b) *Sciurus aestuans*; (c) *Coendou prehensilis*; (d) *Microcavia australis*; (e) *Galea musteloides*; (f) *Cavia aperea*; (g) *Dasyprocta punctata bolivae*; (h) *Chinchilla lanigera*; (i) *Lagidium viscacia*; (j) *Agouti paca*; (k) *Lagostomus maximus* (♂ in front, ♀ behind); (l) *Dolichotis patagonum*; (m) *Dolichotis salinicola*.

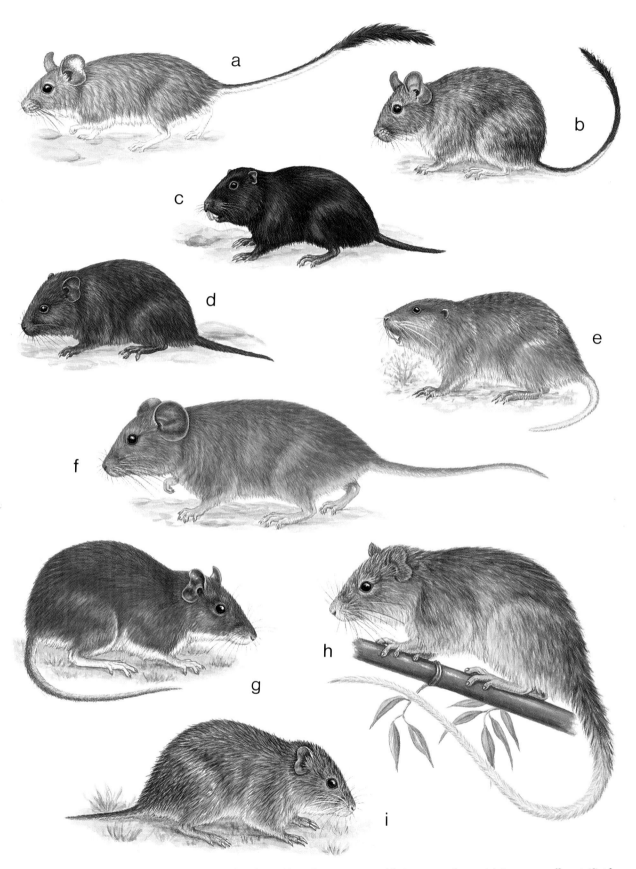

Plate 17: (a) *Octodontomys gliroides*; (b) *Octodon degus*; (c) *Spalacopus cyanus*; (d) *Aconaemys fuscus*; (e) *Ctenomys colburni*; (f) *Abrocoma bennetti*; (g) *Proechimys longicaudatus*; (h) *Kannabateomys amblyonyx*; (i) *Euryzygomatomys spinosus*.

10 cm

30 cm

Plate 18: (a) *Lepus capensis*; (b) *Oryctolagus cuniculus*; (c) *Rattus rattus*; (d) *Mus musculus*; (e) *Ondatra zibethicus*; (f) *Castor cana-densis*; (g) *Sus scrofa*; (h) *Axis axis*; (i) *Dama dama*; (j) *Cervus elaphus*.

13   Mammalian Community Ecology
in Southern South America

Some Generalizations

In this volume we are considering an extremely complex geographical area. The southern end of South America is constricted as one proceeds toward the South Pole; the Andes dominate the western rim of the continent and impede the moisture-laden westerlies. The Tropic of Capricorn cuts through the northern portions of Paraguay, Argentina, and Chile. Xeric areas alternate with mesic areas, and one could scarcely seek a more heterogeneous area in terms of rainfall and temperature than the "southern cone" (see chap. 1).

Let us attempt a few first-order comparisons. The southern end, exemplified by the Falkland (Malvinas) Islands, is dominated by marine mammals. Hamilton (1952) lists fourteen species of cetaceans recorded from this area. Pine, Angle, and Bridge (1978) list only eight species of terrestrial, native mammals for the southern tip of South America, compared with over fourteen species of cetaceans found by Goodall (1978). Clearly, at the southern end of the continent terrestrial mammals are yielding to the dominance of the oceanic forms (see chaps. 7 and 8).

In contrast to the cooler regions of the south are the tropical regions of the north. In the province of Misiones, Argentina, Crespo (1982) completed a survey of the mammal fauna of Iguazú National Park (Argentine side). He and his co-workers recorded fifty-two species of mammals, of which seven were bats, and suspected that twenty-eight more species probably occurred within the combined Brazilian and Argentine administrative area, thus raising the potential total to eighty species. Massoia (1980) lists ninety-five species of native mammals, including twenty-six bats, from the province of Misiones, in which this park is situated. In the northwest of Argentina, the provinces of Jujuy and Salta include lowland tropical areas as well as steep elevational gradients. The result is a "wedding" of Andean and lowland tropical elements. Michael Mares (unpubl. data) recorded a total of 106 species, of which nineteen were bats.

One outstanding fact for readers of volume 1 of *Mammals of the Neotropics* (Eisenberg 1989) is that bat species richness declines rapidly as one proceeds south of 23°30' south latitude, and rodents begin to dominate as one moves into temperate South America. The same principle is true if one climbs and traps over an altitudinal gradient. For example, in the floodplains of the Mato Grosso region of Brazil, Schaller (1983) recorded sixty-seven species of mammals, of which twenty-one were bats and eleven were rodents. Myers (1982) in eastern Paraguay, noted fifteen species of rodents and fifteen species of bats. Massoia (1970) in Formosa province, Argentina, noted seventy species of mammals, of which fourteen were bats and sixteen were rodents. Christie (1984), at Parque Nacional Nahuel Huapi, Argentina, noted thirty-four species of mammals, of which five were bats and sixteen were rodents. At this latter locale in the foothills of the Andes, we are definitely in the Temperate Zone; and in terms of small mammals, rodents dominate. Finally, all the way to the south in Chilean Patagonia, Iriarte (1988b) recorded thirty native mammal species with fifteen rodents and three bats.

If we confine our discussion to small (less than 500 g), trappable mammals (exclusive of bats), some additional generalizations come to light. In the Atlantic rain forest of Brazil, Fonseca and Kierulff (1989) and Stallings (1989) trapped a maximum of fourteen small-mammal species. In their situation, marsupials constituted five species. As one proceeds south, the marsupials diminish and the rodents dominate. Small-mammal communities in Mediterranean or semixeric conditions range from five to seven species—usually with one species of marsupial and the rest rodents (Uruguay Stations 1 and 16, Barlow 1969; Argentina, Monte, Mares 1980b; Chile, Fulk 1975; Meserve et al. 1982; Jaksić, Greene, and Yáñez 1981). In the mosaic of the Chaco in Paraguay, Myers (1982) recorded fifteen species of cricetine rodents and three small marsupials. The temperate rain forests of Argentina and Chile present a slightly different picture. In the south temperate rain forests of Argentina, Pearson (1983) studied a small-mammal community comprising nine rodents and one marsupial. On the other side of the Andes, Patterson, Meserve, and Lang (1989) found seven rodents and two marsupials. Glanz (1977a,b), in the moist temperate forests of Cuesta la Dormida, Chile, studied a small-mammal community with nine rodents and one marsupial.

The conclusion of this review is straightforward. The faunal interchange between North and South America at the close of the Pliocene had negative effects on both the immigrants and the "old endemics." Currently, temperate South America has many ecological niches only partially occupied by mammals when compared with North America (Eisenberg and Redford 1982). Rodents and carnivores dominate in the southern cone, and mammals recently introduced by Europeans, such as rodents, cervids, and lagomorphs, have been very successful (see chap. 14). Rodents dominate in many herbivore niches and have further adapted to insectivore and omnivore niches usually occupied by other taxa on the more contiguous continents. Referring to volume 1 of *Mammals of the Neotropics* (Eisenberg 1989) will quickly show the shifts in faunal dominance. Whereas about 50% of the species in volume 1 were bats, in volume 2 the rodents are numerically dominant in species richness. South America is a case study in catastrophic extinctions. The southern continents may yet have much to teach us about the rate and circumstances of extinction (Eisenberg and Harris 1989).

Some Puzzles in the Temperate Zone of South America

The larger mammals of southern South America seem dominated today by North American immigrants and Old World introductions. The puma (*Puma concolor*) is the major predator on the guanaco (*Lama guanicoe*) in southern Chile and Argentina (Iriarte 1988a). The native Cervidae of the southern cone are clearly of relatively recent North American origin (< 1.5×10^6 ybp) (see Hershkovitz 1982 and Eisenberg 1987 for reviews).

If we turn from the larger mammals to the small,

trappable, nonvolant species of the moist Temperate Zone, then Pearson and Pearson (1982) and Pearson (1983) offer considerable food for thought. Six or seven species of mammals constitute the community they worked on. The tree forms of these forests are dominated by the genus *Nothofagus*. It appears that most of the productivity of these trees is diverted to fungi. There is one marsupial, *Dromiciops*, and there are no hystricognath rodents. All other species are sigmodontine rodents; most if not all of these small mammals do not hibernate, and most are characterized by a low reproductive rate. Noticeably absent from the fauna are mammalian arboreal frugivores and folivores. The biomass of small mammals averaged less than 1,000 g/ha. Eisenberg and Redford (1982) remarked on the absence of squirrel-like forms in these forests. It remains a puzzle, since the *Araucaria* group has seeds with considerable endosperm, though the species of *Nothofagus* have less. Perhaps the sciurids of lowland Jujuy and Misiones provinces have been under long-term selection for tropical forests and have not yet "turned a hand" to harvesting the southern, temperate rain forests.

A Brief Snyopsis of Some Mammal Communities in Southern South America

The Monte Desert of Argentina

Mares (1973, 1975, 1976) worked in both the Monte Desert of Argentina and the Arizona Sonoran Desert in the United States. The Monte Desert exhibits a lower mammalian species richness than Arizona (twenty-eight versus sixty-four). In the Monte there is a noticeable lack of granivorous rodents and large herbivores. Some rodents found in the Monte exhibit adaptations for desert life that are convergent with those of North American rodents adapted to similar niches. These include *Eligmodontia typus*, *Ctenomys fulvus*, and *Octomys mimax*. Clearly the caviormorph rodents have had a longer history of selection for adaptation to xeric conditions.

The most common nonvolant small mammal in this community was *Microcavia australis*. In the Monte they exploit riparian situations and areas supporting shrubs. In the latter habitat, burrows are placed under *Prosopis* or *Bulnesia* bushes. *Oryzomys longicaudatus* is also confined to more mesic habitats, as are *Akodon varius* and *Phyllotis griseoflavus*. *Eligmodontia typus*, while resembling Old World gerbils, is not tied to permanent water and in fact avoids such habitats. It is strongly associated with low shrubs and well-drained fine-grained soils.

The fossorial *Ctenomys fulvus* is most common in *Larrea* flats. It may venture a short distance from the burrow entrance at night, mainly to forage on annuals.

Although the culpeo (*Dusicyon culpaeus*) is present, *D. griseus* is the common canid predator. This fox is an omnivore but does prey on native rodents. The mustelid *Lyncodon patagonicus* may be a far more important predator on *Microcavia* and *Ctenomys* (Mares 1973).

The Pampas as a Grassland Community

The early European explorers of the grasslands in Argentina and Uruguay remarked on the vast numbers of ungulates and rodents. The guanaco (*Lama guanicoe*), pampas deer (*Ozotoceros*), Patagonian cavy (*Dolichotis*), and viscacha (*Lagostomus*) are commented on in writings by Darwin, Hudson, and Azara. As in North America, these temperate grasslands were radically modified by transplanted European agricultural techniques and livestock (Crosby 1986). Today this ecosystem has been virtually eliminated, and only the smaller mammals are accessible for study (Crespo 1966; Cabrera and Yepes 1960). Rood (1972) and Dalby (1974) have offered us a glimpse into this remnant ecosystem. The caviomorph genera *Microcavia*, *Galea*, and *Cavia* have clearly evolved in grasslands and can dominate under appropriate conditions. The subterranean *Ctenomys* species are suitably represented in well-drained soils. Avian, reptilian, and mammalian predators can and do inflict a toll on the mammalian primary consumers of the vegetation (Jaksić and Simonetti 1987). When comparing the nonvolant small-mammal fauna of the pampas with that of the xeric Monte, one recognizes a similarity in composition. Three small cricetine rodents dominated on the trapping grids established by Dalby (1974): *Akodon azarae*, *Oryzomys nigripes*, and *Oxymycterus rutilans*. The more mesic-adapted, insectivorous genus *Oxymycterus* appears on the scene as distinctly different from the more robust, xeric-adapted species and genera in the list compiled by Mares (1973).

Pine, Dalby, and Matson (1985) have made an interesting set of hypotheses concerning the didelphid *Monodelphis dimidiata*, a component of the small-mammal fauna where Dalby worked. This small terrestrial didelphid may be a near approximation to a semelparous mammal, as reviewed for Australia by Lee, Woolley, and Braithwaite (1982), who studied *Antechinus stuartii* and its relatives. Students of South American mammals should consider the case of the Temperate Zone marsupial and consider the consequences of an "in depth" study for our understanding of the selective processes underlying the evolution of life history strategies (see also Eisenberg 1988).

The Patagonian Steppe

The third large open ecosystem to be considered is the Patagonian steppe. Generally defined as lying south of the Río Negro and east of the Andes, it articulates to the west with the Patagonian *Nothofagus* forests. The Patagonian steppe is characterized by scattered low bushes, usually less than 1 m tall, mixed with bunchgrass and considerable bare ground. The shrub *Mulinum spinosus* is a useful indicator species. The steppe—generally cold, flat, and dry—can be viewed in some ways as a southward extension of the high puna of Peru, as reflected in the distribution of such mammalian genera as *Ctenomys*, *Lagidium*, camelids, and *Eligmodontia* (Pearson and Pearson 1982; Pearson 1987).

Over much of this vegetation type there has been little work done on the mammal fauna. There are two exceptions. The first is the work of Oliver Pearson and colleagues (Pearson 1987; Pearson and Christie 1985; Pearson and Pearson 1982). Although not considering only the fauna of the steppe, Pearson has typified the small-mammal fauna of this area as consisting of six forms found largely or exclusively in steppe habitats: *Eligmodontia typus*, *Ctenomys haigi*, *Akodon xanthorhinus*, *Reithrodon auritus*, *Phyllotis darwini*, and *Lestodelphys halli*. *Ctenomys* occurs in open areas with sandy soil, *A. xanthorhinus* prefers bunchgrass areas, *Reithrodon* is a green-grass eater of open habitats, *P. darwini* prefers rocky areas, and *Eligmodontia*, a true desert form restricted to open habitats with bare soil, is the form most commonly trapped. An additional three species are found broadly dispersed from Andean forests down into the steppe: *Auliscomys micropus*, *Akodon longipilis*, and *Oryzomys longicaudatus*.

The second area where research has been done on the steppe mammal fauna is the vicinity of Península Valdés, Chubut province. Taber (Taber and Macdonald 1984; Taber, in prep.) has completed a comprehensive study of the mara (*Dolichotis patagonum*), a quintessential member of the open-habitat mammal fauna of Argentina. Other researchers have examined different aspects of the terrestrial mammal fauna of the area (e.g., Daciuk 1977), though little comprehensive work has been published.

The best-studied component of the Península Valdés mastofauna comprises the marine mammals. For many years researchers from Argentina and other countries have been conducting research on elephant seals (Daciuk 1973), orcas (Lopez and Lopez

1985), various dolphins (Würsig and Bastida 1986), and particularly southern right whales (Payne et al. 1983; Payne and Guinee 1983). The incredibly rich waters combined with the shelters of the peninsula make this an excellent place for such work.

Despite the low diversity of mammals because of low temperatures and low structural diversity of the habitat, there are many interesting projects to be done on the mammals of the Patagonian steppe.

The Mediterranean Ecosystem

Mooney and Dunn (1970) and Mooney, Solbrig, and Simpson (1977) consolidated a great amount of data on climate, soils, and vegetation derived from studies in southern California and Chile. In brief, they underscored the contention that these two areas shared a remarkable similarity in climate and the physiognomy of the vegetation. Strips of trees along watercourses are separated by areas of grassland and low, woody vegetation termed scrub. This ecosystem has received considerable attention from numerous investigators (Glanz 1977a,b; Meserve, Martín, and Rodríguez 1984; Jaksić, Schlatter, and Yáñez 1980; Meserve 1981a,b; Glanz and Meserve 1982; Jaksić and Simonetti 1987).

Although the larger terrestrial mammals and the volant bats (Chiroptera) are vigorous colonizers, the small rodents are a less mobile group and offer interesting insights into biogeography and the distribution of mammals. Caviedes and Iriarte (1988) note that the Andes and the Atacama Desert have presented considerable difficulty for colonization by cricetine (= sigmodontine) rodents in Chile. Thus the Chilean sigmodontine rodent fauna has some distinctive characteristics compared with that of Argentina to the east.

That the Chilean-Mediterranean communities have been altered by humans is not to be disputed (Jaksić and Simonetti 1987; see also chap. 14). In spite of the disastrous effects of European agricultural practices and domestic introductions, the "old endemics" and some Recent (Pliocene) immigrants are holding their own. *Dusicyon culpaeus* still harvests rodent prey thirty-five miles east of Santiago, Chile (Iriarte, Contreras, and Jaksić 1989). Leaving aside the larger mammals, the marsupial and rodent community of the region is distinctive. The small marsupials include only *Marmosa elegans*. This species may exhibit seasonal torpor, and it accumulates considerable general body fat as well as storing fat in its tail before overwintering. Once again our colonizing rodent genera are present (refer to the Monte and pampas discussion): *Oryzomys longicaudadus*, *Akodon olivaceus*, and *Phyllotis darwini*. Some rather unusual caviomorph genera inhabit this re-

gion. The genus *Ctenomys* has been replaced by the ecologically equivalent genus *Spalacopus*. The larger-bodied, ratlike genera are represented by the caviomorphs *Octodon* and *Abrocoma*. In this habitat *Octodon* is numerically dominant; *Abrocoma* may well be in an almost commensal relationship with *Octodon* (see also the species accounts).

The Tropical Communities: Salta and Misiones Provinces, Argentina, and Eastern Paraguay

Tropical vegetation covers much of eastern Paraguay and creeps into western Paraguay, northern Argentina, and northern Uruguay via the riverine gallery forests. An inspection of the range maps in the species accounts will adequately document this contention. As outlined in Eisenberg (1989) and Hall (1981), the phyllostomid bats decrease drastically in species richness as one proceeds toward the poles. In the tropical portion of southern South America, rainfall patterns regulate the timing of reproduction because ultimately rainfall controls the primary productivity of plants. Thus some seasonality prevails in the mammalian life history patterns.

Crespo (1982) has offered us a view of the mammal community in Iguazú National Park. Students of the northern Neotropical forests will recognize the similarities in faunal composition. Arboreal frugivores and folivores dominate in the canopy (e.g., *Bradypus* and *Alouatta*). Frugivorous phyllostomid bats are reaching their southern limit of distribution, but *Artibeus*, *Sturnira*, and *Vampyrops* are still plentiful. Terrestrial frugivores such as *Dasyprocta* and *Agouti* are abundant.

A comparison of the species composition recorded by Crespo (1982) and by Ojeda (1985) reveals the distinctiveness (at the species level) of the mammalian fauna from the tropical portions of western and eastern Argentina. The tropical communities in the east exhibit affinities with those of southeastern Brazil. The western tropical forests and their fauna derive from Bolivia and show affinities with the upper Amazon. In central Argentina and western Paraguay, the xeric Chaco separates these mesic, tropical regions.

Paraguay: A Country of Contrasts

The Río Paraguay divides Paraguay into two radically different ecological communities. To the west lies the Chaco, a vast alluvial plain of high temperatures, little rainfall, and low-stature vegetation. On the other side of the river lies the varied eastern 40% of Paraguay, covered with forests ranging from high tropical to subtropical and broken by savannas. The faunas of the two areas are drastically different, with at least forty-one species of bats found in eastern Paraguay and only twenty-seven in the Chaco. Of these twenty-seven, just one is found only in the Chaco and not in eastern Paraguay. Sixteen species of cricetines are found in eastern Paraguay, nine of which rarely or never cross the Río Paraguay. The Chaco harbors fifteen species, eight of them restricted to the west (Myers 1982).

Eastern Paraguay is home to tropical forest species like *Chironectes*, *Alouatta*, and *Mazama rufina*. The Chaco has arid-adapted forms like *Catagonus*, *Dolichotis salinicola*, and *Cabassous*. In fact, the Chaco has South America's richest armadillo fauna, with eight species. It is also rich in carnivores, with at least seven canids, five procyonids and mustelids, and eight cats.

Predator Influences in Temperate South America

Introduction

Southern South America has a great many mammalian predators, with six canids, two procyonids, nine mustelids, and ten felids. Jaksić and Simonetti (1987) have reviewed the studies on vertebrate predation in the southern cone of South America. Raptors, snakes, and mammalian carnivores are included in their review. Foxes of the genus *Dusicyon* are among the best studied. Dietary selection by foxes strongly tracks the availability of resources, since foxes are omnivorous, readily switching seasonally from vertebrate prey to fruit. Stricter carnivores, such as the mustelid *Galictis*, may shift their ranges to areas showing local abundance of their preferred prey, such as *Microcavia* (Rood 1970b). The diets of predators in the same community may exhibit broad overlap, and some sort of scramble competition is

surely operative; however, predators are highly mobile and adept at switching from prey species to another, thus ameliorating the problem of direct competition. Seasonally, carnivory may become secondary for some species as they include more fruits in their diets.

An Analysis of South American Canid Diets

Members of the family Canidae (exclusive of the domestic dog, *Canis familiaris*) are distributed over a great range of habitats throughout the South American continent. Some species, such as *Speothos venaticus*, have an extremely wide range but appear to be quite rare. *Atelocynus microtis* has a narrower range and appears to be equally rare within it. Other species, such as *Cerdocyon thous*, are widely distributed in a range of habitats over South America and are extremely abundant. It follows, then, that a species like the latter will have been well studied and will provide an excellent data base for ecological inferences. Species that are rare or locally distributed may have been little studied and may for that reason be excluded from certain parts of the analysis. Although both of us have worked with one or two species within the South American assemblage, they have not been a major focus of our fieldwork. Hence this is a compilation from the rich literature, augmented with a few of our own ideas derived from field experiences.

The Species of South American Canids

Table 13.1 summarizes some salient facts concerning the weight class, distribution, and predominant vegetation types used by the South American canids. One species, *Urocyon cinereoargenteus*, is found in northern South America, but its major cen-

Table 13.1   Weight, Distribution, and Predominant Vegetation Types for Species of South American Canids

Species	Adult Weight (kg)	Distribution	Preferred Vegetation Types
Chrysocyon brachyurus	23.30	Northeastern Argentina, Paraguay, Bolivia, Brazil	Cerrado, chaco, open woodland
Cerdocyon thous	6.04	Northern Argentina to Venezuela	Drier forests, savannas
Dusicyon culpaeus	7.37	Tierra del Fuego along Andes to Colombia	Arid, semiarid cool vegetation
D. griseus	3.99	Tierra del Fuego to northern Argentina and Chile	Temperate forests to grasslands
D. gymnocercus	4.49	Argentina, north of Río Negro through Paraguay and Uruguay to southern Brazil and eastern Bolivia	Grasslands, pampas, and open woodlands
D. microtis	ca. 7.00	Amazonian South America	Tropical forests
D. sechurae	2.20	Northwestern Peru and southwestern Ecuador	Coastal desert
D. vetulus	—	Central Brazilian highlands	Cerrado
Speothos venaticus	5.50	Forested areas from Brazil north to Panama	Dry and wet forests
Urocyon cinereoargenteus	3.99	Northern Colombia and Venezuela north to northern North America	Dense cover, drier brush to woodland

Note: References from text.

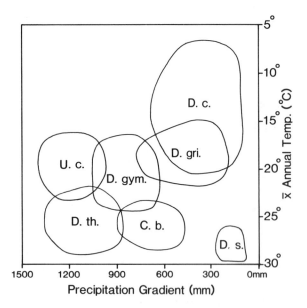

Figure 13.1. Climate diagram for the distribution of South American canids.

ter of distribution is Central America and southeastern North America. It is included, however, for the sake of completeness.

Figure 13.1 is a climate diagram; each species of canid for which we have appropriate data has been plotted indicating the temperature extremes and average precipitation it experiences over its range. *Speothos venaticus* and *Atelocynus microtis* are tropical lowland forms, often occurring in areas of relatively high precipitation. *Speothos*, however, can extend its range into drier, more savannalike forma-

tions, though it usually is tied to permanent gallery forests in association with major watercourses. *Dusicyon culpaeus* experiences some of the lowest temperatures over its range because it is often found in montane habitats and extends very far south in Chile and Argentina. *Dusicyon sechurae* is clearly the most arid adapted of the foxes, occupying areas of very low precipitation and very high mean annual temperature.

The principal components of a species' diet will vary according to such factors as precipitation and mean annual temperature. We know from studies on *Cerdocyon thous* and *Dusicyon culpaeus* that there can be pronounced seasonal differences in diet. These foxes tend to track the most abundant food item, whose availability will vary throughout the annual cycle. In full recognition of this variation, we have attempted to use large samples in our analyses with an eye toward the degree of variation.

The quantitative results (in percentage frequency) and references are included in table 13.2.

Dusicyon culpaeus

We consulted eight studies from Peru, Chile, and Argentina. This data set included analyses of 886 scats and 122 stomachs. Stomach contents correlated rather well with scat analyses: 91% of the stomachs contained mammalian prey, a value approximated by most of the scat analyses. For this reason we combined stomach and scat analyses expressed as percentages of the total sample and concluded that 87% of the diet included mammals, 9% birds, 3% reptiles, 4% invertebrates, and 10% vegetable material

Table 13.2   Food Habits of South American Canids

Species	Number of Studies	Wta (kg) (n)	Number of Stomachs	Number of Scats	Number of Vertebrate Prey	Sheep Carrion	Other Carrion	Hares	Rabbits
Dusicyon culpaeus	9	7.37 (256)	122	886	1,058	3.4	2.7	5.8	15.6
D. griseus	7	3.99 (4)	268	426	616	15.2	0.8	—	2.7
D. gymnocercus	2	4.49 (323)	230	—	270	12.6	3.3	28.1	—
D. sechurae	1	2.20 (?)	—	135	24	—	—	—	—
D. thous	9	6.04 (30)	141	57	302	—	—	—	0.3
Urocyon cinereoargenteus	1	3.99 (171)	111	—	194	—	—	—	25.8
Chrysocyon brachyurus	2	23.30 (9)	68	740	873	—	Anteater	—	2.3

Source: Text, and Asa and Wallace (1990).

(mainly fruits). In some parts of its range *D. culpaeus* may prey on sheep, but a significant proportion of its diet derives from hares, rabbits, and caviomorph rodents.

Dusicyon griseus

We consulted seven studies from Argentina and Chile. The sample included 258 stomachs and 426 scats: 83% of the sample contained mammalian remains, and 43% of this included caviomorph rodents. Birds were found in 15.3% of the sample, invertebrates in 22.9%, and plant material in 38.5%. Clearly this smaller fox takes a higher proportion of invertebrates and fruits than the larger *D. culpaeus*.

Dusicyon gymnocercus

Two studies examined 230 stomachs and 270 scats. Although the remains of sheep and larger ungulates were found in 16% of the stomachs, clearly these were scavanged from carcasses. Of the stomachs, 28% contained lagomorphs, 23% contained cricetine rodents, and only 5% contained remains of caviomorph rodents. Bird remains were found in 29% of the stomachs and reptiles in only 2%. Plant material (fruits) was found in 46.5% of the sample. This small fox has clearly shifted toward smaller prey and is including more fruit in its diet.

Dusicyon sechurae

We consulted a single study for three different departments of Peru. It included a sample of 135 scats. Surprisingly, 10% of the scats contained remains of fish, 25% of reptiles, 41% of birds, 22% of

invertebrates, and 98% of fruit. This small fox has clearly shifted toward arthropods and fruits and away from larger vertebrate prey.

Cerdocyon thous

This widely distributed species has been the subject of nine separate studies at nine sites. A total of 141 stomachs and 57 scats were analyzed. Scavenging carrion involved only 3% of the sample. In the mesic environments it was not surprising that 21% of the sample included frogs: 76% included invertebrates, mainly land crabs, and 75% included fruits. Reptiles were found in 38% of the sample, and mammalian prey, in particular small cricetine rodents, was noted in 25%.

Urocyon cinereoargenteus

This species has been the subject of one study in Venezuela and of many studies in the southeastern United States. Selecting three studies in the United States and combining them with the one from Venezuela, a total of 111 stomachs were sampled: 65% of the sample included vertebrate prey, 15% fruits, and 15% other plant material. Arthropods constituted only 5% of the sample.

Chrysocyon brachyurus

This large species has been the subject of two studies where prey was noted, including 68 stomachs and 740 scats. Once again, scat and stomach samples parallel each other closely. Approximately 70% of the sample included small mammals, and birds composed about 17%, but surprisingly, fruits

Table 13.2   (*continued*)

Caviomorphs	Cricetids	Unidentified Rodents	Total Mammals	Birds	Reptiles	Fish	Percentage of Scats with Invertebrates	Percentage of Scats with Vertebrate Prey	Percentage of Scats with Plants
39.1	13.3	7.7	87.6	9.2	3.2	—	4.1	86.8	10.6
43	9.4	11.5	82.6	15.3	2.1	—	22.9	75.4	38.5
4.8	17.8	—	68.9	9.2	1.9	—	12.2	80.9	46.5
—	8.3	Unidentified vertebrate 16.7	8.3	41.7	25.0	8.3	21.5	15.6	98.5
3.7	12.1	—	24.6	8.4	38.4; frogs 21.9	3.7	76.0	66.4	75.0
—	58.7	1	89.6	7.8	2.6	—	5.0	65.0	15.0
—	—	Unidentified vertebrate 0.3	70.7	28.3	0.7	0.3	14.0	85.0	57.6

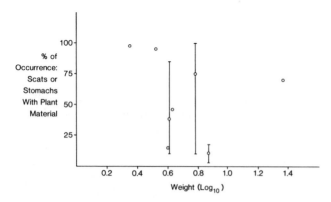

Figure 13.2. Correlation of body size with diet for South American canids.

composed 82%. The high incidence of fruits is because *Chrysocyon* actively feeds on the large, fleshy fruits of *Solanum lycocarpum*.

To account for some of the trends noted in the results, we attempted to see if there was any relation between latitude and the percentage of scats or stomachs containing vertebrate remains. We detected no effect. Similarly, we attempted to analyze what percentage occurrence of plant material in scats or stomachs might be a function of latitude. We detected no trend.

Since the size of the foxes varies greatly, from *Chrysocyon* with a mean weight of 25 kg to *Dusicyon sechurae* with a mean weight of 2.2 kg, we decided to examine any influence of body size on diet. When percentage occurrence of plant material in scats or stomach was plotted against mean body weight, the within-sample variance was so high that only a suggested trend toward a decrease in plant material was discernible. This occurred because of the low frequency of plant material in the diet of *Dusicyon culpaeus* (fig. 13.2). On the other hand, when percentage of scats or stomachs containing vertebrate prey was plotted against body size, a very positive correlation was demonstrable between carnivory and increasing size (fig. 13.3). Even *Chrysocyon*, including *Solanum* in its diet, though not continuing on the regression line, is still high. This suggests that the smaller species, in addition to including fruits in their diets, also eat more arthropods.

Discussion

Two of the most fascinating species, *Atelocynus microtis* and *Speothos venaticus*, could not be included in this study. No detailed field studies of these animals have ever been carried out. Observations in captivity, however, suggest that *Speothos* and *Atelocynus* are highly carnivorous. Scanty field data also corroborate this conclusion for *Speothos*

(see species account). As for the remaining species of South American canids, most are opportunistic feeders tracking abundant resources seasonally. Larger body size apparently permits or mandates taking larger vertebrate prey. The inclusion of arthropods in the diet is common in the small species of foxes, especially where arthropods are seasonally abundant in grasslands, such as the cerrado and llanos habitats. The type of arthropod prey varies depending on how moist the habitat is. Land crabs predominate in extremely mesic areas, whereas orthopterans constitute a good portion of arthropod prey in more xeric areas. Most of the South American canids that have been studied have been found to opportunistically use fruit seasonally, and *Chrysocyon brachyurus* is somewhat unusual in its exploitation of *Solanum*.

The "Harvest" of Southern South American Endemic Mammals and the Effects of Exotic Introductions

The unregulated exploitation of furbearers in southern South America has as its counterpart the disastrous impact of unregulated trapping on furbearers in North America (Novak et al. 1987; Iriarte and Jaksić 1986; Ojeda and Mares 1982). Two processes were at work: exploitation of the native fauna as the Europeans found it, and introductions of exotics in the hope of developing alternative harvests. Temperate South America was vulnerable to exotic introductions by the European colonizers. The horse (*Equus caballus*) occupied the pampas of Argentina after the original settlement by Europeans was abandoned (Crosby 1986). The introduction of the European hare (*Lepus capensis*) and rabbit (*Oryctolagus cuniculus*) has not only opened opportunities for trappers, but also rendered certain habitats vulnerable to degradation (see also chap. 14).

Returning to the impact on endemic species, it is convenient to consider the marine mammals sepa-

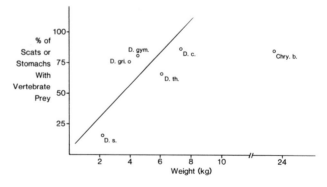

Figure 13.3. Correlation of body size with carnivory for South American canids.

rately from their terrestrial counterparts—both groups were vulnerable to exploitation. The decline of the great whales and seals is rather well documented (see Iriarte and Jaksić 1986 for a review), but the story concerning terrestrial furbearers is less well known. For example, Chile heavily exploited the chinchilla (*Chinchilla lanigera*) in the late nineteenth and early twentieth centuries. The result was the loss of the chinchilla as a renewable resource. The private breeding of the species has never achieved economic viability, despite numerous private efforts.

The coypu or nutria (*Myocastor coypus*) has undergone an interesting twist of fate. The pelt is marketable and commands the same interest among furriers as the pelt of the North American muskrat (*Ondatra zibethica*). The coypu has been widely introduced in Europe and North America, and the muskrat has been released in southern South America (see chap. 14). As Lowery (1974) has so eloquently documented, the introduction of *Myocastor* to the state of Louisiana had some interesting and provocative consequences. Parts of the puzzle will never be untangled, since its solution involves the role of the alligator (*Alligator mississippiensis* as the major predator on both species (the endemic and the introduced; Lowery 1974).

Community Ecology of the Chiroptera

The southern cone of South America presents some interesting opportunities for students of the order Chiroptera. As one proceeds from the tropical regions of Paraguay and northern Argentina south into the Temperate Zone, bat species richness declines until the Verspertilionidae dominate. The same trend can be discerned if one investigates an altitudinal gradient in the Andes.

Table 13.3 lists the genera of bats within the region covered by this volume and classifies their feeding habits according to Wilson (1973). Frugivorous and nectarivorous bats are mainly tropical in their distribution. Insectivorous bats cover the entire range of habitats, but some are migratory while others apparently are able to persist with minimal migratory movements even when their prey base is reduced. Of special interest are those insectivorous genera of the family Vespertilionidae in temperate South America.

In the Northern Hemisphere vespertilionid bats may be either migratory or rather sedentary, but they almost always exhibit torpor during the winter when their insect prey is absent. Pearson and Pearson (1989) studied *Histiotus* and *Myotis* in southern Argentina with the object of clarifying the annual cycle of vespertilionid bats in the southern cone. The

Table 13.3   Feeding Strategies of the Bats of Southern South America

Family and Genus	Feeding Strategy
Emballonuridae	
Peropteryx	Aerial insectivore
Noctilionidae	
Noctilio leporinus	Piscivore
Noctilio albiventris	Aerial insectivore
Phyllostomidae	
Chrotopterus	Carnivore
Macrophyllum	Aerial insectivore
Phyllostomus	Insectivore/frugivore (in part)
Tonatia	Frugivore
Anoura	Nectarivore
Glossophaga	Nectarivore
Carollia	Frugivore
Sturnira	Frugivore
Artibeus	Frugivore
Pygoderma	Frugivore
Vampyressa	Frugivore
Vampyrops	Frugivore
Desmodontidae	
Desmodus	Sanguivore
Furipteridae	
Amorphochilus	Aerial insectivore
Vespertilionidae	
Eptesicus	Aerial insectivore
Histiotus	Aerial insectivore
Lasiurus	Aerial insectivore
Myotis	Aerial insectivore
Molossidae	
Eumops	Aerial insectivore
Molossops	Aerial insectivore
Molossus	Aerial insectivore
Mormopterus	Aerial insectivore
Tadarida	Aerial insectivore
Nyctinomops	Aerial insectivore
Promops	Aerial insectivore

Pearsons found evidence that the reproductive cycles of *Histiotus montanus, H. macrotis,* and *Myotis chiloensis* are similar to those of north temperate vespertilionid bats. Namely, reproduction involves delayed fertilization and is seasonally synchronized to ensure birth of the young when insects are relatively abundant. Although the bats of the extreme Temperate Zone of the southern cone may not hibernate, they certainly adjust their reproductive cycle to the availability of their prey.

Within the range covered by this volume, the frugivorous and nectarivorous bats are predictably confined to the tropics of northern Argentina and Paraguay. The species richness of these bat communities is vastly reduced compared with that of a bat community in the true tropics, such as Panama (Eisenberg 1989). Insectivorous bats that are not cold tolerant, such as species belonging to the families Emballonuridae and Molossidae, are also only marginally represented in the southern cone.

Given these comments on the climatic limitations on bat distributions in southern South America, table 13.3 becomes more informative. One can scan the trophic specializations and note from the range maps that the frugivores and nectarivores are confined to the north. The aerial insectivores as a trophic category do not exhibit the same constraints except when one pays attention to phylogenetic affinity. The way the southern vespertilionids survive the temperate winters remains an active area for future study.

The vampire bats (*Desmodus*) pursue their prey base (larger mammals) wherever they can find them, but vampires are not cold tolerant (McNab 1982). Seasonal movements allow temporary extreme southern distributions.

Plant-Animal Interactions in Temperate South America

As Bucher (1987) documents so well, the grasslands of temperate South America have had a unique history. In the early Pliocene the large mammalian herbivores included the now extinct famlies Mylodontidae, Toxodontidae, Megatheriidae, Glyptodontidae, and Macroucheriidae. The completion of the Pliocene land bridge allowed colonization by the larger North American herbivores, including the Gomphatheriidae, Equidae, Tayassuidae, Tapiridae, Cervidae, and Camelidae. Only the last four families survive. In the era following the discoveries of Christopher Columbus, Europeans began to introduce livestock into the southern cone. Sheep (*Ovis*), goats (*Capra*), swine (*Sus*), horses (*Equus*), and cattle (*Bos*) often became feral and radically altered the vegetation. This account does not include exotic introductions of smaller herbivores such as the European rabbit and hare (*Oryctolagus* and *Lepus*) (see chap. 14).

At present temperate South America has very few endemic large herbivores. Leaving aside the more omnivorous Tayassuidae, we must consider only three families: the Camelidae, with two species clearly temperate adapted; the Tapiridae, with one species confined to tropical climates; and the Cervidae, with eight species occupying a range of habitats in both latitude and elevation. All these native large herbivores have been severely persecuted by humans, and their current influence on native ecosystems has been irreparably mofidied.

Concerning tropical grasslands such as the Chaco, Bucher (1987) remarks on the ubiquity of leaf-cutting ants (*Atta*) and their dominance as herbivores. Some species of these ants are considered to be the main competitors of domestic cattle for herbaceous vegetation (Bucher 1980).

In temperate grasslands Franklin (1983) has assessed the adaptations of the vicuna and guanaco to their plant communities. Both species of camelids may be adversely affected by the grazing of introduced livestock. Indeed, hunting of the larger mammalian herbivores and the grazing of domestic livestock can allow smaller herbivorous rodents (*Dolichotis* and *Lagostomus*) to increase greatly in numbers (Bucher 1980). The grass and scrub habitats of southern South America offer great promise for future research. Extinctions before European occupancy and introductions of livestock cloud the picture, but they offer challenges to the ecologists of the future.

References

Asa, C. S., and M. P. Wallace. 1990. Diet and activity patterns of the Sechuran desert fox (*Dusicyon sechurae*). *J. Mammal.* 71(1): 69–72.

Barlow, J. C. 1969. Observations on the biology of rodents in Uruguay. *Life Sci. Contrib. Roy. Ontario Mus.* 75:1–59.

Bucher, E. H. 1980. Ecología de la fauna chacqueña: Una revisión. *Ecosur* (Argentina) 7:111–59.

———. 1987. Herbivory in arid and semi-arid regions of Argentina. *Rev. Chil. Hist. Nat.* 60: 265–73.

Cabrera, A., and J. Yepes. 1960. *Mamíferos sudamericanos*. 2 vols. Buenos Aires: Ediar.

Caviedes, C. N., and A. Iriarte. 1988. Migration and distribution of rodents in central Chile since the Pleistocene: Polygraphic evidence. *J. Biogeogr.* 16:181–87.

Christie, M. I. 1984. Inventario de la fauna de vertebrados del Parque Nacional Nahuel Huapi. *Rev. Mus. Argent. Cienc. Nat. "Bernardino Rivadavia," Zool.* 13(1–60): 523–34.

Crespo, J. A. 1966. Ecología de una comunidad de roedores silvestres en el Partido de Rojas, provincia de Buenos Aires. *Rev. Mus. Argent. Cienc. Nat. "Bernardino Rivadavia," Ecol.* 1(3): 79–134.

———. 1982. Ecología de la comunidad de mamíferos del Parque Nacional Iguazú, Misiones. *Rev. Mus. Argent. Cienc. Nat. "Bernardino Rivadavia," Ecol.* 3(2): 45–162.

Crosby, A. W. 1986. *Ecological imperialism: The biological expansion of Europe, 900–1900.* New York: Cambridge University Press.

Daciuk, J. 1973. Notas faunisticas y bioecológicas de península Valdés y Patagonia. 10. Estudio cuantitativo y observaciones del comportamiento de la población del elefante marino del sur *Mirounga leonina* (Linne) en sus apostaderos de la provincia de Chubut (república Argentina). *Physis* 32(85): 403–22.

———. 1977. Notas faunisticas y bioecológicas de península Valdés y Patagonia. 20. Presencia de *Histiotus montanus montanus* (Philippi y Landbeck), 1861 en la península Valdés (Chiroptera, Vespertilionidae). *Neotrópica* 23(69): 45–46.

Dalby, P. L. 1974. Ecology and population dynamics of pampa rodents near Balcarce, Argentina. Ph.D. diss., Michigan State University.

Eisenberg, J. F. 1987. Evolutionary history of the Cervidae with special reference to the South American radiation. In *Biology and management of the Cervidae,* ed. C. Wemmer, 60–64. Washington, D.C.: Smithsonian Institution Press.

———. 1988. Reproduction in polyprotodont marsupials and similar-sized eutherians with a speculation concerning the evolution of litter size in mammals. In *Evolution of life histories of mammals,* ed. M. S. Boyce, 291–311. New Haven: Yale University Press.

———. 1989. *Mammals of the Neotropics.* Vol. 1. *Mammals of the northern Neotropics: Panama, Colombia, Venezuela, Guyana, Suriname, French Guiana.* Chicago: University of Chicago Press.

Eisenberg, J. F., and L. H. Harris. 1989. Conservation: A consideration of evolution, population, and life history. In *Conservation for the twenty-first century,* ed. D. Western and M. C. Pearl, 99–108. New York: Oxford University Press.

Eisenberg, J. F., and K. H. Redford. 1982. Comparative niche structure and evolution of mammals of the Nearctic and southern South America. In *Mammalian biology in South America,* ed. M. A. Mares and H. H. Genoways, 77–84. Pymatuning Symposia in Ecology 6. Special Publication Series. Pittsburgh: Pymatuning Laboratory of Ecology, University of Pittsburgh.

Fonseca, G. A. B., and M. C. M. Kierulff. 1989. Biology and natural history of Brazilian Atlantic forest small mammals. *Bull. Florida State Mus.* 34(1): 99–152.

Franklin, W. L. 1983. Contrasting socioecologies of South America's wild camelids, the vicuna and the guanaco. In *Advances in the study of mammalian behavior,* ed. J. F. Eisenberg and D. G. Kleiman,

573–629. Special Publication 7. Shippensburg, Pa.: American Society of Mammalogists.

Fulk, G. W. 1975. Population ecology of rodents in the semiarid shrublands of Chile. *Occas. Pap. Mus., Texas Tech Univ.* 33:1–40.

Glanz, W. E. 1977a. Comparative ecology of small mammal communities in California and Chile. Ph.D. diss., University of California, Berkeley.

———. 1977b. Small mammals. In *Chile-California Mediterranean scrub atlas: A comparative analysis,* ed. N. J. W. Thrower and D. E. Bradbury, 232–37. Stroudsburg, Pa.: Dowden, Hutchinson and Ross.

Glanz, W. E., and P. L. Meserve. 1982. An ecological comparison of small mammal communities in California and Chile. In *Dynamics and management of Mediterranean-type ecosystems,* ed. C. E. Conrad and W. C. Oechel, 220–26. *U.S. Forest* Service, Pacific Southwest Forest Range Experiment Station, General Technical Report PSW 58:1–637.

Goodall, R. N. P. 1978. Report on the small cetaceans stranded on the coasts of Tierra del Fuego. *Sci. Rep. Whales Res. Inst.* 30:197–230.

Hall, E. R. 1981. *The mammals of North America.* 2 vols. New York: John Wiley.

Hamilton, J. E. 1952. Cetacea of the Falkland Islands. *Comun. Zool. Mus. Hist. Nat. Montevideo* 4(66): 1–6.

Hershkovitz, P. 1982. Neotropical deer (Cervidae). Part 1. Pudus, genus *Pudu* Gray. *Fieldiana: Zool.* 11:1–86.

Iriarte, J. A. 1988a. Feeding ecology of the Patagonian puma (*Felis concolor*) in Torres del Paine National Park, Chile. M.A. thesis, University of Florida.

———. 1988b. The mammalian fauna of Torres del Paine National Park, Chile. *Latinamericanist* 24(1): 1–15.

Iriarte, J. A., L. C. Contreras, and F. M. Jaksić. 1989. A long-term study of a small-mammal assemblage in the central Chilean matorral. *J. Mammal.* 70:79–87.

Iriarte, J. A., and F. M. Jaksić. 1986. The fur trade in Chile: An overview of seventy-five years of export data (1910–1984). *Biol. Conserv.* 38: 243–53.

Jaksić, F. M., H. W. Greene, and J. L. Yáñez. 1981. The guild structure of a community of predatory vertebrates in central Chile. *Oecologia* 49:21–28.

Jaksić, F. M., R. P. Schlatter, and J. L. Yáñez. 1980. Feeding ecology of central Chilean foxes, *Dusicyon culpaeus* and *Dusicyon griseus. J. Mammal.* 61:254–60.

Jaksić, F. M., and J. A. Simonetti. 1987. South American terrestrial predation. *Rev. Chil. Hist. Nat.* 60:221–44.

Lee, A. K., P. Woolley, and R. W. Braithwaite. 1982. Life history strategies of disyurid marsupials. In *Carnivorous marsupials*, ed. M. Archer, 1:1–11. Chipping Norton, New South Wales: Surrey Beatty.

Lopez, J. C., and D. Lopez. 1985. Killer whales (*Orcinus orca*) of Patagonia, and their behavior of intentional stranding while hunting near shore. *J. Mammal.* 66:181–83.

Lowery, G. H., Jr. 1974. *The mammals of Louisiana and its adjacent waters.* Baton Rouge: Louisiana State University Press.

McNab, B. 1982. Evolutionary alternatives in the physiological ecology of bats. In *Ecology of bats*, ed. T. H. Kunz, 151–200. New York: Plenum.

Mares, M. A. 1973. Climates, mammalian communities and desert rodent adaptations: An investigation into evolutionary convergence. Ph.D. diss., University of Texas at Austin.

———. 1975. Observations of Argentine desert rodent ecology, with emphasis on water relations of *Eligmodontia typus*. In *Rodents in desert environments*, ed. I. Prakash and P. K. Ghosh, 155–75. The Hague: W. Junk.

———. 1976. Convergent evolution of desert rodents: Multivariate analysis and zoogeographic implications. *Paleobiology* 2:39–63.

———. 1980a. Desert mammals. *Carnegie Mag.* 54(3):15–22.

———. 1980b. Convergent evolution among desert rodents: A global perspective. *Bull. Carnegie Mus. Nat. Hist.* 16:1–51.

Massoia, E. 1970. Contribución al conocimiento de los mamíferos de Formosa con noticias de los que habitan zonas vinaleras. *IDIA* 276:55–63.

———. 1980. Mamalia de Argentina. 1. Los mamíferos silvestres de la provincia de Misiones. *Iguazú* 1(1):15–43.

Meserve, P. L. 1981a. Resource partitioning in a Chilean semi-arid small mammal community. *J. Anim. Ecol.* 40:747–57.

———. 1981b. Trophic relationships among small mammals in a Chilean semiarid thorn scrub community. *J. Mammal.* 62:304–14.

Meserve, P. L., R. E. Martín, and J. Rodríguez. 1984. Comparative ecology of the caviomorph rodent *Octodon degus* in two Chilean Mediterranean-type communities. *Rev. Chil. Hist. Nat.* 57:79–89.

Meserve, P. L., R. Murúa, O. Loppetegui N., and J. R. Rau. 1982. Observations on the small mammal fauna of a primary temperate rain forest in southern Chile. *J. Mammal.* 63:315–17.

Mooney, H. A., and E. L. Dunn. 1970. Convergent evolution of Mediterranean-climate evergreen sclerophyll shrubs. *Evolution* 24:292–303.

Mooney, H. A., O. T. Solbrig, and B. B. Simpson. 1977. Phenology, morphology, physiology. In *Mesquite: Its biology in two desert shrub ecosystems.* US/IBD Synthesis Series 4. Stroudsburg, Pa.: Dowden, Hutchinson, and Ross.

Myers, P. 1982. Origins and affinities of the mammal fauna of Paraguay. In *Mammalian biology in South America*, ed. M. A. Mares and H. H. Genoways, 85–94. Pymatuning Symposia in Ecology 6. Special Publication Series. Pittsburgh: Pymatuning Laboratory of Ecology, University of Pittsburgh.

Novak, M., J. A. Baker, M. E. Obbard, and B. Malloch, eds. 1987. *Wild furbearer management and conservation in North America.* Ontario: Ministry of Natural Resources.

Ojeda, R. A. 1985. A biogeographic analysis of the mammals of Salta province, Argentina: Patterns of community assemblage in the Neotropics. Ph.D. diss., University of Pittsburgh.

Ojeda, R. A., and M. A. Mares. 1982. Conservation of South American mammals: Argentina as a paradigm. In *Mammalian biology in South America*, ed. M. A. Mares and H. H. Genoways, 505–22. Pymatuning Symposia in Ecology 6. Special Publication Series. Pittsburgh: Pymatuning Laboratory of Ecology, University of Pittsburgh.

Patterson, B. D., P. L. Meserve, and B. K. Lang. 1989. Distribution and abundance of small mammals along an elevational transect in temperate rainforests of Chile. *J. Mammal.* 70:67–78.

Payne, R., O. Brazier, E. M. Dorsey, J. S. Perkins, V. J. Rountree, and A. Titus. 1983. External features in southern right whales (*Eubalaena australis*) and their use in identifying individuals. In *Communication and behavior of whales*, ed. R. Payne, 371–445. AAAS Selected Symposia, ser. 76. Boulder, Colo.: Westview Press.

Payne, R., and L. N. Guinee. 1983. Humpback whale (*Megaptera novaeangliae*) songs as an indicator of "stocks." In *Communication and behavior of whales*, ed. R. Payne, 333–58. AAAS Selected Symposia, ser. 76. Boulder, Colo.: Westview Press.

Pearson, O. P. 1983. Characteristics of a mammalian fauna from forests in Patagonia, southern Argentina. *J. Mammal.* 64:476–92.

———. 1987. Mice and the postglacial history of the Traful Valley of Argentina. *J. Mammal.* 68:469–78.

Pearson, O. P., and M. I. Christie. 1985. Los tuco-tucos (género *Ctenomys*) de los Parques Nacionales Lanín y Nahuel Huapi, Argentina. *Hist. Nat.* 5(37): 337–43.

Pearson, O. P., and A. K. Pearson. 1982. Ecology and biogeography of the southern rainforests of Argentina. In *Mammalian biology in South America*, ed. M. A. Mares and H. H. Genoways, 129–42. Pymatuning Symposia in Ecology 6. Special Publication Series. Pittsburgh: Pymatuning Laboratory of Ecology, University of Pittsburgh.

———. 1989. Reproduction of bats in southern Argentina. In *Advances in Neotropical mammalogy*, ed. K. H. Redford and J. F. Eisenberg, 549–66. Gainesville, Fla.: Sandhill Crane Press.

Pine, R. H., J. P. Angle, and D. Bridge. 1978. Mammals from the sea, mainland and islands at the southern tip of South America. *Mammalia* 42(1): 105–14.

Pine, R. H., P. L. Dalby, and J. O. Matson. 1985. Ecology, postnatal development, morphometrics, and taxonomic status of the short-tailed opossum, *Monodelphis dimidiata*, an apparently semelparous annual marsupial. *Ann. Carnegie Mus.* 54(6): 195–231.

Rood, J. P. 1970a. Notes on the behavior of the pygmy armadillo. *J. Mammal.* 51:179.

———. 1970b. Ecology and social behavior of the desert cavy (*Microcavia australis*). *Amer. Midl. Nat.* 83:415–54.

———. 1972. Ecological and behavioral comparisons in three genera of Argentine cavies. *Anim. Behav. Monogr.* 5:67–83.

Schaller, G. B. 1983. Mammals and their biomass on a Brazilian ranch. *Arq. Zool., São Paulo* 31(1): 1–36.

Stallings, J. R. 1989. Small mammal inventories in an eastern Brazilian park. *Bull. Florida State Mus. Biol. Sci.* 34(4): 153–200.

Taber, A. B., and D. W. Macdonald. 1984. Scent dispensing papillae and associated behavior of the mara, *Dolichotis patagonum* (Rodentia: Caviomorpha). *J. Zool.* 203:298–301.

Wilson, D. E. 1973. Bat faunas: A trophic comparison. *Syst. Zool.* 22:14–29.

Würsig, B., and R. Bastida. 1986. Long-range movement and individual associations of two dusky dolphins (*Lagenorhynchus obscurus*) off Argentina. *J. Mammal.* 67:773–74.

14   The Effects of Humans on the Mammalian
Fauna of Southern South America

The activities of *Homo sapiens* in the past 50,000 years have had dramatic effects on the planet earth (cf. Western and Pearl 1989; Hoage 1985). When humans first arrived in South America is a subject of considerable debate, but most workers agree that by 10,000 B.P. human settlements were thriving in the Western Hemisphere. Human beings may have reached Chile 13,000 years ago, or perhaps as early as 34,000 years ago (Bray 1988). The domestic dog (*Canis familiaris*) was brought into South America about 7,000 years ago (Wing 1986).

Southern South America, with its diversity of climates and ecological zones, contained many different Indian groups. Steward and Faron (1959) have named four culture areas from southern South America: the "nomadic hunter-gatherers" of the pampas, Patagonia, and the Chaco; the "southern Andean farmers and pastoralists" of central Chile; the "tropical forest village farmers" of Misiones province, Argentina, and eastern Paraguay; and the "irrigation civilization" of far northern Chile.

The hunter-gatherers inhabited regions in which farming was impossible. In Chile, south of the island of Chiloé, groups such as the Yahgan relied on hunting small land mammals and marine mammals (such as porpoises and sea otters) and on gathering the abundant invertebrate marine life. The Ona tribe, inhabitants of the island of Tierra del Fuego, relied largely on the guanaco, making their houses, clothes, and carrying implements of materials derived from this species. Apparently the second most important mammalian game species was *Ctenomys*. In the "Gran Chaco," which means "great hunting," Indians practiced small-scale agriculture but relied primarily on hunting and gathering (Steward and Faron 1959).

The dense Indian populations in central Chile practiced sophisticated agriculture, and some groups herded llamas. Maize was the principal crop, and

a significant proportion of potential farmland was under cultivation (Miller 1980). Both of the other two culture groups also practiced agriculture, in northern Chile using irrigation systems, and in the tropical forests of Misiones province, Argentina, and eastern Paraguay using the traditional slash-and-burn method (Steward and Faron 1959).

Before the arrival of the Europeans, over most of southern South America the indigenous peoples had not had a great impact on the mammalian fauna. Humans lived at relatively low densities, moved frequently, and were limited in their hunting and farming by native technology. The exception to this occurred in central Chile, where Miller (1980) makes an elegant case for severe indigenous human impact on native mammals. The absence of wild mammals in central Chile, noted even by the early Spanish chroniclers, was due to extensive conversion of forest to cropland. Miller states that the current disjunct distributions of several forest-dwelling rodent species may be due to this early human disturbance.

The arrival of Europeans brought major changes to southern South America. Portions of the "southern cone" were suitable for European agriculture and animal husbandry techniques, but the funding of enterprises in the development of South America was in the hands of bankers, and the explorer was at risk if he borrowed funds and could not repay. Consider the fate of Christopher Colombus. Thus the early exploration was motivated less by the desire to found new agricultural settlements than by an effort to establish bases for extracting valuable natural resources such as gold, silver, and gems, and later hides, forest products, and sea mammals.

Spanish settlement in Argentina resulted from geographical ignorance. The explorers believed that the mineral riches of Peru extended to the southeast across the continent and that they could be more easily reached from the eastern coast. The usual route for the mineral wealth pouring out of Peru and Bolivia was by boat to Panama, by land across the Isthmus, and then again by boat (Weil et al. 1974). After the early explorations of the southern portion of South America about 1500, in an attempt to shorten this route, the Spaniards founded Buenos Aires. The colonists had anticipated drawing supplies from the indigenous peoples, but the Amerindians near Buenos Aires were not horticulturists. Thus the colony was forced to move in 1538 and establish a new settlement at what is now Asunción, Paraguay. Here the Indians practiced some horticulture, and it was possible for the colonists to develop a trading relationship that ensured the permanence of this settlement (Lockhart and Schwartz 1983). However, during the move some horses and cattle were left behind, and the horses became the original stock for the population of the pampas. Meanwhile, events in Bolivia were to strongly influence a further development in what is now Argentina and Paraguay. In 1545 silver was discovered in abundance, and by 1547 a settlement had been founded at what is now known as Potosí. There was a desperate need for horses and mules to be used as draft animals in mining silver. Supplying Bolivia through Peru necessitated long voyages around Cape Horn. Thus settlements began to be established across northern Argentina to provide an overland supply route to Potosí. For example, Tucumán was founded in 1560. Eventually a string of settlements including Salta and Córdoba could connect up with Asunción, and Buenos Aires was refounded in 1580. Throughout this period there was brisk trade in horses and mules, and no one knows how many escaped. But by 1699 cattle by the millions were roaming free on the pampayan grasslands, and horses had become so abundant that the Patagonian Indians had domesticated them and founded a "horse culture" parallel to the one developed by the Plains Indians of North America, again based on feral horses. In the late 1700s, Azara remarked on the abundance of horses and their remarkable increase since the founding stock was introduced as early as 1535. By the end of the seventeenth century the trade in cattle hides between Argentina, Paraguay, and Europe was already well established. A chronicler wrote in 1691 that oxen, cows, and horses roamed the plains in such prodigious numbers that "in some places the fields are covered with them as far as your eyes will reach." At that time the cattle were so numerous that one had only to ride out into the countryside and round them up. They were in fact no individual's property. It is noteworthy that coincident with this period, jaguars (*Panthera onca*) became very numerous, and

there are many records of human-jaguar conflict (Caraman 1976).

By the time Charles Darwin arrived in the nineteenth century (Darwin 1962) he offered the following comments: "According to the principles laid down by Mr. Lyell, few countries have undergone more remarkable changes, since the year 1535, when the first colonist landed with seventy-two horses. The countless herds of horses, cattle, and sheep not only have altered the whole aspect of the vegetation, but they have almost banished the guanaco, deer, and ostrich (*Rhea*). Numberless other changes must likewise have taken place; the wild pig in some parts probably replaces the peccary; packs of wild dogs may be heard howling on the wooded banks of the less frequented streams; and the common cat, altered into a large and fierce animal, inhabits rocky hills" (p. 120).

Azara (1978) noted that by the mid-1700s Buenos Aires and Montevideo were exporting 800,000 to 1,000,000 cattle hides per year. Miller (1980) points out parallel trends in Chile. Once the Spanish settlements were established at what is now Santiago, tremendous numbers of livestock could be noted in central Chile during the seventeenth century. Without a doubt, extensive grazing by livestock accounted for enormous changes in plant community composition.

During the early centuries of colonial occupation of southern South America, human effects on the local mammals were largely incidental. The Europeans were primarily occupied with survival and with extracting mineral wealth. In the late 1700s this began to change and an eager eye was turned to the animal wealth. The first native mammal to be commercially exploited in Chile was the fur seal (*Arctocephalus philippii*), which inhabited the Juan Fernández Islands. Approximately 3.5 million pelts were taken between 1791 and 1809, and the population by the early 1970s was only 500 to 700. Also severely affected was the southern elephant seal, which in Chile was hunted to extinction for its oil (Miller 1980), and the sea lion (*Otaria byronia*), hunted for its pelt (Iriarte and Jaksić 1986).

Whales were also extensively hunted. In Chilean waters whaling began between 1785 and 1790 and was primarily directed at *Balaena glacialis* (Sielfeld 1983). Several of the whale species were hunted to near extinction, a practice that increased in intensity with the much later advent of the explosive harpoon and the factory ship.

Other species were intensively hunted as well. The pampas deer (*Ozotoceros*) was decimated between 1869 and 1879, when an estimated 2 million skins were exported from Buenos Aires alone (Jackson and Langguth 1987). Chinchillas were heavily exploited in the late 1800s, and the target species (*Chinchilla brevicaudata*) is largely extinct in the wild.

In a well-constructed analysis Roig (1989) demonstrated that the geographical ranges of many native Argentine mammals have been substantially altered both directly through destruction of local populations and indirectly through extensive habitat destruction, particularly severe in semiarid woodland areas.

Introductions

European inhabitants of temperate South America further affected the native fauna by introducing game species from their native countries. The European rabbit and hare (*Oryctolagus cuniculus; Lepus capensis*), the European red deer (*Cervus elaphus*), the European wild swine (*Sus scrofa*), and the Asiatic spotted deer (*Axis axis*) are a few of the game species successfully introduced into southern South America. Some Northern Hemisphere furbearing mammals have also been introduced, with varying success. They include the beaver (*Castor canadensis*), the muskrat (*Ondatra zibethica*), and the mink (*Mustela vison*). Some of the more common introductions are illustrated in plate 18. Domestic livestock and exotic game species are not described in the text, but this brief section should bring home the point that some exotics are well established and may continue to be an important component of the fauna of southern South America (table 14.1).

A remarkable fourteen species of ungulates (wild and domestic; table 14.1) have been introduced or have gone feral in southern South America. Their effects are only beginning to be understood, but pioneering studies on red deer (cf. Ramírez et al. 1981; Eldridge 1983) have shown that exotic ungulates may compete with native species, like the pudu, and may affect vegetation succession.

Bonner (1984) notes that the sub-Antarctic islands did not escape introductions (see table 14.2). The early explorers were all too aware that in case of shipwreck islands without fauna could be lands of starvation. Thus it was not uncommon that livestock would be put ashore to "seed the islands." When whaling and sealing operations became an important item of commerce, it became all the more important to provision these outposts with a supply of meat. Thus sheep and reindeer were in later years released to provide food. The inadvertent introduction of European rats and mice also led to the introduction of domestic cats as a form of rodent control.

Table 14.1   Major Mammal Introductions in Selected Countries in Southern South America

Species	Country		
	Chile	Argentina	Uruguay
Lagomorpha			
Lepus	+	+	+
Oryctolagus cuniculus	+	+	+
Rodentia			
Ondatra zibethica	+	+	−
Castor canadensis	+	+	−
Carnivora			
Nasua nasua	+	−	−
Mustela vison	+	+	−
Artiodactyla			
Sus scrofa	+	+	−
Cervus elaphus	+	+	+
Dama dama	+	+	+
Capra hircus	+	+	−
Ovis musimon	+	−	−
Axis axis	−	+	−
Cervus canadensis	−	+	−
Odocoileus hemionus	−	+	−
Rangifer tarandus	−	+	−
Antilope cervicapra	−	+	−
Total	11	14	4

Sources: Miller and Rottmann: DeVos Manville, and Van Gelder 1956; Barlow 1965; Olrog and Lucero 1981; J. A. Iriarte, unpublished data.
Note: Feral populations exclusive of managed domestics such as sheep, dogs, cats, cattle, and the ubiquitous rodents of the genera *Mus* and *Rattus*.

Table 14.2   Mammals Introduced on the Islands in Antarctic Waters

Species	KI	CI	MI	SG	MaI
Rattus rattus	+	+	+	−	−
R. norvegicus	−	−	−	+	−
Mus musculus	+	+	+	+	+
Oryctolagus cuniculus	+	+	+	−	−
Rangifer rangifer	+	+	+	+	−
Ovis aries	+	+	+	+	−
O. ammon	+	−	−	−	−
Felis cattus	+	+	+	−	+

Source: Bonner (1984).
Note: KI = Kerguelen Islands; CI = Crozet Islands; MI = Macquarie Island; SG = South Georgia; MaI = Marion Island.

Having brought up the European rodents that are commensal with humans, we must consider the rodents that rode the vessels of the Spanish, Portuguese, and French during the initial encounter between 1492 and 1500. *Mus musculus*, *Rattus rattus*, and *Rattus norvegicus* are well established in South America and elsewhere. They can be an important component of the mammalian fauna in disturbed habitats and in agricultural settings. Their importance and our need to identify them result in their inclusion in the species accounts of chapter 11.

The introduction of the European hare (*Lepus*) and rabbit (*Oryctolagus*) is also an instructive experiment. The hare, *Lepus capensis*, was introduced into Argentina between 1883 and 1897 (Grigera and Rapoport 1983). Current estimates of the harvest of hares in Argentina range from 5 to 10 million a year. They now occupy practically all of Argentina except for the tropical portions and the higher portions of the Andes. It is noteworthy that upon introduction the animals extended their range on average 19 km per year.

The rabbit (*Oryctolagus*) was introduced first into the islands of the Beagle Channel in about 1880 but subsequently transported to the Chilean side of the strait in 1939. By 1953 their numbers in Chile were estimated to exceed 30 million (Jaksić and Yáñez 1983). Howard and Amaya (1975) note that *Oryctolagus* may have been introduced into Chile by the mid-eighteenth century. In any event, it appeared along the Argentine border between 1945 and 1959.

The introduction of exotics into the southern cone of South America has presented a number of problems. As shown in the previous paragraphs, the vegetation was altered in a significant manner after the introduction of horses, cattle, and sheep. Some of the introductions have had an economic benefit through the trade in hides. For example, Jackson (1986) estimates that at least 6 million hares (*Lepus*) are shot annually, with an average worth of U.S. $24 million. The introduction of the beaver into Tierra del Fuego in 1946 has resulted in a robust population that is harvestable. Sielfeld and Venegas (1980) estimate that beaver are now at a density of 4.2 to 6.7/km² on Navarino Island.

The Present

The impact of European technology is felt throughout the southern cone. Ojeda and Mares (1984) estimate that between 1976 and 1979 U.S. $90 million of wildlife products were exported annually from Buenos Aires. During this time 14 million kg of meat was exported, almost all derived from exotic species, mostly the hare but also axis deer (*Axis axis*) and black buck (*Antilope cervicapra*). Trade in native foxes continues to be very important, with 3,612,459 *Dusicyon* sp. exported from Argentina between 1976 and 1979. Fur exports between 1976 and 1979 totaled U.S. $225 million. A controlled legal harvest of guanacos in Argentina harvested 86,000 animals in 1979. In short, the effects of commercial hunting are pronounced in the southern cone (see Redford and Robinson 1991). Wild mammals have always served and continue to serve as important sources of food for subsistence hunters and ranch workers in rural Argentina. Some indigenous groups continue to hunt extensively, and the Ache tribe of eastern Paraguay has been recorded as taking 1,105 individual mammals over approximately a two-year period (Hawkes et al. and Hill, in Redford and Robinson 1987).

The result of this hunting, combined with extreme habitat destruction and degradation of other habitats through logging and ranching, has been to constrict the natural ranges of many native carnivores, including jaguars, otters, seals, and foxes (Mares and Ojeda 1984). Native rodents with valuable fur, such as the chinchilla and coypu, are also adversely affected. Guanacos have been eliminated from much of their former range (Cajal 1980, 1984), and the pampas deer is virtually extinct in the wild (Jackson and Langguth 1987). Uruguay has been hit most severely, with extinctions of at least jaguars, pumas, collared peccaries, marsh deer, giant anteaters, and possibly tapirs (Tálice 1969). The combined pressures of introduced exotics and human exploitation have so vastly altered some of the biomes of the southern cone that we can scarcely imagine their appearance before the advent of Europeans.

References

Azara, F. de. 1978. *Apuntamientos para la historia natural de los cuadrúpedos del Paraguay y Río La Plata*. New York: Arno Press. Originally published 1802.

Barlow, J. C. 1965. Land mammals from Uruguay: Ecology and zoogeography. Ph.D. diss., University of Kansas.

Bonner, W. M. 1984. Introduced mammals. In *Antarctic ecology*, vol. 1, ed. R. M. Laws, 237–78. New York: Academic Press.

Bray, W. 1988. How old are the Americans? *Américas* 40(3): 50–55.

Cajal, J. L. 1980. *Situación del guanaco en la república Argentina*. Buenos Aires: Subsecretaría de Ciencia y Tecnología.

———. 1984. El guanaco. In *Mamíferos fauna argentina*, vol. 1. Buenos Aires: Centro Editor de América Latina.

Caraman, P. 1976. *The lost paradise: The Jesuit republic in South America*. New York: Seabury Press.

Darwin, C. 1962. *The voyage of the "Beagle."* Garden City, N.Y.: Doubleday.

De Vos, A., R. H. Manville, and R. G. Van Gelder. 1956. Introduced mammals and their influence on native biota. *Zoologica* 41(4): 163–94.

Eldridge, W. 1983. *Impacto ambiental, alimentación y conducta social del ciervo rojo y dama en el sur de Chile.* Boletín Técnico 9. Chile: Corporación Nacional Forestral.

Grigera, D. E., and E. H. Rapoport. 1983. Status and distribution of the European hare in South America. *J. Mammal.* 64:163–66.

Hoage, R., ed. 1985. *Animal extinctions: What everyone should know.* Washington, D.C.: Smithsonian Institution Press.

Howard, W. E., and J. N. Amaya. 1975. European rabbit invades western Argentina. *J. Wildl. Manage.* 39(4): 757–61.

Iriarte, J. A., and F. M. Jaksić. 1986. The fur trade in Chile: An overview of seventy-five years of export data (1910–1984). *Biol. Conserv.* 38:243–53.

Jackson, J. E. 1986. The hare trade in Argentina. *Traffic Bull.* 7(5): 72.

Jackson, J. E., and A. Langguth. 1987. Ecology and status of the pampas deer (*Ozotoceros bezoarticus*) in the Argentinian pampas and Uruguay. In *Biology and management of the Cervidae,* ed. C. M. Wemmer, 402–9. Washington, D.C.: Smithsonian Institution Press.

Jaksić, F. M., and J. L. Yáñez. 1983. Rabbit and fox introductions in Tierra del Fuego: History and assessment of the attempts at biological control of the rabbit infestation. *Biol. Conserv.* 26:367–74.

Lockhart, J., and S. B. Schwartz. 1983. *Early Latin America.* New York: Cambridge University Press.

Mares, M. A., and R. A. Ojeda. 1984. Faunal commercialization and conservation in South America. *BioScience* 34(9): 580–84.

Miller, S. 1980. Human influences on the distribution and abundance of wild Chilean mammals: Prehistoric–present. Ph.D. diss., University of Washington.

Miller, S., and J. Rottmann. 1976. *Guía para el reconocimiento de mamíferos chilenos.* Santiago: Editora Nacional Gabriela Mistral.

Ojeda, R. A., and M. A. Mares. 1984. La degradación de los recursos naturales y la fauna silvestre en Argentina. *Interciencia* 9(1): 21–26.

Olrog, C. C., and M. M. Lucero. 1981. *Guía de los mamíferos argentinos.* Tucumán, Argentina: Ministerio de Cultura y Educación, Fundación Miguel Lillo.

Ramírez, C., R. Godoy, W. Eldridge, and N. Pacheco. 1981. Impacto ecológico del ciervo rojo sobre el bosque de Olivillo en Osorno, Chile. *Ann. Mus. Hist. Nat.* (Valparaíso) 14:197–215.

Redford, K. H., and J. G. Robinson. 1987. The game of choice: Patterns of Indian and colonist hunting in the Neotropics. *Amer. Anthropol.* 89: 650–67.

———. 1991. Commercial and subsistence uses of wildlife in Latin America. In *Neotropical wildlife use and conservation,* ed. J. G. Robinson and K. H. Redford. Chicago: University of Chicago Press.

Roig, V. G. 1989. Desertificatión y distribución geográfica de mamíferos en la república Argentina. In *Detección y control de la desertificatión,* ed. F. A. Roig, 263–78. Mendoza, Argentina: Centro Regional de Investigaciones Cientificas y Tecnológicas Mendoza (CRICYT), Area de Investigaciones de las Zonas Aridas (IADIZA).

Sielfeld, W. 1983. *Mamíferos marinos de Chile.* Santiago: Ediciones de la Universidad de Chile.

Sielfeld, W., and C. Venegas. 1980. Poblamiento e impacto ambiental de *Castor canadensis* Kuhl, en Isla Navarino, Chile. *Anal. Inst. Patagonia, Punta Arenas* (Chile) 11:247–57.

Steward, J. H, and L. C. Faron. 1959. *Native peoples of South America.* New York: McGraw-Hill.

Tálice, R. V. 1969. Mamíferos autóctonos. *Nuestra Tierra* 5:1–68.

Weil, T. E., J. K. Black, H. I. Blutstein, H. J. Hoyer, K. T. Johnston, and D. S. McMorris. 1974. *Area handbook for Argentina.* Washington, D.C.: Government Printing Office.

Western, D., and M. C. Pearl, eds. 1989. *Conservation biology for the twenty-first century.* New York: Oxford University Press.

Wing, E. S. 1986. Domestication of Andean mammals. In *High altitude tropical biogeography,* ed. F. Vuilleumier and M. Monasterio, 246–64. New York: Oxford University Press.

Index of Scientific Names

Index of Common Names